Structural Engineering Solved Problems

Comprehensive Practice for the Structural Engineering (SE) and Civil PE Exams

Sixth Edition

C. Dale Buckner, PhD, PE, SECB

The Power to Pass®
www.ppi2pass.com

Professional Publications, Inc. • Belmont, California

STRUCTURAL ENGINEERING SOLVED PROBLEMS

Sixth Edition

Current printing of this edition: 2

Printing History

date	edition number	printing number	update
Nov 2014	5	4	Minor corrections. Copyright update.
Aug 2015	6	1	New edition. Copyright update. Code updates. Major updates.
Feb 2016	6	2	Minor corrections.

Printed in the United States of America.

PPI
1250 Fifth Avenue
Belmont, CA 94002
(650) 593-9119
ppi2pass.com

ISBN: 978-1-59126-500-9

Library of Congress Control Number: 2015942042

F E D C B A

Table of Contents

Preface and Acknowledgments

Significant changes have occurred in structural design criteria since I wrote the first edition of *Structural Engineering Solved Problems* (previously titled *246 Solved Structural Engineering Problems*). Design specifications for steel, concrete, timber, and masonry structures have been revised multiple times to reflect new research findings and to correct deficiencies. These changes make it necessary to periodically revise the problems and solutions in this book to reflect the latest design expressions, allowable stresses, and code citations.

For this sixth edition of *Structural Engineering Solved Problems*, I have updated all code and specification references to the editions in use by both the 16-hour structural engineering (SE) exam and the structural depth section of the civil PE exam, which are prepared by the National Council of Examiners for Engineering and Surveying (NCEES). The problems in this book will help you in your preparations for either exam.

This book has had a long journey, and many organizations and individuals have contributed along the way to it. In particular, I would like to thank Alan Williams, who technically reviewed the extensive revisions incorporated into the fourth edition, and Ralph Arcena, EIT, who performed a calculation check of the revisions in this edition.

I am also grateful to the staff at PPI, including Sam Webster, editorial manager; Serena Cavanaugh, acquisitions editor; Heather Turbeville, associate project manager; Sierra Cirimelli-Low, lead editor; David Chu, Ellen Nordman, and Ian A. Walker, copy editors; Tom Bergstrom, production associate and technical illustrator; Kate Hayes, production associate; Cathy Schrott, production services manager; and Sarah Hubbard, director of product development and implementation.

While I had much help in creating this sixth edition, any mistakes you find are mine alone. If you find a mistake, please submit it through PPI's errata webpage at **ppi2pass.com/errata**. A reply will be posted and the error revised accordingly in the next printing.

C. Dale Buckner, PhD, PE, SECB

Introduction

HOW TO USE THIS BOOK

Structural Engineering Solved Problems is intended to be used as part of your exam review for either the eight-hour civil PE exam or the 16-hour structural engineering (SE) exam, both prepared by the National Council of Examiners for Engineering and Surveying (NCEES). As such, it provides a comprehensive set of 100 solved problems, which are categorized into chapters that encompass the broad categories found on these exams. These categories are structural analysis, structural concrete design, structural steel design, seismic design, foundations and retaining structures, timber design, and masonry design.

The chapters may be worked in any order, but it is recommended that you work only one chapter at a time. Problems should be worked sequentially, as they are arranged in order of increasing complexity. As a rough guide, the first quarter of a chapter consists of two types of problems. The first type includes qualitative problems that require specific short answers rather than calculations or sketches. These problems cover a broad range of topics and may be used to prepare for the multiple-choice problems of the civil PE structural depth exam and SE exam. The second type includes basic quantitative problems that will require 30 to 45 minutes to solve. The middle half of the chapter consists of more difficult quantitative problems that will require 1 hour to solve. The last quarter of a chapter consists of in-depth quantitative problems that will require $1^1/_2$ to 2 hours to solve. These latter problems are similar in scope to those found on the depth sections of the SE exam.

The solutions to the problems in this book show the steps necessary to arrive at the final answers. However, many of the design problems require assumptions that will lead to unique answers. Therefore, the final answer provided in this book may not necessarily be the exact answer you derived. This is OK. The solutions to such problems are intended to serve as a basis of comparison for you. That is, it is more important that you understand the process by which to solve a problem than to attain the exact answer printed. By comparing this book's solutions to your own solutions, you will be able to see if there are specific areas that require further study.

Before attempting to solve any problems, make sure to read this Introduction's information about the civil PE and SE exams. This information is provided so that you obtain a good understanding of the exams' formats and contents, as well as the rules and regulations for taking the exams. Gaining familiarity with the nature of the exams is an essential preparation step.

This book's "Codes, Standards, and References Used in This Book" section lists the codes, design standards, and other publications that the solutions in this book are based on. For your exam, use the codes and design standards that the current exam is based on. Check PPI's website at **ppi2pass.com/civil** or **ppi2pass.com/structural** for current code information and answers to frequently asked questions (FAQs) about the civil PE and SE exams.

ABOUT THE EXAMS

The civil PE and SE exams are standardized tests prepared by NCEES to ensure that only qualified candidates are allowed to legally practice as engineers. Passing one of these tests is required in most states for licensure as an engineer.

Both the civil PE and SE exams are open book. Therefore, as part of your exam preparation, you should select, prepare, and review your exam reference materials to become intimately familiar with the contents of each selected book. Your exam reference materials should include books that are specific to the structural subjects on the exam. PPI publishes several books for this purpose. These include *Structural Engineering Reference Manual*, *Structural Depth Reference Manual for the Civil PE Exam*, *Bridge Design for the Civil and Structural PE Exams*, *Timber Design for the Civil and Structural PE Exams*, *Concrete Design for the Civil PE and Structural SE Exams*, and *Seismic Design of Building Structures*.

Although both exams are prepared by NCEES, they are administered by the licensing boards of the various states, which usually also require candidates to have attained a certain level of education and experience before being allowed to take the exams. For information regarding these requirements in the state where you plan to become licensed, and to apply for licensure, contact your state board. Current contact information for each state board may be obtained at **ppi2pass.com/stateboards**.

The Civil Structural PE Exam

The NCEES PE exam in civil engineering consists of two four-hour sections, separated by a one-hour lunch period. Each section contains 40 multiple-choice problems, and you must answer all problems in each section to receive full credit.

The breadth section is taken in the morning by all examinees. In the afternoon, you will take one of five depth sections: water resources and environmental, geotechnical, transportation, construction, or structural. The structural depth section covers a range of structural engineering topics including loadings, analysis, mechanics of materials, materials, member design, design criteria, and other topics. Table 1 provides a detailed analysis of the subjects covered by the civil PE structural depth exam.

To pass the civil PE exam, you must obtain an acceptable score on both the breadth and the depth sections of the exam. You will typically receive your results 8–10 weeks after taking the exam.

The Structural Engineering (SE) Exam

The SE exam is offered in two parts. The first part—vertical forces (gravity/other) and incidental lateral—takes place on a Friday. The second part—lateral forces (wind/earthquake)—takes place on a Saturday. Each part comprises a breadth section and a depth section, as outlined in Table 2.

The breadth sections in the morning are each four hours and contain 40 multiple-choice problems that cover a range of structural engineering topics specific to vertical and lateral forces. The depth sections in the afternoon are also each four hours, but instead of multiple-choice problems, they contain essay (design) problems. You may choose either the bridges or the buildings depth section, but you must work the same depth section across both parts of the exam. That is, if you choose to work buildings for the lateral forces part, you must also work buildings for the vertical forces part. Both breadth and depth sections use customary U.S. units.

According to NCEES, the vertical forces (gravity/other) and incidental lateral depth section in buildings covers loads, lateral earth pressures, analysis methods, general structural considerations (e.g., element design), structural systems integration (e.g., connections), and foundations and retaining structures. The depth section in bridges covers gravity loads, superstructures, substructures, and lateral loads other than wind and seismic. It may also require pedestrian bridge and/or vehicular bridge knowledge.

Table 1 *NCEES Civil PE Structural Depth Exam Specifications*

Analysis of Structures (35%)

Loads and load applications (10%): dead and live loads construction loads, wind loads, seismic loads, moving loads (e.g., vehicular, cranes), snow, rain, ice, impact loads, earth pressure and surcharge loads, load paths (e.g., lateral and vertical), load combinations, tributary areas

Forces and load effects (25%): diagrams (e.g., shear and moment), axial (e.g., tension and compression), shear, flexure, deflection, special topics (e.g., torsion, buckling, fatigue, progressive collapse, thermal deformation, bearing)

Design and Details of Structures (50%)

Materials and material properties (12.5%): concrete (e.g., plain, reinforced, cast-in-place, precast, pretensioned, post-tensioned), steel (e.g., structural, reinforcing, cold-formed), timber, masonry (e.g., brick veneer, CMU)

Component design and detailing (37.5%): horizontal members (e.g., beams, slabs, diaphragms), vertical members (e.g., columns, bearing walls, shear walls), systems (e.g., trusses, braces, frames, composite construction), connections (e.g., bearing, bolted, welded, embedded, anchored), foundations (e.g., retaining walls, footings, combined footings, slabs, mats, piers, caissons, drilled shafts)

Codes and Construction (15%)

Codes, standards, and guidance documents (10%): IBC, ACI 318, ACI 530, PCI, AISC, NDS, AASHTO, ASCE/SEI7, SDPWS, OSHA 1910, and OSHA 1926

Temporary structures and other topics (5%): special inspections, submittals, formwork, falsework and scaffolding, shoring and reshoring, concrete maturity and early strength evaluation, bracing, anchorage, OSHA regulations, safety management

---

Table 2 NCEES SE Exam Specifications

Friday: vertical forces (gravity/other) and incidental lateral	

breadth section
4 hours
40 multiple-choice problems
	analysis of structures (30%)
	    loads (10%)
	    methods (20%)
	design and details of structures (65%)
	    general structural considerations (7.5%)
	    structural systems integration (2.5%)
	    structural steel (12.5%)
	    light gage/cold-formed steel (2.5%)
	    concrete (12.5%)
	    wood (10%)
	    masonry (7.5%)
	    foundations and retaining structures (10%)
	construction administration (5%)
	    procedures for mitigating noncomforming work (2.5%)
	    inspection methods (2.5%)

depth section[a]
4 hours
essay (design) problems
	buildings[b]
	    steel structure (1-hour problem)
	    concrete structure (1-hour problem)
	    wood structure (1-hour problem)
	    masonry structure (1-hour problem)
	bridges
	    concrete superstructure (1-hour problem)
	    steel superstructure (2-hour problem)
	    other elements of bridges (e.g., culverts, abutments, retaining walls) (1-hour problem)

Saturday: lateral forces (wind/earthquake)	

breadth section
4 hours
40 multiple-choice problems
	analysis of structures (37.5%)
	    lateral forces (10%)
	    lateral force distribution (22.5%)
	    methods (5%)
	design and detailing of structures (60%)
	    general structural considerations (7.5%)
	    structural systems integration (5%)
	    structural steel (10%)
	    light gage/cold-formed steel (2.5%)
	    concrete (12.5%)
	    wood (7.5%)
	    masonry (7.5%)
	    foundations and retaining structures (7.5%)
	construction administration (2.5%)
	    structural observation (2.5%)

depth section[a]
4 hours
essay (design) problems
	buildings[c]
	    steel structure (1-hour problem)
	    concrete structure (1-hour problem)
	    wood and/or masonry structure (1-hour problem)
	    general analysis (e.g., existing structures, secondary structures, nonbuilding structures, and/or
	        computer verification) (1-hour problem)
	bridges
	    columns (1-hour problem)
	    footings (1-hour problem)
	    general analysis (e.g., seismic and/or wind) (2-hour problem)

[a]Afternoon sections focus on a single area of practice. You must choose *either* the buildings or bridges depth section, and you must work the same depth section across both parts of the exam.
[b]Problems may contain a multistory building, and may contain a foundation.
[c]At least two problems will include seismic content with a seismic design category of D or above. At least one problem will include wind content with a base wind speed of at least 110 mph. Problems may include a multistory building and/or a foundation.

The lateral forces (wind/earthquake) depth section in buildings covers lateral forces, lateral force distribution, analysis methods, general structural considerations (e.g., element design), structural systems integration (e.g., connections), and foundations and retaining structures. The depth section in bridges covers gravity loads, superstructures, substructures, and lateral forces. It may also require pedestrian bridge and/or vehicular bridge knowledge.

To pass the SE exam, you must obtain acceptable results on both eight-hour parts. However, you are not required to obtain acceptable results on both parts in a single exam administration period. That is, you may sit for and obtain an acceptable result for one part one year and then sit for and obtain an acceptable result for the second part at a later date. Some state boards may impose a five-year window for obtaining acceptable results on both parts.

You will typically receive your results 10–12 weeks after taking the exam.

Codes, Standards, and References Used in This Book

CODES AND STANDARDS

ACI 318: *Building Code Requirements for Structural Concrete.* Farmington Hills, MI: American Concrete Institute, 2011.

ACI 530/530.1 (TMS 402/602): *Building Code Requirements and Specification for Masonry Structures* (and related commentaries). Boulder, CO: The Masonry Society, 2011.

AISC: *Seismic Design Manual,* 2nd ed. Chicago, IL: American Institute of Steel Construction, 2012.

AISC: *Steel Construction Manual,* 14th ed. Chicago, IL: American Institute of Steel Construction, 2011.

AITC: *Timber Construction Manual,* 5th ed. Englewood, CO: American Institute of Timber Construction, 2004.

AITC 104: *Typical Construction Details.* Englewood, CO: American Institute of Timber Construction, 2003.

ANSI/AISC 341: *Seismic Provisions for Structural Steel Buildings.* Chicago, IL: American Institute of Steel Construction, 2012.

ASCE/SEI7: *Minimum Design Loads for Buildings and Other Structures.* Reston, VA: American Society of Civil Engineers, 2010.

FEMA 267: *Interim Guidelines: Evaluation, Repair, Modification and Design of Welded Steel Moment Frame Structures and Interim Guidelines Advisory No. 2: Supplement to FEMA-267 Interim Guidelines.* Sacramento, CA: Federal Emergency Management Agency, 1999.

IBC: *International Building Code,* 5th ed. (without supplements). Falls Church, VA: International Code Council, 2012.

NDS: *National Design Specification for Wood Construction ASD/LRFD* and *National Design Specification Supplement, Design Values for Wood Construction.* Washington, DC: American Forest & Paper Association, 2012.

PCI: *PCI Design Handbook: Precast and Prestressed Concrete,* 7th ed. Chicago, IL: Precast/Prestressed Concrete Institute, 2010.

SDPWS: *Special Design Provisions for Wind and Seismic with Commentary.* Washington, DC: American Forest & Paper Association, 2008.

SEAOC: *SEAOC Blue Book: Seismic Design Recommendations.* Sacramento, CA: Structural Engineers Association of California, 2009.

REFERENCES

American Plywood Association. *APA Engineered Wood Construction Guide.* Tacoma, WA: American Plywood Association.

Becker, R., F. Naeim, and E. J. Teal. "Seismic Design Practice for Steel Buildings." *Steel Tips,* August 1988. Walnut Creek, CA: Steel Committee of California.

Blodgett, O. W. *Design of Welded Structures.* Cleveland, OH: James F. Lincoln Welding Foundation.

Breyer, Donald E., Kenneth J. Fridley, Kelly E. Cobeen, and David G. Pollock, Jr. *Design of Wood Structures: ASD/LRFD.* New York, NY: McGraw Hill Professional.

Darwin, D. *Design of Steel and Composite Beams with Web Openings* (AISC Design Guide 2). Chicago, IL: American Institute of Steel Construction.

Fisher, James M., and Donald R. Buettner. *Light and Heavy Industrial Buildings.* Chicago, IL: American Institute of Steel Construction.

Iwankin, N. "Ultimate Strength Considerations for Seismic Design of the Reduced Beam Section (Internal Plastic Hinge)." *Engineering Journal,* 1st Quarter, 1997. Chicago, IL: American Institute of Steel Construction.

Leet, Kenneth M., Chia-Ming Uang, and Anne Gilbert. *Fundamentals of Structural Analysis.* New York, NY: McGraw-Hill Science/Engineering/Math.

Naeim, Farzad, ed. *The Seismic Design Handbook.* New York, NY: Springer.

Nilson, A. H., D. Darwin, and C. W. Dolan. *Design of Concrete Structures.* New York, NY: McGraw-Hill.

Paz, Mario, and William Leigh. *Structural Dynamics: Theory and Competition.* New York, NY: Springer.

Schnabel, Harry W. *Tiebacks in Foundation Engineering and Construction.* New York, NY: Taylor & Francis.

Young, Warren, Richard G. Budynas, and Ali Sadegh. *Roark's Formulas for Stress and Strain.* New York, NY: McGraw-Hill.

Nomenclature

symbol	definition (units)
a	depth of equivalent rectangular stress block (in)
a	length of cantilevered beam (in)
a_p	component amplification factor (ASCE/SEI7 Sec. 13.3.1)
A	area (in^2)
A_b	area of individual bar (in^2)
A_c	equivalent area of concrete compression zone (in^2)
A_{comp}	cross-sectional area in compression (in^2)
A_{cv}	gross area of concrete section bounded by web thickness and length of section in the direction of shear force (in^2)
A_e	effective net area (in^2)
A_{fn}	net cross-sectional area of flange (in^2)
A_g	gross cross-sectional area (in^2)
A_{gv}	gross cross-sectional area subject to shear (in^2)
A_h	area of shear reinforcement parallel to flexural tension reinforcement (in^2)
A_n	area of reinforcement in bracket or corbel resisting tensile force N_{uc} (in^2)
A_n	net cross-sectional area (in^2)
A_{nv}	net area subject to shear (in^2)
A_{ps}	area of prestressing tendon (in^2)
A_{PL}	cross-sectional area of plate (in^2)
A_s	area of tension reinforcement (in^2)
A'_s	area of compression reinforcement (in^2)
A_{sc}	area of primary tension reinforcement in bracket or corbel (in^2)
A_{sh}	total cross-sectional area of transverse reinforcement (including crossties) within spacing s and perpendicular to dimension h_c (in^2)
$A_{s,l}$	area of steel in longitudinal direction (in^2)
A_{st}	total area of longitudinal steel (in^2)
$A_{s,t}$	area of steel in transverse direction (in^2)
A_{sw}	area of steel in web area of wall (in^2)
A_T	tributary area supported by structural member (in^2)
A_{ten}	cross-sectional area in tension (in^2)
A_v	area of shear reinforcement within spacing s (in^2)
A_{vf}	area of shear-friction reinforcement (in^2)
A_w	cross-sectional area of web (in^2)
b	breadth or width (in)
b_c	least dimension center-to-center of closed hoop (in)
b_e	effective width (in)
b_R	required reduction in flange width (in)
b_w	web width (in)
B	breadth of structural shell (in)
B	width of support or member (in)
B_1, B_2	amplifiers to account for second-order effects (AISC App. 8)

symbol	definition (units)
c	coefficient for doubly symmetric I shapes (1.0)
c	distance from extreme compression fiber to neutral axis (in)
c	size of rectangular or equivalent rectangular column, capital, or bracket (in)
c_b	distance from centroid of section to extreme bottom fiber (in)
c_t	distance from centroid of section to extreme top fiber (in)
C	change in elevation (in)
C	compressive force (lbf)
C_c	compressive force in concrete (lbf)
C_d	deflection amplification factor (ASCE/SEI7 Table 12.2-1)
C_D	load duration factor (NDS Table 2.3.2)
C_f	constant based on stress category (AISC Table A-3.1)
C_F	size factor for sawn lumber (NDS Sec. 4.3.6)
C_g	group action factor (NDS Sec. 10.3.6)
C_L	beam stability factor (NDS Sec. 3.3.3)
C_m	factor relating actual moment diagram to equivalent uniform moment diagram
C_M	wet service factor
C_p	wall pressure coefficients (ASCE/SEI7 Fig. 6-6)
C_P	column stability factor (NDS Eq. 3.7-1)
C_{pe}	external wall pressure coefficients (ASCE/SEI7 Fig. 6-6)
C_{pn}	net wall pressure coefficients (ASCE/SEI7 Fig. 6-6)
C_{pr}	factor to account for probable overstrength of steel section
C_r	repetitive use factor (NDS Sec. 4.3.9)
C_s	effective compressive force in reinforcement (lbf)
C_s	seismic response coefficient
C'_s	compressive force in steel (lbf)
C_t	period coefficient (ASCE/SEI7 Table 12.8-2)
C_{tn}	toenail factor (NDS Sec. 11.5.4)
C_V	volume factor (NDS Sec. 5.3.6)
C_{vx}	vertical distribution factor (ASCE/SEI7 Chap. 12)
d	depth or diameter (in)
d	distance from extreme compression fiber to centroid of longitudinal tension reinforcement (in)
d	toe length (in)
d'	distance from extreme compression fiber to centroid of compression reinforcement (in)
d_b	nominal diameter of bar, wire, or prestressing strand (in)
d_{bl}	diameter of longitudinal bar (in)
d_e	effective depth (in)
D	dead load (lbf or lbf/in^2)

symbol	definition (units)
D	diameter or displacement (in)
D	number of 1/16ths of an inch in fillet weld size
D_{1Q}, D_{2Q}	displacement caused by applied load (in)
D_f	depth of foundation (in)
D_F	temperature coefficient (ft/ft-°F)
e	eccentricity of force (in)
E	modulus of elasticity (lbf/in^2)
E'	adjusted modulus of elasticity (lbf/in^2)
E_c	modulus of elasticity of concrete (lbf/in^2)
E_m	modulus of elasticity of masonry (lbf/in^2)
E_{min}, E'_{min}	modulus of elasticity and adjusted modulus of elasticity for beam and column stability calculations (lbf/in^2)
E_{ps}	modulus of elasticity of prestressing steel (lbf/in^2)
E_s	modulus of elasticity of steel (lbf/in^2)
E_w	modulus of elasticity of wood (lbf/in^2)
EI	flexural stiffness of compression member (in^2-lbf)
f	calculated compressive stress in a slender element (kips/in^2)
f	natural frequency (Hz)
$f_{11}, f_{12}, f_{21}, f_{22}$	flexibility influence coefficient (in/lbf)
f_b	shear flow or stress caused by bending moment (lbf/in)
f_{BM}	stress in base material (lbf/in^2)
f_c	calculated compressive stress in concrete (lbf/in^2)
f'_c	specified compressive strength of concrete (lbf/in^2)
$\sqrt{f'_c}$	square root of specified compressive strength of concrete (lbf/in^2)
$f_{c\perp}$	calculated compressive stress perpendicular to grain (lbf/in^2)
f'_m	specified compressive strength of masonry (lbf/in^2)
f_p	passive earth pressure (lbf/in^2)
f_{pc}	stress in concrete caused by prestress (lbf/in^2)
f_{pe}	effective prestress after all losses (lbf/in^2)
f_{pu}	specified tensile strength of prestressing steel (lbf/in^2)
f_r	resultant shear flow (lbf/in)
f_s	calculated stress in reinforcement at service loads (lbf/in^2)
f'_s	stress in compression reinforcement (lbf/in^2)
f_{si}	initial stress in reinforcement (lbf/in^2)
f_{SR}	actual stress range (lbf/in^2)
f_{st}	transfer stress in reinforcement (lbf/in^2)
f_{sw}	working stress in reinforcement (lbf/in^2)
f_v	required shear stress (lbf/in^2)
f_v	shear flow caused by force (lbf/in)
f_v	skin friction (lbf/in^2)
f_y	specified yield strength of reinforcement (lbf/in^2)
f_{ypr}	probable yield strength (lbf/in^2)
f_{yt}	specified yield strength of transverse reinforcement (lbf/in^2)
F	force (lbf)
F	inches of rise per foot of run (in/ft)
F_b	allowable bending stress (lbf/in^2)
F'_b	adjusted allowable bending stress (lbf/in^2)
F^*_b	partially adjusted allowable bending stress (lbf/in^2)

symbol	definition (units)
F_{bE}	critical buckling stress for bending member (lbf/in^2)
F_c	compression parallel to grain (NDS Table 4A) (lbf/in^2)
$F_{c\perp}$	compression perpendicular to grain (NDS Table 4A) (lbf/in^2)
$F'_{c\perp}$	adjusted compression perpendicular to grain (lbf/in^2)
F'_c	adjusted compression parallel to grain (lbf/in^2)
F^*_c	partially adjusted compression parallel to grain (NDS Sec. 15.3.2)
F_{cE}	critical buckling design value for compression member (lbf/in^2)
F_{cr}	flexural buckling stress (lbf/in^2)
F_e	elastic critical buckling stress (lbf/in^2)
F_{EXX}	electrode classification number (lbf/in^2)
F_p	horizontal seismic design force (lbf/in^2)
F_{PL}	force in plate (lbf)
F_s	allowable steel stress in reinforced masonry (lbf/in^2)
F_{SR}	design stress range (lbf/in^2)
F_t	tension design value parallel to grain (lbf/in^2)
F'_t	adjusted tension design value parallel to grain (lbf/in^2)
F_{TH}	fatigue threshold (lbf/in^2)
F_u	specified minimum tensile strength (lbf/in^2)
F_v	shear parallel to grain (horizontal shear) (lbf/in^2)
F_v	vertical component of prestress force (lbf/in^2)
F'_v	adjusted shear parallel to grain (horizontal shear) (lbf/in^2)
F_w	nominal strength of weld per unit area (lbf/in^2)
F_w	weld electrode strength (lbf/in^2)
F_x, F_y, F_z	force in the x-, y-, or z-direction (lbf)
F_x	lateral force (lbf)
F_y	yield stress (lbf/in^2)
FEM	fixed end moment (in-lbf)
FS	factor of safety
g	transverse center-to-center spacing (gage) between fastener gage lines (in)
G	gust effect factor (ASCE/SEI7 Sec. 26.9)
G	shear modulus (lbf/in^2)
G	stiffness ratio
h	height (in)
h	overall thickness of member (in)
h_c	cross-sectional dimension of column core measured center to center of confining reinforcement (in)
h_c	two times distance from centroid to inside face of compression flange less fillet radius (in)
h_n	height of highest level of structure above base (ft)
h_p	height of parapet (ft)
h_p	two times distance from plastic neutral axis to nearest line of fasteners at compression flange (in)
h_s	overall thickness of slab (in)
h_{sx}	story height (in)
h_w	height of wall from base to top (in)
H	height (in)
H	horizontal force (lbf)
I	moment of inertia (in^4)

symbol	definition (units)
I^*	reduced moment of inertia to account for inelastic behavior (in^4)
I_b	moment of inertia of beam (in^4)
I_c	centroidal moment of inertia (in^4)
I_{cr}	moment of inertia of cracked section transformed to concrete (in^4)
I_e	seismic importance factor
I_g	moment of inertia of girder (in^4)
I_g	moment of inertia of gross concrete section about centroidal axis, neglecting reinforcement (in^4)
I_p	component importance factor (ASCE/SEI7 Sec. 13.1.3)
I_s	moment of inertia of slab (in^4)
I_x	live load deflection (in^4)
j	ratio of distance between centroid of flexural compressive forces and centroid of tensile forces to effective depth $= 1 - k/3$
k	neutral axis depth factor $= \sqrt{2\rho n + (\rho n)^2} - \rho n$
k	spring constant (lbf/in)
k	stiffness (lbf/in)
k_a	coefficient of active earth pressure
k_{des}	distance from outer face of flange to web toe of fillet (in)
K	effective length factor
K	relative rigidity (lbf/in)
K_d	wind directionality factor (ASCE/SEI7 Table 26.6-1)
K_f	column stability coefficient (NDS Sec. 15.3.2)
K_{LL}	live load element factor (ASCE/SEI7 Table 4-2)
K_z	velocity pressure exposure coefficient at elevation z (ASCE/SEI7 Sec. 28.3.1)
K_{zt}	topographic factor (ASCE/SEI7 Sec. 26.8.2)
l_a	additional embedment length at support or point of inflection (in)
l_d	development length (in)
l_{dh}	development length of hooked tension bar (in)
l_e	effective length (in)
l_{ev}	vertical edge distance (in)
l_{ld}	length of lap splice dowels (in)
l_n	length of clear span from face to face of beams or other supports (in)
l_o	overall length of continuous beam (in)
l_t	transfer length (in)
l_u	unsupported span length (in)
l_w	overall length of wall (in)
L, L'	length (in)
L	live load (lbf or lbf/in^2)
L	participation factor (lbf-sec^2/in)
L_b	laterally unsupported length of compression flange (in)
L_e	effective length (in)
L_o	unreduced design live load (lbf/in^2)
L_r	reduced roof live load (lbf/in^2)
L_{trib}	length of collector transferring shear to wall (in)
L_w	length of weld (in)
L_x, L_y	projected length of shell from lowest point to highest point, with subscript indicating direction (in)

symbol	definition (units)
m	mass (lbm or lbf-sec^2/in)
m	member internal moment caused by unit virtual force (units same as virtual force)
m^*	generalized mass (lbf-sec^2/in)
M	moment (in-lbf)
M_0	overturning moment (in-lbf)
M_b	nominal flexural strength at balanced strain condition (in-lbf)
M_c	available flexural strength (in-lbf)
M_{cx}	allowable flexural strength about strong axis or x-axis (in-lbf)
M_{cy}	allowable flexural strength about weak axis or y-axis (in-lbf)
M_d	moment caused by dead load (in-lbf)
M_e	moment caused by earthquake (in-lbf)
M_l	moment caused by live load (in-lbf)
M_{lt}	first-order moment caused by translation (in-lbf)
M_n	nominal flexural strength (in-lbf)
M_n	nominal moment strength at section (in-lbf)
M_{not}	bending moment caused by notational load (in-lbf)
M_{nt}	first-order moment, no lateral translation (in-lbf)
M_o	total factored static moment (in-lbf)
M_p	plastic bending moment (in-lbf)
M_{pb}	probable moment strength of steel beam (in-lbf)
M_{pc}	moment at plastic centroid (in-lbf)
M_{pr}	probable flexural strength (in-lbf)
M_{ps}	moment caused by prestress force (in-lbf)
M_r	required flexural strength (in-lbf)
M_r	resisting moment against overturning (in-lbf)
M_{rx}	required flexural strength or bending moment about strong axis or x-axis (in-lbf)
M_{ry}	required flexural strength or bending moment about weak axis or y-axis (in-lbf)
M_s	static moment of inertia (in-lbf)
M_t	torsional moment (in-lbf)
M_t	inherent torsional moment (in-lbf)
M_u	factored moment at section (in-lbf)
M_y	yield moment about axis of bending (in-lbf)
M_{yc}	yield moment for compression flange (in-lbf)
MCE	maximum considered earthquake
n	modular ratio of elasticity $= E_s/E_c$ or E_s/E_m
n	number of fasteners
n_{bar}	number of bars
n_h	number of fasteners resisting horizontal force
n_v	number of fasteners resisting shear force
N	bearing length (in)
N	net lateral earth pressure (lbf/in^2)
N_i	notational load applied at level i (lbf)
N_{uc}	factored tensile force applied at top of bracket or corbel (lbf)
N_{xy}, N_{yx}	stress resultant in a structural shell, with subscript indicating direction (lbf/in)
p	load per unit of length (lbf/in)
p	member internal force caused by unit virtual force (units same as virtual force)
p_{s30}	simplified design wind pressure (ASCE/SEI7 Sec. 28.6.3)

symbol	definition (units)
P	load or force (lbf)
P	unfactored axial load (lbf)
P_1, P_2, P_3	participation factor
P_0	nominal axis load strength at zero eccentricity (lbf)
P_b	nominal axial compression strength at balanced strain condition (lbf)
P_c	allowable or available axial compressive strength (lbf)
P_c	critical load (lbf)
P_c	design axial compressive strength (lbf)
P_{cW}	axial compression force caused by wind (lbf)
P_d	dead load (lbf)
P_e	earthquake-induced load (lbf)
P_e	effective pressure force (lbf)
P_{e1}	elastic critical buckling resistance (AISC App. 8)
P_l	live load (lbf)
P_{lt}	axial force associated with lateral translation
P_n	nominal strength of cross section subject to compression (lbf)
P_n	tensile strength (lbf)
P_{not}	axial force caused by notational load (lbf)
P_{nt}	first-order axial force assuming no lateral translation of frame (lbf)
P_r	required strength (lbf)
P_{rb}	required horizontal brace force (lbf)
P_s	axial load due to snow (lbf)
P_u	factored axial load at given eccentricity (lbf)
P_u	required axial strength in compression (lbf)
P_{uf}	factored load from tributary floor or roof area (lbf)
P_{uw}	factored weight of wall (lbf)
P_w	axial force caused by wind (lbf)
P_x	total vertical design load (lbf)
P_y	load at yield (lbf)
P_z	uniform loading on a structural shell in the z-direction (lbf/in^2)
q	lateral force (lbf)
q	shear flow in fillet weld (lbf/in)
q_z	velocity pressure at elevation z (ASCE/SEI7 Eq. 29.3-1) (lbf/in^2)
Q	statical moment of area (in^3)
Q	virtual load (lbf)
Q_E	effect of horizontal seismic forces (lbf)
Q_f	surcharge load at base of foundation (lbf/in^2)
r	fraction of force carried by beam
r	radius of gyration of cross section of compression member (in)
r_n	nominal shear strength of one bolt (lbf)
r_{ts}	effective radius of gyration (AISC Eq. F2.4) (in)
R	beam reaction (lbf)
R	response modification coefficient (ASCE/SEI7 Table 12.2-1)
R	shear strength of a group of bolts (lbf)
R_1, R_2	reduction factor (ASCE/SEI7 Sec. 4.8.2)
R_a	required strength (lbf)
R_B	slenderness ratio of bending member
R_M	coefficient to account for influence of P-δ on P-Δ
R_n	nominal strength (lbf or lbf/in)
R_p	component response modification factor (ASCE/SEI7 Sec. 13.3.1)

symbol	definition (units)
s	spacing of members (ft)
s'	adjusted spacing of members (ft)
s_o	maximum spacing of transverse reinforcement (in)
S	elastic section modulus of section (in^3)
S_a	spectral acceleration or pseudo-acceleration (in/sec^2)
S_b	section modulus with respect to bottom (in^3)
S_d	spectral displacement (in)
S_{D1}	design spectral response acceleration parameter at a period of 1 sec (ASCE/SEI7 Chap. 11)
S_{DS}	design spectral response acceleration parameter at short periods (ASCE/SEI7 Chap. 11)
S_g	ground acceleration (in/sec^2)
S_{M1}	maximum considered earthquake (MCE) spectral response acceleration parameter at a period of 1 sec (ASCE/SEI7 Chap. 11)
S_{MS}	maximum considered earthquake (MCE) spectral response acceleration parameter at short periods (ASCE/SEI7 Chap. 11)
S_t	section modulus with respect to top (in^3)
S_w	section modulus of weld (in^2)
S_x	elastic section modulus taken around x-axis (in^3)
t	thickness (in)
t_f	connector flange thickness (in)
t_s	thickness of side member (in)
t_w	connector web thickness (in)
t_w	thickness of wall (in)
t_{we}	equivalent width of wall or web (in)
T	fundamental period (sec)
T	temperature (°F)
T	tension force (lbf)
T_a	approximate fundamental period (sec)
T_L	long-period transition period (sec)
T_n	nominal torsional moment strength (in-lbf)
T_s	$S_{D1}/S_{D2} \times 1$ sec (sec)
T_u	factored torsional moment at section (in-lbf)
T_w	tensile force in web area of wall (lbf)
u	bearing pressure caused by gravity loads plus uplift (lbf/in^2)
U	required strength to resist factored loads or related internal moments and forces (lbf)
U	shear lag factor
v	shear stress per unit length (lbf/in)
V	shear force or shear strength (lbf)
V	vertical force (lbf)
V_c	nominal shear strength provided by concrete (lbf)
V_e	effective shear strength (lbf)
V_E	seismic shear (lbf)
V_{ex}	seismic shear in x-direction (lbf)
V_n	nominal shear strength (lbf)
V_r	design shear (lbf)
V_r'	adjusted design shear (lbf)
V_s	nominal shear strength provided by shear reinforcement (lbf)
V_u	factored shear force at section (lbf)
V_u	ultimate shear force (lbf)
V_W	wind shear (lbf)
V_x	seismic shear force (lbf)
w	length of weld (in)

symbol	definition (units)
w	uniformly distributed weight or load per unit of length (lbf/in)
w_b	weight of beam per unit of length or area (lbf/in or lbf/in^2)
w_c	unit weight of concrete (lbf/in^3)
w_d, w_D	dead load (lbf/in or lbf/in^2)
w_e	equivalent uniform load from prestress (lbf/in or lbf/in^2)
w_E	seismic force (lbf/in^2)
w_L	live load (lbf/in or lbf/in^2)
w_s	weight of slab (lbf/in or lbf/in^2)
w_u	factored load per unit length of beam or per unit area of slab (lbf/in or lbf/in^2)
w_W	force per unit length due to wind (lbf/in)
w_w	unit weight of wood (lbf/in^3)
W	seismic dead load (lbf or lbf/ft)
W	weight or vertical force (lbf)
W	width of structural shell (in)
W	withdrawal force for fastener (lbf)
x	displacement (in)
x	distance along member axis (in)
x	period coefficient (ASCE/SEI7 Table 12.8-2)
$\overline{x}$	connection eccentricity (in)
x_m	distance to center of mass (ft)
y	distance in y-direction (in)
y	longer overall dimension of rectangular part of cross section (in)
$\overline{y}$	distance from reference axis to centroid of section (in)
y_c	distance from center of gravity of concrete to center of gravity of prestress steel at center (in)
y_e	distance from center of gravity of concrete to center of gravity of prestress steel at end (in)
$y_{\max}$	maximum vertical deflection (in)
$Y2$	distance from top of steel beam to concrete flange force (in)
Y_i	gravity load applied at level i (lbf)
Z	lateral design value for a single fastener connection (lbf)
Z'	adjusted lateral design value for a single fastener connection (lbf)
$Z_\parallel$	lateral design value for a single fastener connection with wood members loaded parallel to grain (lbf)
$Z'_\perp$	adjusted lateral design value for a single fastener connection with wood members loaded perpendicular to grain (lbf)
Z_x	plastic section modulus about the major axis (in^3)
α	ASD/LRFD force level adjustment factor
α_f	ratio of flexural stiffness of beam section to flexural stiffness of width of slab
α_x, α_y	angle between horizontal and rib of structural shell, with subscript indicating direction (rad)
β_1	depth factor (ACI 318 Sec. 10.2.7.3)
β_{dns}	ratio of dead load to total load
β_l	ratio of live load to total load
β_{NF}	chord rotation (ratio of deflection to member length) of near end relative to far end (rad)
γ_s	unit weight of soil (lbf/in^3)

symbol	definition (units)
δ	moment magnification factor for frames braced against sidesway
δ_u	deflection due to factored loads (in)
δ_x	deflection at center of mass (ASCE/SEI7 Eq. 12.8-15) (in)
δ_{xe}	deflection at center of mass determined by elastic analysis (ASCE/SEI7 Eq. 12.8-15) (in)
Δ	deflection (in)
Δ	design story drift (ASCE/SEI7 Sec. 12.8.6) (in)
Δ_{de}	elastic deflection of diaphragm (in)
Δ_L	live load deflection (in)
Δ_T	deflection due to temperature change (in)
Δ_{we}	elastic deflection of wall (in)
ϵ_c	ultimate strain in concrete (in/in)
ϵ_{ce}	strain in concrete under effective prestress (in/in)
ϵ_{mu}	ultimate strain in masonry (in/in)
ϵ_{pe}	strain in prestressed steel under effective prestress (in/in)
ϵ_{ps}	strain in prestressed steel at nominal flexural strength (in/in)
ϵ_s	tension steel strain
ϵ'_s	strain in compression steel (in/in)
ϵ_{su}	ultimate steel strain (in/in)
ϵ_y	strain in reinforcement at first yield (in/in)
θ	rotation of joint (rad)
λ	adjustment factor for building height and exposure (ASCE/SEI7 Fig. 28.6-1)
λ	lightweight aggregate concrete factor (0.75)
λ	normal weight aggregate concrete factor (1.0)
λ	sand-lightweight aggregate concrete factor (0.85)
λ	shear
λ	slenderness parameter ($b_f/2t_f$)
λ_p	limiting slenderness for compact element (AISC Chap. F)
λ_{pf}	limiting slenderness for compact flange (AISC Chap. F)
λ_{pw}	limiting slenderness for compact web (AISC Chap. F)
λ_r	limiting slenderness for noncompact element (AISC Chap. F)
λ_{rf}	limiting slenderness for noncompact flange (AISC Chap. F)
μ	coefficient of static friction
ξ	damping ratio
ρ	ratio of non-prestressed tension reinforcement = A_s/bd
ρ_l	ratio of longitudinal wall steel to gross area of wall
ρ_t	ratio of transverse wall steel to gross area of wall
τ	shear stress computer by elastic theory (lbf/in^2)
ϕ	rotation (rad)
ϕ	strength reduction factor
ϕ_F	rotation of far end (rad)
ϕ_i	eigenvector
ϕ_N	rotation of near end (rad)
Φ	eigenvector matrix
ψ	stiffness ratio
ψ_e	factor used to modify development length for bar coating
ψ_s	factor used to modify development length for bar size

symbol	definition (units)
ψ_t	factor used to modify development length for bar position
ω	length of weld (in)
ω	natural frequency (rad/sec)
Ω	safety factor

Subscripts

acc	accidental
act	actual
ave	average
b	beam or bottom
c	concrete
col	column
d	dead load
e	effective
ext	exterior
f	flange or floor
F	far end
horiz	horizontal
int	interior
l	live load
L	left
m	masonry
max	maximum

symbol	definition (units)
min	minimum
n	nominal
N	near end
NF	near end relative to far end
pc	plastic centroid
pr	probable
ps	prestress
Q	virtual force
r	roof
R	right
RBS	reduced beam section
req	required
s	slab or steel
sup	superimposed
t	top or torsional
v	shear
vert	vertical
w	web or weld
$w1, w2$	web region 1, web region 2
x	in x-direction
x	related to strong axis bending
y	in y-direction
y	related to weak axis bending
$\parallel$	parallel to grain
$\perp$	perpendicular to grain

1 Structural Analysis

PROBLEM 1

1.1. Draw the bending moment diagrams for the frames with the loads shown. Indicate the magnitude of the maximum bending moments.

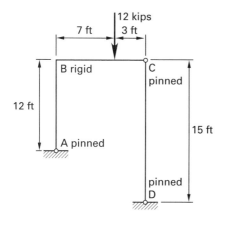

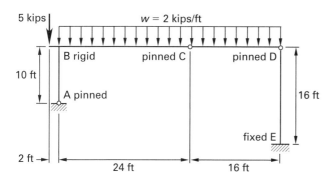

PROBLEM 2

A section of crane rail girder is supported on piles. The piles at A and C are rigid supports. The pile at B was discovered to yield when load was applied. Tests revealed that pile B acts like a spring with a spring constant, k, equal to 600 kips per inch of vertical displacement. A wheel load, P, of 100 kips is placed at B.

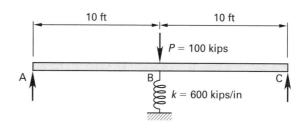

Design Criteria

- assume the force in pile B is zero under girder dead load

- crane girder is 2.5 ft × 2.5 ft, regular weight concrete

- $E_c = 3000$ ksi

2.1. Use the gross cross-sectional area to determine the girder's rigidity.

2.2. Determine the moment and deflection in the girder at point B due to the wheel load.

PROBLEM 3

A schematic section indicating the floor occupancy of an 18-story multiuse building is shown.

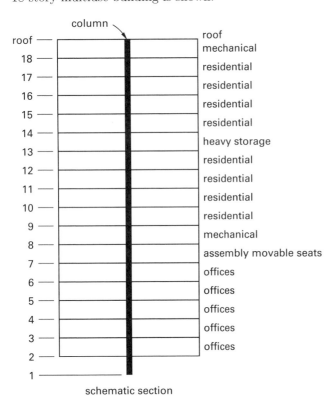

schematic section

Design Criteria

- roof dead load = 80 lbf/ft^2
- floor dead load (including partitions) = 100 lbf/ft^2
- mechanical floor live load = 125 lbf/ft^2
- tributary area to column at each level = 900 ft^2
- column weight neglected

3.1. Determine the dead load plus live load at each story.

PROBLEM 4

A frame is loaded as shown.

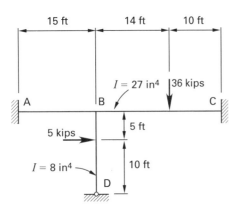

Design Criteria

- $E = 20,000$ ksi

4.1. Determine the moments at each joint.

4.2. Draw the moment diagram.

4.3. Draw the deflected shape.

4.4. Determine the maximum deflection in span AB.

PROBLEM 5

A rigid frame is shown. The right support has settled 6 in.

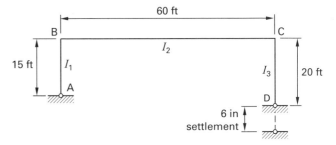

Design Criteria

- $I_1 = 500$ in^4 (at AB)
- $I_2 = 1600$ in^4 (at BC)
- $I_3 = 600$ in^4 (at CD)
- $E = 29,000$ ksi

5.1. Find the required moments at joints B and C due to the settlement.

5.2. Draw the moment diagram.

5.3. Find the vertical and horizontal forces at joints A and D.

PROBLEM 6

All members shown are timber except member DG, which is a steel rod. All connections are pinned.

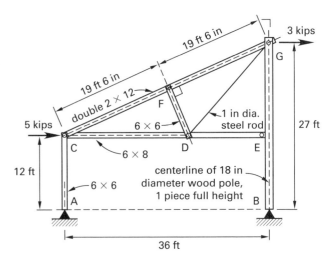

Design Criteria

- $E_w = 1600$ ksi
- $E_s = 29,000$ ksi

6.1. Calculate the external reactions.

6.2. Calculate and draw a diagram of the internal member forces (shear, moment, and axial).

6.3. Calculate the horizontal deflection of point C, neglecting shear distortions within the members. Determine whether the axial or the bending stress contributes more to the overall deflection.

PROBLEM 7

A symmetrical gable frame is shown.

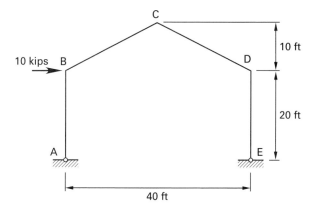

Design Criteria

- $I = 300$ in^4
- $E = 29,000$ ksi
- do not use rigid frame formulas

7.1. Determine the reactions at supports A and E.

7.2. What is the vertical deflection at point C due to the lateral load shown at point B?

7.3. Assume the horizontal load on the frame is removed. Apply a 5 kip vertical load at point C. What is the horizontal deflection at point B?

PROBLEM 8

For the evaluation of moments, shears, and axial loads in high-rise buildings, there are three approximate methods of analysis commonly used. The methods are

- portal method
- cantilever method
- moment distribution method (Hardy Cross)

The frame shown in illustration (a) was analyzed by each of the three methods for overturning forces at the third floor. The results are shown in illustrations (b), (c), and (d), not necessarily respectively.

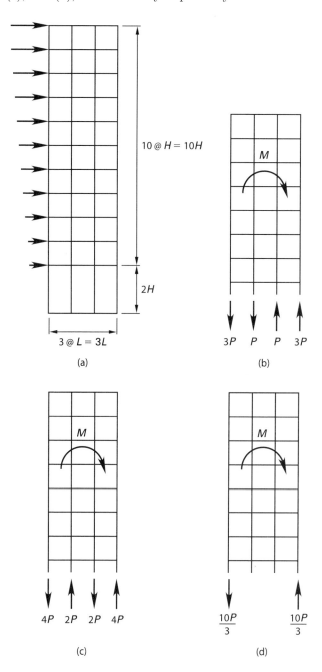

8.1. Identify the method that was used to obtain each result shown in illustrations (b), (c), and (d).

8.2. List the principal assumptions that are used as the basis for each of the three methods.

8.3. For what type and configuration of structural frame is the portal method a better method of analysis than the cantilever method? Explain.

PROBLEM 9

A shear wall and a braced frame are linked as shown. A lateral force of 70 kips is applied. Assume all connections are adequate.

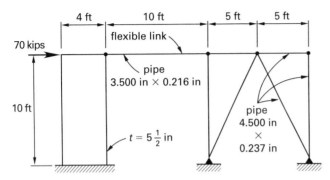

Design Criteria

- $f'_c = 3$ ksi
- $G = 0.4E$
- $F_y = 36$ ksi
- $E_s = 29,000$ ksi
- A500 grade B steel pipe

9.1. Calculate the force resisted by the shear wall.

9.2. Calculate the force resisted by the flexible link.

9.3. Calculate the force resisted by the braced frame.

9.4. Calculate the deflection of the shear wall.

9.5. Calculate the deflection of the braced frame.

PROBLEM 10

The frame shown is fixed at the base and pinned at the top of the columns.

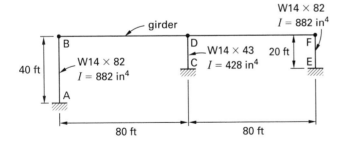

Design Criteria

- change in temperature of girder only is $\Delta T = 60°F$
- temperature coefficient is $D_F = 6 \times 10^{-6}$ in/in-°F
- $E = 30,000$ ksi
- neglect girder shortening induced from column shear
- neglect effect of temperature change on columns

10.1. What are the moments induced at joint A and joint E?

10.2. What are the shear forces at all columns due to the lateral deflection?

10.3. What is the magnitude and direction of deflection at the top of columns?

10.4. What is the bending stress on center column CD due to the lateral deflection?

PROBLEM 11

11.1. Calculate the relative rigidities of the shear-resisting elements in the four structures shown, taking the rigidity in illustration (a) as unity.

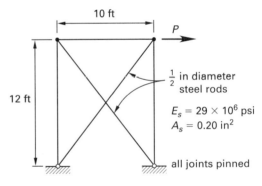

(a) braced frame

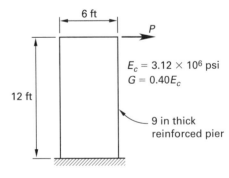

(b) cantilevered wall

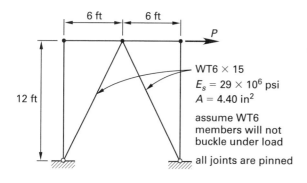

(c) *K*-braced truss

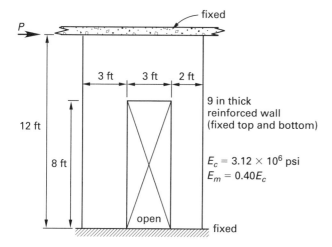

(d) pierced shear wall

PROBLEM 12

The framing condition shown (see *Illustration for Prob. 12*) has been utilized in a 100-unit subdivision. The plans were reviewed and approved prior to construction, and construction was in accordance with this detail.

Design Criteria

• adequate drains and waterproofing provided for the masonry walls

• no connection between 2×6 brace and web members of trusses

• trusses spanning 22 ft, supported by masonry walls that do not retain earth

• horizontal steel in masonry wall not considered

• plan dimensions of garage are 22 ft × 22 ft

standard framing angles	permissible capacity for normal duration loading
type 1	260 lbf
type 2	1000 lbf

12.1. Determine whether the construction detail is adequate. Include substantiating calculations and assumptions. If the detail is inadequate, provide recommendations for remedial repair. Include calculations, assumptions, and suitable sketches to show recommended repairs.

PROBLEM 13

A shoring system has been selected to shore an excavation bank for a new underground structure and support an adjacent warehouse. The system calls for a series of soldier beams dropped into pre-drilled holes at 8 ft on center with timber lagging placed between them. Two tie-backs are placed to resist lateral pressure. A typical section is shown (see *Illustration for Prob. 13*).

Design Criteria

• tie rods provide rigid support

• net pressure is $N = Q_f - 120 D_f$

13.1. Show all lateral pressures acting on the soldier beams. Calculate critical values.

13.2. Calculate the forces at A, B, and C.

13.3. Check the adequacy of the assumed toe length.

13.4. Calculate the grouted anchor length of tie-backs A and B. Draw a sketch to show how length is measured.

13.5. Describe step by step the installation procedure of this system.

Illustration for Prob. 12

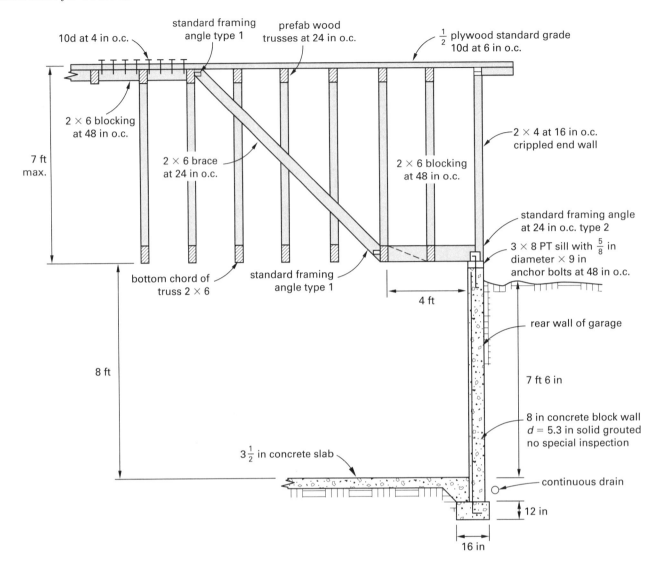

Illustration for Prob. 13

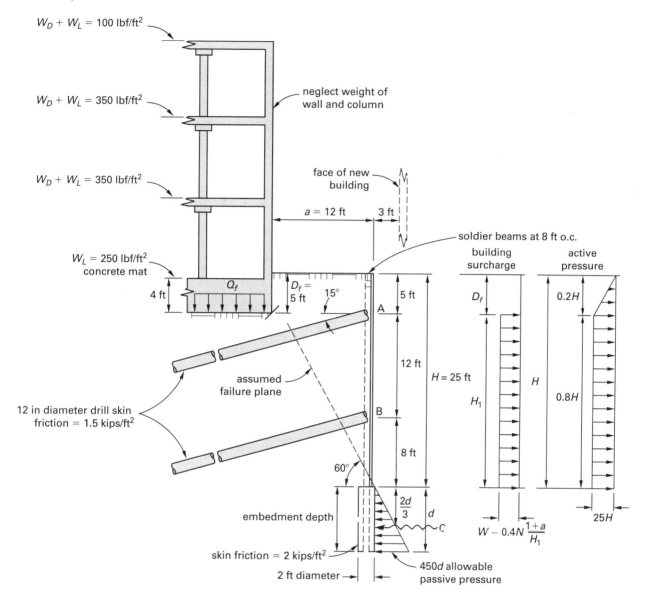

PROBLEM 14

The effective dead loads on each floor of a three-story building are as shown.

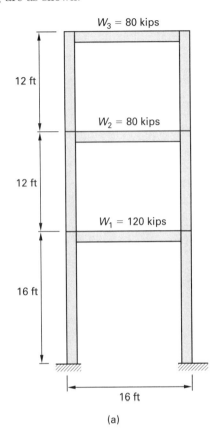

(a)

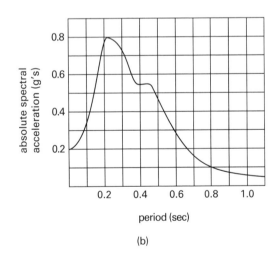

period (sec)

(b)

Design Criteria

- The following dynamic properties of the plane frame are given. The eigenvectors are

$$\Phi = \begin{bmatrix} 1.690 & -1.208 & -0.714 \\ 1.220 & 0.704 & 1.697 \\ 0.572 & 1.385 & -0.948 \end{bmatrix}$$

$$\Phi^T M \Phi = [1]$$

- The eigenvalues, in rad/sec, are

$$\omega = \begin{bmatrix} 8.77 \\ 25.18 \\ 48.13 \end{bmatrix}$$

14.1. Compute the participation factors.

14.2. Check the participation factors.

14.3. Calculate the displacements of each floor based on the spectra shown in illustration (b).

14.4. Calculate the interstory drift for each floor.

PROBLEM 15

For the steel rigid frame shown, the structure has girders that are infinitely stiff compared to the columns, masses concentrated at the plane of the girders, and 7% damping. Neglect axial strains in columns.

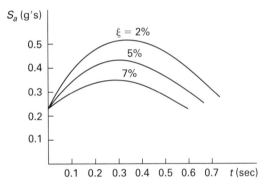

acceleration response spectra

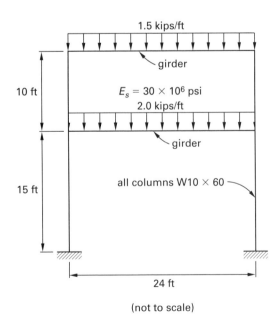

(not to scale)

Design Criteria

- $E = 30{,}000$ ksi

15.1. Calculate the natural frequencies and characteristic shapes of the structure.

15.2. Determine the spectral acceleration and velocity of the first mode.

PROBLEM 16

The following plan and profile show a roof structure supported by cable. The member sizes for the outer rings and inner rings, as well as the plan dimensions, are also shown.

Design Criteria

- lateral force system to be neglected

16.1. What is the horizontal component of cable tension at point A on the cable? (See profile view.) Explain.

16.2. Verify the adequacy of the tension ring indicated for combined stresses. Show calculations.

16.3. Verify the adequacy of the compression ring indicated for combined stresses. Show calculations.

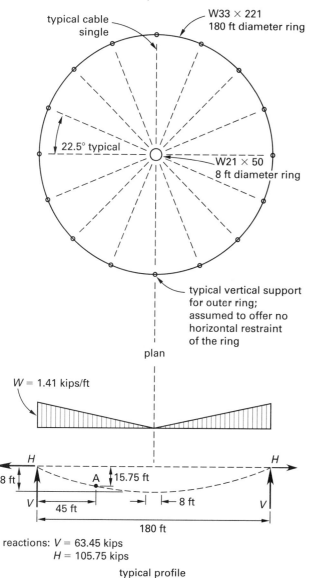

all W shapes A36 steel

typical cable single

W33 × 221
180 ft diameter ring

22.5° typical

W21 × 50
8 ft diameter ring

typical vertical support for outer ring; assumed to offer no horizontal restraint of the ring

plan

$W = 1.41$ kips/ft

H H

18 ft

A 15.75 ft

V V

45 ft   — 8 ft

180 ft

reactions: $V = 63.45$ kips
$H = 105.75$ kips

typical profile

SOLUTION 1

1.1. Construct the shear and moment diagrams for the two frames. For frame (a),

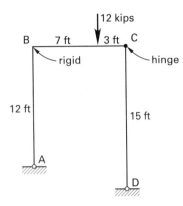

Member CD is a two-force member; therefore, there is only axial force in the member. An appropriate free-body diagram for section ABC is

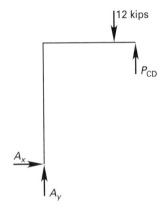

Equilibrium requires

$$\sum M_A = P_{CD}L - Pa = 0$$

$$P_{CD} = \frac{Pa}{L} = \frac{(12 \text{ kips})(7 \text{ ft})}{10 \text{ ft}}$$
$$= 8.4 \text{ kips} \quad [\text{upward}]$$

$$\sum F_y = A_y - P + P_{CD} = 0$$
$$A_y = P - P_{CD} = 12 \text{ kips} - 8.4 \text{ kips}$$
$$= 3.6 \text{ kips} \quad [\text{upward}]$$

$$\sum F_x = A_x = 0$$

Because the horizontal reaction at A is zero, there is no shear or moment in member AB. The only member in frame (a), then, with shear and moment is member BC. The shear and moment diagrams for member BC are

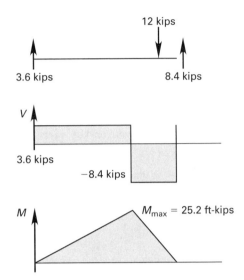

For frame (b), there are five components of the external reactions. The internal hinges at joints C and D provide two equations of conditions (that is, $M_C = M_D = 0$ ft-kips). Thus, there are five independent equations of equilibrium, making the frame statically determinate. Separating the frame at the pin connections at points C and D, and drawing free-body diagrams for each segment, shows

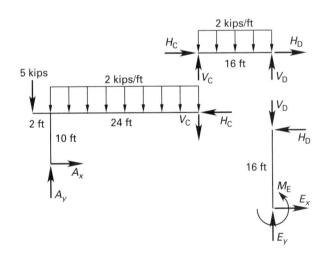

For the free-body diagram of segment CD, equilibrium requires

$$\sum M_C = V_D L - \frac{wL^2}{2} = 0$$

$$V_D = \frac{wL}{2} = \frac{\left(2 \ \frac{\text{kips}}{\text{ft}}\right)(16 \ \text{ft})}{2}$$

$$= 16 \ \text{kips} \quad [\text{upward}]$$

$$\sum F_y = V_C - wL + V_D = 0$$

$$V_C = wL - V_D = \left(2 \ \frac{\text{kips}}{\text{ft}}\right)(16 \ \text{ft}) - 16 \ \text{kips}$$

$$= 16 \ \text{kips} \quad [\text{upward}]$$

$$\sum F_x = H_C + H_D = 0$$

$$H_C = -H_D$$

The 16 kips reaction from member BC acts downward at point C on section ABC. Equilibrium of ABC requires

$$\sum M_A = -V_C L - \frac{wL^2}{2} + H_C h + Pa = 0$$

$$H_C = \frac{V_C L + \frac{wL^2}{2} - Pa}{h}$$

$$= \frac{(16 \ \text{kips})(24 \ \text{ft}) + \frac{\left(2 \ \frac{\text{kips}}{\text{ft}}\right)(24 \ \text{ft})^2}{2} - (5 \ \text{kips})(2 \ \text{ft})}{10 \ \text{ft}}$$

$$= 95 \ \text{kips} \quad [\text{to the left}]$$

$$\sum F_x = A_x - H_C = 0$$

$$A_x = H_C = 95 \ \text{kips} \quad [\text{to the right}]$$

$$\sum F_y = A_y - wL - P - V_C = 0$$

$$A_y = wL + P + V_C$$

$$= \left(2 \ \frac{\text{kips}}{\text{ft}}\right)(24 \ \text{ft}) + 5 \ \text{kips} + 16 \ \text{kips}$$

$$= 69 \ \text{kips} \quad [\text{upward}]$$

The 95 kips axial force at point C transfers through member CD and transfers as a shear at the top of member DE. Equilibrium of DE requires

$$\sum F_x = E_x - H_D = 0$$

$$E_x = H_D = -H_C = -95 \ \text{kips} \quad [\text{to the left}]$$

$$\sum F_y = E_y - V_D = 0$$

$$E_y = V_D = 16 \ \text{kips} \quad [\text{upward}]$$

$$\sum M_E = M_E - H_D h = 0$$

$$M_E = H_D h = (95 \ \text{kips})(16 \ \text{ft})$$

$$= 1520 \ \text{ft-kips} \quad [\text{clockwise}]$$

The shear and moment diagrams for members BC and CD are

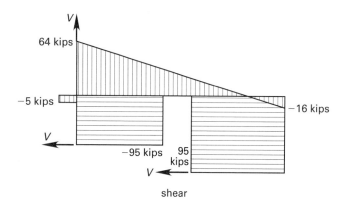

shear

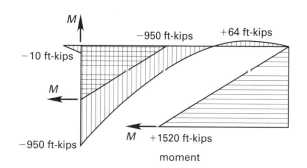

moment

SOLUTION 2

2.1. Compute the rigidity of the girder based on the gross cross-sectional area.

$$E_c I_g = E_c \left(\frac{bh^3}{12}\right)$$

$$= \left(3000 \ \frac{\text{kips}}{\text{in}^2}\right)\left(\frac{(30 \ \text{in})(30 \ \text{in})^3}{12}\right)$$

$$= \boxed{202.5 \times 10^6 \ \text{in}^2\text{-kips}}$$

Flexural Rigidity is:or $\frac{48EI}{L^3}$

Rigidy of beam) Element is EI.

2.2. Take the axial force in the pile as the redundant force, and release the structure

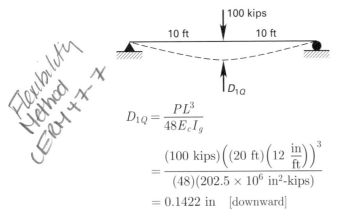

$$D_{1Q} = \frac{PL^3}{48E_cI_g}$$

$$= \frac{(100 \text{ kips})\left((20 \text{ ft})\left(12 \frac{\text{in}}{\text{ft}}\right)\right)^3}{(48)(202.5 \times 10^6 \text{ in}^2\text{-kips})}$$

$$= 0.1422 \text{ in} \quad [\text{downward}]$$

Apply equal and opposite forces of 1 kip each at the release point.

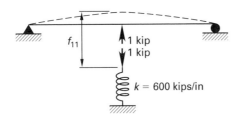

The flexibility influence coefficient is

$$f_{11} = \frac{FL^3}{48E_cI_g} + \frac{1}{k}$$

$$= \frac{(1 \text{ kip})\left((20 \text{ ft})\left(12 \frac{\text{in}}{\text{ft}}\right)\right)^3}{(48)(202.5 \times 10^6 \text{ in}^2\text{-kips})} + \frac{1 \text{ kip}}{600 \frac{\text{kips}}{\text{in}}}$$

$$= 0.003089 \text{ in/kip of load}$$

For consistent displacement,

$$f_{11}B_y + D_{1Q} = 0 \text{ in}$$

$$B_y = \frac{-D_{10}}{f_{11}} = \frac{-(-0.1422 \text{ in})}{0.003089 \frac{\text{in}}{\text{kip}}}$$

$$= 46.0 \text{ kips} \quad [\text{upward}]$$

With the redundant force known, the resultant force acting at point B is $P - B_y$. Use basic equations of solid mechanics to find the moment and deflection at point B.

$$M_B = \frac{(P - B_y)L}{4}$$

$$= \frac{(100 \text{ kips} - 46 \text{ kips})(20 \text{ ft})}{4}$$

$$= \boxed{270 \text{ ft-kips} \quad [\text{positive curvature}]}$$

$$\Delta_B = \frac{(P - B_y)L^3}{48E_cI_g}$$

$$= \frac{(100 \text{ kips} - 46 \text{ kips})\left((20 \text{ ft})\left(12 \frac{\text{in}}{\text{ft}}\right)\right)^3}{(48)(202.5 \times 10^6 \text{ in}^2\text{-kips})}$$

$$= \boxed{0.08 \text{ in} \quad [\text{downward}]}$$

The values of M_B and Δ_B are due only to the wheel load, as required by the problem statement.

SOLUTION 3

3.1. Determine the column load at each story based on the provisions of ASCE/SEI7. For the flat roof's live load, use the minimum load specified in ASCE/SEI7 Table 4-1, $L_o = 20 \text{ lbf/ft}^2$.

Because the roof is flat, the live load can be reduced in accordance with ASCE/SEI7 Sec. 4.8.2.

$$L_r = L_oR_1R_2 \quad [12 \text{ lbf/ft}^2 \leq L_r \leq 20 \text{ lbf/ft}^2]$$

For tributary area $A_T > 600 \text{ ft}^2$, $R_1 = 0.6$. For flat roofs, $R_2 = 1.0$.

$$L_r \geq \begin{cases} L_oR_1R_2 = \left(20 \frac{\text{lbf}}{\text{ft}^2}\right)(0.6)(1.0) \\ \quad\quad = 12 \text{ lbf/ft}^2 \\ 12 \text{ lbf/ft}^2 \end{cases}$$

$$P_L = L_rA_T = \frac{\left(12 \frac{\text{lbf}}{\text{ft}^2}\right)(900 \text{ ft}^2)}{1000 \frac{\text{lbf}}{\text{kip}}} = 11 \text{ kips}$$

For the uppermost mechanical room, the 125 lbf/ft^2 loading (given) cannot be reduced; however, since the column supporting the lower mechanical room is supporting more than two floors, ASCE/SEI7 Sec. 4.7.3 permits a 20% reduction below that floor.

$$L_{\text{upper}} = L = 125 \text{ lbf/ft}^2$$

$$P_L = LA_T = \frac{\left(125 \frac{\text{lbf}}{\text{ft}^2}\right)(900 \text{ ft}^2)}{1000 \frac{\text{lbf}}{\text{kip}}} = 113 \text{ kips}$$

$$L_{\text{lower}} = 0.8L = (0.8)\left(125\ \frac{\text{lbf}}{\text{ft}^2}\right) = 100\ \text{lbf/ft}^2$$

$$P_L = L_{\text{lower}} A_T = \frac{\left(100\ \dfrac{\text{lbf}}{\text{ft}^2}\right)(900\ \text{ft}^2)}{1000\ \dfrac{\text{lbf}}{\text{kip}}} = 90\ \text{kips}$$

In multifamily residential areas, ASCE/SEI7 Table 4-1 requires a minimum live load of 40 lbf/ft^2, which can be reduced per ASCE/SEI7 Sec. 4.7. For the uppermost residential level,

$$L \geq \begin{cases} L_o\left(0.25 + \dfrac{15}{\sqrt{K_{LL}A_T}}\right) \\ \quad = \left(40\ \dfrac{\text{lbf}}{\text{ft}^2}\right)\left(0.25 + \dfrac{15}{\sqrt{(4)(900\ \text{ft}^2)}}\right) \\ \quad = 20\ \text{lbf/ft}^2 \\ 0.4L_o = (0.4)\left(40\ \dfrac{\text{lbf}}{\text{ft}^2}\right) \\ \quad = 16\ \text{lbf/ft}^2 \end{cases}$$

$$P_L = LA_T = \frac{\left(20\ \dfrac{\text{lbf}}{\text{ft}^2}\right)(900\ \text{ft}^2)}{1000\ \dfrac{\text{lbf}}{\text{kip}}} = 18\ \text{kips}$$

For the column supporting two levels of residential loading,

$$L \geq \begin{cases} L_o\left(0.25 + \dfrac{15}{\sqrt{K_{LL}A_T}}\right) \\ \quad = \left(40\ \dfrac{\text{lbf}}{\text{ft}^2}\right)\left(0.25 + \dfrac{15}{\sqrt{(4)((2)(900\ \text{ft}^2))}}\right) \\ \quad = 17\ \text{lbf/ft}^2 \quad [\text{controls}] \\ 0.4L_o = (0.4)\left(40\ \dfrac{\text{lbf}}{\text{ft}^2}\right) \\ \quad = 16\ \text{lbf/ft}^2 \end{cases}$$

$$P_L = LA_T = \frac{\left(17\ \dfrac{\text{lbf}}{\text{ft}^2}\right)(900\ \text{ft}^2)}{1000\ \dfrac{\text{lbf}}{\text{kip}}} = 15\ \text{kips}$$

For the column supporting three or more levels of residential loading,

$$L \geq \begin{cases} L_o\left(0.25 + \dfrac{15}{\sqrt{K_{LL}A_T}}\right) \\ \quad = \left(40\ \dfrac{\text{lbf}}{\text{ft}^2}\right)\left(0.25 + \dfrac{15}{\sqrt{(4)((3)(900\ \text{ft}^2))}}\right) \\ \quad = 15.8\ \text{lbf/ft}^2 \\ 0.4L_o = (0.4)\left(40\ \dfrac{\text{lbf}}{\text{ft}^2}\right) \\ \quad = 16\ \text{lbf/ft}^2 \quad [\text{controls}] \end{cases}$$

$$P_L = LA_T = \frac{\left(16\ \dfrac{\text{lbf}}{\text{ft}^2}\right)(900\ \text{ft}^2)}{1000\ \dfrac{\text{lbf}}{\text{kip}}}$$
$$= 14.4\ \text{kips} \quad [\text{use 15 kips}]$$

For the office areas, ASCE/SEI7 Table 4-1 requires a 50 lbf/ft^2 minimum (excluding corridors).

$$L \geq \begin{cases} L_o\left(0.25 + \dfrac{15}{\sqrt{K_{LL}A_T}}\right) \\ \quad = \left(50\ \dfrac{\text{lbf}}{\text{ft}^2}\right)\left(0.25 + \dfrac{15}{\sqrt{(4)((10)(900\ \text{ft}^2))}}\right) \\ \quad = 16\ \text{lbf/ft}^2 \\ 0.4L_o = (0.4)\left(50\ \dfrac{\text{lbf}}{\text{ft}^2}\right) \\ \quad = 20\ \text{lbf/ft}^2 \quad [\text{controls}] \end{cases}$$

$$P_L = LA_T = \frac{\left(20\ \dfrac{\text{lbf}}{\text{ft}^2}\right)(900\ \text{ft}^2)}{1000\ \dfrac{\text{lbf}}{\text{kip}}} = 18\ \text{kips}$$

For the assembly areas, ASCE/SEI7 Table 4-1 requires $L_o = 100$ lbf/ft^2. ASCE/SEI7 Sec. 4.7.5 does not permit a live load reduction for assembly uses.

$$P_L = LA_T = \frac{\left(100\ \dfrac{\text{lbf}}{\text{ft}^2}\right)(900\ \text{ft}^2)}{1000\ \dfrac{\text{lbf}}{\text{kip}}} = 90\ \text{kips}$$

For heavy storage areas, $L_o = 250$ lbf/ft^2; ASCE/SEI7 Sec. 4.7.3 permits a 20% reduction for columns supporting two or more floors. Thus,

$$L = 0.8L_o = (0.8)\left(250 \,\frac{\text{lbf}}{\text{ft}^2}\right) = 200 \text{ lbf/ft}^2$$

$$P_L = LA_T = \frac{\left(200 \,\dfrac{\text{lbf}}{\text{ft}^2}\right)(900 \text{ ft}^2)}{1000 \,\dfrac{\text{lbf}}{\text{kip}}} = 180 \text{ kips}$$

Dead loads of 80 lbf/ft^2 (at roof) and 100 lbf/ft^2 (at typical floors) give

$$P_{D,r} = w_D A_T = \frac{\left(80 \,\dfrac{\text{lbf}}{\text{ft}^2}\right)(900 \text{ ft}^2)}{1000 \,\dfrac{\text{lbf}}{\text{kip}}} = 72 \text{ kips}$$

$$P_{D,f} = w_D A_T = \frac{\left(100 \,\dfrac{\text{lbf}}{\text{ft}^2}\right)(900 \text{ ft}^2)}{1000 \,\dfrac{\text{lbf}}{\text{kip}}} = 90 \text{ kips}$$

The total load (dead load plus live load) at each story is

level	category	P_L (kips)	P_D (kips)	ΔP (kips)	P (kips)
18-R	roof	11	72	83	83
17-18	mech.	113	90	203	286
16-17	resid.	18	90	108	394
15-16	resid.	15	90	105	499
14-15	resid.	15	90	105	604
13-14	resid.	15	90	105	709
12-13	h. stor.	180	90	270	979
11-12	resid.	15	90	105	1084
10-11	resid.	15	90	105	1189
9-10	resid.	15	90	105	1294
8-9	resid.	15	90	105	1399
7-8	mech.	90	90	180	1579
6-7	assem.	90	90	180	1759
5-6	office	18	90	108	1867
4-5	office	18	90	108	1975
3-4	office	18	90	108	2083
2-3	office	18	90	108	2191
1-2	office	18	90	108	2299

SOLUTION 4

4.1. Analyze the frame using the moment distribution method. Express stiffness factors in terms of the modulus of elasticity, E, which is constant for all members.

$$K_{AB} = K_{BA} = \frac{4E(27 \text{ in}^4)}{15 \text{ ft}} = (7.2 \text{ in}^4/\text{ft})E$$

$$K_{BC} = K_{CB} = \frac{4E(27 \text{ in}^4)}{24 \text{ ft}} = (4.5 \text{ in}^4/\text{ft})E$$

$$K_{BD} = \frac{3E(8 \text{ in}^4)}{15 \text{ ft}} = (1.6 \text{ in}^4/\text{ft})E$$

$$\sum K_B = K_{AB} + K_{BC} + K_{BD}$$

$$= \left(7.2 \,\frac{\text{in}^4}{\text{ft}}\right)E + \left(4.5 \,\frac{\text{in}^4}{\text{ft}}\right)E + \left(1.6 \,\frac{\text{in}^4}{\text{ft}}\right)E$$

$$= (13.3 \text{ in}^4/\text{ft})E$$

The distribution factors are

$$\text{DF}_{AB} = 0 \quad [\text{fixed support}]$$

$$\text{DF}_{BA} = \frac{K_{BA}}{\sum K_B} = \frac{\left(7.2 \,\dfrac{\text{in}^4}{\text{ft}}\right)E}{\left(13.3 \,\dfrac{\text{in}^4}{\text{ft}}\right)E} = 0.541$$

$$\text{DF}_{BD} = \frac{K_{BD}}{\sum K_B} = \frac{\left(1.6 \,\dfrac{\text{in}^4}{\text{ft}}\right)E}{\left(13.3 \,\dfrac{\text{in}^4}{\text{ft}}\right)E} = 0.120$$

$$\text{DF}_{BC} = \frac{K_{BC}}{\sum K_B} = \frac{\left(4.5 \,\dfrac{\text{in}^4}{\text{ft}}\right)E}{\left(13.3 \,\dfrac{\text{in}^4}{\text{ft}}\right)E} = 0.338$$

$$\text{DF}_{CB} = 0 \quad [\text{fixed support}]$$

The carryover factors are

$$\text{CO}_{BD} = 0 \quad [\text{hinged end}]$$

All other carryover factors are $1/2$. The fixed end moments are

$$\text{FEM}_{BC} = \frac{-Pab^2}{L^2}$$

$$= \frac{(-36 \text{ kips})(14 \text{ ft})(10 \text{ ft})^2}{(24 \text{ ft})^2}$$

$$= -87.5 \text{ ft-kips}$$

$$\text{FEM}_{CB} = \frac{Pab^2}{L^2}$$

$$= \frac{(36 \text{ kips})(10 \text{ ft})(14 \text{ ft})^2}{(24 \text{ ft})^2}$$

$$= 122.5 \text{ ft-kips}$$

$$\text{FEM}_{BD} = \left(\frac{Pab}{2L^2}\right)(a + L)$$

$$= \left(\frac{(5 \text{ kips})(10 \text{ ft})(5 \text{ ft})}{(2)(15 \text{ ft})^2}\right)(10 \text{ ft} + 15 \text{ ft})$$

$$= 13.9 \text{ ft-kips}$$

The moment distribution converges in one cycle.

0		0.541	0.120	0.338		0
	co		13.9	−87.5	co	122.5
19.9 ←		39.8	8.8	24.9	→	12.5
19.9		39.8	22.7	−62.6		135.0

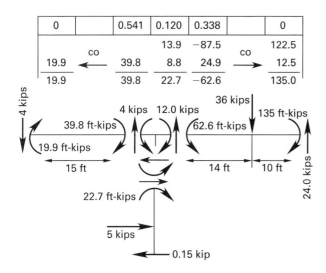

4.2. The moment diagram is as shown.

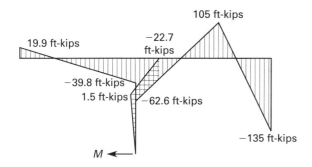

4.3. The deflected shape is drawn based on the moment diagram and the reactions from the positive moment. The positive moment causes compression on the top edge and tension on the bottom of the horizontal members, and causes compression on the left edge and tension on the right edge of the vertical members.

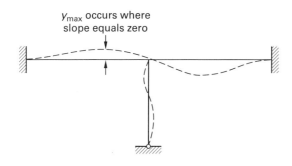

4.4. Calculate the maximum deflection in span AB using the conjugate beam method. Convert moments into units of in-kips and the span length into units of inches to be consistent with the given values of $I = 27 \text{ in}^4$ and $E = 29{,}000$ ksi.

$$M_A = (19.9 \text{ ft-kips})\left(12 \frac{\text{in}}{\text{ft}}\right) = 239 \text{ in-kips}$$

$$M_B = (39.8 \text{ ft-kips})\left(12 \frac{\text{in}}{\text{ft}}\right) = 478 \text{ in-kips}$$

Since both rotation and deflection are zero at joint A, joint A in the conjugate beam is a free end. Joint B in the structure cannot deflect but it can rotate; therefore, joint B will have a vertical reaction component in the conjugate beam. The conjugate beam is loaded with the M/EI of the real structure.

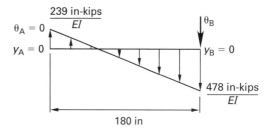

The maximum deflection occurs where θ_x equals zero. By principles of similar triangles,

$$\frac{\dfrac{x}{2}}{239 \text{ in-kips}} = \frac{180 \text{ in}}{239 \text{ in-kips} + 478 \text{ in-kips}}$$

$$x = 120 \text{ in}$$

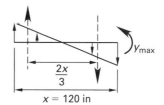

$$\theta = 0.5\left(\frac{239 \text{ in-kips}}{EI}\right)\left(\frac{x}{2}\right)$$

$$y_{max} = \theta\left(\frac{2x}{3}\right) = 0.5\left(\frac{M_A}{EI}\right)\left(\frac{x}{2}\right)\left(\frac{2x}{3}\right)$$

$$= (0.5)\left(\frac{239 \text{ in-kips}}{\left(29{,}000 \dfrac{\text{kips}}{\text{in}^2}\right)(27 \text{ in}^4)}\right)$$

$$\times \left(\frac{120 \text{ in}}{2}\right)\left(\frac{(2)(120 \text{ in})}{3}\right)$$

$$= 0.73 \text{ in} \quad [\text{upward}]$$

Maximum deflection is $\boxed{0.73 \text{ in.}}$

SOLUTION 5

5.1. Analyze the frame using the method of consistent displacements, taking the horizontal reaction at D as the redundant reaction.

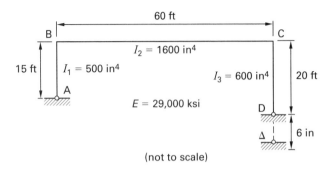

(not to scale)

The structure is one degree statically indeterminate and experiences a 6 in vertical settlement at support D. Release support D.

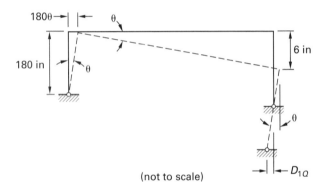

(not to scale)

The released statically determinate structure experiences rigid body rotation about support A. The small displacement theory applies, so $\sin\theta = \tan\theta = \theta$ (in radians).

$$\theta = \frac{\Delta}{L} = \frac{6 \text{ in}}{(60 \text{ ft})\left(12 \frac{\text{in}}{\text{ft}}\right)} = 0.00833 \text{ rad}$$

$$D_{1Q} = h_{AB}\theta - h_{CD}\theta$$
$$= (180 \text{ in})(0.00833 \text{ rad}) - (240 \text{ in})(0.00833 \text{ rad})$$
$$= -0.5 \text{ in} \quad [\text{to the left}]$$

From linear elastic theory, small displacements are calculated with respect to the original geometry (before deformation). Thus,

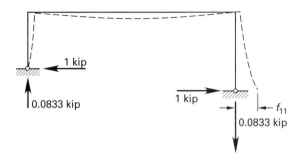

By the dummy load method,

$$f_{11} = \sum_{i=1}^{m} \int_{L_i} \left(\frac{mM\,dx}{EI}\right)_i$$

member	interval	real moment, M	virtual moment, m
AB	$0 \le x \le 180$ in	$+(1 \text{ kip})x$	$+(1^*)x$
BC	$0 \le x \le 720$ in	$+180$ in-kips	$+180$ in-*
		$+(0.0833 \text{ kips})x$	$+(0.0833^*)x$
DC	$0 \le x \le 240$ in	$-(1 \text{ kip})x$	$-(1^*)x$

In the expressions for virtual moments, * denotes an infinitesimal virtual force.

$$E(1^*)f_{11} = \int_{0 \text{ in}}^{180 \text{ in}} \frac{x^2 \text{ in}^2\text{-*kips}}{500 \text{ in}^4}\, dx$$
$$+ \int_{0 \text{ in}}^{720 \text{ in}} \frac{(180 + 0.0833x)^2 \text{ in}^2\text{-*kips}}{1600 \text{ in}^4}\, dx$$
$$+ \int_{0 \text{ in}}^{240 \text{ in}} \frac{(-1x)^2 \text{ in}^2\text{-*kips}}{600}\, dx$$

$$Ef_{11} = \left.\frac{x^3 \text{ in}^3\text{-kips}}{1500 \text{ in}^4}\right|_{0 \text{ in}}^{180 \text{ in}}$$
$$+ \left(\frac{32,400x}{1600 \text{ in}^4} + \frac{30x^2}{3200 \text{ in}^4} \atop + \frac{0.00694x^3}{4800 \text{ in}^4}\right) \text{in}^3\text{-kips} \Bigg|_{0 \text{ in}}^{720 \text{ in}}$$
$$+ \left.\frac{x^3 \text{ in}^3\text{-kips}}{1800 \text{ in}^4}\right|_{0 \text{ in}}^{240 \text{ in}}$$

$$f_{11} = \frac{31,548 \frac{\text{kips}}{\text{in}}}{29,000 \frac{\text{kips}}{\text{in}^2}} = 1.0879 \text{ in} \quad [\text{to the right}]$$

The coefficient f_{11} is the displacement caused by a 1 kip force; therefore, its true dimension is inches per kip. For constant displacements,

$$f_{11}D_x + D_{1Q} = 0 \text{ kips}$$

$$D_x = \frac{-D_{1Q}}{f_{11}} = \frac{-(-0.5 \text{ in})}{1.0879 \dfrac{\text{in}}{\text{kip}}} = 0.46 \text{ kip} \quad [\text{to the right}]$$

From statics,

$$\sum F_x = A_x + D_x = 0$$
$$A_x = -D_x = -0.46 \text{ kip} \quad [0.46 \text{ kip to the left}]$$
$$\sum M_A = D_x(h_2 - h_1) + D_y L = 0$$
$$D_y = \frac{-D_x(h_2 - h_1)}{L}$$
$$= \frac{(-0.46 \text{ kip})(240 \text{ in} - 180 \text{ in})}{720 \text{ in}}$$
$$= -0.038 \text{ kip} \quad [0.038 \text{ kip downward}]$$
$$\sum F_y = A_y + D_y = 0$$
$$D_y = A_y = 0.038 \text{ kip} \quad [\text{upward}]$$

Therefore, the required forces and moments at joints B and C are

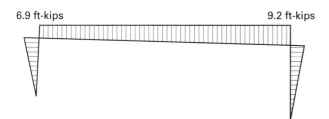

$M_{BA} = M_{BC} = (0.46 \text{ kip})(15 \text{ ft}) = 6.9 \text{ ft-kips}$
$M_{CB} = M_{CD} = (0.46 \text{ kip})(20 \text{ ft}) = 9.2 \text{ ft-kips}$

— 0.046 kip
0.046 kip —
0.038 kip
0.038 kip

5.2. The required moment diagram is

6.9 ft-kips
9.2 ft-kips

5.3. From Sol. 5.1, the horizontal and vertical forces at joints A and D are

$$A_y = 0.038 \text{ kip} \quad [\text{up}]$$
$$A_x = 0.46 \text{ kip} \quad [\text{to the left}]$$
$$D_y = 0.038 \text{ kip} \quad [\text{down}]$$
$$D_x = 0.46 \text{ kip} \quad [\text{to the right}]$$

SOLUTION 6

From the problem statement,

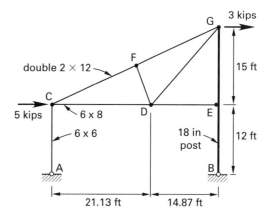

6.1. The external reactions are

$$\sum M_A = B_y L - P_C h_1 - P_G h_2 = 0$$
$$B_y = \frac{P_C h_1 + P_G h_2}{L}$$
$$= \frac{(5 \text{ kips})(12 \text{ ft}) + (3 \text{ kips})(27 \text{ ft})}{36 \text{ ft}}$$
$$= \boxed{3.92 \text{ kips} \quad [\text{upward}]}$$
$$\sum F_y = B_y + A_y = 0$$
$$A_y = -B_y$$
$$= \boxed{-3.92 \text{ kips} \quad [3.92 \text{ kips downward}]}$$
$$\sum F_x = B_x + P_C + P_G = 0$$
$$B_x = -P_C - P_G = -5 \text{ kips} - 3 \text{ kips}$$
$$= \boxed{-8 \text{ kips} \quad [8 \text{ kips to the left}]}$$

6.2. Member BEG is subjected to shear, bending, and axial forces. All others are two-force members, which resist only axial force. A free-body diagram of joint F shows that if forces are summed in a direction perpendicular to member CG, then the force in member DF must be zero. Similarly, since member DF has zero force, a free-body diagram of joint D shows that the force in member DG is also zero.

Taking a free-body diagram of joint C,

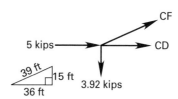

CF
5 kips
CD
39 ft
15 ft
36 ft
3.92 kips

$$\sum F_y = F_{\rm CF} \sin \alpha + A_y = 0$$

$$F_{\rm CF} = \frac{-A_y}{\sin \alpha} = \frac{-(-3.92 \text{ kips})}{\dfrac{15 \text{ ft}}{39 \text{ ft}}}$$

$$= 10.2 \text{ kips} \quad [\text{tension}]$$

$$\sum F_x = 5 \text{ kips} + F_{\rm CF} \cos \alpha + F_{\rm CD} = 0$$

$$F_{\rm CD} = -F_{\rm CF} \cos \alpha - 5 \text{ kips}$$

$$= (-10.2 \text{ kips})\left(\frac{36 \text{ ft}}{39 \text{ ft}}\right) - 5 \text{ kips}$$

$$= -14.4 \text{ kips} \quad [\text{14.4 kips compression}]$$

From inspection of joint F,

$$F_{\rm FG} = F_{\rm CF} = 10.2 \text{ kips} \quad [\text{tension}]$$

From inspection of joint D,

$$F_{\rm DE} = F_{\rm CD} = 14.4 \text{ kips} \quad [\text{compression}]$$

The internal member forces are drawn as follows.

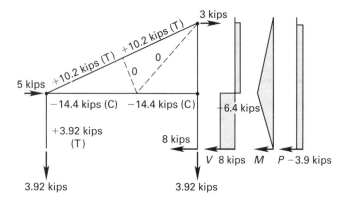

6.3. Compute the horizontal deflection at joint C. Use the virtual work method. Apply a unit virtual force at C.

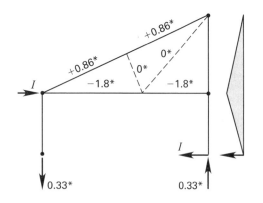

Calculate properties of solid sawn sections using dressed dimensions ($E = 1600$ ksi). For one 6×6 wood member (actual dimensions are 5.5 in $\times$ 5.5 in),

$$EA = \left(1600 \ \frac{\text{kips}}{\text{in}^2}\right)(5.5 \text{ in})^2 = 48{,}400 \text{ kips}$$

For two 2×12 wood members (actual dimensions are 1.5 in $\times$ 11.25 in),

$$EA = \left(1600 \ \frac{\text{kips}}{\text{in}^2}\right)(2)(1.5 \text{ in})(11.25 \text{ in})$$

$$= 54{,}000 \text{ kips}$$

For one 6×8 wood member (actual dimensions are 5.5 in $\times$ 7.5 in),

$$EA = \left(1600 \ \frac{\text{kips}}{\text{in}^2}\right)(5.5 \text{ in})(7.5 \text{ in})$$

$$= 66{,}000 \text{ kips}$$

For one 18 in diameter wood post,

$$EA = E\left(\frac{\pi d^2}{4}\right) = \left(1600 \ \frac{\text{kips}}{\text{in}^2}\right)\left(\frac{\pi(18 \text{ in})^2}{4}\right)$$

$$= 407{,}000 \text{ kips}$$

$$EI = E\left(\frac{\pi d^4}{64}\right) = \left(1600 \ \frac{\text{kips}}{\text{in}^2}\right)\left(\frac{\pi(18 \text{ in})^4}{64}\right)$$

$$= 8{,}240{,}000 \text{ in}^2\text{-kips}$$

The virtual work due to axial forces in the structure is found using $\sum p_i (PL/EA)_i$, and is calculated in the following table.

member	$p(*)$	P (kips)	L (in)	EA (10^3 kips)	pPL/EA
AC	+0.33	3.92	144	48.4	0.00385
CD	−1.80	−14.4	255	66.0	0.10000
DE	−1.80	−14.4	178	66.0	0.07000
CF	+0.86	10.2	234	54.0	0.03800
FG	+0.86	10.2	234	54.0	0.03800
BE	−0.33	−3.92	144	407	0.00050
EG	−0.33	−3.92	180	407	0.00060
$\sum p_i \left(\dfrac{PL}{EA}\right)_i =$					0.251 in

The contribution to the deflection due to axial forces is

$$\Delta_{\text{axial}} = \sum p_i \left(\frac{PL}{EA}\right)_i = 0.251 \text{ in}$$

The contribution due to bending is

$$\Delta_{\text{bending}} = \sum \int \frac{mM\,dx}{EI}$$

$$= \int_{0\text{ in}}^{144\text{ in}} \frac{x(8x)(1\text{ kip})}{EI}\,dx$$

$$+ \int_{0\text{ in}}^{180\text{ in}} \frac{(0.8x)(6.4x)(1\text{ kip})}{EI}\,dx$$

$$= \frac{8x^3(1\text{ kip})}{3EI}\bigg|_{0\text{ in}}^{144\text{ in}} + \frac{5.12x^3(1\text{ kip})}{3EI}\bigg|_{0\text{ in}}^{180\text{ in}}$$

$$= \frac{17{,}915{,}904\text{ in}^3\text{-kips}}{8{,}240{,}000\text{ in}^2\text{-kips}}$$

$$= 2.174\text{ in}$$

Thus,

$$\Delta_{\text{C,horiz}} = \Delta_{\text{axial}} + \Delta_{\text{bending}} = 0.251\text{ in} + 2.174\text{ in}$$

$$= 2.424\text{ in} \quad [\text{to the right}]$$

$$\frac{\Delta_{\text{bending}}}{\Delta_{\text{C,horiz}}} \times 100\% = \frac{2.174\text{ in}}{2.424\text{ in}} \times 100\%$$

$$= 89.6\% \quad (90\%)$$

> 90% of the horizontal deflection is due to bending. Therefore, the bending load contributes more to the structure's overall deflection.

SOLUTION 7

7.1. Since rigid frame formulas may not be used, the best way to compute the reactions is the column analogy method (see Blodgett, Sec. 6.1). Release the horizontal reaction at E and determine the bending moment diagrams for members of the resulting statically determinate structure.

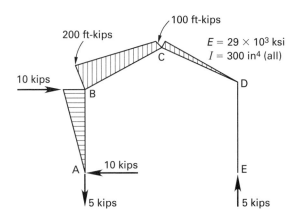

Apply the M/EI of the statically determinate structure as axial loads on the analogous column.

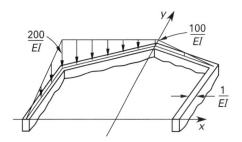

At hinged bases, $1/EI$ approaches infinity; therefore, the area and moment of inertia about the y-axis also approach infinity, and the x-axis is at the bottom of the column. Compute the moment of inertia about the x-axis in terms of the constant rigidity, EI.

$$L_{\text{rafter}} = \sqrt{a^2 + b^2} = \sqrt{(20\text{ ft})^2 + (10\text{ ft})^2} = 22.36\text{ ft}$$

$$I_x = 2(I_{\text{col}} + I_{\text{rafter}})$$

$$= 2\left(\frac{bh^3}{3} + I_{c,\text{rafter}} + A_{\text{rafter}}d^2\right)$$

$$= (2)\left(\left(\frac{1}{EI}\right)\left(\frac{(20\text{ ft})^3}{3}\right) + \frac{bL_{\text{rafter}}a^2}{12} + bL_{\text{rafter}}d^2\right)$$

$$= (2)\left(\begin{array}{l}\left(\dfrac{1}{EI}\right)\left(\dfrac{(20\text{ ft})^3}{3}\right) + \dfrac{\left(\dfrac{1}{EI}\right)(22.36\text{ ft})(10\text{ ft})^2}{12} \\[2ex] + \left(\dfrac{1}{EI}\right)(22.36\text{ ft})(25\text{ ft})^2\end{array}\right)$$

$$= \frac{33{,}656\text{ ft}^3}{EI}$$

Compute the moment of the M/EI load about the x-axis. (Other stress components in the column analogy are zero because both I_y and the area approach infinity.) Treat the trapezoidal load on member BC as the sum of a uniform load, 100 ft-kips/EI, whose centroid is $\overline{y} = 20\text{ ft} + 5\text{ ft} = 25\text{ ft}$ from the x-axis. Account also for a triangle load that varies from 100 ft-kips/EI at point B to zero at point C, and whose centroid is located

$\overline{y} = 20 \text{ ft} + (10 \text{ ft}/3) = 23.33 \text{ ft}$ from the x-axis. The moment of the M/EI loading is

$$M_x = \sum A\overline{y}_i = 0.5bh\left(\frac{2h}{3}\right) + bL\left(h + \frac{a}{3}\right)$$

$$+ 0.5bL\left(h + \frac{2a}{3}\right) + 0.5bL\left(h + \frac{2a}{3}\right)$$

$$= (0.5)\left(\frac{200 \text{ ft-kips}}{EI}\right)(20 \text{ ft})\left(\frac{(2)(20 \text{ ft})}{3}\right)$$

$$+ \left(\frac{100 \text{ ft-kips}}{EI}\right)(22.36 \text{ ft})\left(20 \text{ ft} + \frac{10 \text{ ft}}{3}\right)$$

$$+ (0.5)\left(\frac{100 \text{ ft-kips}}{EI}\right)(22.36 \text{ ft})$$

$$\times \left(20 \text{ ft} + \frac{(2)(10 \text{ ft})}{3}\right)$$

$$+ (0.5)\left(\frac{100 \text{ ft-kips}}{EI}\right)(22.36 \text{ ft})$$

$$\times \left(20 \text{ ft} + \frac{(2)(10 \text{ ft})}{3}\right)$$

$$= \frac{138,467 \text{ ft}^3\text{-kips}}{EI}$$

Obtain the final end moments by superimposing the stresses in the analogous column on the static moments in the released structure. $1/EI$ is common to both the I_x and M_x terms and cancels out in the expression for bending stresses, $M_x c/I_x$. Thus,

$$M_{AB} = M_{s,AB} + \frac{M_x c_{AB}}{I_x}$$

$$= 0 \text{ ft-kips} + \frac{(-138,467 \text{ ft}^3\text{-kips})(0 \text{ ft})}{33,656 \text{ ft}^3}$$

$$= 0 \text{ ft-kips}$$

$$M_{BA} = M_{s,BA} + \frac{M_x c_{BA}}{I_x}$$

$$= 200 \text{ ft-kips} + \frac{(-138,467 \text{ ft}^3\text{-kips})(20 \text{ ft})}{33,656 \text{ ft}^3}$$

$$= 117.7 \text{ ft-kips}$$

$$M_{CB} = M_{s,CB} + \frac{M_x c_{CB}}{I_x}$$

$$= 100 \text{ ft-kips} + \frac{(-138,467 \text{ ft}^3\text{-kips})(30 \text{ ft})}{33,656 \text{ ft}^3}$$

$$= -23.3 \text{ ft-kips}$$

$$M_{DC} = M_{s,DC} + \frac{M_x c_{DC}}{I_x}$$

$$= 0 \text{ ft-kips} + \frac{(-138,467 \text{ ft}^3\text{-kips})(20 \text{ ft})}{33,656 \text{ ft}^3}$$

$$= -82.3 \text{ ft-kips}$$

In the column analogy method, positive moments act counterclockwise, and negative moments are clockwise about the location. With the end moments known, the required reactions are computed using statics.

$$A_x = \frac{M_{BA}}{h} = \frac{117.7 \text{ ft-kips}}{20 \text{ ft}} = \boxed{5.9 \text{ kips} \quad [\text{to the left}]}$$

$$E_x = \frac{M_{DE}}{h} = \frac{82.3 \text{ ft-kips}}{20 \text{ ft}} = \boxed{4.1 \text{ kips} \quad [\text{to the left}]}$$

$$A_y = \boxed{5 \text{ kips} \quad [\text{downward}]}$$

$$E_y = \boxed{5 \text{ kips} \quad [\text{upward}]}$$

7.2. Find the vertical deflection at point C. With the end moments determined, use slope-deflection equations to find the unknown deflection, $\Delta_{C,\text{vert}}$.

$$M_{NF} = \left(\frac{2EI}{L}\right)(2\phi_N + \phi_F - 3\beta_{NF}) \pm \text{FEM}_{NF}$$

Only the members BC and CD need to be considered. The fixed end moments for the two members are zero. Moments and rotations are positive in the clockwise sense at member ends.

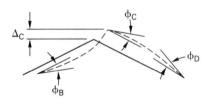

In terms of Δ_C,

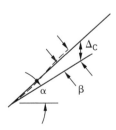

$$\Delta = \Delta_C \cos\alpha = \Delta_C \left(\frac{20 \text{ ft}}{22.36 \text{ ft}}\right)$$

$$\beta = \frac{\Delta}{L} = \frac{\Delta_C \cos\alpha}{L} = \Delta_C \left(\frac{20 \text{ ft}}{(22.36 \text{ ft})^2}\right)$$

$$\frac{2EI}{L} = \frac{(2)\left(29{,}000 \ \frac{\text{kips}}{\text{in}^2}\right)(300 \text{ in}^4)}{(22.36 \text{ ft})\left(12 \ \frac{\text{in}}{\text{ft}}\right)^2} = 5400 \text{ ft-kips}$$

$$M_{NF} = \left(\frac{2EI}{L}\right)(2\phi_N + \phi_F - 3\beta_{NF}) \pm \text{FEM}_{NF}$$

$$M_{BC} = (5400 \text{ ft-kips})(2\phi_B + \phi_C + 3\beta)$$

$$M_{CB} = (5400 \text{ ft-kips})(\phi_B + 2\phi_C + 3\beta)$$

$$M_{CD} = (5400 \text{ ft-kips})(2\phi_C + \phi_D - 3\beta)$$

$$M_{DC} = (5400 \text{ ft-kips})(\phi_C + 2\phi_D - 3\beta)$$

Solving the simultaneous equations for β,

$$\beta = 0.0014 \text{ rad}$$

$$\Delta_C = \frac{\beta L}{\cos\alpha} = \frac{(0.0014 \text{ rad})(22.36 \text{ ft})\left(12 \ \frac{\text{in}}{\text{ft}}\right)}{\dfrac{20 \text{ ft}}{22.36 \text{ ft}}}$$

$$= \boxed{0.42 \text{ in} \quad [\text{upward}]}$$

7.3. The simplest way to find the horizontal deflection at point B is to apply Betti's reciprocity law. The external work done by the loads in system I that act through displacements in system II, will equal the external work done by the loads of system II that act through displacement of system I.

$$(W_{\text{external}})_{\text{I}\to\text{II}} = (W_{\text{external}})_{\text{II}\to\text{I}}$$

$$F_{B,x}\Delta_{B,x} = F_{C,y}\Delta_{C,y}$$

$$(10 \text{ kips})\Delta_{B,x} = (5 \text{ kips})(-0.42 \text{ in})$$

$$\Delta_{B,x} = \boxed{-0.21 \text{ in} \quad [0.21 \text{ in to the left}]}$$

SOLUTION 8

8.1. The analysis method used to obtain the result shown in illustration (b) must be the cantilever method because the column axial forces vary linearly with the distance from the centroid of the column group.

The analysis method used to obtain the result shown in illustration (c) must be the moment distribution method because the couple formed by the axial forces in the interior columns is opposite to the couple formed by axial forces in the exterior columns. This is inconsistent with the force distribution obtained by either of the other two methods.

The analysis method used to obtain the result shown in illustration (d) must be the portal analysis because the shear in the exterior girder is equal and opposite to the shear in the interior girder. This results in zero axial force in the interior columns.

8.2. The principal assumptions are listed as follows for each method.

Portal method

- Interior columns carry twice as much shear as exterior columns.

- The lateral force is distributed in proportion to the relative tributary area of each column.

- Inflection points occur at midspan of girders and midheight of columns.

Cantilever method

- Inflection points occur at midspan of girders and midheight of columns.

- Axial forces in columns resist the overturning moments due to lateral forces.

- Axial forces at a particular level vary linearly as the distance from the centroid of column areas.

Moment distribution method (Hardy Cross)

- Behavior is linear elastic.

- Only bending deformation is considered.

8.3. Of the two approximate methods, the portal method is a better method of analysis with buildings that have low height-to-width ratios. The cantilever method is better with buildings that have large height-to-width ratios, where sway due to column axial deformation can be significant.

SOLUTION 9

For the concrete shear wall,

$$G = 0.4E_c = (0.4)\left(3100 \ \frac{\text{kips}}{\text{in}^2}\right) = 1240 \text{ ksi}$$

From AISC Table 1-14, for 3.500 in × 0.216 in steel pipe,

$$A = 2.08 \text{ in}^2$$
$$I = 2.85 \text{ in}^4$$

For 4.500 in × 0.237 in steel pipe,

$$A = 2.97 \text{ in}^2$$

Check the Euler buckling strength of the 10 ft long 3.500 in × 0.216 in pipe. The pinned pinned end condition gives

$$KL = L = (10 \text{ ft})\left(12 \frac{\text{in}}{\text{ft}}\right) = 120 \text{ in}$$

$$P_E = \frac{\pi^2 EI}{(KL)^2} = \frac{\pi^2\left(29{,}000 \frac{\text{kips}}{\text{in}^2}\right)(2.85 \text{ in}^4)}{(120 \text{ in})^2}$$
$$= 56.6 \text{ kips}$$

Thus, the 3.500 in × 0.216 in pipe can transfer a maximum of 56.6 kips, which should be less than the amount transferred between the shear wall and braced bent. Let the force in this link member be the unknown, and release the indeterminate structure.

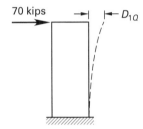

Properties of the 5.5 in × 48 in wall are

$$A = bd = (5.5 \text{ in})(48 \text{ in}) = 264 \text{ in}^2$$

$$I = \frac{bd^3}{12} = \frac{(5.5 \text{ in})(48 \text{ in})^3}{12} = 50{,}700 \text{ in}^4$$

Consider both shear and bending deformation in the wall.

$$D_{1Q} = \frac{Ph^3}{3E_cI} + \frac{1.2Ph}{AG}$$
$$= \frac{(70 \text{ kips})(120 \text{ in})^3}{(3)\left(3100 \frac{\text{kips}}{\text{in}^2}\right)(50{,}700 \text{ in}^4)}$$
$$+ \frac{(1.2)(70 \text{ kips})(120 \text{ in})}{(264 \text{ in}^2)\left(1240 \frac{\text{kips}}{\text{in}^2}\right)}$$
$$= 0.288 \text{ in}$$

Next, apply equal and opposite unit forces at the released points.

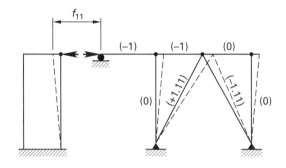

Use the real work method.

$$f_{11} = \Delta_{\text{wall}} + \Delta_{\text{frame}}$$
$$\frac{(1 \text{ kip})\Delta_{\text{frame}}}{2} = \sum_{i=1}^{\text{mem}} \frac{P_i^2 L_i}{2A_i E_s}$$

The length of each diagonal member is

$$L = \sqrt{a^2 + b^2} = \sqrt{(60 \text{ in})^2 + (120 \text{ in})^2} = 134 \text{ in}$$

$$\Delta_{\text{frame}} = \frac{(-1)(-1 \text{ kip})(120 \text{ in})}{(2.08 \text{ in}^2)\left(29{,}000 \frac{\text{kips}}{\text{in}^2}\right)}$$
$$+ \frac{(-1)(-1 \text{ kip})(60 \text{ in})}{(2.97 \text{ in}^2)\left(29{,}000 \frac{\text{kips}}{\text{in}^2}\right)}$$
$$+ \frac{(1.11)(1.11 \text{ kips})(134 \text{ in})}{(2.97 \text{ in}^2)\left(29{,}000 \frac{\text{kips}}{\text{in}^2}\right)}$$
$$+ \frac{(-1.11)(-1.11 \text{ kips})(134 \text{ in})}{(2.97 \text{ in}^2)\left(29{,}000 \frac{\text{kips}}{\text{in}^2}\right)}$$
$$= 0.0065 \text{ in}$$

$$\Delta_{\text{wall}} = \frac{(1 \text{ kip})(120 \text{ in})^3}{(3)\left(3100 \frac{\text{kips}}{\text{in}^2}\right)(50{,}700 \text{ in}^4)}$$
$$+ \frac{(1.2)(1 \text{ kip})(120 \text{ in})}{\left(1240 \frac{\text{kips}}{\text{in}^2}\right)(264 \text{ in}^2)}$$
$$= 0.0041 \text{ in}$$

$$f_{11} = \Delta_{\text{wall}} + \Delta_{\text{frame}}$$
$$= 0.0041 \text{ in} + 0.0065 \text{ in}$$
$$= 0.0106 \text{ in}$$

For consistent displacement,

$$P_{\text{link}} f_{11} + D_{1Q} = 0 \text{ in}$$
$$P_{\text{link}} \left(0.0106 \frac{\text{in}}{\text{kip}} \right) + (-0.288 \text{ in}) = 0 \text{ in}$$
$$P_{\text{link}} = 27.2 \text{ kips}$$

9.1. The force resisted by the shear wall is

$$F_{\text{wall}} = P - P_{\text{link}}$$
$$= 70 \text{ kips} - 27.2 \text{ kips}$$
$$= \boxed{42.8 \text{ kips}}$$

9.2. The force transferred by the flexible link is

$$P_{\text{link}} = \boxed{27.2 \text{ kips}}$$

9.3. The force resisted by the braced frame is

$$F_{\text{truss}} = P_{\text{link}} = \boxed{27.2 \text{ kips}}$$

9.4. The shear wall deflection is

$$\Delta_{\text{wall}} = \left(\frac{F_{\text{wall}}}{P} \right) D_{1Q}$$
$$= \left(\frac{42.8 \text{ kips}}{70 \text{ kips}} \right) (0.288 \text{ in})$$
$$= \boxed{0.18 \text{ in} \quad [\text{to the right}]}$$

9.5. To find the truss deflection, multiply the previously calculated deflection due to 1 kip by 27.2 kips.

$$\Delta_{\text{truss}} = (27.2 \text{ kips}) \left(\frac{\dfrac{(1)^2 (60 \text{ in})}{(2.97 \text{ in}^2) \left(29{,}000 \frac{\text{kips}}{\text{in}^2} \right)}}{ } \right.$$
$$\left. + \frac{(1.11)^2 (134 \text{ in})}{(2.97 \text{ in}^2) \left(29{,}000 \frac{\text{kips}}{\text{in}^2} \right)} \right.$$
$$\left. + \frac{(-1.11)^2 (134 \text{ in})}{(2.97 \text{ in}^2) \left(29{,}000 \frac{\text{kips}}{\text{in}^2} \right)} \right)$$
$$= \boxed{0.12 \text{ in} \quad [\text{to the right}]}$$

SOLUTION 10

10.1. Solve for the reactions in the structure. The structure is two degrees statically indeterminate. Using the method of consistent displacements, treat M_A and M_E as redundant components.

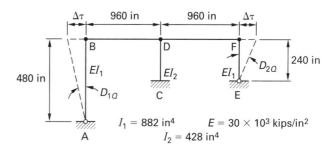

Compute the rotations at joint A and joint E of the released structure due to a 60°F temperature change. With the moments released, members AB and EF can rotate freely. Small displacement theory applies; therefore, $\sin\theta = \tan\theta = \theta$.

$$\Delta_T = D_F (\Delta T) L$$
$$= \left(0.000006 \frac{\text{in}}{\text{in-°F}} \right) (60°\text{F})(960 \text{ in})$$
$$= 0.346 \text{ in}$$
$$D_{1Q} = \frac{\Delta_T}{h} = \frac{0.346 \text{ in}}{480 \text{ in}} = 0.00072 \text{ rad}$$
$$D_{2Q} = \frac{\Delta_T}{h} = \frac{0.346 \text{ in}}{240 \text{ in}} = 0.00144 \text{ rad}$$

Compute the flexibility influence coefficients. First, apply a unit couple clockwise at joint A.

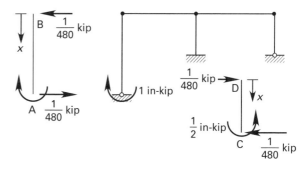

For member AB (taking the origin at B),

$$M_x = \frac{x}{480} \text{ in-kips} \quad [\text{for } 0 \text{ in} < x < 480 \text{ in}]$$

For member CD (taking the origin at D),

$$M_x = -\frac{x}{480} \text{ in-kips} \quad [\text{for } 0 \text{ in} < x < 240 \text{ in}]$$

Apply a unit couple clockwise at joint E. For member CD,

$$M_x = -\frac{x}{240} \text{ in-kips} \quad [\text{for } 0 \text{ in} < x < 240 \text{ in}]$$

For member EF,

$$M_x = \frac{x}{240} \text{ in-kips} \quad [\text{for } 0 \text{ in} < x < 240 \text{ in}]$$

Using the dummy load method, the dummy load moments are identical to the real moments. The flexibility coefficients represent the rotation in radians per in-kip of applied moment. In the following equations, the symbol * denotes the unit of an infinitesimal virtual force in the dummy moment.

$$(1 \text{ in-*})f_{11} = \sum \int_L \left(\frac{mM}{EI}\right) dx$$

$$= \int_0^{480 \text{ in}} \left(\frac{\left(\frac{x}{480}\right)^2 (1 \text{ kip-*})}{EI_1}\right) dx$$

$$+ \int_0^{240 \text{ in}} \left(\frac{\left(\frac{x}{480}\right)^2 (1 \text{ kip-*})}{EI_2}\right) dx$$

$$= \left.\frac{x^3 (1 \text{ kip-*})}{(480)^2 (3EI_1)}\right|_0^{480 \text{ in}} + \left.\frac{x^3 (1 \text{ kip-*})}{(480)^2 (3EI_2)}\right|_0^{240 \text{ in}}$$

$$= \frac{(480 \text{ in})^3 (1 \text{ kip-*})}{(480 \text{ in})^2 (3)\left(30{,}000 \frac{\text{kips}}{\text{in}^2}\right)(882 \text{ in}^4)}$$

$$+ \frac{(240 \text{ in})^3 (1 \text{ kip-*})}{(480 \text{ in})^2 (3)\left(30{,}000 \frac{\text{kips}}{\text{in}^2}\right)(428 \text{ in}^4)}$$

$$f_{11} = 0.0000076 \text{ rad} \quad [\text{per in-kip}]$$

$$(1 \text{ in-*})f_{21} = \int_0^{240 \text{ in}} \left(\frac{\left(\frac{x}{480}\right)^2 \left(-\frac{x}{240}\right)(1 \text{ kip-*})}{EI_2}\right) dx$$

$$= \left.\frac{-x^3 (1 \text{ kip-*})}{(480 \text{ in})(240 \text{ in})(3EI_2)}\right|_0^{240 \text{ in}}$$

$$= \frac{-(240 \text{ in})^3 (1 \text{ kip-*})}{(480 \text{ in})(240 \text{ in})(3)\left(30{,}000 \frac{\text{kips}}{\text{in}^2}\right)(428 \text{ in}^4)}$$

$$f_{21} = -0.0000031 \text{ rad} \quad [\text{per in-kip}]$$

By Maxwell's law, $f_{12} = f_{21}$.

$$(1 \text{ in-*})f_{22} = \int_0^{240 \text{ in}} \left(\frac{\left(\frac{x}{240}\right)^2 (1 \text{ kip-*})}{EI_1}\right) dx$$

$$+ \int_0^{240 \text{ in}} \left(\frac{\left(\frac{x}{240}\right)^2 (1 \text{ kip-*})}{EI_2}\right) dx$$

$$= \left.\frac{x^3 (1 \text{ kip-*})}{(3)(240)^2 (3EI_1)}\right|_0^{240 \text{ in}}$$

$$+ \left.\frac{x^3 (1 \text{ kip-*})}{(3)(240)^2 (3EI_2)}\right|_0^{240 \text{ in}}$$

$$= \frac{(240 \text{ in})^3 (1 \text{ kip-*})}{(3)(240 \text{ in})^2 \left(30{,}000 \frac{\text{kips}}{\text{in}^2}\right)(882 \text{ in}^4)}$$

$$+ \frac{(240 \text{ in})^3 (1 \text{ kip-*})}{(3)(240 \text{ in})^2 \left(30{,}000 \frac{\text{kips}}{\text{in}^2}\right)(428 \text{ in}^4)}$$

$$f_{22} = 0.0000093 \text{ rad} \quad [\text{per in-kip}]$$

For consistent displacements, rotations at A and E must be zero. Thus,

$$f_{11} M_A + f_{12} M_E = D_{1Q}$$

$$\left(0.0000076 \ \frac{\text{rad}}{\text{in-kip}}\right) M_A + \left(0.0000031 \ \frac{\text{rad}}{\text{in-kip}}\right) M_E$$
$$= 0.00072 \text{ rad}$$

$$f_{21} M_A + f_{22} M_E = D_{2Q}$$

$$\left(0.0000031 \ \frac{\text{rad}}{\text{in-kip}}\right) M_A + \left(0.0000093 \ \frac{\text{rad}}{\text{in-kip}}\right) M_E$$
$$= -0.00144 \text{ rad}$$

Solving these two linear equations simultaneously gives

$$M_E = -216 \text{ in-kips} \quad \boxed{[216 \text{ in-kips counterclockwise}]}$$

$$M_A = +183 \text{ in-kips} \quad \boxed{[183 \text{ in-kips clockwise}]}$$

10.2. The column shears in AB and CD are found by summing moments about the bases. Thus,

$$\sum M_B = M_A - V_{AB}h = 0$$

$$= 183 \text{ in-kips} - V_{AB}(480 \text{ in})$$

$$V_{AB} = \boxed{0.38 \text{ kip} \quad [\text{to the right}]}$$

$$\sum M_{EF} = M_E - V_{EF}h = 0$$

$$= -216 \text{ in-kips} - V_{EF}(240 \text{ in})$$

$$V_{EF} = \boxed{-0.90 \text{ kip} \quad [0.90 \text{ kip to the left}]}$$

$$\sum F_x = V_{AB} + V_{EF} + V_{CD} = 0$$

$$= 0.38 \text{ kip} - 0.90 \text{ kip} + V_{CD}$$

$$V_{CD} = \boxed{0.52 \text{ kip} \quad [\text{to the right}]}$$

10.3. With the column shears known, the deflections at the top of columns are calculated as for a cantilevered beam loaded by concentrated forces at its tips.

$$\Delta_B = \frac{V_{AB}h^3}{3EI_1}$$

$$= \frac{(0.38 \text{ kip})(480 \text{ in})^3}{(3)\left(30{,}000 \dfrac{\text{kips}}{\text{in}^2}\right)(882 \text{ in}^4)}$$

$$= \boxed{0.530 \text{ in} \quad [\text{to the left}]}$$

$$\Delta_F = \frac{V_{EF}h^3}{3EI_1}$$

$$= \frac{(0.90 \text{ kip})(240 \text{ in})^3}{(3)\left(30{,}000 \dfrac{\text{kips}}{\text{in}^2}\right)(882 \text{ in}^4)}$$

$$= \boxed{0.157 \text{ in} \quad [\text{to the left}]}$$

A simple check on the solution is made by equating the elongation of the girder to $\Delta_B + \Delta_F = 0.69$.

$$\Delta L = D_F(\Delta T)L$$

$$= \left(0.000006 \dfrac{\text{in}}{\text{in-°F}}\right)(60°\text{F})(960 \text{ in} + 960 \text{ in})$$

$$= 0.69 \text{ in} \quad [\text{OK}]$$

10.4. The maximum bending stress in column CD occurs at the base.

$$M_C = V_{CD}h = (0.52 \text{ kip})(240 \text{ in}) = 125 \text{ in-kips}$$

For a W14 × 43, $S_x = 62.6 \text{ in}^3$.

$$f_b = \frac{M_C}{S_x} = \frac{125 \text{ in-kips}}{62.6 \text{ in}^3} = \boxed{2.0 \text{ ksi}}$$

SOLUTION 11

11.1. For each of the illustrated structures shown in the problem statement, calculate the force needed to create a deflection of 1 in at the top of the structure. Then compare these four values, P_A, P_B, P_C, and P_D, in the following solution to obtain the relative rigidities.

For the braced frame in illustration (a),

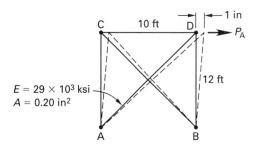

The length of member AD is

$$L_{AD} = \sqrt{(10 \text{ ft}^2 + 12 \text{ ft}^2)} = 15.62 \text{ ft}$$

Draw a free-body diagram of joint D.

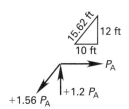

From the free-body diagram,

$$\sum F_x = P_A - P_{AD}\cos\theta = 0$$

$$= P_A - P_{AD}\left(\frac{10 \text{ ft}}{15.62 \text{ ft}}\right)$$

$$P_{AD} = 1.56P_A$$

By the real work method,

$$\frac{P_A\Delta}{2} = \sum \frac{P_i^2 L_i}{2A_iE}$$

$$\frac{P_A(1 \text{ in})}{2} = \frac{P_{AD}^2 L_{AD}}{2A_{AD}E}$$

$$= \frac{(1.56P_A)^2(15.62 \text{ ft})\left(12 \dfrac{\text{in}}{\text{ft}}\right)}{(2)\left(29{,}000 \dfrac{\text{kips}}{\text{in}^2}\right)(0.20 \text{ in}^2)}$$

$$P_A = 12.7 \text{ kips}$$

For the cantilevered wall in illustration (b), consider both shear and bending deformation.

$$G = 0.4E_c = (0.4)\left(3120 \ \frac{\text{kips}}{\text{in}^2}\right) = 1250 \text{ ksi}$$

$$I = \frac{bd^3}{12} = \frac{(9 \text{ in})(72 \text{ in})^3}{12} = 280{,}000 \text{ in}^4$$

$$A = bd = (9 \text{ in})(72 \text{ in}) = 648 \text{ in}^2$$

$$\Delta = \frac{P_{\text{B}}h^3}{3E_cI} + \frac{1.2P_{\text{B}}h}{GA}$$

$$1 \text{ in} = \frac{P_{\text{B}}(144 \text{ in})^3}{(3)\left(3120 \ \frac{\text{kips}}{\text{in}^2}\right)(280{,}000 \text{ in}^4)}$$

$$+ \frac{1.2P_{\text{B}}(144 \text{ in})}{\left(1250 \ \frac{\text{kips}}{\text{in}^2}\right)(648 \text{ in}^2)}$$

$$P_{\text{B}} = 739 \text{ kips}$$

For the K-braced truss in illustration (c),

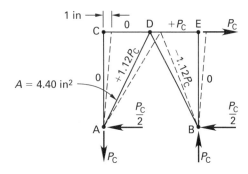

$$L_{\text{AD}} = \sqrt{(12 \text{ ft}^2 + 6 \text{ ft}^2)} = 13.42 \text{ ft}$$

For equilibrium of joint A,

$$\sum F_x = -0.5P_{\text{C}} + P_{\text{AD}}\cos\theta = 0$$

$$= -0.5P_{\text{C}} + P_{\text{AD}}\left(\frac{6 \text{ ft}}{13.42 \text{ ft}}\right)$$

$$P_{\text{AD}} = 1.12P_{\text{C}}$$

From symmetry,

$$P_{\text{DB}} = -P_{\text{AD}} = -1.12P_{\text{C}}$$

Using the real work method,

$$\tfrac{1}{2}P_{\text{C}}(1 \text{ in}) = \sum_{i=1}^{m}\tfrac{1}{2}\left(\frac{P_i^2 L_i}{A_i E}\right)$$

$$P_{\text{C}} = \frac{(1.12P_{\text{C}})^2(13.42 \text{ ft})\left(12 \ \frac{\text{in}}{\text{ft}}\right)}{\left(29{,}000 \ \frac{\text{kips}}{\text{in}^2}\right)(4.4 \text{ in}^2)}$$

$$+ \frac{(-1.12P_{\text{C}})^2(13.42 \text{ ft})\left(12 \ \frac{\text{in}}{\text{ft}}\right)}{\left(29{,}000 \ \frac{\text{kips}}{\text{in}^2}\right)(4.4 \text{ in}^2)}$$

$$= 316 \text{ kips}$$

For the pierced shear wall in illustration (d),

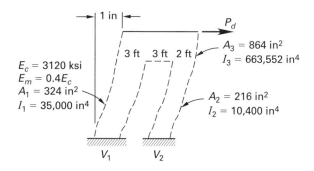

Proportion the force P_{D} to the lower walls on the basis of their translational stiffness. Assume zero rotation at top and bottom. Thus,

$$K_1 = \frac{1}{\dfrac{h^3}{12E_cI_1} + \dfrac{1.2h}{GA_1}}$$

$$= \frac{1}{\dfrac{(96 \text{ in})^3}{(12)\left(3120 \ \frac{\text{kips}}{\text{in}^2}\right)(35{,}000 \text{ in}^4)}}$$

$$+ \dfrac{(1.2)(96 \text{ in})}{(0.4)\left(3120 \ \frac{\text{kips}}{\text{in}^2}\right)(324 \text{ in}^2)}$$

$$= 1042 \text{ kips/in}$$

$$K_2 = \cfrac{1}{\cfrac{h^3}{12E_cI_1} + \cfrac{1.2h}{GA_1}}$$

$$= \cfrac{1}{\cfrac{(96 \text{ in})^3}{(12)\left(3120 \, \cfrac{\text{kips}}{\text{in}^2}\right)(10{,}400 \text{ in}^4)} + \cfrac{(1.2)(96 \text{ in})}{(0.4)\left(3120 \, \cfrac{\text{kips}}{\text{in}^2}\right)(216 \text{ in}^2)}}$$

$$= 370 \text{ kips/in}$$

$$V_1 = \left(\frac{K_1}{K_1 + K_2}\right)P_D$$

$$= \left(\cfrac{1042 \, \cfrac{\text{kips}}{\text{in}^2}}{1042 \, \cfrac{\text{kips}}{\text{in}^2} + 370 \, \cfrac{\text{kips}}{\text{in}^2}}\right)P_D$$

$$= 0.738P_D$$

The stiffness can be calculated using the shear in the left wall segment.

$$\Delta = 1 \text{ in}$$

$$= \frac{V_1 h^3}{12E_cI_1} + \frac{1.2V_1 h}{0.4E_cA_1}$$

$$\quad + \frac{P_D h_3^3}{3E_cI_3} + \frac{1.2P_D h_3}{0.4E_cA_3}$$

$$= \cfrac{0.738P_D(96 \text{ in})^3}{(12)\left(3120 \, \cfrac{\text{kips}}{\text{in}^2}\right)(35{,}000 \text{ in}^4)}$$

$$\quad + \cfrac{(1.2)(0.738P_D)(96 \text{ in})}{(0.4)\left(3120 \, \cfrac{\text{kips}}{\text{in}^2}\right)(324 \text{ in}^2)}$$

$$\quad + \cfrac{P_D(48 \text{ in})^3}{(3)\left(3120 \, \cfrac{\text{kips}}{\text{in}^2}\right)(663{,}552 \text{ in}^4)}$$

$$\quad + \cfrac{1.2P_D(48 \text{ in})}{(0.4)\left(3120 \, \cfrac{\text{kips}}{\text{in}^2}\right)(864 \text{ in}^2)}$$

$$P_D = 1282 \text{ kips}$$

The rigidities of the elements relative to the rigidity of the X-braced frame in illustration (a), which is taken as unity, are

$$P_A = 1$$

$$\frac{P_B}{P_A} = \frac{739 \text{ kips}}{12.7 \text{ kips}} = \boxed{58}$$

$$\frac{P_C}{P_A} = \frac{316 \text{ kips}}{12.7 \text{ kips}} = \boxed{25}$$

$$\frac{P_D}{P_A} = \frac{1282 \text{ kips}}{12.7 \text{ kips}} = \boxed{101}$$

SOLUTION 12

12.1. Assume a well-drained, granular backfill for which $k_a = 0.3$ and $\gamma_s = 110 \text{ lbf/ft}^3$. Based on k_a and γ_s values, the equivalent fluid weight is

$$k_a \gamma_s = (0.3)\left(110 \, \frac{\text{lbf}}{\text{ft}^3}\right) = 33 \text{ lbf/ft}^3$$

The wall spans vertically between the sill plate and the slab centerline at approximately 8.2 ft.

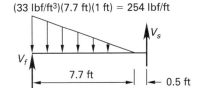

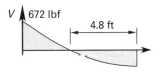

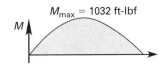

Per foot of wall,

$$w = k_a \gamma_s a(1 \text{ ft})$$

$$= (0.3)\left(110 \, \frac{\text{lbf}}{\text{ft}^3}\right)(7.7 \text{ ft})(1 \text{ ft})$$

$$= 254 \text{ lbf/ft}$$

$$V_s = \cfrac{0.5wa\left(\cfrac{a}{3}\right)}{h}$$

$$= \cfrac{(0.5)\left(254 \, \cfrac{\text{lbf}}{\text{ft}}\right)(7.7 \text{ ft})\left(\cfrac{7.7 \text{ ft}}{3}\right)}{8.2 \text{ ft}}$$

$$= 306 \text{ lbf}$$

$$V_f = 0.5wa - V_s$$
$$= (0.5)\left(254 \ \frac{\text{lbf}}{\text{ft}}\right)(7.7 \text{ ft}) - 300 \text{ lbf}$$
$$= 672 \text{ lbf}$$

The maximum moment occurs where the shear is zero. For simplicity, take positive y as downward from the ground surface.

$$V = V_s - \frac{my^2}{2} = V_s - \left(\frac{w}{a}\right)\left(\frac{y^2}{2}\right) = 0$$

$$306 \text{ lbf} - \left(\frac{254 \ \frac{\text{lbf}}{\text{ft}}}{7.7 \text{ ft}}\right)\left(\frac{y^2}{2}\right) = 0$$

$$y = 4.3 \text{ ft}$$

$$M_{\max} = V_s(y + 0.5 \text{ ft}) - \left(\frac{w}{a}\right)\left(\frac{y^2}{2}\right)\left(\frac{y}{3}\right)$$
$$= (306 \text{ lbf})(4.3 \text{ ft} + 0.5 \text{ ft})$$
$$- \left(\frac{254 \ \frac{\text{lbf}}{\text{ft}}}{7.7 \text{ ft}}\right)\left(\frac{(4.3 \text{ ft})^2}{2}\right)\left(\frac{4.3 \text{ ft}}{3}\right)$$
$$= 1032 \text{ ft-lbf}$$

Check the strength of reinforced masonry wall using the allowable stress method from ACI 530. For the given vertical reinforcement of no. 5 bars spaced 16 in on centers,

$$A_s = \frac{A_b}{s} = \frac{0.31 \text{ in}^2}{\dfrac{16 \text{ in}}{12 \ \frac{\text{in}}{\text{ft}}}} = 0.23 \text{ in}^2/\text{ft}$$

For the unit wall length, $b = 12$ in. Take $f'_m = 1500$ psi and $d = 5.3$ in.

$$n = \frac{E_s}{E_m} = \frac{E_s}{900f'_m} = \frac{29{,}000{,}000 \ \frac{\text{lbf}}{\text{in}^2}}{(900)\left(1500 \ \frac{\text{lbf}}{\text{in}^2}\right)}$$
$$= 21.5$$

$$n\rho = \frac{nA_s}{bd} = \frac{(21.5)\left(0.23 \ \frac{\text{in}^2}{\text{ft}}\right)(1 \text{ ft})}{(12 \text{ in})(5.3 \text{ in})} = 0.0778$$

$$k = \sqrt{(n\rho)^2 + 2n\rho} - n\rho$$
$$= \sqrt{(0.0778)^2 + (2)(0.0778)} - 0.0778$$
$$= 0.324$$

$$j = 1 - \frac{k}{3} = 1 - \frac{0.324}{3} = 0.892$$

$$f_b = \left(\frac{M_{\max}}{bd^2}\right)\left(\frac{2}{jk}\right)$$
$$= \left(\frac{(1032 \text{ ft-lbf})\left(12 \ \frac{\text{in}}{\text{ft}}\right)}{(12 \text{ in})(5.3 \text{ in})^2}\right)\left(\frac{2}{(0.324)(0.892)}\right)$$
$$= 254 \text{ psi}$$

$$f_s = \frac{M_{\max}}{A_s jd} = \frac{(1032 \text{ ft-lbf})\left(12 \ \frac{\text{in}}{\text{ft}}\right)}{\left(0.23 \ \frac{\text{in}^2}{\text{ft}}\right)(0.892)(5.3 \text{ in})(1 \text{ ft})}$$
$$= 11{,}400 \text{ psi}$$

The computed steel stress, f_s, is well below the allowable stress for all grades of reinforcement. The allowable compression stress is

$$F_b = 0.45f'_m = (0.45)\left(1500 \ \frac{\text{lbf}}{\text{in}^2}\right)$$
$$= 675 \text{ psi} \quad [> f_b = 254 \text{ psi, so OK}]$$

Check the shear stress at the wall base.

$$f_v = \frac{V_f}{bt} = \frac{672 \text{ lbf}}{(12 \text{ in})(7.625 \text{ in})} = 7 \text{ psi}$$

$$F_v = 2\sqrt{f'_m} = 2\sqrt{1500 \ \frac{\text{lbf}}{\text{in}^2}}$$
$$= 77 \text{ psi} \quad [> f_v, \text{ so OK}]$$

Therefore, flexural and shear stresses in the masonry are OK, and the $\boxed{\text{wall strength is adequate.}}$

Check the transfer of 306 lbf/ft from wall to sill plates using $^5/_8$ in diameter anchor bolts spaced at 4 ft on centers.

$$V = V_s s = \left(306 \ \frac{\text{lbf}}{\text{ft}}\right)\left(4 \ \frac{\text{ft}}{\text{bolt}}\right)$$
$$= 1220 \text{ lbf/bolt}$$

From NDS Table 11E, the capacity of $^5/_8$ in anchor bolts in nominal 3 in Douglas-fir is

$$Z'_\perp = C_D Z_\perp = (0.9)(610 \text{ lbf})$$
$$= 550 \text{ lbf} \quad [< V = 1220 \text{ lbf/bolt, so no good}]$$

Thus, anchor bolts $\boxed{\text{are not adequate}}$ to transfer force from wall to sill. The type 2 framing angles at 2 ft on centers are OK. (Assume that there are two angles, one on each side, arranged to eliminate cross-grain tension in the plate.)

Several deficiencies occur in the transfer of the lateral force into the roof diaphragm. For example, consider the 2×6 brace at the 7 ft maximum height.

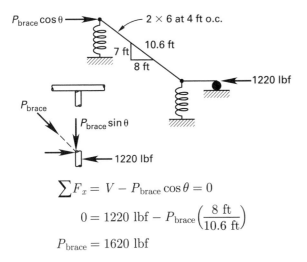

$$\sum F_x = V - P_{\text{brace}} \cos\theta = 0$$

$$0 = 1220 \text{ lbf} - P_{\text{brace}}\left(\frac{8 \text{ ft}}{10.6 \text{ ft}}\right)$$

$$P_{\text{brace}} = 1620 \text{ lbf}$$

First, the 2×6 brace has a weak axis unsupported length, L_e, of 10.6 ft = 127 in. Thus,

$$\frac{L_e}{d} = \frac{(10.6 \text{ ft})\left(12 \frac{\text{in}}{\text{ft}}\right)}{1.5 \text{ in}} = 85$$

The slenderness ratio of the brace exceeds the limit of 50 permitted by NDS. The 2×6 cannot resist the design load.

Second, the type 1 framing angles connecting the brace to the truss chords are not adequate to transfer a force of

$$P_{\text{brace}} \sin\theta = (1620 \text{ lbf})\left(\frac{7 \text{ ft}}{10.6 \text{ ft}}\right) = 1070 \text{ lbf}$$

Third, the vertical component of force in the brace imposes loads on the roof trusses that are probably excessive (e.g., 1080 lbf down on the lower chord, 1080 lbf up on the upper chords where the braces are maximum height).

Remedial action is needed.

One possibility for remedial repair is to create a diaphragm in the plane of the lower chords of the roof trusses.

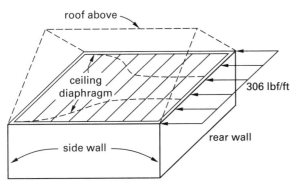

Additional strength must be provided to transfer the 306 lbf/ft into the diaphragm (to supplement the $^5/_8$ in diameter anchor bolts). One method is to install a cleat on the 3×8 sill, which could transfer the force into the sill.

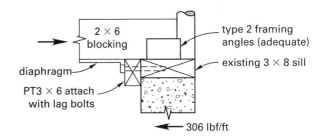

Collector and anchorage elements along the side walls should be investigated and strengthened, if necessary. Front and rear walls plus sills are adequate as chords. The shear transfer through the ceiling diaphragm to the side walls is

$$v = \frac{V_s L}{2B} = \frac{\left(306 \frac{\text{lbf}}{\text{ft}}\right)(22 \text{ ft})}{(2)(22 \text{ ft})} = 153 \text{ lbf/ft}$$

Take the allowable shear for structural panel diaphragms from SDPWS Table 4.2C. Table values are for seismic loading and must be adjusted by $0.9/2.0 = 0.45$ for sustained loads. Try $^3/_8$ in sheathing with 6d common nails at 6 in on center along supported edges (unblocked diaphragm), and at 12 in on center along intermediate supports. The shear capacity is

$$v_{\text{allowable}} = 0.45 v_{\text{tabulated}} = (0.45)\left(250 \frac{\text{lbf}}{\text{ft}}\right)$$

$$= 113 \text{ lbf/ft} \quad [< v = 153 \text{ lbf/ft, so no good}]$$

To correct this, add blocking until v is less than 113 lbf/ft. With blocking (6d at 6 in along blocked edges), the diaphragm capacity is

$$v_{\text{allowable}} = 0.45 v_{\text{tabulated}} = (0.45)\left(370 \frac{\text{lbf}}{\text{ft}}\right)$$

$$= 167 \text{ lbf/ft} \quad [> v, \text{ so OK}]$$

SOLUTION 13

13.1. Determine the critical values of lateral pressure on the soldier beams. The surcharge from the adjacent warehouse is

$$Q_f = w_c h_{\text{footing}} + L_1 + (D+L)_2$$
$$+ (D+L)_3 + (D+L)_{\text{roof}}$$
$$= \left(150\ \frac{\text{lbf}}{\text{ft}^3}\right)(4\ \text{ft}) + 250\ \frac{\text{lbf}}{\text{ft}^2}$$
$$+ 350\ \frac{\text{lbf}}{\text{ft}^2} + 350\ \frac{\text{lbf}}{\text{ft}^2} + 100\ \frac{\text{lbf}}{\text{ft}^2}$$
$$= 1650\ \text{lbf/ft}^2$$

$$N = Q_f - w_{\text{soil}} D_f$$
$$= 1650\ \frac{\text{lbf}}{\text{ft}^2} - \left(120\ \frac{\text{lbf}}{\text{ft}^3}\right)(5\ \text{ft})$$
$$= 1050\ \text{lbf/ft}^2$$

Compute the lateral pressure on the soldier beams given that the beams are spaced 8 ft on centers. The pressure due to the building surcharge is given as uniform over a length of $H_1 = H - 5\ \text{ft} = 25\ \text{ft} - 5\ \text{ft} = 20\ \text{ft}$. The active pressure due to soil increases linearly from the ground surface to a constant pressure of

$$k_a w_{\text{soil}} H = \left(25\ \frac{\text{lbf}}{\text{ft}^3}\right)(25\ \text{ft}) = 625\ \text{lbf/ft}^2$$

The net loading on each beam in pounds-force per foot is

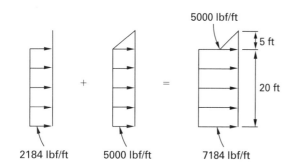

$$\left(25\ \frac{\text{lbf}}{\text{ft}^3}\right)Hs = \left(25\ \frac{\text{lbf}}{\text{ft}^3}\right)(25\ \text{ft})(8\ \text{ft})$$
$$= 5000\ \text{lbf/ft}$$

$$\frac{0.4N(1\ \text{ft} + a)s}{H_1} = \frac{(0.4)\left(1050\ \frac{\text{lbf}}{\text{ft}^2}\right)(1\ \text{ft} + 12\ \text{ft})(8\ \text{ft})}{20\ \text{ft}}$$
$$= 2184\ \text{lbf/ft}$$

Make an approximate analysis of the tie-back forces in order to obtain a trial value for the toe length. The toe will have short embedment, which offers scant resistance to rotation at the base. Therefore, idealize the base as a simple support offering resistance only to vertical and lateral forces. The resultant lateral force on each soldier beam is

$$F_x = 0.5 p_1(0.2H) + p_2(0.8H)$$

$$= \frac{(0.5)\left(5000\ \frac{\text{lbf}}{\text{ft}}\right)(0.2)(25\ \text{ft}) + \left(7184\ \frac{\text{lbf}}{\text{ft}}\right)(0.8)(25\ \text{ft})}{1000\ \frac{\text{lbf}}{\text{kip}}}$$

$$= \boxed{156\ \text{kips}}$$

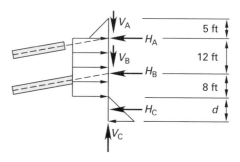

13.2. Calculate the approximate forces at the supports based on tributary length for each loaded segment.

$$H_A \approx 0.5 p_1(0.2H) + p_2(0.4L_1)$$
$$\approx (0.5)\left(5.0\ \frac{\text{kips}}{\text{ft}}\right)(0.2)(25\ \text{ft})$$
$$+ \left(7.184\ \frac{\text{kips}}{\text{ft}}\right)(0.4)(12\ \text{ft})$$
$$\approx 47\ \text{kips}$$

$$H_B \approx p_2(0.6L_1 + 0.75L_2)$$
$$\approx \left(7.184\ \frac{\text{kips}}{\text{ft}}\right)\big((0.6)(12\ \text{ft}) + (0.75)(8\ \text{ft})\big)$$
$$\approx 95\ \text{kips}$$

For horizontal equilibrium,

$$\sum F_x = H_A + H_B + H_C - F_x = 0$$
$$H_C = F_x - H_A - H_B$$
$$= 156\ \text{kips} - 47\ \text{kips} - 95\ \text{kips}$$
$$= 14\ \text{kips}$$

Find the toe length, d, needed to develop 14 kips by passive soil pressure.

$$14 \text{ kips} = 0.5(f_p d)d$$

$$= (0.5)\left(0.45 \frac{\text{kip}}{\text{ft}^2}\right)d^2$$

$$d^2 = 62 \text{ ft}^2$$

$$d \approx 8 \text{ ft}$$

The vertical components of tendon force, V_A and V_B, are resisted by skin friction along the 2 ft diameter toe.

The tendon slopes 15° with respect to the horizontal. Thus,

$$V_A = \left(\frac{H_A}{\cos 15°}\right)\sin 15° = H_A \tan 15°$$

$$= (47 \text{ kips})\tan 15°$$

$$= 12.6 \text{ kips}$$

$$V_B = H_B \tan 15° = (95 \text{ kips})\tan 15° = 25.5 \text{ kips}$$

$$V_C = V_A + V_B = 12.6 \text{ kips} + 25.5 \text{ kips} = 38.1 \text{ kips}$$

Find the skin friction.

$$f_v \pi D d = V_C$$

$$f_v \pi (2 \text{ ft})(8 \text{ ft}) = 38.1 \text{ kips}$$

$$f_v = 0.76 \text{ kip/ft}^2$$

The computed skin friction, 0.76 kip/ft², is well below the allowable value of 2.0 kips/ft²; therefore, an 8 ft embedment at the toe $\boxed{\text{is acceptable.}}$

A more refined analysis of tie-back forces could be made using the trial toe length of 8 ft and treating the horizontal reaction at B, H_B, as the redundant force. However, given the uncertainties in lateral force distribution and support conditions, the approximate analysis is accurate enough.

13.3. To verify adequacy of the 8 ft toe length, both skin friction and passive earth pressure on the embedded shaft must be checked against the given allowable values. The skin friction is computed above to be 0.76 kip/ft², which is below the allowable 2.0 kips/ft². Find the passive pressure.

$$\sum F_x = H_A + H_B + H_C - F_x = 0$$

$$H_C = F_x - H_A - H_B$$

$$= 156 \text{ kips} - 47 \text{ kips} - 95 \text{ kips}$$

$$= 14 \text{ kips}$$

$$0.5 f_p d^2 = H_C$$

$$f_p = \frac{H_C}{0.5 d^2}$$

$$= \frac{14 \text{ kips}}{(0.5)(8 \text{ ft})^2}$$

$$= 0.44 \text{ kip/ft}^2$$

$$f_{\text{allowable}} = 0.45 \text{ kips/ft}^2 \quad [f_p < f_{\text{allowable}}, \text{ so OK}]$$

The passive pressure is less than the allowable pressure, so an 8 ft toe length $\boxed{\text{is OK.}}$

13.4. Compute the grouted anchor lengths. The allowable skin friction on a 12 in diameter drilled shaft is 1.5 kips/ft². The tension force in the rod is

$$T_A = \frac{H_A}{\cos 15°} = \frac{47 \text{ kips}}{\cos 15°} = 48.7 \text{ kips}$$

$$F_f \pi D_{\text{tie rod}} L_A = T_A$$

$$L_A = \frac{T_A}{F_f \pi D_{\text{tie rod}}} = \frac{48.7 \text{ kips}}{\left(1.5 \dfrac{\text{kips}}{\text{ft}^2}\right)\pi(1.0 \text{ ft})}$$

$$= 10.3 \text{ ft}$$

$$T_B = \frac{H_B}{\cos 15°} = \frac{95 \text{ kips}}{\cos 15°} = 98.4 \text{ kips}$$

$$F_f \pi D_{\text{tie rod}} L_B = T_B$$

$$L_B = \frac{T_B}{F_f \pi D_{\text{tie rod}}} = \frac{98.4 \text{ kips}}{\left(1.5 \dfrac{\text{kips}}{\text{ft}^2}\right)\pi(1.0 \text{ ft})}$$

$$= 20.9 \text{ ft}$$

The grouted portion must extend beyond the failure plane, which is at a 60° angle from the horizontal. Sketch the minimum total tendon lengths.

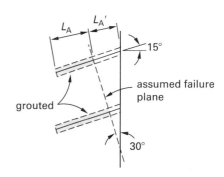

$$\frac{L'_A}{\sin 30°} = \frac{L}{\sin 75°}$$

$$L'_A = \frac{L \sin 30°}{\sin 75°} = \frac{(20 \text{ ft})\sin 30°}{\sin 75°} = 10.4 \text{ ft}$$

$$\frac{L'_B}{\sin 30°} = \frac{L}{\sin 75°}$$

$$L'_A = \frac{L \sin 30°}{\sin 75°} = \frac{(8 \text{ ft})\sin 30°}{\sin 75°} = 4.1 \text{ ft}$$

The minimum length of tendon A is

$$L_A + L'_A = 10.3 \text{ ft} + 10.4 \text{ ft} = \boxed{20.7 \text{ ft}}$$

The minimum length of tendon B is

$$L_B + L'_B = 20.9 \text{ ft} + 4.1 \text{ ft} = \boxed{25.0 \text{ ft}}$$

13.5. The installation procedure is as follows (see Schnabel).

- Drill 2 ft diameter vertical shafts at soldier beam locations.

- Pour concrete in lower (i.e., toe) portion of shaft; insert and align soldier beam.

- Excavate to region below point A, installing lagging planks as work progresses.

- Drill 1 ft diameter shaft inclined 15° between soldier beams at elevation A.

- Grout tendon A (usually by pressure injection rather than by gravity flow).

- Install wales horizontally between soldier beams to transfer tendon tie-back forces to soldier beams (usually two C-sections back to back to form wales).

- Tension tendons to their proof load (typically 80% of ultimate), then anchor the tendons at their working load.

SOLUTION 14

14.1. Calculate the participation factors in accordance with the *SEAOC Blue Book*. The mass at each level and the total mass are

$$m_1 = \frac{W_1}{g} = \frac{80 \text{ kips}}{32.2 \dfrac{\text{ft}}{\text{sec}^2}} = 2.48 \text{ kips-sec}^2/\text{ft}$$

$$m_2 = \frac{W_2}{g} = \frac{80 \text{ kips}}{32.2 \dfrac{\text{ft}}{\text{sec}^2}} = 2.48 \text{ kips-sec}^2/\text{ft}$$

$$m_3 = \frac{W_3}{g} = \frac{120 \text{ kips}}{32.2 \dfrac{\text{ft}}{\text{sec}^2}} = 3.73 \text{ kips-sec}^2/\text{ft}$$

$$m = m_1 + m_2 + m_3$$

$$= 2.48 \frac{\text{kips-sec}^2}{\text{ft}} + 2.48 \frac{\text{kips-sec}^2}{\text{ft}} + 3.73 \frac{\text{kips-sec}^2}{\text{ft}}$$

$$= 8.69 \text{ kips-sec}^2/\text{ft}$$

For the first mode, $\omega_1 = 8.77$ rad/sec, the normalized eigenvector is

$$\{\phi_1\}^T = (1.69, 1.22, 0.572)$$

$$P_1 = \frac{\displaystyle\sum_{i=1}^{3} \phi_{1i} m_i}{\displaystyle\sum_{i=1}^{3} \phi_{1i}^2 m_i}$$

$$= \frac{\begin{array}{l}(1.69)\left(2.48 \dfrac{\text{kips-sec}^2}{\text{ft}}\right) + (1.22) \\[4pt] \times \left(2.48 \dfrac{\text{kips-sec}^2}{\text{ft}}\right) + (0.572)\left(3.73 \dfrac{\text{kips-sec}^2}{\text{ft}}\right)\end{array}}{\begin{array}{l}(1.69)^2\left(2.48 \dfrac{\text{kips-sec}^2}{\text{ft}}\right) + (1.22)^2 \\[4pt] \times \left(2.48 \dfrac{\text{kips-sec}^2}{\text{ft}}\right) + (0.572)^2\left(3.73 \dfrac{\text{kips-sec}^2}{\text{ft}}\right)\end{array}}$$

$$= \boxed{0.780}$$

Similarly, for modes 2 and 3,

$$\{\phi_2\}^T = (-1.208, 0.704, 1.385)$$

$$P_2 = \frac{\displaystyle\sum_{i=1}^{3} \phi_{i2} m_i}{\displaystyle\sum_{i=1}^{3} \phi_{i2}^2 m_i}$$

$$= \frac{\begin{array}{l}(-1.208)\left(2.48 \dfrac{\text{kips-sec}^2}{\text{ft}}\right) + (0.704) \\[4pt] \times \left(2.48 \dfrac{\text{kips-sec}^2}{\text{ft}}\right) + (1.385)\left(3.73 \dfrac{\text{kips-sec}^2}{\text{ft}}\right)\end{array}}{\begin{array}{l}(-1.208)^2\left(2.48 \dfrac{\text{kips-sec}^2}{\text{ft}}\right) + (0.704)^2 \\[4pt] \times \left(2.48 \dfrac{\text{kips-sec}^2}{\text{ft}}\right) + (1.385)^2\left(3.73 \dfrac{\text{kips-sec}^2}{\text{ft}}\right)\end{array}}$$

$$= \boxed{0.327}$$

$$\{\phi_3\}^{\mathrm{T}} = (-0.714, 1.697, -0.948)$$

$$P_3 = \frac{\displaystyle\sum_{i=1}^{3} \phi_{i3} m_i}{\displaystyle\sum_{i=1}^{3} \phi_{i3}^2 m_i}$$

$$= \frac{\begin{array}{c}(-0.714)\left(2.48 \ \frac{\text{kips-sec}^2}{\text{ft}}\right) + (1.697) \\ \times \left(2.48 \ \frac{\text{kips-sec}^2}{\text{ft}}\right) + (-0.948)\left(3.73 \ \frac{\text{kips-sec}^2}{\text{ft}}\right)\end{array}}{\begin{array}{c}(-0.704)^2\left(2.48 \ \frac{\text{kips-sec}^2}{\text{ft}}\right) + (1.697)^2 \\ \times \left(2.48 \ \frac{\text{kips-sec}^2}{\text{ft}}\right) + (-0.948)^2\left(3.73 \ \frac{\text{kips-sec}^2}{\text{ft}}\right)\end{array}}$$

$$= \boxed{-0.094}$$

14.2. As a check, the participation factors should add up to one.

$$P_1 + P_2 + P_3 = 0.780 + 0.327 + (-0.094)$$
$$\approx \boxed{1 \quad [\text{OK}]}$$

14.3. Calculate the displacement at each floor level based on the given response spectrum.

For mode 1,

$$\omega_1 = 8.77 \ \text{rad/sec}$$

$$f_1 = \frac{\omega_1}{2\pi} = \frac{8.77 \ \dfrac{\text{rad}}{\text{sec}}}{2\pi \ \dfrac{\text{rad}}{\text{cycle}}} = 1.4 \ \text{Hz}$$

$$T_1 = \frac{1}{f_1} = \frac{1 \ \text{cycle}}{1.4 \ \text{Hz}} = 0.7 \ \text{sec}$$

From the given response spectrum, for a period of 0.7 sec,

$$S_{a1} = 0.15g = (0.15)\left(32.2 \ \frac{\text{ft}}{\text{sec}^2}\right) = 4.8 \ \text{ft/sec}^2$$

$$S_{d1} = \frac{s_{a1}}{\omega_1^2} = \frac{4.8 \ \dfrac{\text{ft}}{\text{sec}^2}}{\left(8.77 \ \dfrac{\text{rad}}{\text{sec}}\right)^2} = 0.062 \ \text{ft}$$

For mode 2,

$$\omega_2 = 25.18 \ \text{rad/sec}$$

$$f_2 = \frac{\omega_2}{2\pi} = \frac{25.18 \ \dfrac{\text{rad}}{\text{sec}}}{2\pi \ \dfrac{\text{rad}}{\text{cycle}}} = 4.0 \ \text{Hz}$$

$$T_2 = \frac{1}{f_2} = \frac{1 \ \text{cycle}}{4.0 \ \text{Hz}} = 0.25 \ \text{sec}$$

From the given response spectrum, for a period of 0.25 sec,

$$S_{a2} = 0.8g = (0.8)\left(32.2 \ \frac{\text{ft}}{\text{sec}^2}\right) = 25.8 \ \text{ft/sec}^2$$

$$S_{d2} = \frac{S_{a2}}{\omega_2^2} = \frac{25.8 \ \dfrac{\text{ft}}{\text{sec}^2}}{\left(25.18 \ \dfrac{\text{rad}}{\text{sec}}\right)^2} = 0.041 \ \text{ft}$$

$$\omega_2 = 25.18 \ \text{rad/sec}$$
$$f_2 = 4.0 \ \text{Hz}$$
$$T_2 = 0.25 \ \text{sec}$$

For mode 3,

$$\omega_3 = 48.13 \ \text{rad/sec}$$

$$f_3 = \frac{\omega_3}{2\pi} = \frac{48.13 \ \dfrac{\text{rad}}{\text{sec}}}{2\pi \ \dfrac{\text{rad}}{\text{cycle}}} = 7.7 \ \text{Hz}$$

$$T_3 = \frac{1}{f_3} = \frac{1 \ \text{cycle}}{7.7 \ \text{Hz}} = 0.13 \ \text{sec}$$

From the given response spectrum, for a period of 0.13 sec,

$$S_{a3} = 0.4g = (0.4)\left(32.2 \ \frac{\text{ft}}{\text{sec}^2}\right) = 12.9 \ \text{ft/sec}^2$$

$$S_{d3} = \frac{S_{a3}}{\omega_3^2} = \frac{12.9 \ \dfrac{\text{ft}}{\text{sec}^2}}{\left(48.13 \ \dfrac{\text{rad}}{\text{sec}}\right)^2} = 0.0056 \ \text{ft}$$

The displacement at each level is

$$x_{ind} = \phi_{in} P_i S_{dn}$$
$$x_{11d} = (1.690)(0.780)(0.062 \ \text{ft}) = 0.0817 \ \text{ft}$$
$$x_{21d} = (1.220)(0.780)(0.062 \ \text{ft}) = 0.0590 \ \text{ft}$$
$$x_{31d} = (0.572)(0.780)(0.062 \ \text{ft}) = 0.0277 \ \text{ft}$$
$$x_{12d} = (-1.208)(0.327)(0.041 \ \text{ft}) = -0.0162 \ \text{ft}$$
$$x_{22d} = (0.704)(0.327)(0.041 \ \text{ft}) = 0.0094 \ \text{ft}$$
$$x_{32d} = (1.385)(0.327)(0.041 \ \text{ft}) = 0.0186 \ \text{ft}$$
$$x_{13d} = (-0.714)(-0.094)(0.0056 \ \text{ft}) = 0.0004 \ \text{ft}$$
$$x_{23d} = (1.697)(-0.094)(0.0056 \ \text{ft}) = -0.0009 \ \text{ft}$$
$$x_{33d} = (-0.948)(-0.094)(0.0056 \ \text{ft}) = 0.0005 \ \text{ft}$$

Combine the modal maxima using the square root of the sum-of-squares (RSS) method. The displacement for each floor is

$$x_i = \sqrt{\sum_{j=1}^{3} x_{ijd}^2}$$

$$x_1 = \sqrt{\begin{array}{c}(0.0817 \text{ ft})^2 + (-0.0162 \text{ ft})^2 \\ + (0.0004 \text{ ft})^2\end{array}} \left(12 \frac{\text{in}}{\text{ft}}\right)$$

$$= \boxed{1 \text{ in}}$$

$$x_2 = \sqrt{\begin{array}{c}(0.0590 \text{ ft})^2 + (0.0094 \text{ ft})^2 \\ + (-0.0009 \text{ ft})^2\end{array}} \left(12 \frac{\text{in}}{\text{ft}}\right)$$

$$= \boxed{0.7 \text{ in}}$$

$$x_3 = \sqrt{\begin{array}{c}(0.0277 \text{ ft})^2 + (0.0186 \text{ ft})^2 \\ + (0.0005 \text{ ft})^2\end{array}} \left(12 \frac{\text{in}}{\text{ft}}\right)$$

$$= \boxed{0.4 \text{ in}}$$

14.4. The interstory drifts for each floor are

base to level 1: $D_{01} - x_1 = \boxed{0.40 \text{ in}}$

level 1 to level 2: $D_{12} = x_2 - x_1 = 0.7 \text{ in} - 0.40 \text{ in}$
$$= \boxed{0.3 \text{ in}}$$

level 2 to level 3: $D_{23} = x_3 - x_2 = 1.0 \text{ in} - 0.7 \text{ in}$
$$= \boxed{0.3 \text{ in}}$$

SOLUTION 15

15.1. Determine the natural frequencies and mode shapes of the two-story steel frame below. Consider bending deformation only and treat the girders as rigid relative to columns. Lump building masses at the floor levels. For the W10 × 60, use $I = 341 \text{ in}^4$ and $E = 30,000 \text{ ksi}$ per the problem statement.

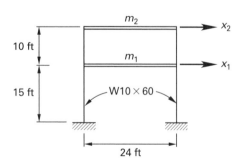

Compute the equivalent mass at the floor levels.

$$W_1 = w_1 L + 2w_{\text{col}}\left(\frac{h_1 + h_2}{2}\right)$$

$$= \left(2000 \frac{\text{lbf}}{\text{ft}}\right)(24 \text{ ft}) + (2)$$

$$\times \left(60 \frac{\text{lbf}}{\text{ft}}\right)\left(\frac{15 \text{ ft} + 10 \text{ ft}}{2}\right)$$

$$= 49{,}500 \text{ lbf}$$

$$m_1 = \frac{W_1}{g} = \frac{49{,}500 \text{ lbf}}{\left(32.2 \frac{\text{ft}}{\text{sec}^2}\right)\left(12 \frac{\text{in}}{\text{ft}}\right)}$$

$$= 128 \text{ lbf-sec}^2/\text{in}$$

$$W_2 = w_2 L + 2w_{\text{col}}\left(\frac{h_2}{2}\right)$$

$$= \left(1500 \frac{\text{lbf}}{\text{ft}}\right)(24 \text{ ft}) + (2)\left(60 \frac{\text{lbf}}{\text{ft}}\right)\left(\frac{10 \text{ ft}}{2}\right)$$

$$= 36{,}600 \text{ lbf}$$

$$m_1 = \frac{W_1}{g} = \frac{36{,}600 \text{ lbf}}{\left(32.2 \frac{\text{ft}}{\text{sec}^2}\right)\left(12 \frac{\text{in}}{\text{ft}}\right)}$$

$$= 95 \text{ lbf-sec}^2/\text{in}$$

Compute the equivalent translational stiffness, k, at each level.

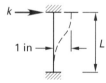

$$k_1 = \frac{12EI}{h^3}$$

$$= \frac{(12)\left(30{,}000{,}000 \frac{\text{lbf}}{\text{in}^2}\right)(341 \text{ in}^4)}{\left((15 \text{ ft})\left(12 \frac{\text{in}}{\text{ft}}\right)\right)^3}$$

$$= 21{,}050 \text{ lbf/in}$$

$$k_2 = \frac{12EI}{h^3}$$

$$= \frac{(12)\left(30{,}000{,}000 \frac{\text{lbf}}{\text{in}^2}\right)(341 \text{ in}^4)}{\left((10 \text{ ft})\left(12 \frac{\text{in}}{\text{ft}}\right)\right)^3}$$

$$= 71{,}040 \text{ lbf/in}$$

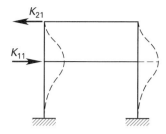

Assemble the stiffness matrix.

$$K_{11} = 2(k_1 + k_2)$$
$$= (2)\left(21{,}050 \ \frac{\text{lbf}}{\text{in}} + 71{,}040 \ \frac{\text{lbf}}{\text{in}}\right)$$
$$= 184{,}183 \ \text{lbf/in}$$

$$K_{21} = -2k_2 = (2)\left(-71{,}040 \ \frac{\text{lbf}}{\text{in}}\right)$$
$$= -142{,}080 \ \text{lbf/in}$$

$$K_{12} = K_{21} = -142{,}080 \ \text{lbf/in}$$

$$K_{22} = 2k_2 = (2)\left(71{,}040 \ \frac{\text{lbf}}{\text{in}}\right)$$
$$= 142{,}080 \ \text{lbf/in}$$

Damping has a negligible effect on natural frequencies and mode shapes. The equations of motion for the frame are

$$[M]\{\ddot{x}\} + [K]\{x\} = \{0\}$$

The solution for the lateral translation, x_i, will be of the form $x_i = a_{\text{I}}\sin(\omega t)$. Substituting and simplifying,

$$\left[[K] - \omega^2[M]\right]\{a\} = \{0\}$$

For a nontrivial solution, the determinant of the coefficient matrix is set to zero.

$$\left|\left[[K] - \omega^2[M]\right]\right| = 0$$

Thus,

$$\begin{vmatrix} \left(184{,}183 \ \dfrac{\text{lbf}}{\text{in}} \\ - \left(128 \ \dfrac{\text{lbf-sec}^2}{\text{in}}\right)\omega^2\right) & -142{,}080 \ \dfrac{\text{lbf}}{\text{in}} \\[4ex] -142{,}080 \ \dfrac{\text{lbf}}{\text{in}} & \left(142{,}080 \ \dfrac{\text{lbf}}{\text{in}} \\ - \left(95 \ \dfrac{\text{lbf-sec}^2}{\text{in}}\right)\omega^2\right) \end{vmatrix} = 0$$

Expanding the determinant yields the characteristic equation (with the common unit lbf^2/in^2 canceled from each term).

$$\begin{pmatrix} \left(184{,}183 \ \dfrac{\text{lbf}}{\text{in}} - \left(128 \ \dfrac{\text{lbf-sec}^2}{\text{in}}\right)\omega^2\right) \\[2ex] \times \left(142{,}080 \ \dfrac{\text{lbf}}{\text{in}} - \left(95 \ \dfrac{\text{lbf-sec}^2}{\text{in}}\right)\omega^2\right) \\[2ex] - \left(142{,}080 \ \dfrac{\text{lbf}}{\text{in}}\right)^2 \end{pmatrix} = 0$$

$$\omega^4(1 \ \text{sec}^4) - 2934.2\omega^2(1 \ \text{sec}^2) + 491{,}900 = 0$$

Solving for ω^2 using the quadratic equation gives

$$\omega^2 = (178.6 \ \text{sec}^{-2}, \ 2755 \ \text{sec}^{-2})$$
$$\omega = \left(13.34 \ \frac{\text{rad}}{\text{sec}}, \ 52.5 \ \frac{\text{rad}}{\text{sec}}\right)$$

Thus,

$$f_1 = \frac{\omega_1}{2\pi \ \dfrac{\text{rad}}{\text{cycle}}} = \frac{13.34 \ \dfrac{\text{rad}}{\text{sec}}}{2\pi \ \dfrac{\text{rad}}{\text{cycle}}} = \boxed{2.1 \ \text{Hz}}$$

$$f_2 = \frac{\omega_2}{2\pi \ \dfrac{\text{rad}}{\text{cycle}}} = \frac{52.5 \ \dfrac{\text{rad}}{\text{sec}}}{2\pi \ \dfrac{\text{rad}}{\text{cycle}}} = \boxed{8.4 \ \text{Hz}}$$

To obtain the mode shapes, arbitrarily set the displacement at level 1 to 1 in, substitute the calculated frequency into the equations of motion, and solve for the displacement at level 2. Thus, for the first mode,

$$\omega_1 = 13.34 \ \text{rad/sec}$$
$$a_{11} = 1 \ \text{in}$$

$$\begin{bmatrix} \left(184{,}183 \ \dfrac{\text{lbf}}{\text{in}} \\ - \left(128 \ \dfrac{\text{lbf-sec}^2}{\text{in}}\right) \\ \times \left(13.34 \ \dfrac{\text{rad}}{\text{sec}}\right)^2\right) & -142{,}080 \ \dfrac{\text{lbf}}{\text{in}} \\[5ex] -142{,}080 \ \dfrac{\text{lbf}}{\text{in}} & \left(142{,}080 \ \dfrac{\text{lbf}}{\text{in}} \\ - \left(95 \ \dfrac{\text{lbf-sec}^2}{\text{in}}\right) \\ \times \left(13.34 \ \dfrac{\text{rad}}{\text{sec}}\right)^2\right) \end{bmatrix}$$

$$\times \begin{Bmatrix} 1 \ \text{in} \\ a_{21} \end{Bmatrix} = \begin{Bmatrix} 0 \ \text{lbf} \\ 0 \ \text{lbf} \end{Bmatrix}$$

$$\left(161{,}405 \ \frac{\text{lbf}}{\text{in}}\right)(1 \text{ in}) - \left(142{,}080 \ \frac{\text{lbf}}{\text{in}}\right)a_{21} = 0 \text{ lbf}$$

$$a_{21} = \boxed{1.136 \text{ in}}$$

For the second mode,

$$
\begin{bmatrix}
\left(\begin{array}{l} 184{,}183 \ \dfrac{\text{lbf}}{\text{in}} \\[2mm] - \left(128 \ \dfrac{\text{lbf-sec}^2}{\text{in}}\right) \\[2mm] \times \left(52.5 \ \dfrac{\text{rad}}{\text{sec}}\right)^2 \end{array}\right) & -142{,}080 \ \dfrac{\text{lbf}}{\text{in}} \\[10mm]
-142{,}080 \ \dfrac{\text{lbf}}{\text{in}} & \left(\begin{array}{l} 142{,}080 \ \dfrac{\text{lbf}}{\text{in}} \\[2mm] - \left(95 \ \dfrac{\text{lbf-sec}^2}{\text{in}}\right) \\[2mm] \times \left(52.5 \ \dfrac{\text{rad}}{\text{sec}}\right)^2 \end{array}\right)
\end{bmatrix}
$$

$$\times \begin{Bmatrix} 1 \text{ in} \\ a_{22} \end{Bmatrix} = \begin{Bmatrix} 0 \text{ lbf} \\ 0 \text{ lbf} \end{Bmatrix}$$

$$\left(-168{,}617 \ \frac{\text{lbf}}{\text{in}}\right)(1 \text{ in}) - \left(142{,}080 \ \frac{\text{lbf}}{\text{in}}\right)a_{22} = 0 \text{ lbf}$$

$$a_{22} = \boxed{-1.187 \text{ in}}$$

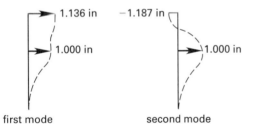

15.2. The acceleration response spectrum for 7% critical damping is the lowest of the three given curves. For the first mode,

$$\omega_1 = 2.1 \text{ Hz}$$

$$T_1 = \frac{1}{\omega_1} = 0.48 \text{ sec} \approx 0.5 \text{ sec}$$

Thus, the spectral acceleration is

$$S_a = 0.28g = (0.28)\left(32.2 \ \frac{\text{ft}}{\text{sec}^2}\right) = \boxed{9.0 \text{ ft/sec}^2}$$

The spectral pseudo-velocity relates directly to S_a.

$$S_v = \frac{S_a}{\omega} = \frac{9.0 \ \dfrac{\text{ft}}{\text{sec}^2}}{13.34 \ \dfrac{\text{rad}}{\text{sec}}} = \boxed{0.7 \text{ ft/sec}}$$

SOLUTION 16

16.1. To compute the horizontal component of cable tension at point A, consider a typical strand as a free-body diagram.

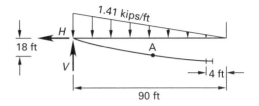

Equilibrium requires

$$\sum F_y = V - \frac{wL}{2} = 0$$

$$= V - \frac{\left(1.41 \ \dfrac{\text{kips}}{\text{ft}}\right)(90 \text{ ft})}{2}$$

$$V = 63.45 \text{ kips} \quad [\text{upward}]$$

Taking moments about the centerline and the line of action of the cable force from the tension ring to the centerline gives

$$\sum M_{\text{centerline}} = Hh - VL + \left(\frac{wL}{2}\right)\left(\frac{2L}{3}\right) = 0$$

$$= H(18 \text{ ft}) - (63.45 \text{ kips})(90 \text{ ft})$$

$$+ \left(\frac{\left(1.41 \ \dfrac{\text{kips}}{\text{ft}}\right)(90 \text{ ft})}{2}\right)\left(\frac{(2)(90 \text{ ft})}{3}\right)$$

$$H = 105.75 \text{ kips} \quad [\text{to the left}]$$

A free-body diagram through point A on the cable shows

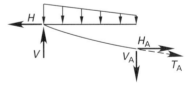

Equilibrium requires

$$\sum F_x = -H + H_A = 0$$

$$= -105.75 \text{ kips} + H_A$$

$$H_A = \boxed{105.75 \text{ kips} \quad [\text{to the right}]}$$

Equilibrium requires that the horizontal component of cable tension is constant at all points along the cable.

16.2. Verify that W21 × 50 of A36 steel is adequate for the tension ring.

The maximum ring tension occurs midway between cable connection points (see Young, Budynas, and Sadegh, Table 17, case 7).

$$T_{max} = \sum H \cos \theta_i$$
$$= (105.75 \text{ kips}) \left(\begin{array}{c} \cos 11.25^\circ + \cos 33.75^\circ \\ + \cos 56.25^\circ + \cos 78.75^\circ \end{array} \right)$$
$$= 271 \text{ kips}$$

$$M^+ = \left(\frac{Hr}{2} \right) \left(\frac{1}{\sin \theta} - \frac{1}{\theta} \right)$$
$$\theta = (11.25^\circ) \left(\frac{\pi}{180^\circ} \right) = 0.19635 \text{ rad}$$

$$M^+ = \frac{(105.75 \text{ kips})(4 \text{ ft}) \left(\frac{1}{\sin 11.25^\circ} - \frac{1}{0.19635 \text{ rad}} \right)}{2}$$
$$= 7.0 \text{ ft-kips}$$

$$M^- = \left(\frac{Hr}{2} \right) \left(\frac{1}{\theta} - \frac{1}{\tan \theta} \right)$$
$$= \frac{(105.75 \text{ kips})(4 \text{ ft}) \left(\frac{1}{0.19635 \text{ rad}} - \frac{1}{\tan 11.25^\circ} \right)}{2}$$
$$= 13.9 \text{ ft-kips}$$

Use ASD. Relevant section properties of the W21 × 50 are $A_g = 14.7 \text{ in}^2$ and $Z_x = 110 \text{ in}^3$. The weak axis bending is negligible. Assume that the cable-to-ring connection is detailed such that the full area of the W21 × 50 resists tension. Check combined axial tension plus bending as per AISC Sec. H1.2.

$$P_c = \frac{F_y A_g}{\Omega} = \frac{\left(36 \frac{\text{kips}}{\text{in}^2} \right)(14.7 \text{ in}^2)}{1.67} = 317 \text{ kips}$$

$$M_{cx} = \frac{F_y Z_x}{\Omega} = \frac{\left(36 \frac{\text{kips}}{\text{in}^2} \right)(110 \text{ in}^3)}{1.67} = 2371 \text{ in-kips}$$

$$P_r = T_{max} = 271 \text{ kips}$$
$$\frac{P_r}{P_c} = \frac{271 \text{ kips}}{317 \text{ kips}} = 0.85 \quad [>0.2]$$

The ratio of required strength to allowable tensile strength exceeds 0.2, so AISC Eq. H1-1a applies.

$$\frac{P_r}{P_c} + \frac{8}{9} \left(\frac{M_{rx}}{M_{cx}} \right) = \frac{271 \text{ kips}}{317 \text{ kips}} + \left(\frac{8}{9} \right) \left(\frac{(13.9 \text{ ft-kips}) \left(12 \frac{\text{in}}{\text{ft}} \right)}{2371 \text{ in-kips}} \right)$$
$$= 0.92 \quad [<1.0]$$

W21 × 50 satisfies the interaction equation and is OK.

16.3. Check the W33 × 221 for the compression ring.

$$C_{max} = T_{max} = 271 \text{ kips} = P_r$$
$$M^+ = \left(\frac{Hr}{2} \right) \left(\frac{1}{\sin \theta} - \frac{1}{\theta} \right)$$
$$\theta = (11.25^\circ) \left(\frac{\pi}{180^\circ} \right) = 0.19635 \text{ rad}$$

$$M^+ = \frac{(105.75 \text{ kips})(90 \text{ ft}) \left(\frac{1}{\sin 11.25^\circ} - \frac{1}{0.19635 \text{ rad}} \right)}{2}$$
$$= 156 \text{ ft-kips}$$

$$M^- = \left(\frac{Hr}{2} \right) \left(\frac{1}{\theta} - \frac{1}{\tan \theta} \right)$$
$$= \frac{(105.75 \text{ kips})(90 \text{ ft}) \left(\frac{1}{0.19635 \text{ rad}} - \frac{1}{\tan 11.25^\circ} \right)}{2}$$
$$= 312 \text{ ft-kips}$$

The negative bending moment of 312 ft-kips controls. Relevant properties of the W33 × 221 are $A_g = 65.3 \text{ in}^2$, $r_x = 14.1 \text{ in}$, $I_x = 12,900 \text{ in}^4$, $Z_x = 857 \text{ in}^3$, $b_f/2t_f = 6.2$, and $h/t_w = 38.5$. To check buckling of the compression ring, calculate an equivalent radial pressure using an available solution (see Young, Budynas, and Sadegh, Table 34, case 8).

$$p' = \frac{3EI_x}{r^3} = \frac{(3) \left(29,000 \frac{\text{kips}}{\text{in}^2} \right)(12,900 \text{ in}^4)}{\left((90 \text{ ft}) \left(12 \frac{\text{in}}{\text{ft}} \right) \right)^3}$$
$$= 0.89 \text{ kip/in}$$

$$P' = p'r = \left(0.89 \frac{\text{kip}}{\text{in}} \right)(90 \text{ ft}) \left(12 \frac{\text{in}}{\text{ft}} \right)$$
$$= 961 \text{ kips}$$

The equivalent effective length from the Euler equation is

$$P_{ex} = \frac{\pi^2 E I_x}{(KL_x)^2} = P'$$

$$KL_x = \sqrt{\frac{\pi^2 E I_x}{P'}}$$

$$= \sqrt{\frac{\pi^2 \left(29{,}000 \ \dfrac{\text{kips}}{\text{in}^2}\right)(12{,}900 \ \text{in}^4)}{961 \ \text{kips}}}$$

$$= 1960 \ \text{in}$$

$$\frac{KL_x}{r_x} = \frac{1960 \ \text{in}}{14.1 \ \text{in}} = 139$$

From AISC Table 4-22,

$$\frac{F_{cr}}{\Omega} = 7.78 \ \text{ksi}$$

$$P_c = \left(\frac{F_{cr}}{\Omega}\right) A_g = \left(7.78 \ \frac{\text{kips}}{\text{in}^2}\right)(65.3 \ \text{in}^2)$$

$$= 508 \ \text{kips}$$

Assuming that the section is adequately braced against weak axis buckling, so that $L_b = 0$, the moment strength of the section is given by AISC Eq. F2.1.

$$M_{cx} = \frac{F_y Z_x}{\Omega} = \frac{\left(36 \ \dfrac{\text{kips}}{\text{in}^2}\right)(857 \ \text{in}^3)}{1.67}$$

$$= 18{,}474 \ \text{in-kips}$$

For axial load only,

$$\frac{P_r}{P_c} = \frac{271 \ \text{kips}}{508 \ \text{kips}} = 0.53 \quad [> 0.2]$$

The ratio is greater than 0.2, so AISC Eq. H1-1a applies.

$$M_{rx} = M^- = (312 \ \text{ft-kips})\left(12 \ \frac{\text{in}}{\text{ft}}\right) = 3744 \ \text{in-kips}$$

$$\frac{P_r}{P_c} + \frac{8}{9}\left(\frac{M_{rx}}{M_{cx}}\right) = \frac{271 \ \text{kips}}{508 \ \text{kips}} + \left(\frac{8}{9}\right)\left(\frac{3744 \ \text{in-kips}}{18{,}474 \ \text{in-kips}}\right)$$

$$= 0.71 \quad [< 1.0]$$

The interaction equation is satisfied, so the W33 × 221 is OK.

2 Structural Concrete Design

PROBLEM 1

1.1. In multistory reinforced concrete construction, (a) when and where would reshoring be permitted to be placed? (b) What determines the required amount of reshoring? (c) When would reshoring be permitted to be removed? (d) What physical property of concrete other than compressive strength is important when considering form removal? Why?

1.2. Name two possible effects on concrete beams if shoring is removed too soon. What measures should be taken to ensure proper scheduling of shore removal?

1.3. A concrete mix is so harsh that it cannot be pumped readily. What are two adjustments that will make the concrete mix more workable without reducing its strength?

1.4. Name three methods of curing concrete.

1.5. List at least six practices that might be included in a good cold weather concreting procedure.

1.6. What are three practical construction procedures that will minimize the harmful effects of hot weather on freshly poured concrete?

1.7. What two factors relating to the handling of fresh concrete might adversely affect the final shrinkage?

1.8. In reinforced concrete high-rise construction, why are columns usually poured and allowed to harden before the floor system is poured?

1.9. In prestressed concrete construction, why is it necessary to check both the jacking force and the tendon elongation during stressing procedures?

1.10. What is the major cause of plastic shrinkage cracks in slabs? List five precautions that can be taken to minimize or prevent plastic shrinkage cracking.

1.11. Due to a mistake in placement, a group of no. 10 grade 60 steel reinforcing bars extends only 18 in from the previous pour. Plans and code require a 48 in lap splice for no. 10 bars. Draw a sketch to show a recommended solution for developing the bar in tension.

1.12. A no. 7 grade 40 steel reinforcing dowel for a jamb bar was inadvertently omitted from a foundation-pour (a continuous footing 12 in wide × 24 in deep) for an 8 in thick concrete block wall. Specified compressive

strength, f'_c, is 3 ksi. Knowing that the block wall has not been installed, how could this problem be corrected? Describe the procedure.

1.13. Name two undesirable effects of excessive vibration on freshly mixed concrete.

1.14. What are two readily identifiable defects of poorly consolidated concrete?

1.15. A horizontal construction joint is needed in a highly stressed concrete shear wall. What treatments should be required for the concrete contact surface?

1.16. When investigating a concrete slab, it is found that it is exhibiting greater than anticipated deflections. The slab was placed and then stripped five days later when the laboratory-cured cylinders reached design strength. Job records are available. (a) List six items other than the design details that should be checked. Explain why. (b) What three design details might have affected deflection of the slab?

1.17. Epoxy adhesives are used in concrete work. List two advantages and two disadvantages to their use.

1.18. When welding reinforcing steel, what is meant by the terms (a) preheat, (b) carbon equivalent, and (c) heat-affected zone?

1.19. If grade 40 bars have a minimum yield of 40 ksi, what is the requirement for maximum yield?

1.20. For what two main purposes is ASTM A706 reinforcing steel used?

1.21. What analysis must be made prior to welding reinforcing steel? Why?

1.22. What two alloying elements used in the manufacture of steel rebar have the greatest effect on electrode selection and preheat in welding?

1.23. What factor will *most* affect the compressive strength of a well-cured concrete of a given age?

1.24. The following are types of admixtures used in concrete mixes: (a) air-entraining agents, (b) retarders, (c) accelerators, and (d) water reducers. For each, briefly discuss its principal benefits and disadvantages, and name the precautions that should be taken when using it.

1.25. High concentrations of carbon dioxide can have a harmful effect on freshly placed concrete. Why?

1.26. List three advantages and three disadvantages of the use of fly ash as an admixture in concrete.

1.27. When designing an exterior beam-column joint in a post-tensioned frame, what factor must be considered in addition to gravity and lateral loads?

1.28. Why isn't mild steel, such as A615 grade 40, used for prestressing?

1.29. Concrete deformations due to creep and shrinkage may be two or more times greater than elastic deformations. Give four factors affecting concrete shrinkage.

1.30. What is creep? Are creep limits normally specified? If so, how?

1.31. In reinforced concrete, why is the required embedment, or splice length, longer for top bars?

1.32. What minimum capacity of a reinforcing steel bar must be developed by a full-welded tension butt splice? What minimum capacity would a weld of a no. 14 grade 60 reinforcing bar be required to develop?

1.33. In a prestressed beam that has been installed in a building, it is discovered that the effective stress in the tendons is 30% less than specified. The beam was designed for zero tension in the concrete at service loads (pretensioned). (a) What would be the loss in ultimate capacity of the beam? (b) Name two other features of the beam that should also be evaluated, and describe how they have been affected by the loss in prestress.

1.34. What three factors other than loads affect the deflection of a flat slab?

1.35. The web reinforcing in a ductile flexural member is based on ultimate moment capacities with a yield strength 25% greater than specified yield and no capacity reduction factor ϕ. Why?

1.36. Why is it important to place part of the negative reinforcement of a T-beam in the flange area?

1.37. Where would construction joints in a reinforced concrete slab-beam-girder system be specified?

1.38. Why is ductile concrete used in seismic design?

1.39. What is the primary purpose of a column spiral?

1.40. In reinforced concrete moment-resisting frame structures, overstrength reinforcement in girders can present a seismic hazard to columns. Why?

1.41. In addition to providing adequate main reinforcing steel, how else is ductility provided in reinforced concrete columns?

1.42. Briefly describe the basic difference between the internal resisting couple of a reinforced concrete beam and that of a prestressed concrete beam.

1.43. Sketch the shape of the concrete stress block that results from the location of the compressive force C for each of the four prestressed beams shown.

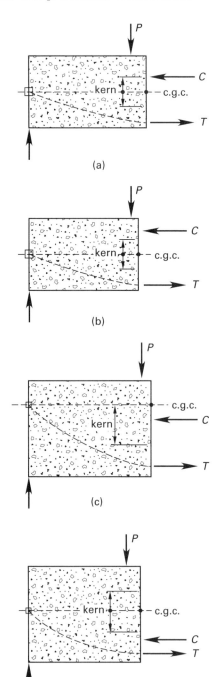

(a)

(b)

(c)

(d)

c.g.c. is the location of the center of gravity for the concrete section.

1.44. It is proposed to convert an existing lamella roof building into a theater. This will require removal of existing tie rods. The roof plan and typical cross section are shown. Local conditions prohibit additional construction beyond the property line. The interior clearance needed for use as a theater is indicated in the section. (See *Illustration for Prob. 1.44*.) Describe and briefly discuss three reasonable methods that could be used to resist the lateral thrust from dead load plus live load, while maintaining the given clearance. Calculations are not required.

1.45. A reinforced concrete high-rise is planned for a site six miles from an active earthquake fault that has the capacity to generate a magnitude 7.5 earthquake (Richter). To minimize story heights, the structure's lateral force resisting system will consist of moment-resisting frames located at the periphery of the building. A typical elevation of one of the moment-resisting frames is shown. (See *Illustration for Prob. 1.45*.) (a) Discuss the relative merits of details D, E, and F under the possible severe seismic overloads associated with a magnitude 7.5 earthquake. (b) Which detail should be selected? Why? (c) In the detail that is selected, what is the most important pitfall to avoid when proportioning the concrete dimensions and the amounts of principal reinforcement?

Illustration for Prob. 1.44

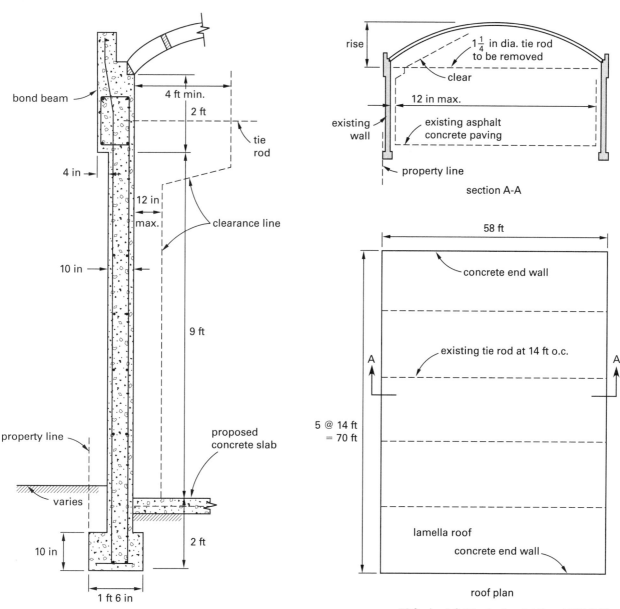

span 58 ft, rise 9 ft 8 in, horizontal thrust 870 lbf/ft

Illustration for Prob. 1.45

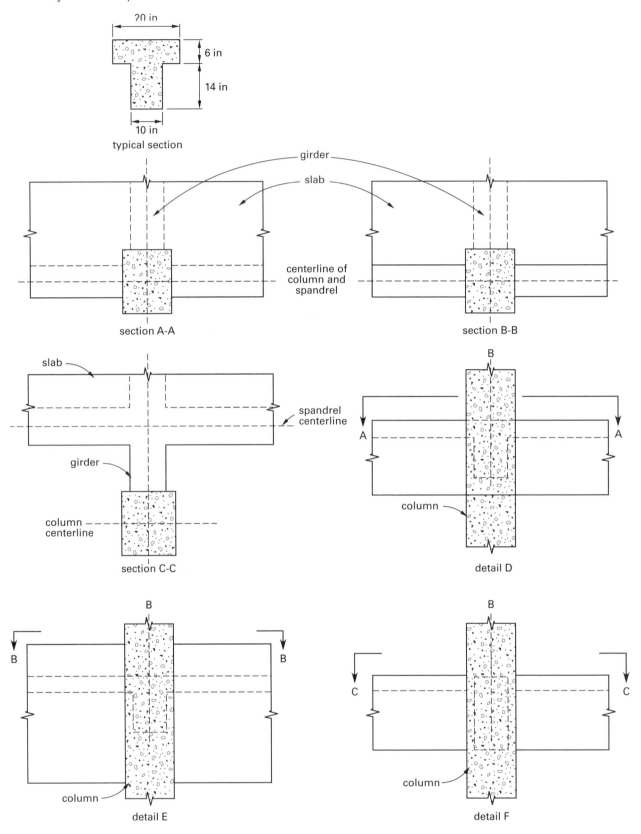

PROBLEM 2

A partial plan and a cross section of a typical floor for a multistory building are shown. The columns are 24 in square and the beams are shown in section A-A. Assume that the beams and columns are adequate.

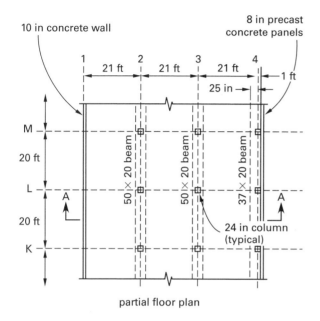

partial floor plan

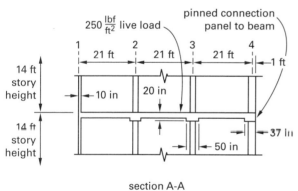

section A-A

Design Criteria

- use ACI 318
- design as one-way slab
- $f'_c = 4$ ksi stone aggregate concrete
- $f_y = 40$ ksi reinforcing steel
- live load $= 250$ lbf/ft^2 (do not reduce)
- fire rating does not constrain slab thickness

2.1. What slab thickness should be established for the design? Show calculations to support the development.

2.2. Determine the maximum slab moments.

2.3. Design and sketch the slab reinforcement at the critical line. Indicate the size of bars, spacing, clearances, placing, and cutoff points. (Do not use truss bars.)

PROBLEM 3

A short concrete column is shown in cross section. It is 21 in wide × 17 in deep with four no. 9 bars and two no. 8 bars placed as shown. The column carries combined axial and bending loads about the Y-Y axis.

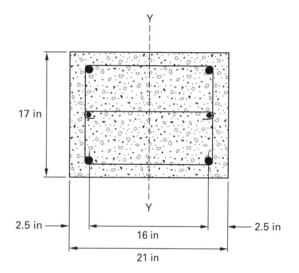

Design Criteria

- $P_u = 700$ kips
- $M_u = 330$ ft-kips
- $f'_c = 4.5$ ksi
- $f_y = 60$ ksi
- do not use column tables

3.1. Plot the ultimate strength interaction diagram for axial compression and bending moment.

3.2. Check adequacy of column for $P_u = 700$ kips and $M_u = 330$ ft-kips.

PROBLEM 4

Shown is a typical interior span of a cast in place one way beam-slab system. The beam span is 36 ft. Beams are spaced 22 ft o.c. The typical beam depth is 36 in with a web width of 18 in. Beam loading and end moment are given as ultimate loads and moments. All principal reinforcement is to be no. 8 bars. Neglect column width and any unbalanced loading.

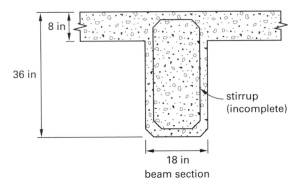

beam section

ultimate negative moment (M)
at D = 1190 ft-kips
ultimate negative end moment
at E = 1080 ft-kips

total ultimate uniform
load = 13.3 kips/ft

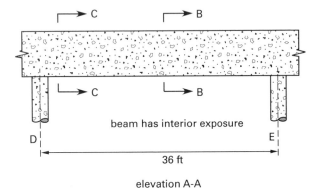

beam has interior exposure

elevation A-A

Design Criteria

- $f'_c = 4$ ksi hard rock aggregate

- $f_y = 60$ ksi for principal reinforcement

- $f_y = 40$ ksi for stirrups (use no. 3 or no. 4 bars)

4.1. Design the maximum positive and the maximum negative reinforcement.

4.2. Find the location where 50% of the positive reinforcement may be terminated. Dimension the required bar lengths.

4.3. Design stirrups from the left support to the beam centerline in 4 ft increments. Show any additional reinforcement that may be required.

4.4. Detail reinforcement at sections B-B and C-C. Draw a beam elevation, and show bar lengths for positive reinforcement. Show stirrup size and spacing. Show all calculations; do not use graphs or charts.

PROBLEM 5

Precast concrete beams are supported on concrete corbels as shown

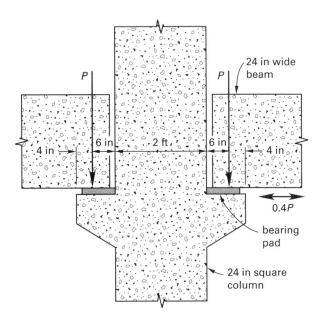

Design Criteria

- $f'_c = 4$ ksi

- $f_y = 60$ ksi

- $P_{\text{dead load}} = 48$ kips

- $P_{\text{live load}} = 32$ kips

- friction coefficient between concrete beam and bearing pad = 0.40

- width of corbel = 24 in

5.1. Design the corbels using shear friction concept.

5.2. Draw a sketch large enough to clearly indicate all reinforcement and dimensions for your design.

PROBLEM 6

An existing two-story office building is to be converted to a warehouse. Determine the maximum allowable live load that the second floor will safely support. A typical interior bay is shown.

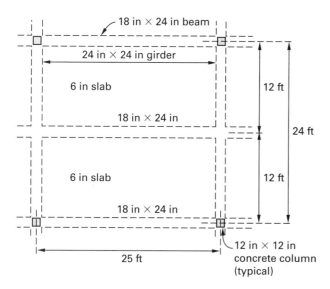

Design Criteria

- $f'_c = 3.5$ ksi (from core samples) hardrock concrete

- $f_y = 40$ ksi

- slab reinforcement ($3/4$ in clearance):

    no. 4 at 12 in o.c., bottom

    no. 4 at 8 in o.c., top at supports

- beam reinforcement ($1^1/2$ in clearance to stirrups):

    four no. 9, bottom

    four no. 10, top to supports

    eight no. 3 stirrups at 8 in o.c., each end (starting 4 in from support)

6.1. What is the maximum unit live load the second floor slab and beam will support?

6.2. List other items that should be checked for which no information has been given.

PROBLEM 7

A column cross section is shown.

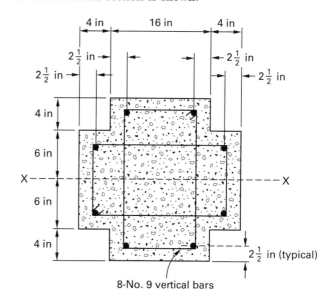

8-No. 9 vertical bars

Design Criteria

- $f'_c = 5$ ksi

- $f_y = 60$ ksi

- $Kl_u = 18$ ft

- $C_m = 1.0$

7.1. Draw the interaction diagram for the column cross section shown. Locate at least three critical points.

7.2. Determine the maximum allowable factored total load moment about the X-X axis for the column when a factored axial total load of 100 kips is applied. (Assume that the ratio of factored dead load moment to total load moment is 0.6.) Do not use column design tables. Show all calculations.

PROBLEM 8

A short reinforced concrete tied column is shown in cross section. Table values are not acceptable; calculations are required. Use ASCE/SEI7.

Design Criteria

The column carries vertical loads and moments about the center line X-X as follows ($N=$ vertical load, $M=$ moment). The loads are unfactored, design-level values.

- $f'_c = 4$ ksi
- $f_y = 50$ ksi
- service dead load only, $N = 150$ kips, $M = 100$ ft-kips
- service live load (office occupancy) only, $N = 70$ kips, $M = 20$ ft-kips
- earthquake only (design level), $N = 0$, $M = 450$ ft-kips
- seismic design category D, redundancy factor $= 1$; $S_{DS} = 0.5$

8.1. Verify the adequacy of the column design. What comments could be offered on this analysis?

8.2. What effect would vertical acceleration have on this column during an earthquake?

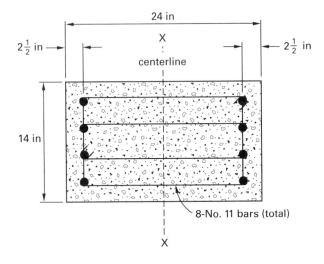

PROBLEM 9

An elevation of a concrete shear wall is shown. The seismic lateral loads for which the wall is to be designed are shown to the right of the wall.

Design Criteria

- neglect vertical dead and live loads
- assume wall is fixed at base
- assume given lateral loads are design level

9.1. Analyze the wall and draw a series of free-body diagrams which show the forces acting on each resisting element of the wall. Assume that the floor slab acts as a rigid diaphragm and distributes the concentrated lateral force at each level among the walls.

9.2. Draw an elevation of the wall and show all additional boundary and trim reinforcement needed to resist lateral forces. It is not necessary to completely detail the reinforcement. Show the amount of reinforcement needed, approximate lengths, and placing. Do not show typical wall reinforcement.

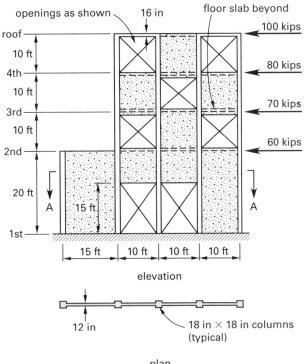

PROBLEM 10

A prestressed concrete pile is shown in cross section, along with the stress and strain relationship for a 270 ksi stress-relieved strand.

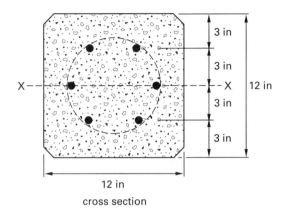

cross section

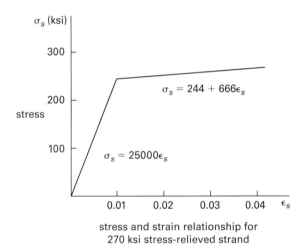

stress and strain relationship for
270 ksi stress-relieved strand

Design Criteria

- $f'_c = 5$ ksi at 28 days
- $f_{pu} = 270$ ksi prestress strands
- $A_s = 0.115$ in^2
- $A_c = 144$ in^2
- $E_s = 27 \times 10^6$ psi
- fabricator will use $0.75f_{pu}$ as jacking stress
- assume 10% loss in prestress at transfer
- make a reasonable assumption about further loss of prestress before member is put into use

10.1. Calculate the allowable service load for the pile under axial load only.

10.2. Due to the soil condition, the pile has to resist both axial load and bending moment during an earthquake. Determine the maximum allowable bending moment about the X-X axis. Use the following values.

$$P_{\text{dead}+\text{live}+\text{seismic}} = 150 \text{ kip factored design load}$$

$$\phi = 0.65 \text{ for both axial and bending}$$

PROBLEM 11

A cast-in-place reinforced concrete building is being designed with exterior bearing and shear walls. A typical cross section is shown. (See *Illustration for Prob. 11.*) Dimensions and member sizes are shown. Minimum wall reinforcement satisfies the requirements for lateral force design. Assume that the foundation provides fixity to exterior walls at the first floor line, and that interior column stiffness is negligible.

Design Criteria

- $f'_c = 3$ ksi hard rock concrete
- ASTM A615 grade 40 reinforcing steel
- $f_y = 40$ ksi
- floor slab design live load $= 175$ lbf/ft^2
- floor slab dead load including ceiling $= 120$ lbf/ft^2

11.1. Determine the critical second floor slab loading for design of joint A.

11.2. For the loading condition in Prob. 11.1, draw a complete moment and shear diagram of the second floor. Indicate maximum moments and locate points of inflection.

11.3. Check the adequacy of the reinforcement at joint A.

Illustration for Prob. 11

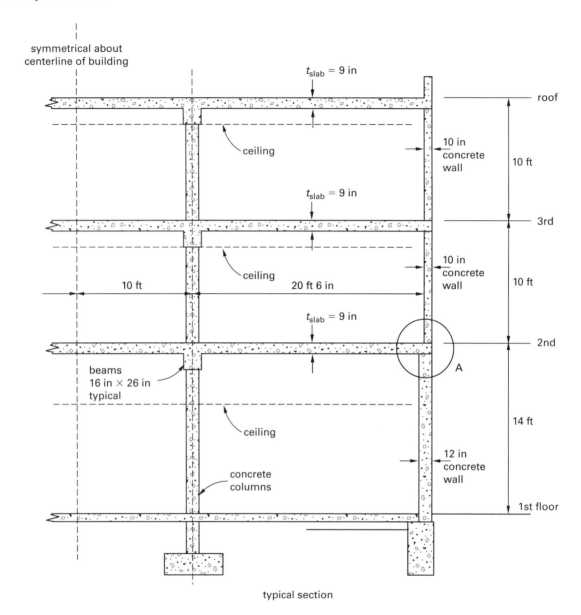

typical section

11.4. Determine the size, spacing, and required length of additional reinforcement needed at joint A. (Do not consider the slab to be pin-connected at joint A.) How should the detail be revised?

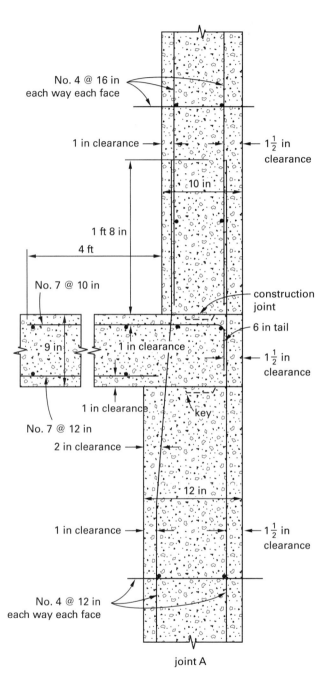

joint A

PROBLEM 12

A precast prestressed girder with a depth of 42 in is shown, as is a deck slab that is poured in place to form a T-section. No other dead loads are involved.

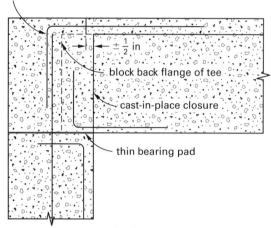

cast-in-place closure

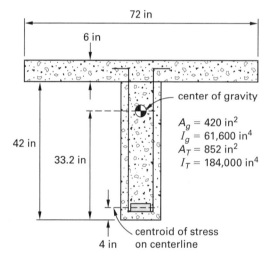

poured-in-place slab

$A_g = 420 \text{ in}^2$
$I_g = 61{,}600 \text{ in}^4$
$A_T = 852 \text{ in}^2$
$I_T = 184{,}000 \text{ in}^4$

Design Criteria

- 65 ft span
- 6 ft slab width
- 125 lbf/ft^2 live load
- $f'_c = 5$ ksi hard rock concrete
- prestressed steel:

$$f'_s = 260 \text{ ksi} \quad \text{[ultimate]}$$
$$f_{st} = 0.7f'_s = 182 \text{ ksi} \quad \text{[transfer]}$$
$$\text{losses} = 35 \text{ ksi}$$
$$f_{sw} = 147 \text{ ksi} \quad \text{[working]}$$
$$f_{si} = 167 \text{ ksi} \quad \text{[initial]}$$

12.1. Calculate the prestress force required. Do not select a prestressing system.

12.2. Calculate the stresses at top and bottom at transfer. Temporary tensile stresses in excess of $6\sqrt{f'_c}$ may be resisted by mild steel reinforcement.

12.3. What types of bars and hooks should be used?

12.4. Calculate top and bottom stresses (girder) at dead load and live load.

12.5. In the design of the joint at the girder support, what are the expected effects of shrinkage, creep, and temperature on the assumptions for fixity?

PROBLEM 13

A precast concrete girder is shown with a cast-in-place slab that is intended to work compositely with the girder.

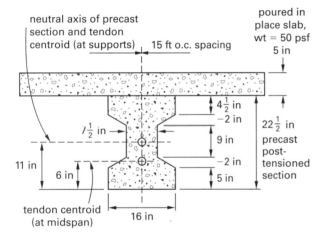

Design Criteria

- beam span = 40 ft 10 in

- dead load of beam = 210 lbf/ft

- precast concrete beam:

  $A = 265$ in^2

  $S_t = 1200$ in^3

  $S_b = 1250$ in^3

- composite section:

  $S_t = 13{,}500$ in^3

  $S_b = 2350$ in^3

- no continuity assumed at the columns

- tendon force = 200 kips effective (after all losses have occurred)

- beam and slab have the same modulus of elasticity

Construction Procedure

The precast concrete beam is precast and post-tensioned in shop. Beam is hauled to job site and placed over column supports. Temporary support is installed at beam midspan after girder is in place. Formwork is placed and slab is poured over beams. Weight of formwork may be neglected in computations. Temporary support is removed after slab concrete reaches its strength.

13.1. Calculate stresses at top and bottom of precast section at midspan when effective tendon force is acting with dead weight of beam.

13.2. Calculate stresses at top and bottom of precast section at midspan when the slab is poured.

13.3. Calculate stresses at top and bottom of precast section at midspan when temporary support is removed.

13.4. Calculate stresses at top and bottom of precast section at midspan when a live load of 50 lbf/ft^2 is applied (non-reduced).

PROBLEM 14

An elevation is shown for a reinforced concrete shear wall.

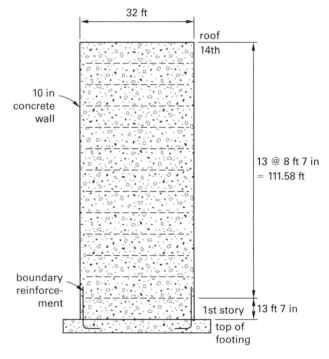

wall elevation

Design Criteria

- service dead load at base = 21 kips/ft

- service live load at base = 2.7 kips/ft

- seismic shear at base = 456 kips

- seismic overturning moment at base = 28,270 ft-kips

- seismic design category D
- importance factor $= 1.0$
- redundancy factor $= 1.0$
- $S_{DS} = 0.50$
- dual system
- $f'_c = 3$ ksi
- $f_y = 60$ ksi

14.1. Calculate the shear reinforcement for the first story wall. What type of bars should be used?

14.2. Determine required boundary reinforcement.

14.3. Calculate anchorage length for the boundary reinforcement.

14.4. Detail the placement of the shear and boundary reinforcement.

PROBLEM 15

The reinforced concrete beam shown is part of a special reinforced concrete moment frame in seismic design category D. (See *Illustration for Prob. 15*.) The frames

have been designed to resist seismic forces in accordance with ASCE/SEI7. Assume that factored axial force is less than $A_g f'_c / 20$ and that vertical acceleration effects are negligible.

Design Criteria

- original design calls for

  $f'_c = 3$ ksi
  ASTM A615 grade 40 reinforcing steel for longitudinal reinforcement and stirrups

- actual testing of material during construction shows

  $f'_c = 3.4$ ksi at 28 days
  longitudinal reinforcement has actual yield strength of 72 ksi
  stirrups were not tested

15.1. What effect, if any, does this increase in material strength have on the ultimate shear requirement of the member shown? Substantiate with calculations.

15.2. Using these increased material stresses, determine the minimum spacing of the no. 3 stirrups adjacent to the support.

Illustration for Prob. 15

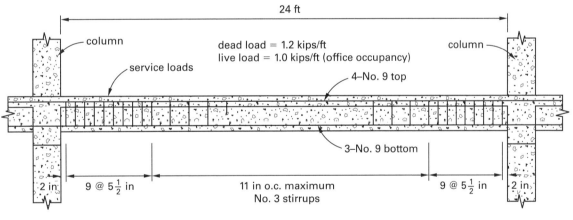

elevation

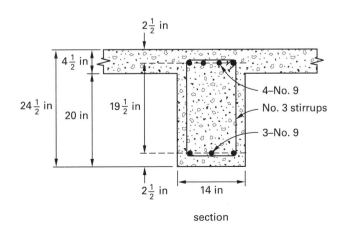

section

PROBLEM 16

A roof plan and a cross section through a one-story building are shown. The 8 in roof slab is supported along the building centerline by a wide, but shallow, reinforced concrete roof beam.

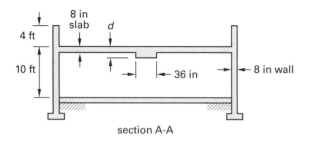

section A-A

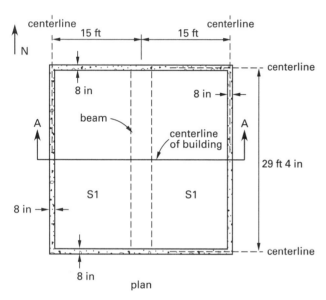

plan

Design Criteria

- $f'_c = 3$ ksi concrete
- $f_y = 40$ ksi reinforced steel
- $w_D = 20$ lbf/ft^2 plus weight of concrete
- $w_L = 50$ lbf/ft^2
- neglect unbalanced live load
- neglect effect of width of support
- assume gross section of roof beam for deflections

16.1. Assume a depth of 30 in for the roof beam. Design the reinforcing steel for the 8 in roof slab (as indicated by S1) at section A-A. Use continuous reinforcement with splices at appropriate locations. Draw a section to show the design. Do not detail the wall steel.

16.2. Assume a depth of 15 in for the roof beam. Prepare a design similar to that called for in Prob. 16.1.

PROBLEM 17

The existing concrete beam shown is to be used to support a new concrete floor.

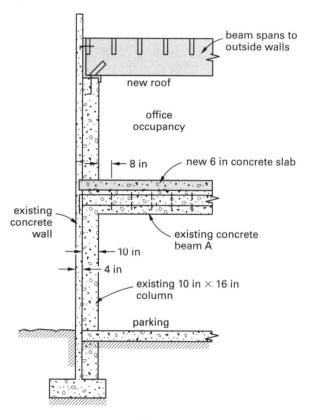

section A-A

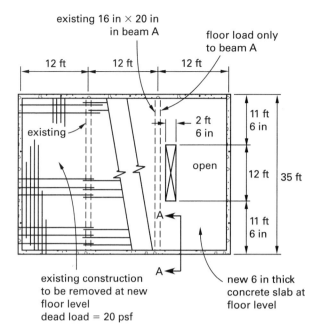

floor plan

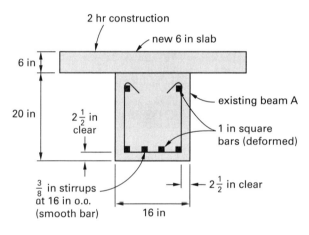

typical section

Design Criteria

- existing concrete properties

  $f'_c = 2$ ksi

  $f_y = 40$ ksi

  $E = 3.1 \times 10^6$ psi

  measured deflection $= \frac{3}{8}$ in

- new concrete: $f'_c = 3$ ksi (normal weight)

- office occupancy

17.1. For the new condition, draw the shear and moment diagram for the composite girder.

17.2. Describe how a composite girder for the new condition could be designed. Assume shoring is provided. Show calculations.

17.3. Sketch an elevation and typical cross section of the composite girder showing details of the new design.

17.4. What also must be checked per ACI 318?

PROBLEM 18

The plan of a cantilevered canopy is shown, along with the adjacent part of the roof structure. (See *Illustration for Prob. 18.*) The owner proposes to remove the cantilevered canopy and to cut it off at the face of the girder shown on line 3. The cantilever to the east of line 4 is to remain.

Design Criteria

- $f'_c = 3$ ksi lightweight concrete

- dead load of concrete $= 115$ lbf/ft^2

- grade 40 reinforcing steel

- assume no roofing

- use ASCE/SEI7

- structure first erected in 1960

- disregard lateral forces

- present structure (before modification) is adequate for dead and live loads

- assume central beam is loaded uniformly

- assume the roof is simply supported

18.1. Determine whether the altered structure will be able to resist the increased moment.

18.2. Assume that some modification is required to strengthen the altered structure. What structural modification should be considered and recommended? (For this part, do not add columns, and no calculations are required.)

18.3. Describe a construction sequence to accomplish the modifications identified in Prob. 18.2.

Illustration for Prob. 18

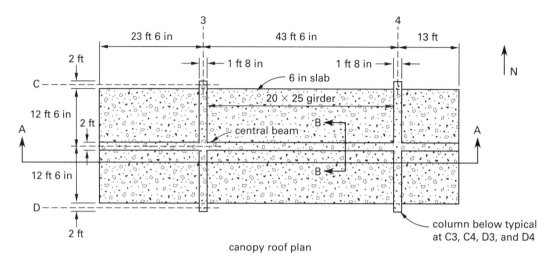

canopy roof plan

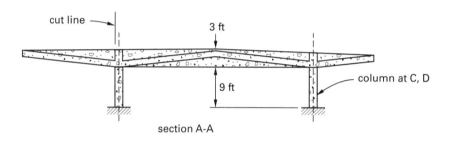

section A-A

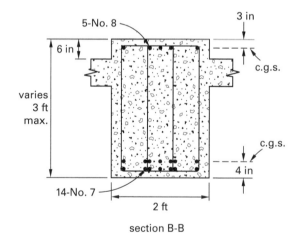

section B-B

PROBLEM 19

The following illustration shows an idealization of a typical design condition. (See *Illustration for Prob. 19.*) For simplicity, the reinforcement and column details are not given, and the strength of the column is assumed to be accurately represented by the interaction curve. The system is assumed to be laterally braced, and column strength reduction factors may be neglected. Anchorage of the beam reinforcement in the column is adequate. Assume the column and beam are laterally held against buckling. Neglect the effect of the two no. 4 bars that act as compression steel.

Design Criteria

- $f'_c = 3$ ksi hard rock aggregate concrete

- $f_y = 40$ ksi reinforcing steel

- show all calculations

19.1. What is the safe working load on the beam? Assume the column is adequate.

19.2. At what load will the tension steel in the beam yield? Assume the column is adequate.

19.3. What is the ultimate load for the beam? Assume the column is adequate.

19.4. Calculate ductility for the beam section, and calculate the ductility ratio of the beam.

19.5. Describe the expected mode of failure of the beam.

19.6. Based on the given interaction curve, is the column adequate? Explain. Should secondary stresses be investigated in the column? Explain.

19.7. Describe the expected mode of failure of the beam if the top reinforcement in the beam were increased from two no. 8 bars to two no. 11 bars. Assume the column is adequate.

Illustration for Prob. 19

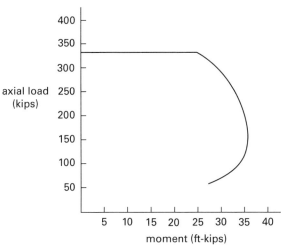

interaction curve for column
(ultimate capacity)

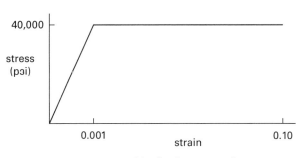

idealized stress-strain
curve for reinforcing steel

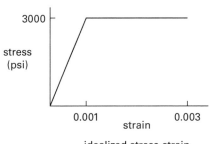

idealized stress-strain
curve for concrete

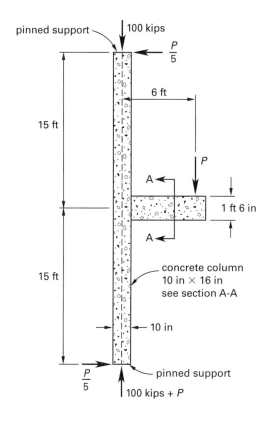

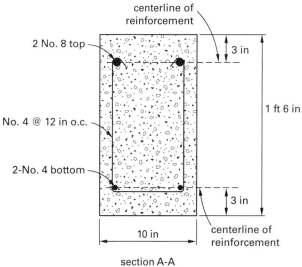

section A-A

PROBLEM 20

A roof structure is composed of a four-quadrant hyperbolic paraboloid that joins the straight ribs (edge members) as shown.

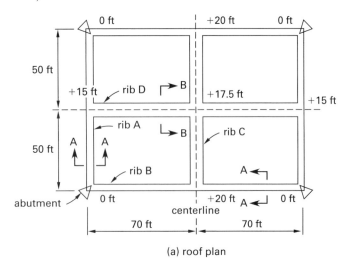

(a) roof plan

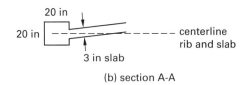

(b) section A-A

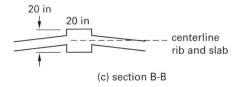

(c) section B-B

Design Criteria

- thickness of slab, or membrane, is 3 in
- all ribs have a 20 in × 20 in cross section (see sections A-A and B-B)
- all concrete is lightweight (110 lbf/ft^3 air dry with $f'_c = 3$ ksi)
- vertical support provided by abutments at four corners
- assume perfect hinge condition at juncture of ribs to various abutments
- weight of ceiling $= 3$ lbf/ft^2
- weight of roofing $= 3$ lbf/ft^2
- roof live load $= 12$ lbf/ft^2 (not reducible)
- grade 60 reinforcing steel

20.1. What is the maximum axial load in rib A?

20.2. What is the maximum axial load in rib B?

20.3. What is the maximum axial load in rib C?

20.4. What is the maximum axial load in rib D?

20.5. Show a plan view of the placing of the membrane reinforcement with respect to the ribs. Provide a suitable bar size and spacing.

20.6. Is the structure stable as shown? Is a column required at the center? If so, what would be the total column load?

SOLUTION 1

1.1. (a) Reshoring is used after the formwork is removed and before the floor system is capable of supporting the loads that will be placed on it. In multistory construction, the loads come primarily from the weight of the forms and wet concrete for the higher floors still to be cast. Usually, several lower floors adjacent to the one to be constructed are reshored, so that this weight is distributed safely among all these floors.

(b) The amount of reshoring depends on the construction load and the strength and relative stiffness of the supporting floors below.

(c) Shores can be removed after the floor above gains sufficient strength so that its formwork can be removed and it becomes self-supporting.

(d) An important property to consider, besides compressive strength, is the modulus of elasticity. This property determines the rigidity of the floors, which then determines how many floors should be reshored.

1.2. Two possible effects of early shore removal in reinforced concrete are increased deflection (the concrete's modulus of elasticity is lower than intended) and increased crack width (the curvature is greater than intended). To establish a schedule for shore removal, it is necessary to set a minimum compressive strength at which the structure is satisfactorily self-supporting. Proper curing (moisture retention and temperature control) must be maintained to ensure adequate strength gain. Field-cured test specimens provide a more dependable gauge of in-place strength than does the usual practice of specifying age at removal.

1.3. Two ways to make a harsh mixture more workable (pumpable) are

- increasing the amount of fines and cement paste without changes to water-cement ratio

- adding mineral admixture, such as fly ash

1.4. Three methods of curing concrete are

- continuous saturation or spraying

- covering with a plastic sheet or wet sand or burlap

- sealing with a curing compound

1.5. Good cold weather concrete practices include

- protecting the subgrade from freezing

- maintaining proper concrete temperature during placement

- determining curing procedures that will ensure strength gain and prevent freezing

- considering the use of an accelerating admixture

- maintaining temperature records during construction and cure

- heating the water and aggregates to be used in the mix

1.6. Good hot weather concrete practices include

- cooling the aggregates and water to be used in the mix

- moistening the forms

- erecting temporary sunshades

1.7. Two factors related to handling fresh concrete that might affect shrinkage are (1) the addition of water (for example, to make the concrete more workable), and (2) the use of admixtures such as calcium chloride that increase the demand for mix water.

1.8. It is difficult to place and consolidate concrete in columns if the floor system reinforcement is in place.

1.9. If only the jacking force is measured during strand tensioning, it is possible that extraordinary friction loss will occur at some point and prevent adequate stress over a significant length of strand. If only the elongation is measured, it is possible that a strand will slip or fracture and go undetected. It is thus necessary to check both.

1.10. Plastic shrinkage cracking is caused by rapid evaporation of water near the slab surface. Cracking is particularly severe during hot weather, low relative humidity, and windy conditions. Some methods to alleviate the problem are

- moistening the subgrade and forms

- moistening any aggregates that are dry

- erecting temporary wind breaks

- erecting temporary sunshades

- protecting the surface with a temporary covering such as a plastic sheet

- protecting the surface immediately after finishing, such as with fog spray, wet burlap, or wet sawdust

1.11. There are several methods that can be used to splice a no. 10 grade 60 rebar with only an 18 in projection (ACI 439). Special devices and tools are required, and the decision depends on access, clearance between bars, cost, and whether the splice is for tension or compression. One possibility, which will develop the bar in either tension or compression, is the mortar-filled sleeve (ACI 439 Sec. R-3), shown as follows.

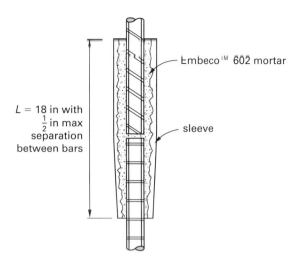

1.12. To install a no. 7 dowel into an existing grade beam ($f'_c = 3$ ksi, $f_y = 40$ ksi), as shown, the following procedure should be used.

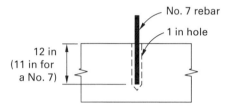

- Drill a hole 1 in in diameter by 11 in deep, using a handheld pneumatic drill. If necessary, chip the top surface to the depth of the main steel so that the hole will miss the beam's reinforcement.

- Blow out the hole with compressed air.

- Fill the hole with high modulus epoxy gel using a caulking gun with polyethylene tube extension.

- Insert the dowel, working up and down to ensure complete embedment.

- Maintain the dowel's alignment in the center of hole until the epoxy sets.

1.13. Excessive vibration can create excessive form-work pressure and can cause segregation of the mortar and coarse aggregate.

1.14. Poorly consolidated concrete will show *honey-comb* (voids left in the concrete where the cement has not completely flowed into the spaces among the pieces of aggregate, visible when they occur at surfaces and edges) and variation in color. In extreme cases, cold joints may be evident from the color difference between portions placed at different times.

1.15. To prepare a concrete surface for a horizontal construction joint,

- deliberately roughen the contact surface (such as by raking the fresh concrete with metal tines)

- remove the laitance from the surface of the hardened concrete (for example, by wetting the surface and blasting with compressed air)

- moisten the contact surface before placing the fresh concrete

1.16. (a) The following six items should be checked.

- Curing conditions: Were temperature and moisture controlled to allow the slab to gain strength?

- As-built dimensions: Is the slab thickness correct?

- Rebars: Were correct bar sizes and positions used?

- Construction loads: Was the slab overloaded by temporary forces?

- Falsework: Did settlement occur while concrete was plastic?

- In-place strength: Was water added to the mix after the cylinders were cast?

(b) The following 3 design details should be checked.

- Was variation in flexural rigidity, *EI*, properly computed?

- Was assumed joint rigidity (that is, end condition) achieved?

- Were deformations of supporting members considered?

1.17. Advantages of epoxy adhesive include

- excellent bond to most structural materials (it is generally insensitive to moisture)

- fast setting with high strength

Disadvantages include

- a short pot life (typically less than 30 min)

- fumes during mixing and placing that can be irritating

1.18. (a) *Preheat* is the localized application of heat prior to welding (minimum of 3 in to each side of the weld). (b) To ascertain the weldability of steel rebars, the chemical composition based on mill analysis must be known. Elements in the steel (both impurities and alloy elements) are expressed in percentages. The percentages of the various elements are multiplied by factors and summed to obtain a quantity called the *carbon equivalent*. For the rebar to be weldable without preheat, it must have a carbon equivalent less than 0.55. (c) The heat-affected zone is the base metal adjacent to a weld over which the temperature during welding becomes high enough that too rapid cooling (quenching) might cause hardening and embrittlement.

1.19. There is no requirement. Table 2 in ASTM A615 establishes minimum yield and tensile strengths for grade 40 rebar, but no maximum value for maximum yield strength is specified.

1.20. ASTM A706 covers low-alloy bars intended for special applications where welding, bending, or both are of importance. (See ACI 318 Sec. 3.5.3.1.)

Better quality control of chemical composition helps ensure that yield strength does not deviate excessively from the minimum specified. This makes A706 the preferred reinforcement when seismic loadings govern. (See ACI 318 Sec. 21.1.5.)

1.21. A mill test analysis of the chemical composition must be available. Elements in the steel are converted to carbon equivalents for determining weld electrode and preheat.

1.22. The two alloying elements that most affect weld electrode selection and preheat are manganese and nickel, both of which yield relatively high carbon equivalents.

1.23. The concrete's compressive strength is most affected by its water-cement ratio.

1.24. (a) Air entrainment is used primarily to increase the resistance of hardened concrete to freeze-thaw cycles. The entrainment also improves workability. The primary disadvantage of an air-entraining admixture is that the compressive strength of the mix is lowered. Precautions include testing the air content of fresh concrete to ensure that the proper air content is obtained through batching and mixing.

(b) Retarders are used to delay the set of fresh concrete. They might be needed to compensate for hot weather or to avoid cold joints in pours of massive elements. The disadvantage to their use is that strength gain is slowed, which might cause delays in subsequent construction phases. Also, the amount of shrinkage in the hardened concrete is less predictable. As a precaution, acceptance tests should be conducted under actual job conditions.

(c) An accelerator is used to accelerate strength gain. Calcium chloride is the most common accelerator. The primary disadvantage to calcium chloride is that the Cl^- ion speeds metal corrosion in the hardened concrete. The precaution to be taken in using calcium chloride is that it should be added to the mix water first, dissolved, and only then added to the batch. This prevents high local concentrations that might occur if flakes are added dry to the batch.

(d) Water reducers (also called superplasticizers) are added to improve workability without sacrificing strength. For normal strength ranges (3000 psi to 5000 psi), the admixture is generally used to produce high-slump, cohesive concrete that is flowable. In the normal slump range (1 in to 3 in), the water reducer is usually added to produce high-strength concrete. The primary disadvantage of water reducers is their cost. Precautions to take include understanding that the

effect of the admixture is short-lived, typically about 30 min.

1.25. CO_2 can form a weak carbonic acid, which attacks some compounds in the cement.

1.26. Advantages of adding fly ash to concrete include

* better workability (lower water demand, less segregation, greater pumpability, and so on)

* lowered heat of hydration

* lowered permeability, leading to greater durability (for example, better resistance to sulfate attack)

Disadvantages include

* slower strength gain

* longer curing period needed

* more material to be stored and handled in batching

1.27. It is important to consider the effects of volume change caused by creep, shrinkage, and change in temperature. Confinement that restrains the concrete against these effects, such as spiral reinforcement or hoops and crossties, can create stresses of the same magnitude as those caused by gravity loads.

1.28. Mild steel cannot be used to prestress concrete because creep and shrinkage strains in concrete will practically eliminate the prestress.

1.29. Four factors that influence shrinkage are

* time

* the environment (especially the relative humidity)

* the shape of the element (the volume-to-surface ratio)

* the curing condition (moist versus accelerated)

1.30. Creep is time-dependent strain under constant stress. Creep limits are usually not specified, but are established indirectly through limits on long-term deformation (deflection, camber, or axial shortening).

1.31. In fresh concrete, excess mix water and air "bleed" upward and accumulate under the horizontal rebar. This weakens the concrete and reduces the bond.

1.32. The splice must develop 125% of the minimum specified yield strength. So, the splice for a no. 14 grade 60 bar would need to develop

$$T = 1.25 f_y A_b$$
$$= (1.25)\left(60 \ \frac{\text{kips}}{\text{in}^2}\right)(2.25 \ \text{in}^2)$$
$$= 169 \ \text{kips}$$

1.33. (a) The 30% additional loss of prestress has practically no effect on the ultimate moment strength. In properly designed flexural members, the elastic strain corresponding to effective prestress is generally small relative to ultimate strain. (b) The lowered effective prestress can create serviceability problems, such as additional deflection and cracking of tension zones.

1.31. Three factors that affect deflection in the flat slab are

- variation in flexural rigidity, EI

- camber (specified or actual)

- support conditions

1.35. Under seismic lateral forces, plastic hinges should form in the girders near the column faces. To ensure ductile behavior, it is important that shear and development strengths exceed flexural strength. Because the yield strength of rebars is generally higher than their minimum specified yield stress, the flexural capacity of the member is proportionally higher. This gives rise to higher shears in the girder than if the bars yielded at or below the minimum yield strength.

1.36. Part of the negative reinforcement is placed in the flange area for crack control. (See ACI 318 Sec. 10.6.)

1.37. Construction joints should be located near midspan, where shear transfer across the joint is relatively low.

1.38. The design philosophy is that the structure should have enough toughness to respond inelastically to severe ground motion.

1.39. Column spirals have two purposes:

- to restrain the longitudinal steel in order to maintain its alignment during construction and prevent it from buckling at design load

- to confine concrete in the core, making it stronger and more ductile

Which purpose is primary depends on the situation. Spirals are often used for convenience in circular columns where the benefit of added ductility is of little practical importance. Frequently, however, spirals are used where energy absorption—toughness—is of prime importance, such as in ductile frames governed by earthquake loading.

1.40. During an earthquake, several cycles of inelastic deformation are expected. In a ductile frame, the design intent is usually to force plastic hinges to form in the girders rather than columns. (See FEMA 267.) If the girder reinforcement has higher yield strength than expected, the moment capacity will also be higher than expected, and the hinges might form in the columns instead.

1.41. Ductility is achieved by using an appropriate amount of confinement steel and limiting the steel ratio.

1.42. For a conventional reinforced concrete beam, the lever arm of the internal couple remains essentially constant as bending moment is increased. Equilibrium is maintained by increases in the internal forces $C = T = f_s A_s$.

For a prestressed concrete beam, the internal couple increases with the bending moment by increasing the internal lever arm, with the internal forces $C = T$ remaining essentially constant.

1.43. The concrete stress distribution for each beam is as follows.

(a) Resultant C on upper kern; zero stress on bottom fiber.

(b) Resultant C above upper kern; tension in bottom fiber (assuming tensile stress is less than the modulus of rupture).

(c) Resultant C is within the kerns and below the c.g.c.; the section is in compression throughout, with higher stress at the bottom than at the top.

(d) Resultant C below bottom kern; this is the opposite condition to case (b).

1.44. *Method 1:* Provide transfer girders to span horizontally 70 ft between new tie rods located at the end walls. The tie rods at the ends would furnish strength equivalent to that of the existing rods to be removed. Given the limited depth of the girders, a post-tensioned member would probably be needed. The post-tensioning can be controlled so that the tension from the existing rods is transferred before they are removed. This would minimize the horizontal deflection at the eave. Careful attention would need to be given to details so that the girder could deform independently of the main structure. Long-term deformation would also need to be studied.

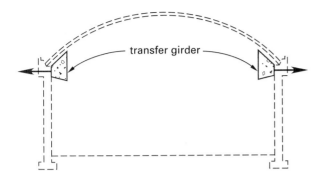

Method 2: Construct rigid frames within the clearance envelope and install new tie rods below the finish floor. The frames can be designed so that they will tilt inward when erected, then deflect outward so that they become vertical when the tension is transferred to them.

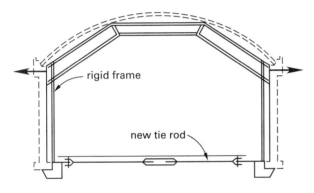

Method 3: A third possibility—at least from a structural standpoint—is to use a truss or frame constructed over the existing roof to resist the tension. For example,

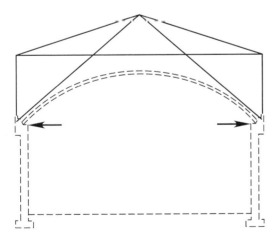

Connection details would be difficult, but are feasible. The construction sequence can be controlled to permit force transfer without excessive horizontal deflection.

1.45. (a) Detail D has the disadvantage that the top layers of longitudinal steel in intersecting members conflict; however, the structural advantages outweigh this consideration.

Detail E is bad in several respects. The dimensions do not satisfy ACI 318 Sec. 21.5.1.3, which requires that the width-to-depth ratio must be no more than 0.3. More importantly, the design philosophy for the ductile moment-resisting frame is that plastic hinges should form in the beams rather than the columns. It would be difficult to satisfy this objective using the relative depths of beam and column shown in detail E. Also, the spandrel would be difficult to construct. The top position would probably have to be formed and poured separately sometime after the floor system was placed.

Detail F is impractical for a moment-resisting exterior frame. The moment transfer from spandrel to columns would have to be through torsion in the girder, which would be difficult to accomplish. Also, the girder then has to transfer the additional vertical load from the spandrel beams. The benefit of confinement from intersecting members is also lost in the column-girder joint.

(b) The preferred detail is D, as explained above.

(c) The primary pitfall is in the joint. This creates construction problems both in rebar placement and in placement and consolidation of concrete.

SOLUTION 2

2.1. Determine the required slab thickness for the floor system.

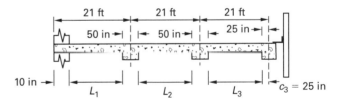

Use the strength design method from ACI 318, with $f'_c = 4$ ksi and $f_y = 40$ ksi. The slab is integral with the 10 in wall at the left end, and integral with the spandrel beam at the right end. The approximate method of ACI 318 Sec. 8.3 applies, with clear spans calculated as follows.

$$l_1 = L - h_w - \frac{b_w}{2}$$

$$= 21 \text{ ft} - \frac{10 \text{ in}}{12 \ \frac{\text{in}}{\text{ft}}} - \frac{50 \text{ in}}{(2)\left(12 \ \frac{\text{in}}{\text{ft}}\right)}$$

$$= 18.1 \text{ ft}$$

$$l_2 = L - \frac{b_{w,\text{left}}}{2} - \frac{b_{w,\text{right}}}{2}$$

$$= 21 \text{ ft} - \frac{50 \text{ in}}{(2)\left(12 \frac{\text{in}}{\text{ft}}\right)} - \frac{50 \text{ in}}{(2)\left(12 \frac{\text{in}}{\text{ft}}\right)}$$

$$= 16.8 \text{ ft}$$

$$l_3 = L - \frac{b_{w,\text{left}}}{2} - c_3$$

$$= 21 \text{ ft} - \frac{50 \text{ in}}{(2)\left(12 \frac{\text{in}}{\text{ft}}\right)} - \frac{25 \text{ in}}{12 \frac{\text{in}}{\text{ft}}}$$

$$= 16.8 \text{ ft}$$

Fire rating does not constrain slab thickness. For serviceability, ACI 318 Table 9.5(a) requires that the minimum thickness be

$$h \geq \begin{cases} \dfrac{l_n}{28} = \dfrac{(18.1 \text{ ft})\left(12 \frac{\text{in}}{\text{ft}}\right)}{28} \\ \qquad = 7.75 \text{ in} \quad \text{[for both ends continuous]} \\ \dfrac{l_n}{24} = \dfrac{(16.8 \text{ ft})\left(12 \frac{\text{in}}{\text{ft}}\right)}{24} \\ \qquad = 8.4 \text{ in} \quad \text{[for one end continuous]} \end{cases}$$

Per the footnote to Table 9.5(a), a reduction factor could be applied to the computed slab thickness to account for f_y less than 60 ksi. However, the values computed above are conservative and reasonable; therefore, try an 8.5 in slab. Check the shear.

$$w_u = 1.2 w_c h + 1.6 w_L$$

$$= \frac{(1.2)\left(150 \frac{\text{lbf}}{\text{ft}^3}\right)(8.5 \text{ in})}{12 \frac{\text{in}}{\text{ft}}} + (1.6)\left(250 \frac{\text{lbf}}{\text{ft}^3}\right)$$

$$= 528 \text{ lbf/ft}^2$$

Design on a per-foot-width basis: $b = 1 \text{ ft} = 12 \text{ in}$. Allow $3/4$ in cover and $1/2$ in additional distance to the centroid of main steel.

$$d = h - \text{cover} - \frac{d_b}{2}$$

$$= 8.5 \text{ in} - 0.75 \text{ in} - \frac{1 \text{ in}}{2}$$

$$= 7.25 \text{ in}$$

$$\phi V_c = 2\phi\lambda\sqrt{f_c'}\,b_w d$$

$$= (2)(0.75)(1.0)\sqrt{4000 \frac{\text{lbf}}{\text{in}^2}}(12 \text{ in})(7.25 \text{ in})$$

$$= 8250 \text{ lbf}$$

$$V_u \geq \begin{cases} \dfrac{w_u l_1}{2} - w_u d = \dfrac{\left(528 \frac{\text{lbf}}{\text{ft}}\right)(18.1 \text{ ft})}{2} \\ \qquad\qquad - \dfrac{\left(528 \frac{\text{lbf}}{\text{ft}}\right)(7.25 \text{ in})}{12 \frac{\text{in}}{\text{ft}}} \\ \qquad = 4460 \text{ lbf} \\ \dfrac{1.15 w_u l_3}{2} - w_u d = \dfrac{(1.15)\left(528 \frac{\text{lbf}}{\text{ft}}\right)(16.8 \text{ ft})}{2} \\ \qquad\qquad - \dfrac{\left(528 \frac{\text{lbf}}{\text{ft}}\right)(7.25 \text{ in})}{12 \frac{\text{in}}{\text{ft}}} \\ \qquad = 4780 \text{ lbf} \quad \text{[controls]} \end{cases}$$

ϕV_c is greater than V_u, so the shear strength is more than adequate. Use an $\boxed{8.5 \text{ in}}$ one-way slab.

2.2. To determine maximum slab moments, use ACI 318 Sec. 8.3.

$$M_{u,1L}^- = \frac{w_u l_n^2}{16} = \frac{\left(528 \frac{\text{lbf}}{\text{ft}}\right)(18.1 \text{ ft})^2}{16}$$

$$= \boxed{10{,}800 \text{ ft-lbf}}$$

$$M_{u,1}^+ = \frac{w_u l_n^2}{14} = \frac{\left(528 \frac{\text{lbf}}{\text{ft}}\right)(18.1 \text{ ft})^2}{14}$$

$$= \boxed{12{,}400 \text{ ft-lbf}}$$

$$M_{u,1R}^- = \frac{w_u l_n^2}{10} = \frac{\left(528 \frac{\text{lbf}}{\text{ft}}\right)\left(\frac{18.1 \text{ ft} + 16.8 \text{ ft}}{2}\right)^2}{10}$$

$$= \boxed{16{,}100 \text{ ft-lbf}}$$

$$M_{u,2}^+ = \frac{w_u l_n^2}{16} = \frac{\left(528 \frac{\text{lbf}}{\text{ft}}\right)(16.8 \text{ ft})^2}{16}$$

$$= \boxed{9300 \text{ ft-lbf}}$$

$$M_{u,3L}^- = \frac{w_u l_n^2}{10} = \frac{\left(528 \frac{\text{lbf}}{\text{ft}}\right)\left(\frac{16.8 \text{ ft} + 16.8 \text{ ft}}{2}\right)^2}{10}$$

$$= \boxed{14{,}900 \text{ ft-lbf}}$$

$$M_{u,3}^+ = \frac{w_u l_n^2}{11} = \frac{\left(528 \; \frac{\text{lbf}}{\text{ft}}\right)(16.8 \; \text{ft})^2}{11}$$

$$= \boxed{13{,}500 \; \text{ft-lbf}}$$

$$M_{u,3R}^- = \frac{w_u l_n^2}{24} = \frac{\left(528 \; \frac{\text{lbf}}{\text{ft}}\right)(16.8 \; \text{ft})^2}{24}$$

$$= \boxed{6200 \; \text{ft-lbf}}$$

2.3. Design slab reinforcement (per foot of width). For example, at the left end of span 1,

$$A_{s,\text{min}} = 0.002bh = (0.002)(12 \; \text{in})(8.5 \; \text{in})$$

$$= 0.20 \; \text{in}^2$$

Use no. 4 bars at 12 in centers for temperature and shrinkage steel.

For flexural steel ($f_y = 40$ ksi, $f_c' = 4$ ksi),

$$M_u = \phi M_n = \phi \rho b d^2 f_y \left(1 - 0.59\rho\left(\frac{f_y}{f_c'}\right)\right)$$

$$(10{,}800 \; \text{ft-lbf})\left(12 \; \frac{\text{in}}{\text{ft}}\right)$$

$$= 0.9\rho(12 \; \text{in})(7.25 \; \text{in})^2\left(40{,}000 \; \frac{\text{lbf}}{\text{in}^2}\right)$$

$$\times \left(1 - 0.59\rho\left(\frac{40{,}000 \; \frac{\text{lbf}}{\text{in}^2}}{4000 \; \frac{\text{lbf}}{\text{in}^2}}\right)\right)$$

$$\rho = 0.00591$$

$$A_s = \rho bd = (0.00591)(12 \; \text{in})(7.25 \; \text{in}) = 0.52 \; \text{in}^2$$

The calculated area of steel is per foot of slab width; therefore, use no. 5 bars at 7 in centers at the left end of span 1.

Similarly, for $M_u = 12{,}400$ ft-lbf, $A_s = 0.59$ in^2/ft.

Use no. 5 bars at 6 in centers.

For $M_u = 16{,}100$ ft-lbf, $A_s = 0.78$ in^2/ft.

Use no. 6 bars at 6 in centers.

For $M_u = 9300$ ft-lbf, $A_s = 0.44$ in^2/ft.

Use no. 5 bars at 8 in centers.

For $M_u = 14{,}950$ ft-lbf, $A_s = 0.72$ in^2/ft.

Use no. 6 bars at 7 in centers.

For $M_u = 13{,}500$ ft-lbf, $A_s = 0.65$ in^2/ft.

Use no. 5 bars at 6 in centers.

For $M_u = 6200$ ft-lbf, $A_s = 0.29$ in^2/ft.

Use no. 5 bars at 12 in centers.

Determine locations for bar cutoffs (see Nilson et al., 2008, Graph A.3). For negative steel at the left end of span 1,

$$L = 0.15l_1 = (0.15)(18.1 \; \text{ft}) = 2.7 \; \text{ft}$$

For negative steel at the right end of span 1,

$$L = 0.3l_1 = (0.3)(18.1 \; \text{ft}) = 5.4 \; \text{ft}$$

For negative steel on interior spans,

$$L = 0.22l_2 = (0.22)(16.8 \; \text{ft}) = 3.7 \; \text{ft}$$

For negative steel at the left end of span 3,

$$L = 0.3l_3 = (0.3)(16.8 \; \text{ft}) = 5.0 \; \text{ft}$$

For negative steel at the right end of span 3,

$$L = 0.1l_3 = (0.1)(16.8 \; \text{ft}) = 1.7 \; \text{ft}$$

Per ACI 318 Sec. 12.12.3, at least one-third of the top flexural steel must extend beyond the cutoff points to a distance of at least $12d_b$ or $1/16$ the clear span, whichever is greater.

$$12d_b = (12)(0.75 \; \text{in})$$

$$= 9 \; \text{in}$$

$$\frac{\text{clear span}}{16} = \frac{(18.1 \; \text{ft})\left(12 \; \frac{\text{in}}{\text{ft}}\right)}{16}$$

$$= 13.6 \; \text{in}$$

To be conservative, extend alternate bars 13.6 in (1.13 ft) beyond the cutoff locations.

For positive steel, extend alternate bars 6 in into supports, and terminate other bars at points of inflection. Thus, for the exterior spans,

$$L' = \sqrt{\frac{8M_u^+}{w_u}}$$

$$= \sqrt{\frac{(8)(12{,}600 \; \text{ft-lbf})}{538 \; \frac{\text{lbf}}{\text{ft}}}}$$

$$= 13.7 \; \text{ft}$$

For the interior span,

$$L' = \sqrt{\frac{8M_u^+}{w_u}}$$

$$= \sqrt{\frac{(8)(9500 \; \text{ft-lbf})}{538 \; \frac{\text{lbf}}{\text{ft}}}}$$

$$= 11.9 \; \text{ft}$$

The small diameter reinforcing bars used in the slab satisfy the minimum embedment required by ACI 318 Sec. 12.11.

The section through the slabs may be illustrated as follows.

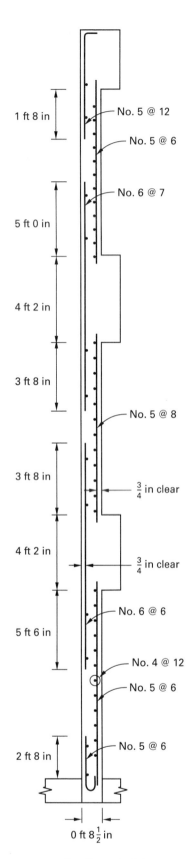

section through slabs
(not to scale)

SOLUTION 3

3.1. Plot the interaction diagram for the given column.

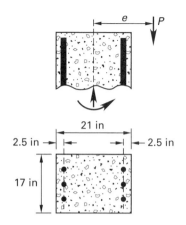

Given two no. 9 bars and one no. 8 bar on each face,

$$A_s = A'_s = 2A_{b,9} + A_{b,8}$$
$$= (2)(1.00 \text{ in}^2) + 0.79 \text{ in}^2$$
$$= 2.79 \text{ in}^2$$
$$A_{st} = 2A_s = (2)(2.79 \text{ in}^2) = 5.58 \text{ in}^2$$
$$A_g = bh = (17 \text{ in})(21 \text{ in}) = 357 \text{ in}^2$$
$$f'_c = 4500 \text{ psi}$$
$$\beta_1 = 0.825 \quad [\text{per ACI 318 Sec. 10.2.7.3}]$$
$$f_y = 60,000 \text{ psi}$$

For the concentric load case $(M = 0)$,

$$P_0 = 0.85f'_c(A_g - A_{st}) + f_yA_{st}$$
$$= (0.85)\left(4500 \ \frac{\text{lbf}}{\text{in}^2}\right)(357 \text{ in}^2 - 5.58 \text{ in}^2)$$
$$\quad + \left(60,000 \ \frac{\text{lbf}}{\text{in}^2}\right)(5.58 \text{ in}^2)$$
$$= 1,679,000 \text{ lbf} \quad (1679 \text{ kips})$$

For pure bending $(P = 0)$, iterate to find the neutral axis position. Start by assuming $f'_s = 0$ psi.

$$C_c = 0.85f'_cba = T = A_sf_y$$
$$a = \frac{A_sf_y}{0.85f'_cb} = \frac{(2.79 \text{ in}^2)\left(60,000 \ \frac{\text{lbf}}{\text{in}^2}\right)}{(0.85)\left(4500 \ \frac{\text{lbf}}{\text{in}^2}\right)(17 \text{ in})}$$
$$= 2.57 \text{ in}$$
$$c = \frac{a}{\beta_1} = \frac{2.57 \text{ in}}{0.825} = 3.12 \text{ in}$$

From similar triangles on the strain diagram,

$$\frac{\epsilon'_s}{c - d'} = \frac{\epsilon_c}{c}$$

$$\epsilon'_s = \frac{(c - d')\epsilon_c}{c}$$

$$= \frac{(3.12 \text{ in} - 2.5 \text{ in})(0.003)}{3.12 \text{ in}}$$

$$= 0.0006$$

$$\epsilon_y = \frac{f_y}{E_s} = \frac{60{,}000 \, \frac{\text{lbf}}{\text{in}^2}}{29{,}000{,}000 \, \frac{\text{lbf}}{\text{in}^2}} = 0.00207 \quad [\epsilon_y > \epsilon'_s]$$

$$f'_s = E_s\epsilon'_s = \left(29{,}000{,}000 \, \frac{\text{lbf}}{\text{in}^2}\right)(0.0006)$$

$$= 17{,}300 \text{ psi}$$

The assumption that f'_s is zero leads to a calculated stress in the compression side steel of 17,300 psi. A second trial should assume a value for f'_s smaller than 17,300 psi because including stress in the compression steel will decrease the neutral axis depth and so give a lower calculated stress. Assume $f'_s = 10{,}000$ psi.

$$C_c = 0.85f'_c ba = T - C'_s = A_s f_y - (f'_s - 0.85f'_c)A'_s$$

$$a = \frac{A_s f_y - (f'_s - 0.85f'_c)A'_s}{0.85f'_c b}$$

$$= \frac{(2.79 \text{ in}^2)\left(60{,}000 \, \frac{\text{lbf}}{\text{in}^2}\right)}{}$$

$$\frac{- \left(10{,}000 \, \frac{\text{lbf}}{\text{in}^2} - (0.85)\left(4500 \, \frac{\text{lbf}}{\text{in}^2}\right)\right)(2.79 \text{ in}^2)}{(0.85)\left(4500 \, \frac{\text{lbf}}{\text{in}^2}\right)(17 \text{ in})}$$

$$- 2.31 \text{ in}$$

$$c = \frac{a}{\beta_1} = \frac{2.31 \text{ in}}{0.825} = 2.80 \text{ in}$$

From similar triangles,

$$\frac{\epsilon'_s}{c - d'} = \frac{\epsilon_c}{c}$$

$$\epsilon'_s = \frac{(2.80 \text{ in} - 2.5 \text{ in})(0.003)}{2.80 \text{ in}}$$

$$= 0.00032$$

$$\epsilon_y = \frac{f_y}{E_s} = \frac{60{,}000 \, \frac{\text{lbf}}{\text{in}^2}}{29{,}000{,}000 \, \frac{\text{lbf}}{\text{in}^2}}$$

$$= 0.00207 \quad [\epsilon_y > \epsilon'_s]$$

$$f'_s = E_s\epsilon'_s = \left(29{,}000{,}000 \, \frac{\text{lbf}}{\text{in}^2}\right)(0.00032)$$

$$= 9300 \text{ lbf/in}^2$$

The stress calculated at this step is close enough to the assumed value of 10,000 psi, so take the compression

stress as 10 ksi. Sum moments about the tension steel to determine the nominal moment strength.

$$M_n = C_c\left(d - \frac{a}{2}\right) + C'_s(d - d')$$

$$= 0.85f'_c ba\left(d - \frac{a}{2}\right) + (f'_s - 0.85f'_c)A'_s(d - d')$$

$$= \frac{\substack{(0.85)\left(4500 \, \frac{\text{lbf}}{\text{in}^2}\right)(17 \text{ in})(2.31 \text{ in}) \\ \times \left(18.5 \text{ in} - \frac{2.31 \text{ in}}{2}\right) \\ + \left(10{,}000 \, \frac{\text{lbf}}{\text{in}^2} - (0.85)\left(4500 \, \frac{\text{lbf}}{\text{in}^2}\right)\right) \\ \times (2.79 \text{ in}^2)(18.5 \text{ in} - 2.5 \text{ in})}}{\left(12 \, \frac{\text{in}}{\text{ft}}\right)\left(1000 \, \frac{\text{lbf}}{\text{kip}}\right)}$$

$$= 240 \text{ ft-kips}$$

For balanced strain conditions,

$$\epsilon_y = \frac{f_y}{E_s} = \frac{60{,}000 \, \frac{\text{lbf}}{\text{in}^2}}{29{,}000{,}000 \, \frac{\text{lbf}}{\text{in}^2}} = 0.00207$$

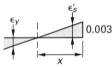

By similar triangles,

$$\frac{\epsilon_y + \epsilon_c}{d} = \frac{\epsilon_c}{c}$$

$$c = \frac{\epsilon_c d}{\epsilon_y + \epsilon_c}$$

$$= \frac{(0.003)(18.5 \text{ in})}{0.00207 + 0.003}$$

$$= 10.93 \text{ in}$$

$$\frac{\epsilon'_s}{c - d'} = \frac{\epsilon_c}{c}$$

$$\epsilon'_s = \frac{(10.93 \text{ in} - 2.5 \text{ in})(0.003)}{10.93 \text{ in}}$$

$$= 0.0023 \quad [\epsilon_y > \epsilon'_s \text{ so compression steel yields}]$$

$$f'_s = f_y = 60{,}000 \text{ psi}$$

$$T = A_s f_y = (2.79 \text{ in}^2)\left(60{,}000 \, \frac{\text{lbf}}{\text{in}^2}\right) = 167{,}400 \text{ lbf}$$

$$C'_s = (f'_s - 0.85f'_c)A'_s$$

$$= \left(60{,}000 \, \frac{\text{lbf}}{\text{in}^2} - (0.85)\left(4500 \, \frac{\text{lbf}}{\text{in}^2}\right)\right)(2.79 \text{ in}^2)$$

$$= 156{,}700 \text{ lbf}$$

$a = \beta_1 c = (0.825)(10.93 \text{ in}) = 9.02 \text{ in}$

$C_c = 0.85 f'_c ba$

$= (0.85)\left(4500 \ \dfrac{\text{lbf}}{\text{in}^2}\right)(17 \text{ in})(9.02 \text{ in})$

$= 586{,}000 \text{ lbf}$

$P_b = C_c + C'_s - T$

$= 586{,}000 \text{ lbf} + 156{,}700 \text{ lbf} - 167{,}400 \text{ lbf}$

$= 575{,}300 \text{ lbf} \quad (575 \text{ kips})$

$M_b = C_c\left(\dfrac{h}{2} - \dfrac{a}{2}\right) + C'_s\left(\dfrac{h}{2} - d'\right) + T\left(d - \dfrac{h}{2}\right)$

$$= \frac{(586{,}000 \text{ lbf})\left(\dfrac{21 \text{ in}}{2} - \dfrac{9.02 \text{ in}}{2}\right)}{}$$

$$+ (156{,}700 \text{ lbf})\left(\dfrac{21 \text{ in}}{2} - 2.5 \text{ in}\right)$$

$$= \frac{+ (167{,}400 \text{ lbf})\left(18.5 \text{ in} - \dfrac{21 \text{ in}}{2}\right)}{\left(12 \ \dfrac{\text{in}}{\text{ft}}\right)\left(1000 \ \dfrac{\text{lbf}}{\text{kip}}\right)}$$

$= 509 \text{ ft-kips}$

Calculate some intermediate points to define the shape of the interaction curve. The simplest point corresponds to $f_s = 0$.

$c = h - d' = 21 \text{ in} - 2.5 \text{ in} = 18.5 \text{ in}$

$\epsilon'_s > \epsilon_y$

Therefore, $f'_s = f_y = 60{,}000 \text{ psi}$. The tension force, T, is zero for this position of the neutral axis, so

$C'_s = (f'_s - 0.85 f'_c) A'_s$

$= \left(60{,}000 \ \dfrac{\text{lbf}}{\text{in}^2} - (0.85)\left(4500 \ \dfrac{\text{lbf}}{\text{in}^2}\right)\right)(2.79 \text{ in}^2)$

$= 156{,}700 \text{ lbf}$

$a = \beta_1 c = (0.825)(18.5 \text{ in}) = 15.26 \text{ in}$

$C_c = 0.85 f'_c ba$

$= (0.85)\left(4500 \ \dfrac{\text{lbf}}{\text{in}^2}\right)(17 \text{ in})(15.26 \text{ in})$

$= 992{,}000 \text{ lbf}$

$P = C_c + C'_s - T$

$= 992{,}000 \text{ lbf} + 156{,}700 \text{ lbf} - 0 \text{ lbf}$

$= 1{,}149{,}000 \text{ lbf} \quad (1150 \text{ kips})$

$M_b = C_c\left(\dfrac{h}{2} - \dfrac{a}{2}\right) + C'_s\left(\dfrac{h}{2} - d'\right) + T\left(d - \dfrac{h}{2}\right)$

$$= \frac{(992{,}000 \text{ lbf})\left(\dfrac{21 \text{ in}}{2} - \dfrac{15.26 \text{ in}}{2}\right)}{}$$

$$+ (156{,}700 \text{ lbf})\left(\dfrac{21 \text{ in}}{2} - 2.5 \text{ in}\right)$$

$$= \frac{+ (0 \text{ lbf})\left(18.5 \text{ in} - \dfrac{21 \text{ in}}{2}\right)}{\left(12 \ \dfrac{\text{in}}{\text{ft}}\right)\left(1000 \ \dfrac{\text{lbf}}{\text{kip}}\right)}$$

$= 342 \text{ ft-kips}$

For this case, the computed axial force is larger than P_b, so the point is on the compression branch of the interaction curve. Choose another neutral axis location that is on the tension branch of the curve (that is, any value of c less than c_b). A convenient choice is that point corresponding to a tension-controlled failure, for which ϵ_s equals 0.005.

$\dfrac{\epsilon_y + \epsilon_c}{d} = \dfrac{\epsilon_c}{c}$

$c = \dfrac{\epsilon_c d}{\epsilon_y + \epsilon_c}$

$= \dfrac{(0.003)(18.5 \text{ in})}{0.005 + 0.003}$

$= 6.94 \text{ in}$

$a = \beta_1 c = (0.825)(6.94 \text{ in}) = 5.73 \text{ in}$

$\epsilon'_s = \left(\dfrac{c - d'}{c}\right)\epsilon_c$

$= \left(\dfrac{6.94 \text{ in} - 2.5 \text{ in}}{6.94 \text{ in}}\right)(0.003)$

$= 0.0019 \quad [\epsilon'_s < \epsilon_y]$

$f'_s = E_s \epsilon'_s = \left(29{,}000{,}000 \ \dfrac{\text{lbf}}{\text{in}^2}\right)(0.0019) = 55{,}100 \text{ psi}$

$T = A_s f_y = (2.79 \text{ in}^2)\left(60{,}000 \ \dfrac{\text{lbf}}{\text{in}^2}\right) = 167{,}400 \text{ lbf}$

$C'_s = (f'_s - 0.85 f'_c) A'_s$

$= \left(55{,}100 \ \dfrac{\text{lbf}}{\text{in}^2} - (0.85)\left(4500 \ \dfrac{\text{lbf}}{\text{in}^2}\right)\right)(2.79 \text{ in}^2)$

$= 143{,}000 \text{ lbf}$

$C_c = 0.85 f'_c ba$

$= (0.85)\left(4500 \ \dfrac{\text{lbf}}{\text{in}^2}\right)(17 \text{ in})(5.73 \text{ in})$

$= 372{,}600 \text{ lbf}$

$$P = C_c + C'_s - T$$
$$= 372{,}600 \text{ lbf} + 143{,}000 \text{ lbf} - 167{,}400 \text{ lbf}$$
$$= 348{,}200 \text{ lbf} \quad (348 \text{ kips})$$

$$M = C_c\left(\frac{h}{2} - \frac{a}{2}\right) + C'_s\left(\frac{h}{2} - d'\right) + T\left(d - \frac{h}{2}\right)$$

$$= \frac{\begin{aligned}(372{,}600 \text{ lbf})&\left(\frac{21 \text{ in}}{2} - \frac{5.73 \text{ in}}{2}\right) \\ + (143{,}000 \text{ lbf})&\left(\frac{21 \text{ in}}{2} - 2.5 \text{ in}\right) \\ + (167{,}400 \text{ lbf})&\left(18.5 \text{ in} - \frac{21 \text{ in}}{2}\right)\end{aligned}}{\left(12 \frac{\text{in}}{\text{ft}}\right)\left(1000 \frac{\text{lbf}}{\text{kip}}\right)}$$

$$= 444 \text{ ft-kips}$$

For design values, apply $\phi = 0.65$ for compression failure, for tension-controlled failure use $\phi = 0.9$, and between the limits ϕ varies linearly from 0.65 to 0.9. For practical purposes, ACI 318 takes the steel strain corresponding to compression-controlled failure as 0.002, which is essentially the yield strain for grade 60 reinforcement. Thus, for tension steel strains, ϵ_s, varying from 0.002 to 0.005,

$$\phi = 0.65 + (83.3)(\epsilon_s - 0.002)$$

The interaction diagram for the column is

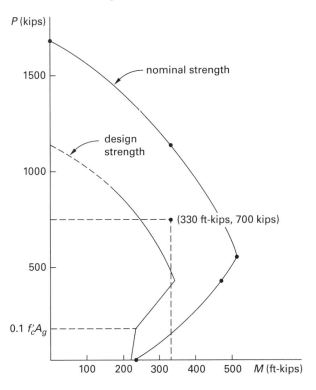

3.2. Check adequacy of the column for the given loading. The point $(M_u = 330 \text{ ft-kips}, P_u = 700 \text{ kips})$ falls outside the design interaction curve for the column; therefore, the column is $\boxed{\text{inadequate}}$ for this loading.

SOLUTION 4

Given: $f'_c = 4000$ psi, $f_y = 60{,}000$ psi for longitudinal steel, and $f_y = 40{,}000$ psi for stirrups. Use no. 8 bars for the main steel, and use the given values of factored moments and shears (do not redistribute moments or reduce moments to the face of support). Also, since unbalanced loadings are neglected, the given shears and moments occur under the factored uniformly distributed load, 13.3 kips/ft.

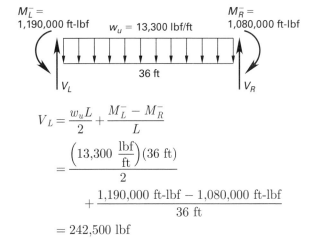

$$V_L = \frac{w_u L}{2} + \frac{M_L^- - M_R^-}{L}$$
$$= \frac{\left(13{,}300 \frac{\text{lbf}}{\text{ft}}\right)(36 \text{ ft})}{2}$$
$$\quad + \frac{1{,}190{,}000 \text{ ft-lbf} - 1{,}080{,}000 \text{ ft-lbf}}{36 \text{ ft}}$$
$$= 242{,}500 \text{ lbf}$$

Maximum positive bending moment occurs where $V = 0$.

$$V = V_L - w_u x$$
$$x = \frac{V_L - V}{w_u}$$
$$= \frac{242{,}500 \text{ lbf} - 0 \text{ lbf}}{13{,}300 \frac{\text{lbf}}{\text{ft}}}$$
$$= 18.2 \text{ ft}$$
$$M_u^+ = V_L x - \frac{w_u x^2}{2} - M_L^-$$
$$= (242{,}500 \text{ lbf})(18.2 \text{ ft})$$
$$\quad - \frac{\left(13{,}300 \frac{\text{lbf}}{\text{ft}}\right)(18.2 \text{ ft})^2}{2} - 1{,}190{,}000 \text{ ft-lbf}$$
$$= 1{,}020{,}000 \text{ ft-lbf}$$

4.1. Design the positive and negative flexural reinforcement using the strength design method of ACI 318. For the positive reinforcement, use $M_u^+ = 1{,}020{,}000$ ft lbf. The section resisting positive flexure is a T-beam.

$$d^+ = 36 \text{ in} - 2.5 \text{ in} = 33.5 \text{ in}$$

$$b \leq \begin{cases} \dfrac{L}{4} = \dfrac{(36 \text{ ft})\left(12 \frac{\text{in}}{\text{ft}}\right)}{4} = 108 \text{ in} \quad \text{[controls]} \\[2mm] b_w + 16h = 18 \text{ in} + (16)(8 \text{ in}) = 146 \text{ in} \\[2mm] s = (22 \text{ ft})\left(12 \frac{\text{in}}{\text{ft}}\right) = 264 \text{ in} \end{cases}$$

$$M_u = \phi M_n = \phi \rho b d^2 f_y \left(1 - 0.59\rho\left(\frac{f_y}{f_c'}\right)\right)$$

$$(1{,}020{,}000 \text{ ft-lbf})\left(12 \frac{\text{in}}{\text{ft}}\right)$$

$$= 0.9\rho(108 \text{ in})(33.5 \text{ in})^2\left(60{,}000 \frac{\text{lbf}}{\text{in}^2}\right)$$

$$\times \left(1 - 0.59\rho\left(\frac{60{,}000 \frac{\text{lbf}}{\text{in}^2}}{4000 \frac{\text{lbf}}{\text{in}^2}}\right)\right)$$

$$\rho = 0.0019$$

$$A_s = \rho b d = (0.0019)(108 \text{ in})(33.5 \text{ in}) = 6.87 \text{ in}^2$$

Check the neutral axis depth.

$$c = \frac{a}{\beta_1} = \frac{A_s f_y}{0.85 f_c' b \beta_1}$$

$$= \frac{(6.87 \text{ in}^2)\left(60{,}000 \frac{\text{lbf}}{\text{in}^2}\right)}{(0.85)\left(4000 \frac{\text{lbf}}{\text{in}^2}\right)(108 \text{ in})(0.85)}$$

$$= 1.32 \text{ in}$$

The compression region is well within the slab thickness, so the beam is a T-beam as assumed. Steel strain is well above 0.005, so the section is tension controlled with $\phi = 0.9$. Check the minimum steel required.

$$A_{s,\min} \geq \begin{cases} \dfrac{200 b_w d}{f_y} = \dfrac{\left(200 \frac{\text{lbf}}{\text{in}^2}\right)(18 \text{ in})(33.5 \text{ in})}{60{,}000 \frac{\text{lbf}}{\text{in}^2}} \\[2mm] \qquad = 2.01 \text{ in}^2 \quad \text{[does not control]} \\[4mm] \dfrac{3\sqrt{f_c'} b_w d}{f_y} = \dfrac{3\sqrt{4000 \frac{\text{lbf}}{\text{in}^2}}(18 \text{ in})(33.5 \text{ in})}{60{,}000 \frac{\text{lbf}}{\text{in}^2}} \\[2mm] \qquad = 1.91 \text{ in}^2 \end{cases}$$

Try no. 8 bars ($A_b = 0.79 \text{ in}^2$).

$$n_{\text{bar}} = \frac{A_{s,\text{req}}}{A_b} = \frac{6.87 \text{ in}^2}{0.79 \text{ in}^2} = 8.7 \quad \text{[use 9]}$$

Bars may need to be bundled to fit the 18 in web width. Determine negative reinforcement for the left end.

$$M_u^- = \phi M_n = \phi \rho b_w d^2 f_y \left(1 - 0.59\rho\left(\frac{f_y}{f_c'}\right)\right)$$

$$(1{,}190{,}000 \text{ ft-lbf})\left(12 \frac{\text{in}}{\text{ft}}\right)$$

$$= 0.9\rho(18 \text{ in})(33.5 \text{ in})^2\left(60{,}000 \frac{\text{lbf}}{\text{in}^2}\right)$$

$$\times \left(1 - 0.59\rho\left(\frac{60{,}000 \frac{\text{lbf}}{\text{in}^2}}{4000 \frac{\text{lbf}}{\text{in}^2}}\right)\right)$$

$$\rho = 0.0151$$

$$A_s = \rho b d = (0.0151)(18 \text{ in})(33.5 \text{ in}) = 9.11 \text{ in}^2$$

Check the neutral axis depth.

$$c < 0.375 d = (0.375)(33.5 \text{ in}) = 12.7 \text{ in}$$

$$c = \frac{a}{\beta_1} = \frac{A_s f_y}{0.85 f_c' b \beta_1}$$

$$= \frac{(9.11 \text{ in}^2)\left(60{,}000 \frac{\text{lbf}}{\text{in}^2}\right)}{(0.85)\left(4000 \frac{\text{lbf}}{\text{in}^2}\right)(18 \text{ in})(0.85)}$$

$$= 10.5 \text{ in} < 12.7 \text{ in}$$

The section is tension controlled, with the required A_s well above $A_{s,\min}$.

$$n_{\text{bar}} = \frac{A_{s,\text{req}}}{A_b} = \frac{9.11 \text{ in}^2}{0.79 \text{ in}^2} = 11.5 \quad \text{[use 12]}$$

Use 12 no. 8 bars at the top left end.

For the right end,

$$M_u^- = \phi M_n = \phi \rho b_w d^2 f_y \left(1 - 0.59\rho\left(\frac{f_y}{f_c'}\right)\right)$$

$$(1{,}080{,}000 \text{ ft-lbf})\left(12 \frac{\text{in}}{\text{ft}}\right)$$

$$= 0.9\rho(18 \text{ in})(33.5 \text{ in})^2\left(60{,}000 \frac{\text{lbf}}{\text{in}^2}\right)$$

$$\times \left(1 - 0.59\rho\left(\frac{60{,}000 \frac{\text{lbf}}{\text{in}^2}}{4000 \frac{\text{lbf}}{\text{in}^2}}\right)\right)$$

$$\rho = 0.0135$$

$$A_s = \rho b d = (0.0135)(18 \text{ in})(33.5 \text{ in}) = 8.14 \text{ in}^2$$

The steel area required is smaller than at the left end; therefore, the section is tension controlled.

$$n_{\text{bar}} = \frac{A_{s,\text{req}}}{A_b} = \frac{8.14 \text{ in}^2}{0.79 \text{ in}^2} = 10.3 \quad \text{[use 11]}$$

Use 11 no. 8 bars at the top right end.

4.2. Find the locations where 50% of A_s^+ can terminate. Theoretically, 50% of the maximum A_s is

$$0.5A_{s,max} = (0.5)(6.86 \text{ in}^2) = 3.43 \text{ in}^2$$

So,

$$a = \frac{A_s f_y}{0.85 f'_c b} = \frac{(3.43 \text{ in}^2)\left(60{,}000 \frac{\text{lbf}}{\text{in}^2}\right)}{(0.85)\left(4000 \frac{\text{lbf}}{\text{in}^2}\right)(108 \text{ in})}$$

$$= 0.56 \text{ in}$$

$$\phi M_n = \phi A_s f_y \left(d - \frac{a}{2}\right)$$

$$= \frac{(0.9)(3.43 \text{ in}^2)\left(60{,}000 \frac{\text{lbf}}{\text{in}^2}\right)\left(33.5 \text{ in} - \frac{0.56 \text{ in}}{2}\right)}{\left(12 \frac{\text{in}}{\text{ft}}\right)\left(1000 \frac{\text{lbf}}{\text{kip}}\right)}$$

$$= 513 \text{ ft-kips}$$

Locate the region over which the bending moment is less than 513 ft-kips.

$$\frac{w_u (L')^2}{8} = 513 \text{ ft-kips}$$

$$L' = \sqrt{\frac{(8)(513 \text{ ft-kips})}{13.3 \frac{\text{kips}}{\text{ft}}}}$$

$$= 17.6 \text{ ft}$$

The distance L' centers about the point of M_{max}, 18.2 ft from the left support, which is practically at midspan. Per ACI 318 Sec. 12.10, bars must extend beyond the theoretical cutoff point by at least

$$\text{extension} \geq \begin{cases} d - \dfrac{33.5 \text{ in}}{12 \frac{\text{in}}{\text{ft}}} \\ \qquad = 2.8 \text{ ft} \quad \text{[controls]} \\ 12 d_b = \dfrac{12 \text{ in}}{12 \frac{\text{in}}{\text{ft}}} = 1 \text{ ft} \end{cases}$$

$$x = \frac{L - \left(L' + 2(\text{extension})\right)}{2}$$

$$= \frac{36 \text{ ft} - 17.6 \text{ ft} - (2)(2.8 \text{ ft})}{2}$$

$$= 6.4 \text{ ft}$$

This cutoff is for the left end, but is conservative for the right end.

> Cut off four no. 8 bottom bars at 6 ft 4 in from the center of each column.

4.3. Calculate the stirrup spacing at 4 ft intervals. For the critical locations at a distance d from the face of support,

$$V_u = V_L - w_u d$$

$$= 242{,}500 \text{ lbf} - \frac{\left(13{,}300 \frac{\text{lbf}}{\text{ft}}\right)(33.5 \text{ in})}{12 \frac{\text{in}}{\text{ft}}}$$

$$= 205{,}400 \text{ lbf}$$

$$V_c = 2\lambda \sqrt{f'_c} b_w d$$

$$= (2)(1.0)\sqrt{4000 \frac{\text{lbf}}{\text{in}^2}}(18 \text{ in})(33.5 \text{ in})$$

$$= 76{,}300 \text{ lbf}$$

The design shear is approximately three times larger than the shear resistance of concrete. Stirrups are required. Try double U-stirrups.

$$A_v = 4A_b = (4)(0.20 \text{ in}^2) = 0.80 \text{ in}^2$$

$$V_s = \frac{V_u}{\phi} - V_c$$

$$= \frac{205{,}400 \text{ lbf}}{0.75} - 76{,}300 \text{ lbf}$$

$$= 197{,}600 \text{ lbf}$$

$$4\sqrt{f'_c} b_w d = 2V_c = (2)(76{,}300 \text{ lbf})$$

$$= 152{,}600 \text{ lbf}$$

$$8\sqrt{f'_c} b_w d = 4V_c = (4)(76{,}300 \text{ lbf})$$

$$= 305{,}200 \text{ lbf}$$

$$4\sqrt{f'_c} b_w d < V_s < 8\sqrt{f'_c} b_w d$$

Because $4\sqrt{f'_c} b_w d < V_s$, the spacing is

$$s \leq \begin{cases} \dfrac{d}{4} = \dfrac{33.5}{4} = 8.4 \text{ in} \\ \dfrac{A_v f_y}{50 b_w} = \dfrac{(0.80 \text{ in}^2)\left(40{,}000 \frac{\text{lbf}}{\text{in}^2}\right)}{\left(50 \frac{\text{lbf}}{\text{in}^2}\right)(18 \text{ in})} \\ \qquad = 36 \text{ in} \quad \left[0.75\sqrt{f'_c} < 50 \text{ for } f'_c < 4400 \text{ psi}\right] \\ \dfrac{A_v f_y d}{V_s} = \dfrac{(0.80 \text{ in}^2)\left(40{,}000 \frac{\text{lbf}}{\text{in}^2}\right)(33.5 \text{ in})}{205{,}400 \text{ lbf}} \\ \qquad = 5.2 \text{ in} \quad \text{[controls]} \end{cases}$$

Use no. 4 U-stirrups at 5 in on centers from 0 ft to 4 ft from the column centerline.

For the region $4 \text{ ft} \leq x \leq 8 \text{ ft}$,

$$V_u = V_L - w_u r$$

$$= 242{,}500 \text{ lbf} - \left(13{,}300 \ \frac{\text{lbf}}{\text{ft}}\right)(4 \text{ ft})$$

$$= 189{,}300 \text{ lbf}$$

$$V_s = \frac{V_u}{\phi} - V_c$$

$$= \frac{189{,}300 \text{ lbf}}{0.75} - 76{,}300 \text{ lbf}$$

$$= 176{,}100 \text{ lbf}$$

$$4\sqrt{f'_c}\, b_w d < V_s < 8\sqrt{f'_c}\, b_w d$$

$$s \leq \begin{cases} \dfrac{d}{4} = \dfrac{33.5}{4} = 8.4 \text{ in} \\[2ex] \dfrac{A_v f_y}{50 b_w} = \dfrac{(0.80 \text{ in}^2)\left(40{,}000 \ \frac{\text{lbf}}{\text{in}^2}\right)}{\left(50 \ \frac{\text{lbf}}{\text{in}^2}\right)(18 \text{ in})} = 36 \text{ in} \\[3ex] \dfrac{A_v f_y d}{V_s} = \dfrac{(0.80 \text{ in}^2)\left(40{,}000 \ \frac{\text{lbf}}{\text{in}^2}\right)(33.5 \text{ in})}{176{,}100 \text{ lbf}} \\[2ex] \qquad = 6.0 \text{ in} \quad [\text{controls}] \end{cases}$$

Use no. 4 U-stirrups at 6 in on centers from 4 ft to 8 ft.

For the region $8 \text{ ft} \leq x \leq 12 \text{ ft}$,

$$V_u = V_L - w_u x$$

$$= 242{,}500 \text{ lbf} - \left(13{,}300 \ \frac{\text{lbf}}{\text{ft}}\right)(8 \text{ ft})$$

$$= 136{,}100 \text{ lbf}$$

$$V_s = \frac{V_u}{\phi} - V_c$$

$$= \frac{136{,}100 \text{ lbf}}{0.75} - 76{,}300 \text{ lbf}$$

$$= 105{,}200 \text{ lbf}$$

$$V_s < 4\sqrt{f'_c}\, b_w d$$

$$s \leq \begin{cases} \dfrac{d}{2} = \dfrac{33.5}{2} = 16.8 \text{ in} \\[2ex] \dfrac{A_v f_y}{50 b_w} = \dfrac{(0.80 \text{ in}^2)\left(40{,}000 \ \frac{\text{lbf}}{\text{in}^2}\right)}{\left(50 \ \frac{\text{lbf}}{\text{in}^2}\right)(18 \text{ in})} = 36 \text{ in} \\[3ex] \dfrac{A_v f_y d}{V_s} = \dfrac{(0.80 \text{ in}^2)\left(40{,}000 \ \frac{\text{lbf}}{\text{in}^2}\right)(33.5 \text{ in})}{105{,}200 \text{ lbf}} \\[2ex] \qquad = 10.1 \text{ in} \quad [\text{controls}] \end{cases}$$

Use no. 4 U-stirrups at 10 in on centers from 8 ft to 12 ft.

For the region $12 \text{ ft} \leq x \leq 16 \text{ ft}$,

$$V_u = V_L - w_u x$$

$$= 242{,}500 \text{ lbf} - \left(13{,}300 \ \frac{\text{lbf}}{\text{ft}}\right)(12 \text{ ft})$$

$$= 82{,}900 \text{ lbf}$$

$$V_s = \frac{V_u}{\phi} - V_c = \frac{82{,}900 \text{ lbf}}{0.75} - 76{,}300 \text{ lbf}$$

$$= 34{,}200 \text{ lbf}$$

$$V_s < 4\sqrt{f'_c}\, b_w d$$

$$s \leq \begin{cases} \dfrac{d}{2} = \dfrac{33.5}{2} = 16.8 \text{ in} \quad [\text{controls}] \\[2ex] \dfrac{A_v f_y}{50 b_w} = \dfrac{(0.80 \text{ in}^2)\left(40{,}000 \ \frac{\text{lbf}}{\text{in}^2}\right)}{\left(50 \ \frac{\text{lbf}}{\text{in}^2}\right)(18 \text{ in})} = 36 \text{ in} \\[3ex] \dfrac{A_v f_y d}{V_s} = \dfrac{(0.80 \text{ in}^2)\left(40{,}000 \ \frac{\text{lbf}}{\text{in}^2}\right)(33.5 \text{ in})}{34{,}200 \text{ lbf}} = 31.3 \text{ in} \end{cases}$$

Minimum spacing controls. Use only a single U-stirrup at this location.

$$A_v = 2A_b = (2)(0.20 \text{ in}^2) = 0.40 \text{ in}^2$$

Use no. 4 U-stirrups at 16 in on centers from 12 ft to 16 ft.

From 16 ft to 20 ft, minimum spacing governs, so $s \leq d/2 = 16.8 \text{ in}$; use 16 in spacing. Beyond 20 ft, the shear increases in magnitude and the stirrup pattern closely mirrors the pattern over the left half. Use the same pattern (slightly conservative).

See the following elevation for a summary of stirrup spacings.

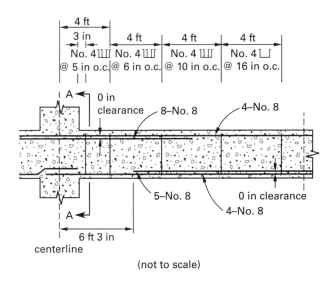

(not to scale)

4.4. For detail reinforcement (ACI 318 Sec. 10.6.6), distribute the top flexural reinforcement over a flange width of one-tenth the span of 3.6 ft—say, 3 ft 6 in. Because the overall depth is less than 36 in, intermediate web reinforcement is not required (ACI 318 Sec. 10.6.7).

The beam is on the interior and is exempt from the structural integrity requirements of ACI 318 Sec. 7.13.2.3. The cutoff location for four of the nine no. 8 bars (to be conservative) was previously calculated to be 6 ft 4 in from the support centerline. The point of inflection occurs at

$$x = \frac{L - \sqrt{\dfrac{8M_u^+}{w_u}}}{2}$$

$$= \frac{36 \text{ ft} - \sqrt{\dfrac{(8)(1017 \text{ ft-kips})}{13.3 \dfrac{\text{kips}}{\text{ft}}}}}{2}$$

$$= 5.6 \text{ ft}$$

Check provisions of ACI 318 Sec. 12.11.3. For the no. 8 bottom bars,

$$l_d = \left(\frac{f_y \psi_t \psi_e \lambda}{20\sqrt{f_c'}}\right) d_b$$

$$= \left(\frac{\left(60{,}000 \dfrac{\text{lbf}}{\text{in}^2}\right)(1)(1)(1)}{20\sqrt{4000 \dfrac{\text{lbf}}{\text{in}^2}}}\right)\left(\frac{1.0 \text{ in}}{12 \dfrac{\text{in}}{\text{ft}}}\right)$$

$$= 4.0 \text{ ft}$$

$$V_u = V_L - w_u x$$

$$= 242{,}400 \text{ lbf} - \left(13{,}300 \frac{\text{lbf}}{\text{ft}}\right)(5.6 \text{ ft})$$

$$= 168{,}000 \text{ lbf}$$

$$\phi M_n = (513 \text{ ft-kips})\left(1000 \frac{\text{lbf}}{\text{kip}}\right) = 513{,}000 \text{ ft-lbf}$$

$$M_n = \frac{513{,}000 \text{ ft-lbf}}{\phi} = \frac{513{,}000 \text{ ft-lbf}}{0.9} = 570{,}000 \text{ ft-lbf}$$

$$l_a \leq \begin{cases} 12d_b = (12)(1 \text{ in}) = 12 \text{ in} \\[4pt] d = \dfrac{33.5 \text{ in}}{12 \dfrac{\text{in}}{\text{ft}}} \\[8pt] = 2.8 \text{ ft} \quad [\text{controls}] \end{cases}$$

$$l_d \leq \frac{M_n}{V_u} + l_a$$

$$= \frac{569{,}000 \text{ ft-lbf}}{168{,}000 \text{ lbf}} + 2.8 \text{ ft}$$

$$= 6.1 \text{ ft}$$

The no. 8 bottom bars have adequate embedment beyond the points of inflection. For top bars, let four no. 8 bars be continuous to provide support for stirrups. Extend the remaining eight no. 8 bars beyond the points of inflection per ACI 318 Sec. 12.12.3.

From ACI 318 Sec. 12.2.4, the development length of the top bars is calculated to be $(1.3)(4 \text{ ft}) = 5.2 \text{ ft}$. Thus, the terminating top bars have adequate development from the face of supports. An elevation and cross section of the beam are shown.

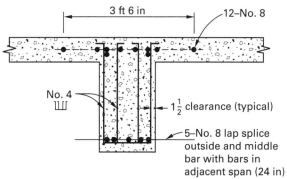

section A-A
slab reinforcement omitted for clarity
(not to scale)

SOLUTION 5

5.1. Use the strength design method from ACI 318 to design the corbel. Given: $f'_c = 4000$ psi and $f_y = 60,000$ psi. Use the prescriptive criteria in ACI 318 Sec. 11.8. To simplify the formwork, let $b =$ column width $= 24$ in. Use a trial-and-error procedure to find h. Try 13 in.

$$d = h - \text{cover} - \frac{d_b}{2}$$
$$= 13 \text{ in} - 0.75 \text{ in} - \frac{1 \text{ in}}{2}$$
$$= 11.8 \text{ in}$$
$$\frac{a}{d} = \frac{6 \text{ in}}{11.8 \text{ in}}$$
$$= 0.5 \quad [< 1; \text{ACI 318 Sec. 11.8 criteria apply}]$$
$$V_u = 1.2 P_D + 1.6 P_L$$
$$= (1.2)(48,000 \text{ lbf}) + (1.6)(32,000 \text{ lbf})$$
$$= 109,000 \text{ lbf}$$

According to ACI 318 Sec. 11.8.3.4, the normal force, N_{uc}, is to be treated as a live load; therefore, apply a load factor of 1.6 to the given gravity dead and live loads.

$$N_{uc} = \mu_s \big(1.6(P_D + P_L)\big)$$
$$= (0.4)\big((1.6)(48,000 \text{ lbf} + 32,000 \text{ lbf})\big)$$
$$= 51,000 \text{ lbf}$$
$$M_u = V_u a + N_{uc}(h - d)$$
$$= (109,000 \text{ lbf})(6 \text{ in}) + (51,000 \text{ lbf})(13 \text{ in} - 11.8 \text{ in})$$
$$= 715,000 \text{ in-lbf}$$

Approximate the area of flexural steel using an internal lever arm equal to $0.9d$. (ACI 318 Sec. 11.8.3.1 requires $\phi = 0.75$ for all calculations.)

$$A_s = \frac{M_u}{\phi f_y (0.9d)}$$
$$= \frac{715,000 \text{ in-lbf}}{(0.75)\left(60,000 \ \frac{\text{lbf}}{\text{in}^2}\right)(0.9)(11.8 \text{ in})}$$
$$= 1.50 \text{ in}^2$$

Check the approximation.

$$a = \frac{A_s f_y}{0.85 f'_c b}$$
$$= \frac{(1.50 \text{ in}^2)\left(60,000 \ \frac{\text{lbf}}{\text{in}^2}\right)}{(0.85)\left(4000 \ \frac{\text{lbf}}{\text{in}^2}\right)(24 \text{ in})}$$
$$= 1.1 \text{ in}$$

$$d - \frac{a}{2} = 11.8 \text{ in} - \frac{1.1 \text{ in}}{2}$$
$$= 11.3 \text{ in}$$
$$> 0.9d = (0.9)(11.8 \text{ in}) = 10.6 \text{ in}$$

Thus, the calculated A_{vf} is conservative. The additional steel needed to resist the direct tension is

$$A_n = \frac{N_{uc}}{\phi f_y} = \frac{51,000 \text{ lbf}}{(0.75)\left(60,000 \ \frac{\text{lbf}}{\text{in}^2}\right)} = 1.13 \text{ in}^2$$

The total area of main steel must resist combined flexure and axial force.

$$A_{sc} \geq A_s + A_n$$
$$\geq 1.50 \text{ in}^2 + 1.13 \text{ in}^2$$
$$\geq 2.63 \text{ in}^2$$

Check shear transfer.

$$V_u \leq \phi V_n$$
$$\leq \phi A_{vf} f_y \mu$$
$$A_{vf} \geq \frac{V_u}{\phi f_y \mu}$$
$$\geq \frac{109,000 \text{ lbf}}{(0.75)\left(60,000 \ \frac{\text{lbf}}{\text{in}^2}\right)(1.4)}$$
$$\geq 1.73 \text{ in}^2$$

Per ACI 318 Sec. 11.8.3.2.1,

$$V_n \leq \begin{cases} 0.2 f'_c bd = (0.2)\left(4000 \ \frac{\text{lbf}}{\text{in}^2}\right)(24 \text{ in}) \\ \qquad \times (11.8 \text{ in}) \\ \qquad = 226,000 \text{ lbf} \quad [\text{controls}] \\[4pt] \left(\begin{array}{c} 480 \ \frac{\text{lbf}}{\text{in}^2} \\ + 0.08 f'_c \end{array}\right) bd = \left(\begin{array}{c} 480 \ \frac{\text{lbf}}{\text{in}^2} \\ + (0.08)\left(4000 \ \frac{\text{lbf}}{\text{in}^2}\right) \end{array}\right) \\ \qquad \times (24 \text{ in})(11.8 \text{ in}) \\ \qquad = 226,500 \text{ lbf} \\[4pt] \left(1600 \ \frac{\text{lbf}}{\text{in}^2}\right) bd = \left(1600 \ \frac{\text{lbf}}{\text{in}^2}\right)(24 \text{ in})(11.8 \text{ in}) \\ \qquad = 453,120 \text{ lbf} \end{cases}$$

The shear capacity, V_n, is well below the limiting values. Distribute the shear friction reinforcement as required by ACI 318 Sec. 11.8.3.

$$A_{sc} \geq \left(\frac{2A_{vf}}{3} + A_n\right) = \frac{(2)(1.73 \text{ in}^2)}{3} + 1.13 \text{ in}^2 = 2.28 \text{ in}^2$$

Check the minimum steel requirement of ACI 318 Sec. 11.8.

$$A_{sc,\min} = 0.04\left(\frac{f'_c}{f_y}\right)bd$$

$$= (0.04)\left(\frac{4000 \ \frac{\text{lbf}}{\text{in}^2}}{60,000 \ \frac{\text{lbf}}{\text{in}^2}}\right)(24 \text{ in})(11.8 \text{ in})$$

$$= 0.76 \text{ in}^2$$

Use $A_{sc} = 2.63 \text{ in}^2$ as required for flexure and axial force.

Use two no. 7 and two no. 8 top bars.

Stirrups are required to furnish the additional shear strength (ACI 318 Sec. 11.8.4).

$$A_h = 0.5(A_{sc} - A_n)$$

$$= (0.5)(2.63 \text{ in}^2 - 1.13 \text{ in}^2)$$

$$= 0.75 \text{ in}^2$$

Use two no. 4 closed stirrups, $A_h = 0.80 \text{ in}^2$. The detail is as shown.

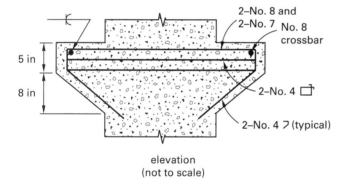

5 in

8 in

2–No. 8 and
2–No. 7 No. 8
crossbar

2–No. 4 □

2–No. 4 ⊐ (typical)

elevation
(not to scale)

SOLUTION 6

6.1. Compute the service live load capacity based on the strength of the one-way slab and beam. Use the ACI 318 strength design method, with $f'_c = 3500$ psi and $f_y = 40,000$ psi.

For the typical interior spans, ACI 318 Sec. 8.3 gives

$$M_u^+ = \frac{w_u l_n^2}{16}$$

$$M_u^- = \frac{w_u l_n^2}{11}$$

$$V_u = \frac{w_u l_n}{2}$$

For the 6 in slab,

$$d = h_s - \text{cover} - \frac{d_b}{2}$$

$$= 6 \text{ in} - 0.75 \text{ in} - \frac{0.5 \text{ in}}{2}$$

$$= 5.0 \text{ in}$$

Design on a unit width basis, with $b = 12$ in. In the negative moment region,

$$A_s^- = \left(\frac{12 \text{ in}}{s}\right)A_b = \left(\frac{12 \text{ in}}{8 \text{ in}}\right)(0.20 \text{ in}^2)$$

$$= 0.30 \text{ in}^2 \quad \text{[per foot width]}$$

$$l_n = L - b_w = 12 \text{ ft} - 1.5 \text{ ft} = 10.5 \text{ ft}$$

$$a = \frac{A_s f_y}{0.85 f'_c b} = \frac{(0.30 \text{ in}^2)\left(40,000 \ \frac{\text{lbf}}{\text{in}^2}\right)}{(0.85)\left(3500 \ \frac{\text{lbf}}{\text{in}^2}\right)(12 \text{ in})} = 0.34 \text{ in}$$

$$c = \frac{a}{\beta_1} = \frac{0.34 \text{ in}}{0.85} = 0.40 \text{ in}$$

$$< 0.375d = (0.375)(5.0 \text{ in})$$

$$= 1.875 \text{ in} \quad [\text{therefore, } \phi = 0.9]$$

$$\phi M_n = \phi A_s f_y\left(d - \frac{a}{2}\right)$$

$$= \frac{(0.9)(0.30 \text{ in}^2)\left(40,000 \ \frac{\text{lbf}}{\text{in}^2}\right)\left(5.0 \text{ in} - \frac{0.34 \text{ in}}{2}\right)}{12 \ \frac{\text{in}}{\text{ft}}}$$

$$= 4350 \text{ ft-lbf}$$

$$M_u^- = \frac{w_u l_n^2}{11} \le \phi M_n$$

$$w_u \le \frac{11(\phi M_n)}{l_n^2}$$

$$\le \frac{(11)(4350 \text{ ft-lbf})}{(10.5 \text{ ft})^2}$$

$$\le 434 \text{ lbf/ft}^2$$

For the positive moment region, no. 4 bars at 12 in o.c. give 0.20 in² per foot width.

$$a = \frac{A_s f_y}{0.85 f'_c b} = \frac{(0.20 \text{ in}^2)\left(40,000 \ \frac{\text{lbf}}{\text{in}^2}\right)}{(0.85)\left(3500 \ \frac{\text{lbf}}{\text{in}^2}\right)(12 \text{ in})} = 0.22 \text{ in}$$

Slab positive region is tension controlled.

$$\phi M_n = \phi A_s f_y \left(d - \frac{a}{2} \right)$$

$$= \frac{(0.9)(0.20 \text{ in}^2) \left(40{,}000 \, \frac{\text{lbf}}{\text{in}^2} \right) \left(5.0 \text{ in} - \frac{0.22 \text{ in}}{2} \right)}{12 \, \frac{\text{in}}{\text{ft}}}$$

$$= 2930 \text{ ft-lbf}$$

$$M_u^+ = \frac{w_u l_n^2}{16} \le \phi M_n$$

$$w_u \le \frac{16(\phi M_n)}{l_n^2}$$

$$\le \frac{(16)(2930 \text{ ft-lbf})}{(10.5 \text{ ft})^2}$$

$$\le 425 \text{ lbf/ft} \quad \text{[per foot width]}$$

Find slab shear strength.

$$\phi V_c = \phi 2 \lambda \sqrt{f'_c} \, b_w d$$

$$= (0.75)(2)(1.0) \sqrt{3500 \, \frac{\text{lbf}}{\text{in}^2}} (12 \text{ in})(5.0 \text{ in})$$

$$= 5320 \text{ lbf}$$

$$V_u = \frac{w_u l_n}{2} - w_u d$$

$$= \frac{w_u(10.5 \text{ ft})}{2} - \frac{w_u(5.0 \text{ in})}{12 \, \frac{\text{in}}{\text{ft}}}$$

$$= (4.83 \text{ ft}) w_u$$

$$V_u \le \phi V_c$$

$$(4.83 \text{ ft}) w_u \le 5320 \text{ lbf}$$

$$w_u \le 1100 \text{ lbf/ft} \quad \text{[per foot width]}$$

Because the beam depth is the same as the girder depth, 24 in, take the clear span of the beams as the face-to-face distance between columns. For the 18 in × 24 in beam, find the negative bending.

$$b = b_w = 18 \text{ in}$$

$$d^- = h - \text{cover} - \frac{d_b}{2}$$

$$= 24 \text{ in} - 1.875 \text{ in} - \frac{1.27 \text{ in}}{2}$$

$$= 21.5 \text{ in}$$

$$\approx d^+$$

$$l_n = L - h_{\text{column}} = 25 \text{ ft} - 2 \text{ ft} = 23 \text{ ft}$$

$$A_s^- = 4 A_b = (4)(1.27 \text{ in}^2) = 5.08 \text{ in}^2$$

$$u = \frac{A_s f_y}{0.85 f'_c b} = \frac{(5.08 \text{ in}^2) \left(40{,}000 \, \frac{\text{lbf}}{\text{in}^2} \right)}{(0.85) \left(3500 \, \frac{\text{lbf}}{\text{in}^2} \right) (18 \text{ in})}$$

$$= 3.80 \text{ in}$$

$$c = \frac{a}{\beta_1} = \frac{3.80 \text{ in}}{0.85} = 4.47 \text{ in}$$

$$< 0.375 d \quad \text{[therefore, } \phi = 0.9 \text{]}$$

$$\phi M_n = \phi A_s f_y \left(d - \frac{a}{2} \right)$$

$$= \frac{(0.9)(5.08 \text{ in}^2) \left(40{,}000 \, \frac{\text{lbf}}{\text{in}^2} \right) \left(21.5 \text{ in} - \frac{3.80 \text{ in}}{2} \right)}{12 \, \frac{\text{in}}{\text{ft}}}$$

$$= 298{,}700 \text{ ft-lbf}$$

$$M_u^- = \frac{w_u l_n^2}{11} \le \phi M_n$$

$$w_u \le \frac{11(\phi M_n)}{l_n^2}$$

$$\le \frac{(11)(298{,}700 \text{ ft-lbf})}{(23 \text{ ft})^2}$$

$$= 6211 \text{ lbf/ft}$$

Per 12 ft of spacing,

$$w_{u,\text{per 12 ft}} = \frac{6211 \, \frac{\text{lbf}}{\text{ft}}}{12 \text{ ft}} = 517 \text{ lbf/ft}^2$$

Find the positive moment for the T-beam.

$$b_{\text{eff}} \le \begin{cases} \dfrac{L}{4} = \left(\dfrac{23 \text{ ft}}{4} \right) \left(12 \, \dfrac{\text{in}}{\text{ft}} \right) \\ \qquad = 69 \text{ in} \quad \text{[controls]} \\ \text{beam spacing} = (12 \text{ ft}) \left(12 \, \dfrac{\text{in}}{\text{ft}} \right) = 144 \text{ in} \\ b_w + 16 h_s = 18 \text{ in} + (16)(6 \text{ in}) \\ \qquad = 114 \text{ in} \end{cases}$$

$$A_s^+ = n_{\text{bar}} A_b = (4)(1.00 \text{ in}^2) = 4.00 \text{ in}^2$$

$$a = \frac{A_s f_y}{0.85 f'_c b} = \frac{(4.00 \text{ in}^2) \left(40{,}000 \, \frac{\text{lbf}}{\text{in}^2} \right)}{(0.85) \left(3500 \, \frac{\text{lbf}}{\text{in}^2} \right) (69 \text{ in})} = 0.78 \text{ in}$$

$$c = \frac{a}{\beta_1} = \frac{0.78 \text{ in}}{0.85} = 0.92 \text{ in}$$

$$< 0.375 d \quad \text{[therefore, } \phi = 0.9 \text{]}$$

$$\phi M_n = \phi A_s f_y \left(d - \frac{a}{2} \right)$$

$$= \frac{(0.9)(4.00 \text{ in}^2)\left(40{,}000 \, \frac{\text{lbf}}{\text{in}^2}\right)\left(21.5 \text{ in} - \frac{0.78 \text{ in}}{2}\right)}{12 \, \frac{\text{in}}{\text{ft}}}$$

$$= 253{,}300 \text{ ft-lbf}$$

$$M_u^+ = \frac{w_u l_n^2}{16} \le \phi M_n$$

$$w_u \le \frac{16(\phi M_n)}{l_n^2}$$

$$\le \frac{(16)(253{,}300 \text{ ft-lbf})}{(23 \text{ ft})^2}$$

$$= 7661 \text{ lbf/ft}$$

Per 12 ft of spacing,

$$w_{u,\text{per } 12 \text{ ft}} = \frac{7661 \, \dfrac{\text{lbf}}{\text{ft}}}{12 \text{ ft}} = 638 \text{ lbf/ft}^2$$

Find the shear for no. 3 stirrups at 8 in o.c.

$$A_v = 2A_b = (2)(0.11 \text{ in}^2) = 0.22 \text{ in}^2$$

$$V_c = 2\lambda \sqrt{f_c'} \, b_w d$$

$$= (2)(1.0)\sqrt{3500 \, \frac{\text{lbf}}{\text{in}^2}}(18 \text{ in})(21.5 \text{ in})$$

$$= 45{,}800 \text{ lbf}$$

$$V_u = \frac{w_u l_n}{2} - w_u d$$

$$= \frac{w_u(23 \text{ ft})}{2} - \frac{w_u(21.5 \text{ in})}{12 \, \frac{\text{in}}{\text{ft}}}$$

$$= (9.7 \text{ ft})w_u$$

$$V_s = \frac{A_v f_y d}{s} = \frac{(0.22 \text{ in}^2)\left(40{,}000 \, \frac{\text{lbf}}{\text{in}^2}\right)(21.5 \text{ in})}{8 \text{ in}}$$

$$= 23{,}700 \text{ lbf}$$

$$V_u = \phi(V_c + V_s) = (0.75)(45{,}800 \text{ lbf} + 23{,}700 \text{ lbf})$$

$$= 52{,}100 \text{ lbf}$$

$$V_u \le \phi(V_c + V_s)$$

$$(9.7 \text{ ft})w_u \le 52{,}100 \text{ lbf}$$

$$w_u \le 5370 \text{ lbf/ft}$$

Per 12 ft of spacing,

$$w_{u,\text{per } 12 \text{ ft}} = \frac{5370 \, \dfrac{\text{lbf}}{\text{ft}}}{12 \text{ ft}} = 448 \text{ lbf/ft}^2$$

Check shear capacity in the central region where there are no stirrups. (ACI 318 limits shear on the unreinforced web to $\phi V_c/2$.)

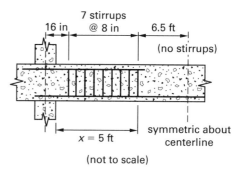

$$V_u = \frac{w_u l_n}{2} - w_u x$$

$$= \frac{w_u(23 \text{ ft})}{2} - w_u(5 \text{ ft})$$

$$= w_u(6.5 \text{ ft}) \le \frac{\phi V_c}{2}$$

$$w_u \le \frac{\phi V_c}{(2)(6.5 \text{ ft})}$$

$$\le \frac{(0.75)(45{,}800 \text{ lbf})}{(2)(6.5 \text{ ft})}$$

$$\le 2640 \text{ lbf/ft}$$

Per 12 ft of spacing,

$$w_{u,\text{per } 12 \text{ ft}} = \frac{2640 \, \dfrac{\text{lbf}}{\text{ft}}}{12 \text{ ft}} = 220 \text{ lbf/ft}^2$$

Thus, the floor system strength is governed by shear capacity in the central region. The design load capacity is

$$w_u = 1.2w_d + 1.6w_L = 220 \text{ lbf/ft}^2$$

Assume that normal weight concrete is the only dead load.

$$w_d = w_c h_s + \frac{w_c(h - h_s)b_w}{s}$$

$$= w_c\left(h_s + \frac{(h - h_s)b_w}{s}\right)$$

$$= \left(150 \, \frac{\text{lbf}}{\text{ft}^3}\right)\left(\frac{6 \text{ in}}{12 \, \frac{\text{in}}{\text{ft}}} + \frac{(24 \text{ in} - 6 \text{ in})(18 \text{ in})}{(12 \text{ ft})\left(12 \, \frac{\text{in}}{\text{ft}}\right)^2}\right)$$

$$= 103 \text{ lbf/ft}^2$$

$$w_u = 1.2w_D + 1.6w_L$$

$$w_L = \frac{w_u - 1.2w_D}{1.6}$$

$$= \frac{220 \ \frac{\text{lbf}}{\text{ft}^2} - (1.2)\left(103 \ \frac{\text{lbf}}{\text{ft}^2}\right)}{1.6}$$

$$= 60 \ \text{lbf/ft}^2$$

Shear on the unreinforced beam web governs. $\boxed{w_L \le 60 \ \text{lbf/ft}^2.}$

6.2. To establish true capacity, it would also be necessary to check

- capacity of girders in shear and flexure
- capacity of column and foundation
- edge and corner bay locations
- special locations, such as around stairwells

Also, for each case, the appropriate building code would likely permit a live load reduction for elements supporting large tributary areas, which could lead to a specified live load that is larger than has been calculated. Serviceability considerations (for example, live load deflection or cracking) might also limit the acceptable loading. Finally, any lateral force effects that might combine with the gravity loads would have to be considered.

SOLUTION 7

7.1. Plot the interaction diagram for the given column using at least three points. $f'_c = 5000$ psi, $\beta_1 = 0.8$, and $f_y = 60,000$ psi.

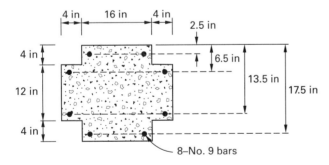

$$A_{st} = n_{bar}A_b = (8)(1.00 \ \text{in}^2) = 8.00 \ \text{in}^2$$

$$A_g = bh - 4a^2$$

$$= (24 \ \text{in})(20 \ \text{in}) - (4)(4 \ \text{in})^2$$

$$= 416 \ \text{in}^2$$

For the concentric load case (M is zero),

$$P_o = 0.85f'_c(A_g - A_{st}) + f_yA_{st}$$

$$= (0.85)\left(5000 \ \frac{\text{lbf}}{\text{in}^2}\right)(416 \ \text{in}^2 - 8.00 \ \text{in}^2)$$

$$+ \left(60,000 \ \frac{\text{lbf}}{\text{in}^2}\right)(8.00 \ \text{in}^2)$$

$$= 2,214,000 \ \text{lbf} \quad (2214 \ \text{kips})$$

The section is symmetrical, so its plastic centroid coincides with its geometric centroid. For balanced conditions,

$$\epsilon_y = \frac{f_y}{E_s} = \frac{60,000 \ \frac{\text{lbf}}{\text{in}^2}}{29,000,000 \ \frac{\text{lbf}}{\text{in}^2}} = 0.00207$$

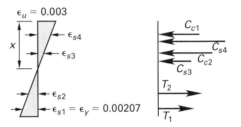

By similar triangles,

$$d_1 = h - 2.5 \ \text{in} = 20 \ \text{in} - 2.5 \ \text{in} = 17.5 \ \text{in}$$

$$d_2 = h - 6.5 \ \text{in} = 20 \ \text{in} - 6.5 \ \text{in} = 13.5 \ \text{in}$$

$$d_3 = h - 13.5 \ \text{in} = 20 \ \text{in} - 13.5 \ \text{in} = 6.5 \ \text{in}$$

$$d_4 = 2.5 \ \text{in}$$

$$\frac{c}{\epsilon_c} = \frac{d_1}{\epsilon_c + \epsilon_y}$$

$$c = \frac{d_1\epsilon_c}{\epsilon_c + \epsilon_y}$$

$$= \frac{(17.5 \ \text{in})(0.003)}{0.003 + 0.00207}$$

$$= 10.3 \ \text{in}$$

$$\frac{\epsilon_4}{c - d_4} = \frac{\epsilon_c}{c}$$

$$\epsilon_4 = \frac{\epsilon_c(c - d_4)}{c}$$

$$= \frac{(0.003)(10.3 \ \text{in} - 2.5 \ \text{in})}{10.3 \ \text{in}}$$

$$= 0.0023$$

$$\frac{\epsilon_3}{c - d_3} = \frac{\epsilon_c}{c}$$

$$\epsilon_3 = \frac{\epsilon_c(c - d_3)}{c}$$

$$= \frac{(0.003)(10.3 \text{ in} - 6.5 \text{ in})}{10.3 \text{ in}}$$

$$= 0.0011$$

$$\frac{\epsilon_2}{c - d_2} = \frac{\epsilon_c}{c}$$

$$\epsilon_2 = \frac{\epsilon_c(c - d_2)}{c}$$

$$= \frac{(0.003)(10.3 \text{ in} - 13.5 \text{ in})}{10.3 \text{ in}}$$

$$= -0.000932 \quad \text{[tension]}$$

The compression strain in the top steel exceeds the yield strain, so $f_{s4} = f_y = 60,000$ psi. The compression strain in steel layer 3 is less than the yield strain, so

$$f_{s3} = \epsilon_{s3} E_s$$

$$= (0.0011)\left(29,000,000 \ \frac{\text{lbf}}{\text{in}^2}\right)$$

$$= 31,900 \text{ psi}$$

The tensile strain in layer 2 is also below the yield strain, giving

$$f_{s2} = \epsilon_{s2} E_s$$

$$= (0.00092)\left(29,000,000 \ \frac{\text{lbf}}{\text{in}^2}\right)$$

$$= 26,700 \text{ psi}$$

The strain in the extreme tension steel is at the yield strain, by the definition of balanced conditions. Thus,

$$T_1 = A_{s1} f_y = (2.00 \text{ in}^2)\left(60,000 \ \frac{\text{lbf}}{\text{in}^2}\right) = 120,000 \text{ lbf}$$

$$T_2 = A_{s2} f_{s2} = (2.00 \text{ in}^2)\left(26,700 \ \frac{\text{lbf}}{\text{in}^2}\right) = 53,400 \text{ lbf}$$

$$C_{s3} = (f_{s3} - 0.85 f'_c) A_{s3}$$

$$= \left(32,100 \ \frac{\text{lbf}}{\text{in}^2} - (0.85)\left(5000 \ \frac{\text{lbf}}{\text{in}^2}\right)\right)(2.00 \text{ in}^2)$$

$$= 56,000 \text{ lbf}$$

$$C_{s4} = (f_{s4} - 0.85 f'_c) A_{s4}$$

$$= \left(60,000 \ \frac{\text{lbf}}{\text{in}^2} - (0.85)\left(5000 \ \frac{\text{lbf}}{\text{in}^2}\right)\right)(2.00 \text{ in}^2)$$

$$= 112,000 \text{ lbf}$$

$$a = \beta_1 c = (0.8)(10.3 \text{ in}) = 8.24 \text{ in}$$

$$C_{c1} = 0.85 f'_c b_1 a_1$$

$$= (0.85)\left(5000 \ \frac{\text{lbf}}{\text{in}^2}\right)(16 \text{ in})(4.0 \text{ in})$$

$$= 272,000 \text{ lbf}$$

$$C_{c2} = 0.85 f'_c b_2 a_2$$

$$= (0.85)\left(5000 \ \frac{\text{lbf}}{\text{in}^2}\right)(24 \text{ in})(4.24 \text{ in})$$

$$= 436,000 \text{ lbf}$$

$$P_b = C_{c1} + C_{c2} + C_{s4} + C_{s3} - T_2 - T_1$$

$$= 272,000 \text{ lbf} + 436,000 \text{ lbf} + 112,000 \text{ lbf}$$

$$+ 56,000 \text{ lbf} - 53,600 \text{ lbf} - 120,000 \text{ lbf}$$

$$= 702,000 \text{ lbf} \quad (702 \text{ kips})$$

$$M_b = C_{c1}\left(\frac{h}{2} - \frac{a_1}{2}\right) + C_{c2}\left(\frac{h}{2} - a_1 - \frac{a_2}{2}\right) + C_{s4}\left(\frac{h}{2} - d_4\right)$$

$$+ C_{s3}\left(\frac{h}{2} - d_3\right) + T_2\left(d_2 - \frac{h}{2}\right) + T_1\left(d_1 - \frac{h}{2}\right)$$

$$= \frac{\begin{aligned} & (272,000 \text{ lbf})\left(\frac{20 \text{ in}}{2} - \frac{4 \text{ in}}{2}\right) + (436,000 \text{ lbf}) \\ & \times\left(\frac{20 \text{ in}}{2} - 4 \text{ in} - \frac{4.24 \text{ in}}{2}\right) + (112,000 \text{ lbf}) \\ & \times\left(\frac{20 \text{ in}}{2} - 2.5 \text{ in}\right) + (56,400 \text{ lbf})\left(\frac{20 \text{ in}}{2} - 6.5 \text{ in}\right) \\ & + (53,600 \text{ lbf})\left(13.5 \text{ in} - \frac{20 \text{ in}}{2}\right) \\ & + (120,000 \text{ lbf})\left(17.5 \text{ in} - \frac{20 \text{ in}}{2}\right) \end{aligned}}{\left(12 \ \frac{\text{in}}{\text{ft}}\right)\left(1000 \ \frac{\text{lbf}}{\text{kip}}\right)}$$

$$= 497 \text{ ft-kips}$$

As a third point, chose the point on the tension failure segment corresponding to a strain in extreme tension steel equal to 0.005.

$$\frac{c}{\epsilon_c} = \frac{d_1}{\epsilon_c + \epsilon_{s1}}$$

$$c = \frac{d_1 \epsilon_c}{\epsilon_c + \epsilon_{s1}} = \frac{(17.5 \text{ in})(0.003)}{0.003 + 0.005} = 6.56 \text{ in}$$

$$\frac{\epsilon_4}{c - d_4} = \frac{\epsilon_c}{c}$$

$$\epsilon_4 = \frac{\epsilon_c(c - d_4)}{c}$$

$$= \frac{(0.003)(6.56 \text{ in} - 2.5 \text{ in})}{6.56 \text{ in}}$$

$$= 0.00186$$

$$\frac{\epsilon_3}{c - d_3} = \frac{\epsilon_c}{c}$$

$$\begin{aligned}
\epsilon_3 &= \frac{\epsilon_c(c - d_3)}{c} \\
&= \frac{(0.003)(6.56 \text{ in} - 6.5 \text{ in})}{6.56 \text{ in}} \\
&= 0.00002 \quad [\approx 0]
\end{aligned}$$

$$\frac{\epsilon_2}{c - d_2} = \frac{\epsilon_c}{c}$$

$$\begin{aligned}
\epsilon_2 &= \frac{\epsilon_c(c - d_2)}{c} \\
&= \frac{(0.003)(6.56 \text{ in} - 13.5 \text{ in})}{6.56 \text{ in}} \\
&= -0.0032 \quad [\text{tension}]
\end{aligned}$$

For this case, the strain (stress) in steel layer 3 is practically zero and can be ignored. The compression strain in layer 4 is less than the yield strain, so

$$\begin{aligned}
f_{s4} = \epsilon_{s4} E_s &= (0.00186)(29{,}000{,}000 \text{ psi}) \\
&= 53{,}900 \text{ psi}
\end{aligned}$$

The tensile strain in the lower layers is above yield, so $T_2 = T_1 = 120{,}000$ lbf. The corresponding point on the interaction diagram is

$$\begin{aligned}
C_{s4} &= (f_{s4} - 0.85 f'_c) A_{s4} \\
&= \left(53{,}900 \, \frac{\text{lbf}}{\text{in}^2} - (0.85)\left(5000 \, \frac{\text{lbf}}{\text{in}^2}\right)\right)(2.00 \text{ in}^2) \\
&= 99{,}300 \text{ lbf}
\end{aligned}$$

$$a = \beta_1 c = (0.8)(6.56 \text{ in}) = 5.25 \text{ in}$$

$$\begin{aligned}
C_{c1} &= 0.85 f'_c b_1 a_1 \\
&= (0.85)\left(5000 \, \frac{\text{lbf}}{\text{in}^2}\right)(16 \text{ in})(4.0 \text{ in}) \\
&= 272{,}000 \text{ lbf}
\end{aligned}$$

$$\begin{aligned}
C_{c2} &= 0.85 f'_c b_2 a_2 \\
&= (0.85)\left(5000 \, \frac{\text{lbf}}{\text{in}^2}\right)(24 \text{ in})(1.25 \text{ in}) \\
&= 127{,}500 \text{ lbf}
\end{aligned}$$

$$\begin{aligned}
P &= C_{c1} + C_{c2} + C_{s4} - T_2 - T_1 \\
&= 272{,}000 \text{ lbf} + 127{,}500 \text{ lbf} + 99{,}300 \text{ lbf} \\
&\quad - 120{,}000 \text{ lbf} - 120{,}000 \text{ lbf} \\
&= 258{,}000 \text{ lbf} \quad (258 \text{ kips})
\end{aligned}$$

$$\begin{aligned}
M &= C_{c1}\left(\frac{h}{2} - \frac{a_1}{2}\right) + C_{c2}\left(\frac{h}{2} - a_1 - \frac{a_2}{2}\right) + C_{s4}\left(\frac{h}{2} - d_4\right) \\
&\quad + T_2\left(d_2 - \frac{h}{2}\right) + T_1\left(d_1 - \frac{h}{2}\right) \\
&= \frac{(272{,}000 \text{ lbf})\left(\frac{20 \text{ in}}{2} - \frac{4 \text{ in}}{2}\right) + (127{,}500 \text{ lbf})}{} \\
&\quad \times \left(\frac{20 \text{ in}}{2} - 4 \text{ in} - \frac{1.25 \text{ in}}{2}\right) + (99{,}300 \text{ lbf}) \\
&\quad \times \left(\frac{20 \text{ in}}{2} - 2.5 \text{ in}\right) + (120{,}000 \text{ lbf}) \\
&\quad \times \left(13.5 \text{ in} - \frac{20 \text{ in}}{2}\right) + (120{,}000 \text{ lbf}) \\
&= \frac{\quad\quad \times \left(17.5 \text{ in} - \frac{20 \text{ in}}{2}\right)}{\left(12 \, \frac{\text{in}}{\text{ft}}\right)\left(1000 \, \frac{\text{lbf}}{\text{kip}}\right)} \\
&= 411 \text{ ft-kips}
\end{aligned}$$

For design values, apply $\phi = 0.65$ for compression failure (that is, above balanced strain) and $\phi = 0.9$ for flexure. ϕ varies linearly from 0.65 to 0.9 as strain in extreme tension steel varies from 0.002 to 0.005. A conservative, lower-bound design interaction curve for this column is shown.

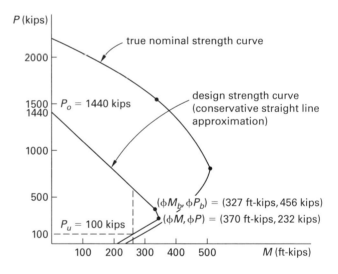

7.2. Compute the maximum value of M_{ux} that can be applied simultaneously with an applied P_u equal to 100 kips. From the interaction diagram for $\phi P_n = 100$ kips, ϕM_n is approximately equal to 250 ft-kips.

Check slenderness effects using the approximate method of ACI 318 Sec. 10.10. Kl_u is given as 18 ft (216 in) and β_{dns} is 0.6. A_g is 416 in^2 as calculated in the first part of this solution. C_{mx} equals C_m, which is given as 1.0, so the column is part of a non-sway frame (that is, ACI 318 Sec. 10.10.6 applies). From ACI 318 Sec. 10.10.1,

$$I_g = \frac{bh_1^3 + 2bh_2^3}{12}$$

$$= \frac{(16 \text{ in})(20 \text{ in})^3 + (2)(4 \text{ in})(12 \text{ in})^3}{12}$$

$$= 11,820 \text{ in}^4$$

$$r = \sqrt{\frac{I_g}{A_g}}$$

$$= \sqrt{\frac{11,820 \text{ in}^4}{416 \text{ in}^2}}$$

$$= 5.33 \text{ in}$$

$$\frac{Kl_u}{r} = \frac{216 \text{ in}}{5.33 \text{ in}}$$

$$= 40.5$$

Thus, KL/r exceeds the limit of ACI 318 Eq. 10-7. Slenderness effects must be considered.

$$E_c = 57,000\sqrt{f'_c}$$

$$= 57,000\sqrt{5000 \ \frac{\text{lbf}}{\text{in}^2}}$$

$$= 4,030,000 \text{ psi}$$

$$EI = \frac{0.4E_c I_g}{1 + \beta_{dns}}$$

$$= \frac{(0.4)\left(4,030,000 \ \frac{\text{lbf}}{\text{in}^2}\right)(11,820 \text{ in}^4)}{1 + 0.6}$$

$$= 11.9 \times 10^9 \text{ in}^2\text{-lbf}$$

$$P_c = \frac{\pi^2(EI)}{(Kl_u)^2}$$

$$= \frac{\pi^2(11.9 \times 10^9 \text{ in}^2\text{-lbf})}{(216 \text{ in})^2}$$

$$= 2,515,000 \text{ lbf}$$

$$\delta = \frac{C_m}{1 - \dfrac{P_u}{0.75P_c}}$$

$$= \frac{1.0}{1 - \dfrac{100,000 \text{ lbf}}{(0.75)(2,515,000 \text{ lbf})}}$$

$$= 1.06$$

$$M_{ux} = \frac{\phi M_n}{\delta}$$

$$= \frac{250 \text{ ft-kips}}{1.06}$$

$$= \boxed{236 \text{ ft-kips}}$$

SOLUTION 8

8.1. Basic load combinations for strength design are calculated by ASCE/SEI7 Sec. 2.3.2, with seismic load as defined in ASCE/SEI7 Sec. 12.4.2.3. For gravity loading, load combination 2 gives

$$P_u = 1.2P_D + 1.6P_L$$

$$= (1.2)(150,000 \text{ lbf}) + (1.6)(70,000 \text{ lbf})$$

$$= 292,000 \text{ lbf}$$

$$M_u = 1.2M_D + 1.6M_L$$

$$= (1.2)(100,000 \text{ ft-lbf}) + (1.6)(20,000 \text{ ft-lbf})$$

$$= 152,000 \text{ ft-lbf}$$

For seismic plus gravity loading, two cases must be considered. Load combination 5 gives

$$P_u = 1.2P_D + 0.2S_{DS}P_D + 0.5P_L + P_E$$

$$= (1.2)(150,000 \text{ lbf}) + (0.2)(0.50)(150,000 \text{ lbf})$$

$$\quad + (0.5)(70,000 \text{ lbf}) + 0 \text{ lbf}$$

$$= 230,000 \text{ lbf}$$

$$M_u = 1.2M_D + 0.2S_{DS}M_D + 0.5M_L + M_E$$

$$= (1.2)(100,000 \text{ ft-lbf}) + (0.2)(0.50)$$

$$\quad \times (100,000 \text{ ft-lbf}) + (0.5)(20,000 \text{ ft-lbf})$$

$$\quad + (450,000 \text{ ft-lbf})$$

$$= 590,000 \text{ ft-lbf}$$

Load combination 7 requires

$$P_u = 0.9P_D - 0.2S_{DS}P_D + P_E$$

$$= (0.9)(150,000 \text{ lbf}) - (0.2)(0.50)$$

$$\quad \times (150,000 \text{ lbf}) + 0 \text{ lbf}$$

$$= 120,000 \text{ lbf}$$

$$M_u = 0.9M_D - 0.2S_{DS}M_D + M_E$$

$$= (0.9)(100,000 \text{ ft-lbf}) - (0.2)(0.50)$$

$$\quad \times (100,000 \text{ ft-lbf}) + 450,000 \text{ ft-lbf}$$

$$= 530,000 \text{ ft-lbf}$$

Thus, there are three loading combinations that might be critical on the given column.

$$(M_u, P_u) = \begin{cases} (152 \text{ ft-kips}, 292 \text{ kips}) \\ (590 \text{ ft-kips}, 230 \text{ kips}) \\ (530 \text{ ft-kips}, 120 \text{ kips}) \end{cases}$$

Plot the interaction diagram for the column. The following are given.

$$A_s = A'_s = 4A_b = (4)(1.56 \text{ in}^2) = 6.24 \text{ in}^2$$

$$A_{st} = 2A_s = (2)(6.24 \text{ in}^2) = 12.48 \text{ in}^2$$

$$A_g = bh = (14 \text{ in})(24 \text{ in}) = 336 \text{ in}^2$$

$$f'_c = 4000 \text{ psi}$$

$$\beta_1 = 0.85$$

$$f_y = 50,000 \text{ psi}$$

For the concentric load case ($M = 0$),

$$P_0 = 0.85 f'_c (A_g - A_{st}) + f_y A_{st}$$

$$= (0.85)\left(4000 \ \frac{\text{lbf}}{\text{in}^2}\right)(336 \text{ in}^2 - 12.48 \text{ in}^2)$$

$$+ \left(50,000 \ \frac{\text{lbf}}{\text{in}^2}\right)(12.48 \text{ in}^2)$$

$$= 1,724,000 \text{ lbf}$$

For balanced strain conditions,

$$\epsilon_y = \frac{f_y}{E_s} = \frac{50,000 \ \dfrac{\text{lbf}}{\text{in}^2}}{29,000,000 \ \dfrac{\text{lbf}}{\text{in}^2}} = 0.0017$$

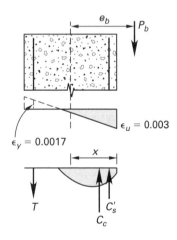

By similar triangles,

$$\frac{\epsilon_y + \epsilon_c}{d} = \frac{\epsilon_c}{c}$$

$$c = \frac{\epsilon_c d}{\epsilon_y + \epsilon_c}$$

$$= \frac{(0.003)(21.5 \text{ in})}{0.0017 + 0.003}$$

$$= 13.7 \text{ in}$$

$$\frac{\epsilon'_s}{c - d'} = \frac{\epsilon_c}{c}$$

$$\epsilon'_s = \frac{(c - d')\epsilon_c}{c}$$

$$= \frac{(13.7 \text{ in} - 2.5 \text{ in})(0.003)}{13.7 \text{ in}}$$

$$= 0.0025 \quad [\epsilon_y > \epsilon'_s, \text{ so compression steel yields}]$$

$$f'_s = f_y = 50,000 \text{ psi}$$

$$T = A_s f_y = (6.24 \text{ in}^2)\left(50,000 \ \frac{\text{lbf}}{\text{in}^2}\right) = 312,000 \text{ lbf}$$

$$C'_s = (f'_s - 0.85 f'_c)A'_s$$

$$= \left(50,000 \ \frac{\text{lbf}}{\text{in}^2} - (0.85)\left(4000 \ \frac{\text{lbf}}{\text{in}^2}\right)\right)(6.24 \text{ in}^2)$$

$$= 291,000 \text{ lbf}$$

$$a = \beta_1 c = (0.85)(13.7 \text{ in}) = 11.65 \text{ in}$$

$$C_c = 0.85 f'_c ba = (0.85)\left(4000 \ \frac{\text{lbf}}{\text{in}^2}\right)(14 \text{ in})(11.65 \text{ in})$$

$$= 555,000 \text{ lbf}$$

$$P_b = C_c + C'_s - T$$

$$= 555,000 \text{ lbf} + 291,000 \text{ lbf} - 312,000 \text{ lbf}$$

$$= 534,000 \text{ lbf}$$

$$M_b = C_c \left(\frac{h}{2} - \frac{a}{2}\right) + C'_s \left(\frac{h}{2} - d'\right) + T\left(d - \frac{h}{2}\right)$$

$$= \frac{\begin{aligned}&(555,000 \text{ lbf})\left(\dfrac{24 \text{ in}}{2} - \dfrac{11.65 \text{ in}}{2}\right)\\[4pt]&+ (291,000 \text{ lbf})\left(\dfrac{24 \text{ in}}{2} - 2.5 \text{ in}\right)\\[4pt]&+ (312,000 \text{ lbf})\left(21.5 \text{ in} - \dfrac{24 \text{ in}}{2}\right)\end{aligned}}{\left(12 \ \dfrac{\text{in}}{\text{ft}}\right)\left(1000 \ \dfrac{\text{lbf}}{\text{kip}}\right)}$$

$$= 763 \text{ ft-kips}$$

The axial force corresponding to balanced conditions (534,000 lbf) is more than twice the maximum factored axial force applied to the column (294,000 lbf). Thus, for this problem only the tension failure of the interaction curve is needed. Calculate two points on this branch of the curve, the points corresponding to pure bending and to tension-controlled failure. For pure bending ($P = 0$), iterate to find the neutral axis position.

Trial 1: Assume f'_s is zero.

$$C_c = T$$

$$0.85f'_c ba = A_s f_y$$

$$a = \frac{A_s f_y}{0.85f'_c b}$$

$$= \frac{(6.24 \text{ in}^2)\left(50{,}000 \ \frac{\text{lbf}}{\text{in}^2}\right)}{(0.85)\left(4000 \ \frac{\text{lbf}}{\text{in}^2}\right)(14 \text{ in})}$$

$$= 6.55 \text{ in}$$

$$c = \frac{a}{\beta_1}$$

$$= \frac{6.55 \text{ in}}{0.85}$$

$$= 7.71 \text{ in}$$

From similar triangles of the strain diagram,

$$\frac{\epsilon'_s}{c - d'} = \frac{\epsilon_c}{c}$$

$$\epsilon'_s = \frac{(c - d')\epsilon_c}{c}$$

$$= \frac{(7.71 \text{ in} - 2.5 \text{ in})(0.003)}{7.71 \text{ in}}$$

$$= 0.0020$$

$$\epsilon_y = \frac{f_y}{E_s}$$

$$= \frac{50{,}000 \ \dfrac{\text{lbf}}{\text{in}^2}}{29{,}000{,}000 \ \dfrac{\text{lbf}}{\text{in}^2}}$$

$$= 0.0017 \quad [\epsilon_y < \epsilon'_s]$$

The assumption that f'_s is zero leads to a calculated stress in the compression side steel that is above yield. A second trial should assume a value of f'_s that is smaller than the yield stress, because including stress in the compression steel will decrease the neutral axis depth and hence give a lower calculated stress. (f'_s could not be yielding, because in that case the force in the compression steel would equal the tension force, leaving zero force in the concrete.)

Trial 2: Assume $f'_s = 0.5f_y = (0.5)(50{,}000 \text{ psi}) = 25{,}000 \text{ psi}$.

$$C_c = T - C'_s$$

$$0.85f'_c ba = A_s f_y - (f'_s - 0.85f'_c)A'_s$$

$$a = \frac{A_s f_y - (f'_s - 0.85f'_c)A'_s}{0.85f'_c b}$$

$$= \frac{(6.24 \text{ in}^2)\left(50{,}000 \ \dfrac{\text{lbf}}{\text{in}^2}\right)}{(0.85)\left(4000 \ \dfrac{\text{lbf}}{\text{in}^2}\right)(14 \text{ in})}$$
$$\frac{- \left(25{,}000 \ \dfrac{\text{lbf}}{\text{in}^2} - (0.85)\left(4000 \ \dfrac{\text{lbf}}{\text{in}^2}\right)\right)(6.24 \text{ in}^2)}{(0.85)\left(4000 \ \dfrac{\text{lbf}}{\text{in}^2}\right)(14 \text{ in})}$$

$$= 3.72 \text{ in}$$

$$c = \frac{a}{\beta_1} = \frac{3.72 \text{ in}}{0.85} = 4.38 \text{ in}$$

From similar triangles,

$$\frac{\epsilon'_s}{c - d'} = \frac{\epsilon_c}{c}$$

$$\epsilon'_s = \frac{(c - d')\epsilon_c}{c}$$

$$= \frac{(4.38 \text{ in} - 2.5 \text{ in})(0.003)}{4.38 \text{ in}}$$

$$= 0.0013 \quad [\epsilon_y > \epsilon'_s]$$

$$f'_s = E_s \epsilon'_s = \left(29{,}000{,}000 \ \frac{\text{lbf}}{\text{in}^2}\right)(0.0013)$$

$$= 37{,}700 \text{ psi}$$

Thus, the stress in the compression steel is between the assumed 25,000 psi and the calculated 37,700 psi.

Trial 3: Assume $f'_s = 30{,}000 \text{ psi}$.

$$a = \frac{A_s f_y - (f'_s - 0.85f'_c)A'_s}{0.85f'_c b}$$

$$= \frac{(6.24 \text{ in}^2)\left(50{,}000 \ \dfrac{\text{lbf}}{\text{in}^2}\right)}{(0.85)\left(4000 \ \dfrac{\text{lbf}}{\text{in}^2}\right)(14 \text{ in})}$$
$$\frac{- \left(30{,}000 \ \dfrac{\text{lbf}}{\text{in}^2} - (0.85)\left(4000 \ \dfrac{\text{lbf}}{\text{in}^2}\right)\right)(6.24 \text{ in}^2)}{(0.85)\left(4000 \ \dfrac{\text{lbf}}{\text{in}^2}\right)(14 \text{ in})}$$

$$= 3.06 \text{ in}$$

$$c = \frac{a}{\beta_1} = \frac{3.06 \text{ in}}{0.85} = 3.60 \text{ in}$$

$$\epsilon'_s = \frac{(c - d')\epsilon_c}{c}$$

$$= \frac{(3.60 \text{ in} - 2.5 \text{ in})(0.003)}{3.60 \text{ in}}$$

$$= 0.00092 \quad [\epsilon_y > \epsilon'_s]$$

$$f'_s = E_s \epsilon'_s = \left(29{,}000{,}000 \ \frac{\text{lbf}}{\text{in}^2}\right)(0.00092)$$

$$= 27{,}000 \text{ psi}$$

The stress calculated at this step is close enough to the assumed value, so take the compression stress as 30,000 psi. Sum moments about the tension steel to determine the nominal moment strength.

$$M_n = C_c\left(d - \frac{a}{2}\right) + C'_s(d - d')$$

$$= 0.85f'_c ba\left(d - \frac{a}{2}\right) + (f'_s - 0.85f'_c)A'_s(d - d')$$

$$(0.85)\left(4000\ \frac{\text{lbf}}{\text{in}^2}\right)(14\text{ in})(3.06\text{ in})$$

$$\times\left(21.5\text{ in} - \frac{3.06\text{ in}}{2}\right) + \left(\begin{array}{c}30,000\ \frac{\text{lbf}}{\text{in}^2} - (0.85)\\ \times\left(4000\ \frac{\text{lbf}}{\text{in}^2}\right)\end{array}\right)$$

$$= \frac{\times(6.24\text{ in}^2)(21.5\text{ in} - 2.5\text{ in})}{\left(12\ \frac{\text{in}}{\text{ft}}\right)\left(1000\ \frac{\text{lbf}}{\text{kip}}\right)}$$

$$= 505\text{ ft-kips}$$

For tension-controlled failure,

$$\frac{\epsilon_s + \epsilon_c}{d} = \frac{\epsilon_c}{c}$$

$$c = \frac{\epsilon_c d}{\epsilon_s + \epsilon_c}$$

$$= \frac{(0.003)(21.5\text{ in})}{0.005 + 0.003}$$

$$= 8.06\text{ in}$$

$$a = \beta_1 c = (0.85)(8.06\text{ in}) = 6.85\text{ in}$$

$$\epsilon'_s = \frac{(c - d')\epsilon_c}{c}$$

$$= \frac{(8.06\text{ in} - 2.5\text{ in})(0.003)}{8.06\text{ in}}$$

$$= 0.00207 > \epsilon_y$$

$$f'_s = f_y = 50,000\text{ psi}$$

$$T = A_s f_y = (6.24\text{ in}^2)\left(50,000\ \frac{\text{lbf}}{\text{in}^2}\right)$$

$$= 312,000\text{ lbf}$$

$$C'_s = (f'_s - 0.85f'_c)A'_s$$

$$= \left(50,000\ \frac{\text{lbf}}{\text{in}^2} - (0.85)\left(4000\ \frac{\text{lbf}}{\text{in}^2}\right)\right)(6.24\text{ in}^2)$$

$$= 290,700\text{ lbf}$$

$$C_c = 0.85f'_c ba$$

$$= (0.85)\left(4000\ \frac{\text{lbf}}{\text{in}^2}\right)(14\text{ in})(6.85\text{ in})$$

$$= 326,000\text{ lbf}$$

$$P = C_c + C'_s - T$$

$$= 320,000\text{ lbf} + 290,700\text{ lbf} - 312,000\text{ lbf}$$

$$= 304,700\text{ lbf}$$

$$M = C_c\left(\frac{h}{2} - \frac{a}{2}\right) + C'_s\left(\frac{h}{2} - d'\right) + T\left(d - \frac{h}{2}\right)$$

$$= \frac{\begin{array}{c}(326,000\text{ lbf})\left(\frac{24\text{ in}}{2} - \frac{6.85\text{ in}}{2}\right)\\ + (290,700\text{ lbf})\left(\frac{24\text{ in}}{2} - 2.5\text{ in}\right)\\ + (312,000\text{ lbf})\left(21.5\text{ in} - \frac{24\text{ in}}{2}\right)\end{array}}{\left(12\ \frac{\text{in}}{\text{ft}}\right)\left(1000\ \frac{\text{lbf}}{\text{kip}}\right)}$$

$$= 710\text{ ft-kips}$$

For design values, apply $\phi = 0.65$ for compression failure and $\phi = 0.9$ for tension-controlled failure. ϕ varies linearly from 0.65 to 0.9 between the limits for tension steel strains that vary from 0.002 to 0.005 ($\phi = 0.65 + (83.3)(\epsilon_s - 0.002)$). The interaction diagram for the column is

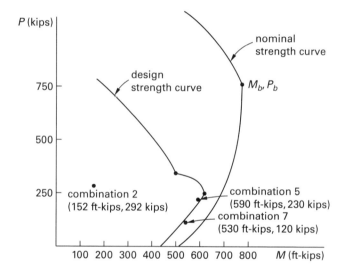

The column is governed by the seismic loading combination corresponding to $P_u = 0.9P_d - 0.2S_{DS}P_d$, and the design is inadequate in that the point corresponding to this set of axial force and moment falls outside the design interaction curve.

8.2. The effect of vertical acceleration downward is generally to increase the likelihood of failure by reducing the design axial force on the column for the short duration loading. This is because the critical loading point is near the tension failure side of the balanced point on the interaction diagram. In this region, axial compression is beneficial in increasing the column's resisting moment strength (in a loose sense, the axial compression "prestresses" the column in this region). Conversely, upward acceleration would increase the moment resistance and be beneficial.

SOLUTION 9

9.1. Analyze the wall and draw free-body diagrams of the resisting elements. The lateral forces are reversible; therefore, all elements act in both compression and tension. The wall panels are significantly stiffer in compression than in tension, so the lateral force is transferred from one level to the level below primarily through compression struts with tension ties.

From the roof to level 4,

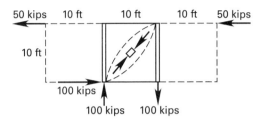

From level 4 to level 3,

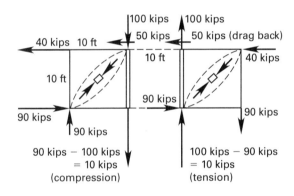

From level 3 to level 2,

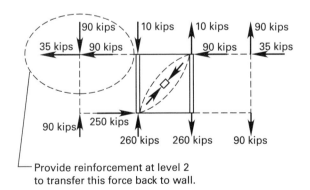

Provide reinforcement at level 2 to transfer this force back to wall.

From level 2 to level 1,

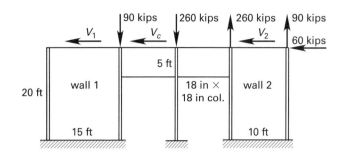

Distribute the base shear, 310 kips, to the wall and column elements from the ground to level 2 on the basis of their translational stiffness. Wall 1 is cantilevered from a fixed base and has a height, H, of 20 ft; include both shear and flexural deformations, and take $G = 0.4E_c$.

For wall 1,

$$A_1 = (l_{w1} + h_c)t_{w1}$$
$$= (15 \text{ ft} + 1.5 \text{ ft})(1 \text{ ft})$$
$$= 16.5 \text{ ft}^2$$

$$I_1 = \frac{(l_{w1} + h_c)^3 t_{w1}}{12} + 2\left(\frac{h_c^4}{12} + h_c^2(0.5l_{w_1})^2\right)$$
$$= \frac{(15 \text{ ft} + 1.5 \text{ ft})^3(1 \text{ ft})}{12}$$
$$\quad + (2)\left(\frac{(1.5 \text{ ft})^4}{12} + (1.5 \text{ ft})^2(0.5)(15 \text{ ft})^2\right)$$
$$= 459 \text{ ft}^4$$

$$\frac{K_1 H^3}{3E_c I_1} + \frac{1.2K_1 H}{GA_1} = 1$$

$$K_1 = \frac{1}{\dfrac{H^3}{3E_c I_1} + \dfrac{1.2H}{(0.4E_c)A_1}}$$
$$= \frac{1}{\dfrac{(20 \text{ ft})^3}{3E_c(459 \text{ ft}^4)} + \dfrac{(1.2)(20 \text{ ft})}{0.4E_c(16.5 \text{ ft}^2)}}$$
$$= 0.106E_c$$

For wall 2,

$$A_2 = (l_{w2} + h_c)t_{w1}$$
$$= (10 \text{ ft} + 1.5 \text{ ft})(1 \text{ ft})$$
$$= 11.5 \text{ ft}^2$$

$$I_2 = \frac{(l_{w2} + h_c)^3 t_{w1}}{12} + 2\left(\frac{h_c^4}{12} + h_c^2(0.5l_w)^2\right)$$

$$= \frac{(10 \text{ ft} + 1.5 \text{ ft})^3 (1 \text{ ft})}{12}$$

$$+ (2)\left(\frac{(1.5 \text{ ft})^4}{12} + (1.5 \text{ ft})^2(0.5)(10 \text{ ft})^2\right)$$

$$= 165 \text{ ft}^4$$

$$K_2 = \cfrac{1}{\cfrac{H^3}{3E_c I_2} + \cfrac{1.2H}{(0.4E_c)A_2}}$$

$$= \cfrac{1}{\cfrac{(20 \text{ ft})^3}{3E_c(165 \text{ ft}^4)} + \cfrac{(1.2)(20 \text{ ft})}{0.4E_c(11.5 \text{ ft}^2)}}$$

$$= 0.047 E_c$$

Consider the column to be fixed top and bottom and only flexural deformations to be significant.

$$\frac{K_c H^3}{12 E_c I_c} = 1$$

$$I_c = \frac{bh^3}{12} = \frac{(1.5 \text{ ft})(1.5 \text{ ft})^3}{12} = 0.42 \text{ ft}^4$$

$$K_c = \frac{12E_c(0.42 \text{ ft}^4)}{(15 \text{ ft})^3} = 0.0015 E_c$$

$$\sum K = K_1 + K_c + K_2$$

$$= 0.106 E_c + 0.0015 E_c + 0.047 E_c$$

$$= 0.155 E_c$$

$$V_1 = \frac{K_1 V}{\sum K} = \frac{0.106 E_c (310 \text{ kips})}{0.155 E_c} = 213 \text{ kips}$$

$$V_c = \frac{K_c V}{\sum K} = \frac{0.0015 E_c (310 \text{ kips})}{0.155 E_c} = 3 \text{ kips}$$

$$V_2 = \frac{K_2 V}{\sum K} = \frac{0.047 E_c (310 \text{ kips})}{0.155 E_c} = 94 \text{ kips}$$

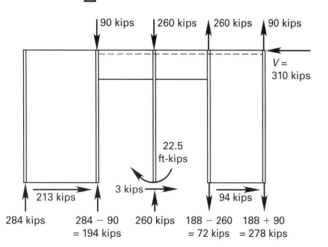

Check the equilibrium of the entire structure.

$$\sum F_x = V - V_1 - V_c - V_2$$

$$= 310 \text{ kips} - 213 \text{ kips} - 3 \text{ kips} - 94 \text{ kips}$$

$$= 0$$

$$\sum F_y = P_1 - P_2 + P_3 - P_4 - P_5$$

$$= 284 \text{ kips} - 194 \text{ kips} + 260 \text{ kips}$$

$$- 72 \text{ kips} - 278 \text{ kips}$$

$$= 0$$

$$\sum M_1 = V_2 h_2 + V_3 h_3 + V_4 h_4 + V_5 h_5$$

$$- P_2 l_2 + P_3 l_3 - M_3 - P_4 l_4 - P_5 l_5$$

$$= (60 \text{ kips})(20 \text{ ft}) + (70 \text{ kips})(30 \text{ ft}) + (80 \text{ kips})$$

$$\times (40 \text{ ft}) + (100 \text{ kips})(50 \text{ ft}) - (194 \text{ kips})$$

$$\times (15 \text{ ft}) + (260 \text{ kips})(25 \text{ ft}) - 22.5 \text{ ft-kips}$$

$$- (72 \text{ kips})(35 \text{ ft}) - (278 \text{ kips})(45 \text{ ft})$$

$$= 38 \text{ ft-kips} \quad [\approx 0 \text{ ft-kips}]$$

The small moment (38 ft-kips) is trivial compared to the total overturning moment, so the equilibrium checks.

9.2. Determine the boundary and trim reinforcement. From the design criteria, neglect gravity loads and assume that the given lateral forces are design level. ACI 318 Sec. 21.3.5.6, requires that columns supporting discontinuous stiff walls must be provided with transverse reinforcement throughout their full lengths below the wall. Check the compressive strength of the 18 in × 18 in columns using the minimum percentage of longitudinal steel.

$$A_{st} = \rho_{\min} bh = (0.01)(18 \text{ in})(18 \text{ in}) = 3.24 \text{ in}^2$$

Four no. 8 bars provide

$$A_{st} = 4A_b = (4)(0.79 \text{ in}^2) = 3.16 \text{ in}^2$$

This is close enough. The design compressive strength is

$$\phi P_{n,\max} = 0.8\phi\left(0.85 f_c'(A_g - A_{st}) + f_y A_{st}\right)$$

$$= (0.8)(0.65)\left(\begin{array}{c} (0.85)\left(4000 \ \dfrac{\text{lbf}}{\text{in}^2}\right) \\ \times (324 \text{ in}^2 - 3.16 \text{ in}^2) \\ + \left(60{,}000 \ \dfrac{\text{lbf}}{\text{in}^2}\right)(3.16 \text{ in}^2) \end{array} \right)$$

$$= 666{,}000 \text{ lbf}$$

The columns have adequate compressive strength for the design loading. The lateral force is reversible, so each vertical boundary element must resist the same magnitude force acting in tension. Calculate the area of longitudinal steel to resist tension force of $P_u = 284$ kips.

$$\phi T_n = \phi f_y A_{st} = P_u$$

$$A_{st} = \frac{P_u}{\phi f_y} = \frac{284{,}000 \text{ lbf}}{(0.9)\left(60{,}000 \, \frac{\text{lbf}}{\text{in}^2}\right)}$$

$$= 5.26 \text{ in}^2 \quad [\text{controls}]$$

Provide six no. 9 longitudinal bars in the outer columns. Calculate similarly for other members.

For $P_u = 278$ kips,

$$A_{st} = \frac{P_u}{\phi f_y} = \frac{278{,}000 \text{ lbf}}{(0.9)\left(60{,}000 \, \frac{\text{lbf}}{\text{in}^2}\right)}$$

$$= 5.15 \text{ in}^2 \quad [\text{say six no. 9}]$$

For $P_u = 260$ kips,

$$A_{st} = \frac{P_u}{\phi f_y} = \frac{260{,}000 \text{ lbf}}{(0.9)\left(60{,}000 \, \frac{\text{lbf}}{\text{in}^2}\right)}$$

$$= 4.81 \text{ in}^2 \quad [\text{say six no. 8}]$$

For $P_u = 194$ kips,

$$A_{st} = \frac{P_u}{\phi f_y} = \frac{194{,}000 \text{ lbf}}{(0.9)\left(60{,}000 \, \frac{\text{lbf}}{\text{in}^2}\right)}$$

$$= 3.59 \text{ in}^2 \quad [\text{say, six no. 7}]$$

For P_u less than 150 kips, the minimum longitudinal steel controls; use four no. 8 bars.

Calculate for horizontal boundary elements. For $P_u = 250$ kips at level 2,

$$A_{st,2} = \frac{P_u}{\phi f_y} = \frac{250{,}000 \text{ lbf}}{(0.9)\left(60{,}000 \, \frac{\text{lbf}}{\text{in}^2}\right)}$$

$$= 4.62 \text{ in}^2 \quad [\text{say four no. 9}]$$

For $P_u = 90$ kips $+ 35$ kips $= 125$ kips at level 3,

$$A_{st,3} = \frac{P_u}{\phi f_y} = \frac{125{,}000 \text{ lbf}}{(0.9)\left(60{,}000 \, \frac{\text{lbf}}{\text{in}^2}\right)}$$

$$= 2.31 \text{ in}^2 \quad [\text{say four no. 7}]$$

For $P_u = 50$ kips at level 4,

$$A_{st,4} = \frac{P_u}{\phi f_y} = \frac{50{,}000 \text{ lbf}}{(0.9)\left(60{,}000 \, \frac{\text{lbf}}{\text{in}^2}\right)}$$

$$= 0.93 \text{ in}^2 \quad [\text{say two no. 7}]$$

Provide transverse confinement steel (ACI 318 Sec. 21.5.3) in the form of closed hoops and crossties for the full height of the center column from first level to second level, and in the vertical boundary elements (ACI 318 Sec. 21.9.6). Provide transverse confinement steel in horizontal boundary elements wherever an opening exists either above or below the element.

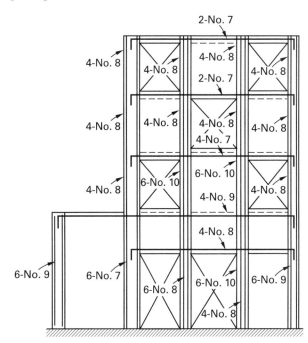

SOLUTION 10

Assume that the given value of $0.75f_{pu}$ is the temporary jacking stress and that the specified loss of 10% in prestress only pertains to strand anchorage, relaxation, and elastic shortening at release. This solution will assume an additional 12% loss to account for creep, shrinkage, and additional strand relaxation before the member is placed in service. Thus, for $f_{pu} = 270{,}000$ psi,

$$f_{pe} = 0.75 f_{pu}(1 - 0.22)$$

$$= (0.75)\left(270{,}000 \, \frac{\text{lbf}}{\text{in}^2}\right)(0.78)$$

$$= 158{,}000 \text{ psi}$$

$$A_{ps} = n_{\text{strands}} A_b = (6)(0.115 \text{ in}^2) = 0.69 \text{ in}^2$$

$$P = f_{pe} A_{ps} = \left(158{,}000 \, \frac{\text{lbf}}{\text{in}^2}\right)(0.69 \text{ in}^2) = 109{,}000 \text{ lbf}$$

$$f_{pc} = \frac{P}{A_c} = \frac{109{,}000 \text{ lbf}}{144 \text{ in}^2} = 757 \text{ psi}$$

10.1. Compute the allowable service load under axial compression (see Sec. 5.9.7 of the *PCI Design Handbook*). Loading is assumed to be due to sustained load (ACI 318 Sec. 18.4.2).

$$f = f_{pc} + \frac{P}{A_g} \le 0.45 f'_c$$

$$P \le A_g(0.45 f'_c - f_{pc})$$

$$= (144 \text{ in}^2)\left((0.45)\left(5000 \frac{\text{lbf}}{\text{in}^2}\right) - 757 \frac{\text{lbf}}{\text{in}^2}\right)$$

$$= \boxed{215{,}000 \text{ lbf}}$$

This value is the structural capacity. In a real situation, the capacity based on skin friction and/or end bearing would also need to be considered, but in this problem the information needed for this calculation isn't given.

10.2. Calculate the moment strength about the x-axis corresponding to an axial force of $P_u = 150$ kips.

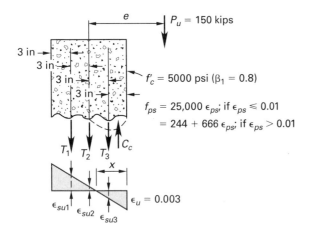

Use trial and error to establish the position of the neutral axis so as to satisfy strain compatibility and equilibrium. First, assume that the depth of the neutral axis is 6 in from the compression edge.

$$\epsilon_{ps1} = \epsilon_{pe} + \epsilon_{ce} + \epsilon_{su1}$$

$$\epsilon_{pe} = \frac{f_{pe}}{E_{ps}} = \frac{158{,}000 \frac{\text{lbf}}{\text{in}^2}}{27{,}000{,}000 \frac{\text{lbf}}{\text{in}^2}} = 0.00585$$

$$E_c = 57{,}000\sqrt{f'_c} = 57{,}000\sqrt{5000 \frac{\text{lbf}}{\text{in}^2}} = 4{,}030{,}000 \text{ psi}$$

$$\epsilon_{ce} = \frac{f_{pc}}{E_c} = \frac{757 \frac{\text{lbf}}{\text{in}^2}}{4{,}030{,}000 \frac{\text{lbf}}{\text{in}^2}} = 0.00019$$

By similar triangles,

$$\frac{\epsilon_{su1}}{d_1 - c} = \frac{\epsilon_c}{c}$$

$$\epsilon_{su1} = \frac{\epsilon_c(d_1 - c)}{c}$$

$$= \frac{(0.003)(9 \text{ in} - 6 \text{ in})}{6 \text{ in}}$$

$$= 0.0015$$

$$\epsilon_{ps1} = \epsilon_{pe} + \epsilon_{ce} + \epsilon_{su1}$$

$$= 0.00585 + 0.00019 + 0.0015$$

$$= 0.00754$$

Similarly,

$$\epsilon_{su2} = \frac{\epsilon_c(d_2 - c)}{c}$$

$$= \frac{(0.003)(6 \text{ in} - 6 \text{ in})}{6 \text{ in}}$$

$$= 0$$

$$\epsilon_{ps2} = \epsilon_{pe} + \epsilon_{ce} + \epsilon_{su1}$$

$$= 0.00585 + 0.00019 + 0$$

$$= 0.00604$$

$$\epsilon_{su3} = \frac{\epsilon_c(d_3 - c)}{c}$$

$$= \frac{(0.003)(3 \text{ in} - 6 \text{ in})}{6 \text{ in}}$$

$$= -0.0015$$

$$\epsilon_{ps3} = \epsilon_{pe} + \epsilon_{ce} + \epsilon_{su3}$$

$$= 0.00585 + 0.00019 - 0.0015$$

$$= 0.00454$$

The strain separating the two regions of the bilinear stress-strain curve is found from the figure.

$$25{,}000\epsilon_s = 244 + 666\epsilon_s$$

$$\epsilon_s = 0.0100$$

The calculated strain for the assumed 6 in depth of neutral axis is less than ϵ_s, so

$$f_{ps1} = E_{ps}\epsilon_{ps1} = \left(27{,}000{,}000 \frac{\text{lbf}}{\text{in}^2}\right)(0.00754)$$

$$= 203{,}600 \text{ psi}$$

$$f_{ps2} = E_{ps}\epsilon_{ps2} = \left(27{,}000{,}000 \frac{\text{lbf}}{\text{in}^2}\right)(0.00604)$$

$$= 163{,}000 \text{ psi}$$

$$f_{ps2} = E_{ps}\epsilon_{ps2} = \left(27{,}000{,}000 \frac{\text{lbf}}{\text{in}^2}\right)(0.00454)$$

$$= 122{,}000 \text{ psi}$$

Neglecting the chamfers,

$$\sum F_{\text{long}} = P_u + f_{ps1}(2A_b) + f_{ps2}(2A_b)$$
$$+ f_{ps3}(2A_b) - 0.85f'_c ba = 0$$

$$a = \frac{P_u + (f_{ps1} + f_{ps2} + f_{ps3})2A_b}{0.85f'_c b}$$

$$= \frac{150,000 \text{ lbf} + \left(\begin{array}{c} 203,600 \ \frac{\text{lbf}}{\text{in}^2} + 163,000 \ \frac{\text{lbf}}{\text{in}^2} \\ + 122,000 \ \frac{\text{lbf}}{\text{in}^2} \end{array}\right)}{}$$

$$= \frac{\times (2)(0.115 \text{ in}^2)}{(0.85)\left(5000 \ \frac{\text{lbf}}{\text{in}^2}\right)(12 \text{ in})}$$

$$= 5.14 \text{ in}$$

$$c = \frac{a}{\beta_1}$$

$$= \frac{5.14 \text{ in}}{0.8}$$

$$= 6.43 \text{ in}$$

The calculated depth of the neutral axis is 10% greater than the trial value; compute it again using a trial value between 6 in and 6.43 in; say 6.3 in.

$$\frac{\epsilon_{su1}}{d_1 - c} = \frac{\epsilon_c}{c}$$

$$\epsilon_{su1} = \frac{\epsilon_c(d_1 - c)}{c}$$

$$= \frac{(0.003)(9 \text{ in} - 6.3 \text{ in})}{6.3 \text{ in}}$$

$$= 0.00129$$

$$\epsilon_{ps1} = \epsilon_{pe} + \epsilon_{ce} + \epsilon_{su1}$$

$$= 0.00585 + 0.00019 + 0.00129$$

$$= 0.00733$$

Similarly,

$$\epsilon_{su2} = \frac{\epsilon_c(d_2 - c)}{c}$$

$$= \frac{(0.003)(6 \text{ in} - 6.3 \text{ in})}{6.3 \text{ in}}$$

$$= -0.00014$$

$$\epsilon_{ps2} = \epsilon_{pe} + \epsilon_{ce} + \epsilon_{su2}$$

$$= 0.00585 + 0.00019 - 0.00014$$

$$= 0.00590$$

$$\epsilon_{su3} = \frac{\epsilon_c(d_3 - c)}{c}$$

$$= \frac{(0.003)(3 \text{ in} - 6.3 \text{ in})}{6.3 \text{ in}}$$

$$= -0.00157$$

$$\epsilon_{ps3} = \epsilon_{pe} + \epsilon_{ce} + \epsilon_{su3}$$

$$= 0.00585 + 0.00019 - 0.00157$$

$$= 0.00447$$

$$f_{ps1} = E_{ps}\epsilon_{ps1} = \left(27,000,000 \ \frac{\text{lbf}}{\text{in}^2}\right)(0.00733)$$

$$= 198,000 \text{ psi}$$

$$f_{ps2} = E_{ps}\epsilon_{ps2} = \left(27,000,000 \ \frac{\text{lbf}}{\text{in}^2}\right)(0.00590)$$

$$= 159,300 \text{ psi}$$

$$f_{ps3} = E_{ps}\epsilon_{ps3} = \left(27,000,000 \ \frac{\text{lbf}}{\text{in}^2}\right)(0.00447)$$

$$= 120,700 \text{ psi}$$

$$\sum F_{\text{long}} = P_u + f_{ps1}2A_b + f_{ps2}2A_b$$
$$+ f_{ps3}2A_b - 0.85f'_c ba = 0$$

$$a = \frac{P_u + (f_{ps1} + f_{ps2} + f_{ps3})2A_b}{0.85f'_c b}$$

$$= \frac{150,000 \text{ lbf} + \left(\begin{array}{c} 198,000 \ \frac{\text{lbf}}{\text{in}^2} + 159,300 \ \frac{\text{lbf}}{\text{in}^2} \\ + 120,700 \ \frac{\text{lbf}}{\text{in}^2} \end{array}\right)}{}$$

$$= \frac{\times (2)(0.115 \text{ in}^2)}{(0.85)\left(5000 \ \frac{\text{lbf}}{\text{in}^2}\right)(12 \text{ in})}$$

$$= 5.10 \text{ in}$$

$$c = \frac{a}{\beta_1}$$

$$= \frac{5.10 \text{ in}}{0.8}$$

$$= 6.4 \text{ in} \quad \text{[close enough]}$$

The calculated depth of neutral axis is close enough to the trial value of 6.3 in; therefore, the process converges. Base the moment strength on a depth of 6.3 in.

$$M_n = 0.85 f'_c ba \left(\frac{h}{2} - \frac{a}{2} \right) + f_{ps1} 2 A_b \left(d_1 - \frac{h}{2} \right)$$

$$+ f_{ps2} 2 A_b \left(d_2 - \frac{h}{2} \right) - f_{ps3} 2 A_b \left(d_3 - \frac{h}{2} \right)$$

$$= \frac{\begin{array}{c} (0.85) \left(5000 \; \frac{\text{lbf}}{\text{in}^2} \right) (12 \; \text{in}) (5.10 \; \text{in}) \\[2mm] \times \left(\frac{12 \; \text{in}}{2} - \frac{5.10 \; \text{in}}{2} \right) + \left(198{,}000 \; \frac{\text{lbf}}{\text{in}^2} \right) (2) \\[2mm] \times (0.115 \; \text{in}^2) \left(9 \; \text{in} - \frac{12 \; \text{in}}{2} \right) + \left(166{,}800 \; \frac{\text{lbf}}{\text{in}^2} \right) \\[2mm] \times (2)(0.115 \; \text{in}^2) \left(6 \; \text{in} - \frac{12 \; \text{in}}{2} \right) + \left(120{,}700 \; \frac{\text{lbf}}{\text{in}^2} \right) \\[2mm] \times (2)(0.115 \; \text{in}^2) \left(3 \; \text{in} - \frac{12 \; \text{in}}{2} \right) \end{array}}{\left(12 \; \frac{\text{in}}{\text{ft}} \right) \left(1000 \; \frac{\text{lbf}}{\text{kip}} \right)}$$

$$= 79.2 \; \text{ft-kips}$$

$$\phi M_n = (0.65)(79.2 \; \text{ft-kips})$$

$$= \boxed{51.5 \; \text{ft-kips}}$$

SOLUTION 11

11.1. Design on a per-linear-foot-width basis. Use the strength design method in ACI 318 with $f'_c = 3000$ psi and $f_y = 40{,}000$ psi. Analyze the live load pattern that produces critical negative bending moment at joint A. Live load reduction does not apply for loads in excess of 100 lbf/ft^2 (ASCE/SEI7 Sec. 4.7.3). The moment of inertia for each member is

$$I_{\text{slab}} = \frac{b h_{\text{slab}}^3}{12} = \frac{(12 \; \text{in})(9 \; \text{in})^3}{12} = 729 \; \text{in}^4$$

$$I_{\text{wall,1}} = \frac{b h_{\text{wall,1}}^3}{12} = \frac{(12 \; \text{in})(10 \; \text{in})^3}{12} = 1000 \; \text{in}^4$$

$$I_{\text{wall,2}} = \frac{b h_2^3}{12} = \frac{(12 \; \text{in})(12 \; \text{in})^3}{12} = 1728 \; \text{in}^4$$

The relative stiffness of each member is

$$K_1 = \frac{4 E I_{\text{wall,1}}}{L_1} = \frac{4 E (1000 \; \text{in}^4)}{10 \; \text{ft}}$$

$$= \left(400 \; \frac{\text{in}^4}{\text{ft}} \right) E$$

$$K_2 = \frac{4 E I_{\text{wall,2}}}{L_2} = \frac{4 E (1728 \; \text{in}^4)}{14 \; \text{ft}}$$

$$= \left(493 \; \frac{\text{in}^4}{\text{ft}} \right) E$$

$$K_3 = \frac{4 E I_{\text{slab}}}{L_3} = \frac{4 E (729 \; \text{in}^4)}{20.8 \; \text{ft}}$$

$$= \left(140 \; \frac{\text{in}^4}{\text{ft}} \right) E$$

$$K_4 = \frac{4 E I_{\text{slab}}}{L_4} = \frac{4 E (729 \; \text{in}^4)}{20 \; \text{ft}}$$

$$= \left(146 \; \frac{\text{in}^4}{\text{ft}} \right) E$$

The ratio of the relative stiffness of each member compared to that of the upper wall is

$$\psi_1 = \frac{K_1}{K_1} = \frac{\left(400 \; \frac{\text{in}^4}{\text{ft}} \right) E}{\left(400 \; \frac{\text{in}^4}{\text{ft}} \right) E} = 1.00$$

$$\psi_2 = \frac{K_2}{K_1} = \frac{\left(493 \; \frac{\text{in}^4}{\text{ft}} \right) E}{\left(400 \; \frac{\text{in}^4}{\text{ft}} \right) E} = 1.23$$

$$\psi_3 = \frac{K_3}{K_1} = \frac{\left(140 \; \frac{\text{in}^4}{\text{ft}} \right) E}{\left(400 \; \frac{\text{in}^4}{\text{ft}} \right) E} = 0.350$$

$$\psi_4 = \frac{K_4}{K_1} = \frac{\left(146 \; \frac{\text{in}^4}{\text{ft}} \right) E}{\left(400 \; \frac{\text{in}^4}{\text{ft}} \right) E} = 0.365$$

The factored load is

$$w_u = 1.2 w_D + 1.6 w_L$$

$$= (1.2) \left(120 \; \frac{\text{lbf}}{\text{ft}} \right) + (1.6) \left(175 \; \frac{\text{lbf}}{\text{ft}} \right)$$

$$= 424 \; \text{lbf/ft}$$

For dead load only, this is

$$w_u = 1.2 w_D = (1.2) \left(120 \; \frac{\text{lbf}}{\text{ft}} \right) = 144 \; \text{lbf/ft}$$

So, the second floor slab loading for joint A is

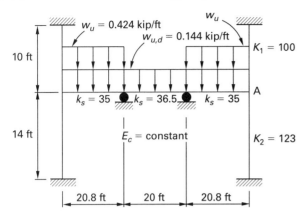

11.2. The moment distribution is given in the following table. (See *Table for Sol. 11.2*.) Moments are given in foot-pounds; moments distributed to walls are omitted.

From this table, the moments may be shown as

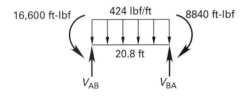

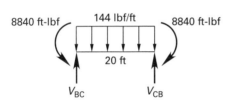

Find the shear forces.

$$\sum M_A = V_{BA} L_{AB} - \frac{w_u L_{AB}^2}{2} + M_{AB} - M_{BA} = 0$$

$$V_{BA} = \frac{\dfrac{w_u L_{AB}^2}{2} - M_{AB} + M_{BA}}{L_{AB}}$$

$$= \frac{\dfrac{\left(424\ \dfrac{\text{lbf}}{\text{ft}}\right)(20.8\ \text{ft})^2}{2} - 16{,}600\ \text{ft-lbf} + 8840\ \text{ft-lbf}}{20.8\ \text{ft}}$$

$$= 4037\ \text{lbf}$$

$$\sum F_y = V_{AB} - w_u L_{AB} + V_{BA} = 0$$
$$V_{AB} = w_u L_{AB} - V_{BA}$$
$$= \left(424\ \frac{\text{lbf}}{\text{ft}}\right)(20.8\ \text{ft}) - 4037\ \text{lbf}$$
$$= 4782\ \text{lbf}$$

By symmetry,

$$V_{BC} = V_{CB} = \frac{w_u L_{BC}}{2} = \frac{\left(144\ \dfrac{\text{lbf}}{\text{ft}}\right)(20\ \text{ft})}{2}$$
$$= 1440\ \text{lbf}$$

Table for Sol. 11.2

walls	A	B	B	C	C	D	walls	
0.39	0.13	0.49	0.51	0.51	0.49	0.13	0.39	DF
0.48	−15,300	15,300	−4800	4800	−15,300	15,300	0.48	FEM
	1990	995						Bal.A,CO
	−2815	−5630	−5865	−2933				Bal.B,CO
		3425	6850	6583	3292			Bal.C,CO
				−1210	−2420			Bal.D,CO
	366	183						Bal.A,CO
	−884	−1768	−1840	−920				Bal.B,CO
		543	1086	1044	522			Bal.C,CO
				−34	−68			Bal.D,CO
	115	58						Bal.A,CO
	−147	−294	−307	−153				Bal.B,CO
		27	78	75	38			Bal.C,CO
					−5			Bal.D,noCO
	19	9						Bal.A,CO
		−18	−18					Bal.B,noCO
$\sum =$	−16,600	8840	−8840	8840	−8840	16,600		**final M**

Therefore, the moment and shear diagram is

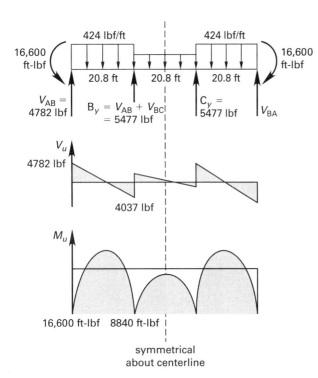

11.3. Check the flexural strength of the slab at the wall.

$$A_s = A_b \left(\frac{12 \text{ in}}{s}\right) = (0.60 \text{ in}^2)\left(\frac{12 \text{ in}}{10 \text{ in}}\right) = 0.72 \text{ in}^2$$

$$a = \frac{A_s f_y}{0.85 f'_c b} = \frac{(0.72 \text{ in}^2)\left(40{,}000 \dfrac{\text{lbf}}{\text{in}^2}\right)}{(0.85)\left(3000 \dfrac{\text{lbf}}{\text{in}^2}\right)(12 \text{ in})} = 0.94 \text{ in}$$

$$d = h - \text{cover} - 0.5 d_b$$
$$= 9 \text{ in} - 1 \text{ in} - (0.5)(0.875 \text{ in})$$
$$= 7.56 \text{ in}$$

$$c = \frac{a}{\beta_1} = \frac{0.94 \text{ in}}{0.85} = 1.11 \text{ in} \quad [c < 0.375d, \text{ so } \phi = 0.9]$$

$$\phi M_n = \phi A_s f_y \left(d - \frac{a}{2}\right)$$
$$= \frac{(0.9)(0.72 \text{ in}^2)\left(40{,}000 \dfrac{\text{lbf}}{\text{in}^2}\right)\left(7.56 - \dfrac{0.94 \text{ in}}{2}\right)}{\left(12 \dfrac{\text{in}}{\text{ft}}\right)\left(1000 \dfrac{\text{lbf}}{\text{kip}}\right)}$$
$$= 15.3 \text{ ft-kips} \quad [\text{per foot of slab width}]$$

The design flexural strength is less than the calculated moment at the centerline of wall. However, the moment is reducible to the face of wall, and ACI 318 Sec. 8.4 permits a redistribution of the negative moment if the strain in the extreme tension steel exceeds 0.0075.

$$\frac{\epsilon_t}{d-c} = \frac{\epsilon_c}{c}$$
$$\epsilon_t = \frac{\epsilon_c(d-c)}{c}$$
$$= \frac{(0.003)(7.56 \text{ in} - 1.11 \text{ in})}{1.11 \text{ in}}$$
$$= 0.017$$

This is greater than 0.0075, so the percentage of redistribution is

$$\%_{\text{redistribution}} \leq \begin{cases} 1000\epsilon_t = (1000)(0.017) \times 100\% \\ \qquad = 17\% \quad [\text{controls}] \\ 20\% \end{cases}$$

The moment at centerline after redistribution is

$$M'_u = \left(1 - \frac{\%_{\text{redistribution}}}{100\%}\right) M_u$$
$$= \left(1 - \frac{17\%}{100\%}\right)(16{,}600 \text{ ft-lbf})$$
$$= 13{,}780 \text{ ft-lbf} \quad [M'_u < \phi M_n]$$

Therefore, $\phi M_n > M_u$ and the slab is OK.

11.4. Check the slab's shear strength.

$$\phi V_c = 2\phi\lambda\sqrt{f'_c}\, b_w d$$
$$= (2)(0.75)(1.0)\sqrt{3000 \dfrac{\text{lbf}}{\text{in}^2}}\,(12 \text{ in})(7.56 \text{ in})$$
$$= 7450 \text{ lbf} \quad [> V_u]$$

$\phi V_c > V_u$ and the slab's shear strength is OK.

Check the moment transfer to the wall.

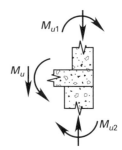

$$M_{u1} = \left(\frac{\psi_1}{\psi_1 + \psi_2}\right) M'_u$$

$$= \left(\frac{1.0}{1.00 + 1.23}\right)(13{,}780 \text{ ft-lbf})$$

$$= 6180 \text{ ft-lbf}$$

Neglecting axial compression in the wall (conservative),

$$A_s = A_b\left(\frac{12 \text{ in}}{s}\right) = (0.20 \text{ in}^2)\left(\frac{12 \text{ in}}{16 \text{ in}}\right) = 0.15 \text{ in}^2$$

$$a = \frac{A_s f_y}{0.85 f'_c b} = \frac{(0.15 \text{ in}^2)\left(40{,}000 \dfrac{\text{lbf}}{\text{in}^2}\right)}{(0.85)\left(3000 \dfrac{\text{lbf}}{\text{in}^2}\right)(12 \text{ in})}$$

$$= 0.20 \text{ in}$$

$$d = h - \text{cover} - 0.5 d_b$$

$$= 10 \text{ in} - 1.5 \text{ in} - (0.5)(0.50 \text{ in})$$

$$= 8.25 \text{ in}$$

$$c = \frac{a}{\beta_1} = \frac{0.20 \text{ in}}{0.85}$$

$$= 0.24 \text{ in} \quad [c < 0.375d, \text{ therefore } \phi = 0.9]$$

$$\phi M_n = \phi A_s f_y\left(d - \frac{a}{2}\right)$$

$$= \frac{(0.9)(0.15 \text{ in}^2)\left(40{,}000 \dfrac{\text{lbf}}{\text{in}^2}\right)\left(8.25 \text{ in} - \dfrac{0.20 \text{ in}}{2}\right)}{12 \dfrac{\text{in}}{\text{ft}}}$$

$$= 3670 \text{ ft-lbf} \quad [\text{per foot of slab width}]$$

> The flexural strength of the wall above the joint is not sufficient to resist its share of the end moment.

Either recompute ϕM_n taking into account the axial force, or increase the flexural steel in proportion to the required additional strength. Try the latter option.

$$M_{u1} = \phi M_n = \phi \rho b d^2 f_y\left(1 - 0.59\rho\left(\frac{f_y}{f'_c}\right)\right)$$

$$(6180 \text{ ft-lbf})\left(12 \frac{\text{in}}{\text{ft}}\right)$$

$$= 0.9\rho(12 \text{ in})(8.25 \text{ in})^2\left(40{,}000 \frac{\text{lbf}}{\text{in}^2}\right)$$

$$\times \phi \rho b d^2 f_y \left(1 - 0.59\rho\left(\frac{40{,}000 \dfrac{\text{lbf}}{\text{in}^2}}{3000 \dfrac{\text{lbf}}{\text{in}^2}}\right)\right)$$

$$\rho = 0.00258$$

$$A_s = \rho b d = (0.00258)(12 \text{ in})(8.25 \text{ in}) = 0.26 \text{ in}^2$$

Use no. 4 bars at 9 in o.c. on the inside face of the wall above the joint.

For the wall below the joint,

$$M_{u2} = \left(\frac{\psi_2}{\psi_1 + \psi_2}\right) M'_u$$

$$= \left(\frac{1.23}{1.00 + 1.23}\right)(13{,}780 \text{ ft-lbf})$$

$$= 7600 \text{ ft-lbf}$$

$$d = h - \text{cover} - 0.5 d_b$$

$$= 12 \text{ in} - 1.5 \text{ in} - (0.5)(0.5 \text{ in})$$

$$= 10.25 \text{ in}$$

$$M_{u2} = \phi M_n = \phi \rho b d^2 f_y\left(1 - 0.59\rho\left(\frac{f_y}{f'_c}\right)\right)$$

$$(7600 \text{ ft-lbf})\left(12 \frac{\text{in}}{\text{ft}}\right)$$

$$= 0.9\rho(12 \text{ in})(10.25 \text{ in})^2\left(40{,}000 \frac{\text{lbf}}{\text{in}^2}\right)$$

$$\times \left(1 - 0.59\rho\left(\frac{40{,}000 \dfrac{\text{lbf}}{\text{in}^2}}{3000 \dfrac{\text{lbf}}{\text{in}^2}}\right)\right)$$

$$\rho = 0.00204$$

$$A_s = \rho b d = (0.00204)(12 \text{ in})(10.25 \text{ in}) = 0.25 \text{ in}^2$$

> Use no. 4 bars at 9 in o.c. on the outside face of wall below the joint. The only revision necessary to the detail provided is to extend the bottom reinforcement 6 in past the face of the wall, rather than the 2 in extension shown.

SOLUTION 12

12.1. Calculate the required prestress force in the composite member. Assume unshored construction and use the given section properties and strand stresses. Use the ACI 318 strength design method, with $f'_c = 5$ ksi and $f_{pu} = 260$ ksi.

$$w_{\text{slab}} = w_c h s = \frac{\left(150 \dfrac{\text{lbf}}{\text{ft}^3}\right)(6 \text{ in})(72 \text{ in})}{\left(12 \dfrac{\text{in}}{\text{ft}}\right)^2}$$

$$= 450 \text{ lbf/ft}$$

$$w_D = w_{\text{slab}} + w_{\text{precast}} = 450 \frac{\text{lbf}}{\text{ft}} + 440 \frac{\text{lbf}}{\text{ft}} = 890 \text{ lbf/ft}$$

$$w_L = ws = \left(125 \frac{\text{lbf}}{\text{ft}^2}\right)(6 \text{ ft}) = 750 \text{ lbf/ft}$$

$$w_u = 1.2 w_D + 1.6 w_L$$

$$= (1.2)\left(890 \frac{\text{lbf}}{\text{ft}}\right) + (1.6)\left(750 \frac{\text{lbf}}{\text{ft}}\right)$$

$$= 2270 \text{ lbf/ft}$$

$$M_u = \frac{w_u L^2}{8} = \frac{\left(2270 \frac{\text{lbf}}{\text{ft}}\right)(65 \text{ ft})^2}{8} = 1,200,000 \text{ ft-lbf}$$

Approximate the required area of prestressing strand by assuming that flexural strength governs design.

$$d_p = h_{\text{girder}} + h_{\text{slab}} - \text{cover}$$

$$= 42 \text{ in} + 6 \text{ in} - 4 \text{ in}$$

$$= 44 \text{ in}$$

$$d_p - \frac{a}{2} \approx 0.95 d_p$$

$$\approx (0.95)(44 \text{ in})$$

$$\approx 41.8 \text{ in}$$

$$A_{ps} = \frac{M_u}{\phi f_{ps}(0.95 d_p)}$$

$$= \frac{(1,200,000 \text{ ft-lbf})\left(12 \frac{\text{in}}{\text{ft}}\right)}{(0.9)\left(260,000 \frac{\text{lbf}}{\text{in}^2}\right)(41.8 \text{ in})}$$

$$= 1.47 \text{ in}^2$$

For $1/2$ in diameter strands,

$$n_{\text{strand}} = \frac{A_{ps}}{A_b} = \frac{1.47 \text{ in}^2}{0.153 \frac{\text{in}^2}{\text{strand}}} = 9.6 \quad [\text{say, 10 strands}]$$

Use ACI 318 Eq. 18-1, assuming a modern strand $(\gamma_p = 0.28)$. Since there is no mild steel flexural reinforcement, $d = 0$, and Eq. 18-1 reduces to

$$f_{ps} = f_{pu}\left(1 - \left(\frac{\gamma_p}{\beta_1}\right)\left(\frac{A_{ps}}{bd_p}\right)\left(\frac{f_{pu}}{f'_c}\right)\right)$$

$$= \left(260,000 \frac{\text{lbf}}{\text{in}^2}\right)\left(\begin{array}{c} 1 - \left(\frac{0.28}{0.8}\right)\left(\frac{1.53 \text{ in}^2}{(72 \text{ in})(44.0 \text{ in})}\right) \\ \times \left(\frac{260,000 \frac{\text{lbf}}{\text{in}^2}}{5000 \frac{\text{lbf}}{\text{in}^2}}\right) \end{array}\right)$$

$$= \boxed{257,700 \text{ psi}}$$

Check.

$$a = \frac{A_{ps} f_{ps}}{0.85 f'_c b} = \frac{(1.53 \text{ in}^2)\left(257,700 \frac{\text{lbf}}{\text{in}^2}\right)}{(0.85)\left(5000 \frac{\text{lbf}}{\text{in}^2}\right)(72 \text{ in})} = 1.29 \text{ in}$$

$$c = \frac{a}{\beta_1} = \frac{1.29 \text{ in}}{0.8} = 1.61 \text{ in}$$

$$\frac{\epsilon_t}{d_p - c} = \frac{\epsilon_c}{c}$$

$$\epsilon_t = \frac{(d_p - c)\epsilon_c}{c}$$

$$= \frac{(44.0 \text{ in} - 1.61 \text{ in})(0.003)}{1.61}$$

$$= 0.079 \quad [> 0.005, \text{ therefore } \phi = 0.9]$$

$$\phi M_n = \phi A_{ps} f_{ps}\left(d_p - \frac{a}{2}\right)$$

$$= \frac{(0.9)(1.53 \text{ in}^2)\left(257,700 \frac{\text{lbf}}{\text{in}^2}\right)\left(44.0 \text{ in} - \frac{1.29 \text{ in}}{2}\right)}{12 \frac{\text{in}}{\text{ft}}}$$

$$= 1,282,000 \text{ ft-lbf} \quad [\phi M_n > M_u]$$

Thus, the flexural strength is adequate; use ten $1/2$ in diameter strands: $A_{ps} = 1.53 \text{ in}^2$.

12.2. Check the stresses at transfer. The critical location occurs at the end of the transfer length $l_t = 50 d_b = (50)(0.5 \text{ in}) = 25 \text{ in} \approx 2 \text{ ft}$. Assume a straight strand pattern with eccentricity, e.

$$e = \frac{h}{2} - \text{cover}$$

$$= \frac{42 \text{ in}}{2} - 4 \text{ in}$$

$$= 17 \text{ in}$$

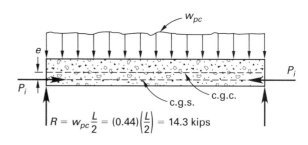

$$P_i = f_{pi} A_{ps} = \left(167,000 \frac{\text{lbf}}{\text{in}^2}\right)(1.53 \text{ in}^2) = 255,500 \text{ lbf}$$

At a distance $x = l_t = 2$ ft from the end of member,

$$M_D = 0.5(w_D L x - w_D x^2)$$

$$= (0.5)\left(\left(440 \ \frac{\text{lbf}}{\text{ft}}\right)(65 \text{ ft})(2 \text{ ft}) - \left(440 \ \frac{\text{lbf}}{\text{ft}}\right)(2 \text{ ft})^2\right)$$

$$= 27{,}720 \text{ ft-lbf} \quad [\text{compression top}]$$

$$M_{ps} = P_i e = (255{,}500 \text{ lbf})\left(\frac{17 \text{ in}}{12 \ \frac{\text{in}}{\text{ft}}}\right)$$

$$= 362{,}000 \text{ in-lbf} \quad [\text{tension top}]$$

Determine stresses at transfer.

$$A = 420 \text{ in}^2$$

$$S = \frac{I_g}{0.5h} = \frac{61{,}600 \text{ in}^4}{(0.5)(42 \text{ in})} = 2933 \text{ in}^3$$

$$f_{\text{top}} = \frac{P_i}{A} + \frac{M_D - M_{ps}}{S}$$

$$= \frac{255{,}500 \text{ lbf}}{420 \text{ in}^2}$$

$$+ \frac{(27{,}720 \text{ ft-lbf} - 362{,}000 \text{ ft-lbf})\left(12 \ \frac{\text{in}}{\text{ft}}\right)}{2933 \text{ in}^3}$$

$$= \boxed{-759 \text{ psi} \quad [\text{tension}]}$$

$$f_{\text{bottom}} = \frac{P_i}{A} - \frac{M_D - M_{ps}}{S}$$

$$= \frac{255{,}500 \text{ lbf}}{420 \text{ in}^2}$$

$$- \frac{(27{,}720 \text{ ft-lbf} - 362{,}000 \text{ ft-lbf})\left(12 \ \frac{\text{in}}{\text{ft}}\right)}{2933 \text{ in}^3}$$

$$= \boxed{1976 \text{ psi} \quad [\text{compression}]}$$

12.3. The strength of concrete at transfer was not specified. ACI 318 limits the compressive stress at transfer to $0.6f'_{ci}$. This requires

$$f'_c = \frac{1974 \ \frac{\text{lbf}}{\text{in}^2}}{0.6} = 3290 \text{ psi}$$

There should be no problem gaining sufficient strength at transfer to satisfy the compressive strength requirement. However, the computed tensile stress at transfer is limited to $6\sqrt{f'_c}$, which cannot be accommodated by high enough strength. There are several options.

- Depress the strands to reduce the end eccentricity.

- Blanket several strands to reduce the prestress near the end of member.

- Add mild steel to resist the total tensile force at the top at transfer.

Any of the options could be chosen. This solution uses the last option: to add mild steel at a working stress of 30 ksi (grade 60 rebars).

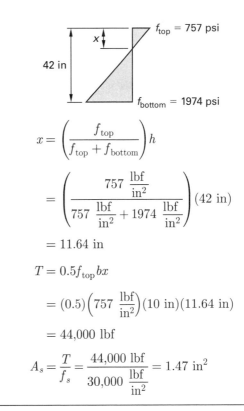

$$x = \left(\frac{f_{\text{top}}}{f_{\text{top}} + f_{\text{bottom}}}\right)h$$

$$= \left(\frac{757 \ \frac{\text{lbf}}{\text{in}^2}}{757 \ \frac{\text{lbf}}{\text{in}^2} + 1974 \ \frac{\text{lbf}}{\text{in}^2}}\right)(42 \text{ in})$$

$$= 11.64 \text{ in}$$

$$T = 0.5 f_{\text{top}} b x$$

$$= (0.5)\left(757 \ \frac{\text{lbf}}{\text{in}^2}\right)(10 \text{ in})(11.64 \text{ in})$$

$$= 44{,}000 \text{ lbf}$$

$$A_s = \frac{T}{f_s} = \frac{44{,}000 \text{ lbf}}{30{,}000 \ \frac{\text{lbf}}{\text{in}^2}} = 1.47 \text{ in}^2$$

$$\boxed{\text{Use two no. 8 top bars with } 90° \text{ hooks at the bottom.}}$$

12.4. Calculate the stresses under dead plus live load, assuming unshored construction.

$$P_e = f_{pe} A_{ps} = \left(147{,}000 \ \frac{\text{lbf}}{\text{in}^2}\right)(1.53 \text{ in}^2) = 225{,}000 \text{ lbf}$$

For the dead load acting alone on the precast section under effective prestress, the critical stresses occur at midspan.

$$M_D = \frac{(w_{\text{slab}} + w_{\text{precast}})L^2}{8}$$

$$= \frac{\left(450 \ \frac{\text{lbf}}{\text{ft}} + 440 \ \frac{\text{lbf}}{\text{ft}}\right)(65 \text{ ft})^2}{8}$$

$$= 470{,}000 \text{ ft-lbf} \quad [\text{compression top}]$$

$$M_{ps} = P_e e$$

$$= \frac{(225{,}000 \text{ lbf})(17 \text{ in})}{12 \ \frac{\text{in}}{\text{ft}}}$$

$$= 319{,}000 \text{ ft-lbf} \quad [\text{tension top}]$$

$$f_{\text{top,dead}} = \frac{P_e}{A} + \frac{M_D - M_{ps}}{S}$$

$$= \frac{225{,}000 \text{ lbf}}{420 \text{ in}^2}$$

$$+ \frac{(470{,}000 \text{ ft-lbf} - 319{,}000 \text{ ft-lbf})\left(12 \dfrac{\text{in}}{\text{ft}}\right)}{2933 \text{ in}^3}$$

$$= 1150 \text{ psi} \quad [\text{compression}]$$

$$f_{\text{bottom,dead}} = \frac{P_e}{A} - \frac{M_D - M_{ps}}{S}$$

$$= \frac{225{,}000 \text{ lbf}}{420 \text{ in}^2}$$

$$- \frac{(470{,}000 \text{ ft-lbf} - 319{,}000 \text{ ft-lbf})\left(12 \dfrac{\text{in}}{\text{ft}}\right)}{2933 \text{ in}^3}$$

$$= -82 \text{ psi} \quad [\text{tension}]$$

Superimpose the live load stresses acting on the composite section.

$$M_L = \frac{w_l L^2}{8}$$

$$= \frac{\left(750 \dfrac{\text{lbf}}{\text{ft}}\right)(65 \text{ ft})^2}{8}$$

$$= 396{,}000 \text{ ft-lbf} \quad [\text{compression top}]$$

$$f_{\text{top,precast}} = \frac{M_L c_{\text{precast}}}{I_{\text{top,comp}}}$$

$$= \frac{(396{,}000 \text{ ft-lbf})\left(12 \dfrac{\text{in}}{\text{ft}}\right)(8.8 \text{ in})}{184{,}000 \text{ in}^4}$$

$$= 227 \text{ psi} \quad [\text{compression top}]$$

$$f_{\text{top,slab}} = \frac{M_L c_{\text{slab}}}{I_{\text{top,comp}}}$$

$$= \frac{(396{,}000 \text{ ft-lbf})(14.8 \text{ in})\left(12 \dfrac{\text{in}}{\text{ft}}\right)}{184{,}000 \text{ in}^4}$$

$$= 382 \text{ psi} \quad [\text{compression}]$$

$$f_{\text{bottom,precast}} = \frac{M_L c_{\text{bottom}}}{I_{\text{bottom,comp}}}$$

$$= \frac{(396{,}000 \text{ ft-lbf})(33.2 \text{ in})\left(12 \dfrac{\text{in}}{\text{ft}}\right)}{184{,}000 \text{ in}^3}$$

$$= -857 \text{ psi} \quad [\text{tension}]$$

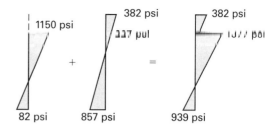

Thus, the calculated stresses are

$$f_{\text{bottom}} = f_{\text{bottom,precast}} + f_{\text{bottom,dead}}$$

$$= -857 \frac{\text{lbf}}{\text{in}^2} - 82 \frac{\text{lbf}}{\text{in}^2}$$

$$= \boxed{-939 \text{ psi} \quad [\text{tension}]}$$

$$f_{\text{top}} = f_{\text{top,precast}} + f_{\text{top,dead}}$$

$$= 227 \frac{\text{lbf}}{\text{in}^2} + 1150 \frac{\text{lbf}}{\text{in}^2}$$

$$= \boxed{1377 \text{ psi} \quad [\text{compression}]}$$

$$f_{\text{top,slab}} = \boxed{382 \text{ psi} \quad [\text{compression}]}$$

(The calculated tension stress exceeds $7.5\sqrt{f'_c}$, but is less than $12\sqrt{f'_c}$. This classifies the member as Class T, requiring that deflections be calculated per ACI 318 Sec. 9.5.4.2.)

12.5. Depending on the age of the precast at the time the slab and closure are cast, the beam will experience long-term deformation that will likely result in shortening and upward deflection (that is, additional camber). This will in turn cause a separation between the precast member and the closure concrete. Consequently, with application of the live load, there will not be significant moment resistance at the connection. Thus, the connection is flexible, behaving essentially as a simple connection.

SOLUTION 13

Compute the top and bottom midspan stresses under the specified loading. Because the beam is given as posttensioned, assume a parabolic strand profile with a sag of $y_c - y_e = 11 \text{ in} - 6 \text{ in} = 5 \text{ in}$.

13.1. Under the dead load plus effective prestress at the midspan,

$$w_e = \frac{8 P_e (\text{sag})}{L^2} = \frac{(8)(200 \text{ kips})(5 \text{ in})}{(40.83 \text{ ft})^2 \left(12 \dfrac{\text{in}}{\text{ft}}\right)} = 0.40 \text{ kip/ft}$$

The strand is concentric at the ends with the precast beam's centroid. Therefore, the equivalent load diagram is

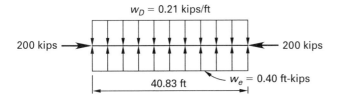

The net loading is

$$w_{\text{net}} = w_e - w_D = 0.40 \; \frac{\text{kip}}{\text{ft}} - 0.21 \; \frac{\text{kip}}{\text{ft}}$$

$$= 0.19 \; \text{kip/ft} \quad [\text{upward}]$$

$$M_{D+ps} = \frac{w_{\text{net}} L^2}{8}$$

$$= \frac{\left(0.19 \; \frac{\text{kip}}{\text{ft}} \right)(40.83 \; \text{ft})^2}{8}$$

$$= 39.6 \; \text{ft-kips} \quad [\text{top fiber in tension}]$$

Values given for the precast section are $A = 265 \; \text{in}^2$, $S_t = 1200 \; \text{in}^3$, and $S_b = 1250 \; \text{in}^3$.

$$f_{\text{top,1}} = \frac{P_e}{A} - \frac{M_{D+ps}}{S_t}$$

$$= \frac{200 \; \text{kips}}{265 \; \text{in}^2} - \frac{(39.6 \; \text{ft-kips})\left(12 \; \frac{\text{in}}{\text{ft}} \right)}{1200 \; \text{in}^3}$$

$$= \boxed{0.359 \; \text{ksi} \quad [\text{compression}]}$$

$$f_{\text{bottom,1}} = \frac{P_e}{A} + \frac{M_{D+ps}}{S_b}$$

$$= \frac{200 \; \text{kips}}{265 \; \text{in}^2} + \frac{(39.6 \; \text{ft-kips})\left(12 \; \frac{\text{in}}{\text{ft}} \right)}{1250 \; \text{in}^3}$$

$$= \boxed{1.135 \; \text{ksi} \quad [\text{compression}]}$$

13.2. With the precast section shored at midspan and subjected to a wet concrete load of $w = w_{\text{slab}}s = (50 \; \text{lbf/ft}^2)(15 \; \text{ft}) = 750 \; \text{lbf/ft}^2$ applied after the shore is positioned, calculate the midspan stresses and superimpose them with the stresses in the previously simply supported member.

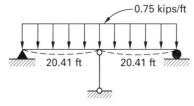

$$M^- = \frac{w(0.5L)^2}{8}$$

$$= \frac{\left(0.75 \; \frac{\text{kip}}{\text{ft}} \right)((0.5)(40.83 \; \text{ft}))^2}{8}$$

$$= 39.1 \; \text{ft-kips} \quad [\text{tension on top fiber}]$$

$$f_{\text{top,2}} = \frac{M^-}{S_t}$$

$$= \frac{(39.1 \; \text{ft-kips})\left(12 \; \frac{\text{in}}{\text{ft}} \right)}{1200 \; \text{in}^3}$$

$$= -0.391 \; \text{ksi} \quad [\text{tension}]$$

$$f_{\text{bottom,2}} = \frac{M^-}{S_b}$$

$$= \frac{(39.1 \; \text{ft-kips})\left(12 \; \frac{\text{in}}{\text{ft}} \right)}{1250 \; \text{in}^3}$$

$$= 0.375 \; \text{ksi} \quad [\text{compression}]$$

Superimpose the stresses.

$$f_{\text{top}} = f_{\text{top,1}} + f_{\text{top,2}}$$

$$= 0.359 \; \frac{\text{kip}}{\text{in}^2} - 0.391 \; \frac{\text{kip}}{\text{in}^2}$$

$$= \boxed{-0.032 \; \text{ksi} \quad [\text{tension}]}$$

$$f_{\text{bottom}} = f_{\text{bottom,1}} + f_{\text{bottom,2}}$$

$$= 1.135 \; \frac{\text{kips}}{\text{in}^2} + 0.375 \; \frac{\text{kip}}{\text{in}^2}$$

$$= \boxed{1.51 \; \text{ksi} \quad [\text{compression}]}$$

13.3. Remove the midspan shore after the concrete hardens and compute the resulting stresses. This is equivalent to applying a concentrated force at the midspan of the now simply supported member equal to the reaction of the shore under the wet load. In all computations for this problem, ideal linear elastic behavior is assumed without adjustments for the effects of creep and shrinkage.

$$R_{\text{shore}} = 2\left(\frac{5w(0.5L)}{8}\right)$$

$$= (2)\left(\frac{(5)\left(0.75\ \dfrac{\text{kip}}{\text{ft}}\right)(0.5)(40.83\ \text{ft})}{8}\right)$$

$$= 19.1\ \text{kips}$$

$$M_{\text{shore}} = \frac{R_{\text{shore}}L}{4}$$

$$= \frac{(19.1\ \text{kips})(40.83\ \text{ft})}{4}$$

$$= 195.3\ \text{ft-kips}\quad[\text{compression on top}]$$

The stresses caused by this bending moment are induced in the composite cross section. Thus, the stresses at the bottom and at the interface of the precast and cast-in-place slab must be calculated using properties of the composite section. The neutral axis location is not given, but it can be found from the given section moduli and dimensions.

$$I = S_t c_t = S_b c_b$$

$$S_t c_t = S_b(h + h_s - c_t)$$

$$c_t = \frac{S_b(h + h_s)}{S_t + S_b}$$

$$= \frac{(2350\ \text{in}^3)(22.5\ \text{in} + 5\ \text{in})}{13{,}500\ \text{in}^3 + 2350\ \text{in}^3}$$

$$= 4.08\ \text{in}$$

$$I = S_t c_t$$

$$= (13{,}500\ \text{in}^3)(4.08\ \text{in})$$

$$= 55{,}080\ \text{in}^4$$

Therefore, the stress induced at the top of the precast section when the shore is removed is

$$f_{\text{top},3} = \frac{M_{\text{shore}}\,y}{I}$$

$$= \frac{(195.5\ \text{ft-kips})\left(12\ \dfrac{\text{in}}{\text{ft}}\right)(5\ \text{in} - 4.08\ \text{in})}{55{,}080\ \text{in}^4}$$

$$= -0.039\ \text{ksi}\quad[\text{tension}]$$

$$f_{\text{bottom},3} = \frac{M_{\text{shore}}}{S_b}$$

$$= \frac{(195.5\ \text{ft-kips})\left(12\ \dfrac{\text{in}}{\text{ft}}\right)}{2350\ \text{in}^3}$$

$$= -0.998\ \text{ksi}\quad[\text{tension}]$$

Superimposing on previous stresses,

$$f_{\text{top}} = f_{\text{top},1} + f_{\text{top},2} + f_{\text{top},3}$$

$$= 0.359\ \frac{\text{kip}}{\text{in}^2} + \left(-0.391\ \frac{\text{kip}}{\text{in}^2}\right) + \left(-0.039\ \frac{\text{kip}}{\text{in}^2}\right)$$

$$= \boxed{-0.071\ \text{ksi}\quad[\text{tension}]}$$

$$f_{\text{bottom}} = f_{\text{bottom},1} + f_{\text{bottom},2} + f_{\text{bottom},3}$$

$$= 1.135\ \frac{\text{kips}}{\text{in}^2} + 0.375\ \frac{\text{kip}}{\text{in}^2} + \left(-0.998\ \frac{\text{kip}}{\text{in}^2}\right)$$

$$= \boxed{0.511\ \text{ksi}\quad[\text{compression}]}$$

13.4. Apply the live load to the composite section.

$$w = w_L s$$

$$= \left(50\ \frac{\text{lbf}}{\text{ft}^2}\right)(15\ \text{ft})$$

$$= 750\ \text{lbf/ft}^2$$

$$M_L = \frac{wL^2}{8}$$

$$= \frac{\left(750\ \dfrac{\text{lbf}}{\text{ft}}\right)(40.83\ \text{ft})^2}{(8)\left(1000\ \dfrac{\text{lbf}}{\text{kips}}\right)}$$

$$= 156.3\ \text{ft-kips}$$

$$f_{\text{top},4} = \frac{M_L y}{I}$$

$$= \frac{(156.3\ \text{ft-kips})\left(12\ \dfrac{\text{in}}{\text{ft}}\right)(5\ \text{in} - 4.08\ \text{in})}{55{,}080\ \text{in}^4}$$

$$= -0.031\ \text{ksi}\quad[\text{tension}]$$

$$f_{\text{bottom},3} = \frac{M_L}{S_b}$$

$$= \frac{(156.3\ \text{ft-kips})\left(12\ \dfrac{\text{in}}{\text{ft}}\right)}{2350\ \text{in}^3}$$

$$= -0.798\ \text{ksi}\quad[\text{tension}]$$

Superimpose on the previously computed stresses.

$$f_{\text{top}} = f_{\text{top},1} + f_{\text{top},2} + f_{\text{top},3} + f_{\text{top},4}$$

$$= 0.359\ \frac{\text{kip}}{\text{in}^2} + \left(-0.391\ \frac{\text{kip}}{\text{in}^2}\right)$$

$$+ \left(-0.039\ \frac{\text{kip}}{\text{in}^2}\right) + \left(-0.031\ \frac{\text{kip}}{\text{in}^2}\right)$$

$$= \boxed{-0.102\ \text{kip/in}^2\quad[\text{tension}]}$$

$$f_{bottom} = f_{bottom,1} + f_{bottom,2} + f_{bottom,3} + f_{bottom,4}$$

$$= 1.135 \frac{kips}{in^2} + 0.375 \frac{kip}{in^2}$$

$$+ \left(-0.998 \frac{kip}{in^2}\right) + \left(-0.798 \frac{kip}{in^2}\right)$$

$$= \boxed{-0.286 \ kip/in^2 \quad [tension]}$$

The stress at the top of the slab is

$$f_{t,s} = \frac{M_l}{S_t}$$

$$= \frac{(156.3 \ ft\text{-}kips)\left(12 \ \frac{in}{ft}\right)}{13,500 \ in^3}$$

$$= 0.138 \ kip/in^2 \quad [compression]$$

SOLUTION 14

14.1. Design the shear reinforcement based on the given 10 in thick by 32 ft long wall.

$$A_{cv} = t_w l_w = (10 \ in)(32 \ ft)\left(12 \ \frac{in}{ft}\right) = 3840 \ in^2$$

Try using the minimum percentages of longitudinal and transverse wall reinforcement (ACI 318 Sec. 21.9.2.1).

$$\rho_t = \rho_l = 0.0025$$

$$V_u = V_e = 456,000 \ lbf$$

$$2A_{cv}\sqrt{f'_c} = (2)(3840 \ in^2)\sqrt{3000 \ \frac{lbf}{in^2}}$$

$$= 420,000 \ lbf \quad [V_u > 2A_{cv}\sqrt{f'_c}]$$

Thus, two curtains of horizontal and vertical steel are required by ACI 318 Sec. 21.9.2.2. Assume that the nominal shear strength of the wall will exceed the shear corresponding to nominal flexural strength of the wall. ACI 318 Sec. 9.3.4 permits $\phi = 0.75$ for this case. The height-to-length ratio of the wall is

$$\frac{h_w}{l_w} = \frac{125.2 \ ft}{32 \ ft} = 3.9$$

Since 3.9 is greater than 2.0, use $\alpha_c = 2.0$ (ACI 318 Sec. 21.9.4.1).

$$\phi V_n = \phi A_{cv}(\alpha_c\sqrt{f'_c} + \rho_t f_y)$$

$$= (0.75)(3840 \ in^2)$$

$$\times \left(2.0\sqrt{3000 \ \frac{lbf}{in^2}} + (0.0025)\left(60,000 \ \frac{lbf}{in^2}\right)\right)$$

$$= \boxed{747,500 \ lbf \quad [\phi V_n > V_u]}$$

Thus, shear strength is adequate with minimum wall reinforcement.

$$A_{s,l} = A_{s,t} = \rho_t t_w$$

$$= (0.0025)\left(12 \ \frac{in}{ft}\right)(10 \ in)$$

$$= 0.30 \ in^2/ft \quad [0.15 \ in^2/ft \ in \ each \ face \ of \ wall]$$

Use no. 4 bars at 16 in on centers vertical and horizontal in two faces.

14.2. Determine the boundary reinforcement per ACI 318 Sec. 21.9.5. There are two loading cases to consider (ASCE/SEI7 Sec. 2.3.2 and Sec. 12.4.2.3). Load combination 5 gives

$$P_{u,5} = 1.2P_D + 0.2S_{DS}P_D + 0.5P_L + 0.2P_S$$

$$= (1.2)\left(21,000 \ \frac{lbf}{ft}\right)(32 \ ft)$$

$$+ (0.2)(0.50)\left(21,000 \ \frac{lbf}{ft}\right)(32 \ ft)$$

$$+ (0.5)\left(2700 \ \frac{lbf}{ft}\right)(32 \ ft) + 0 \ lbf$$

$$= 917,000 \ lbf$$

Load combination 7 requires

$$P_{u,7} = 0.9P_D - 0.2S_{DS}P_D$$

$$= (0.9)\left(21,000 \ \frac{lbf}{ft}\right)(32 \ ft)$$

$$- (0.2)(0.50)\left(21,000 \ \frac{lbf}{ft}\right)(32 \ ft)$$

$$= 538,000 \ lbf$$

The seismic overturning moment, which is $M_u = \pm M_E = \pm 28,270,000 \ ft\text{-}lbf$, acts in combination with axial compressive force in combinations 5 and 7. Determine if special boundary elements are required per ACI 318 Sec. 21.9.6.3 based on the critical compression loading, combination 5. The index compression stress is

$$I_g = \frac{t_w l_w^3}{12} = \frac{(10 \ in)(32 \ ft)^3\left(12 \ \frac{in}{ft}\right)^3}{12}$$

$$= 47.2 \times 10^6 \ in^4$$

$$f_c = \frac{P_{u,5}}{A_{cv}} + \frac{M_u c}{I_g}$$

$$= \frac{917,000 \ lbf}{3840 \ in^2} + \frac{(28,270,000 \ ft\text{-}lbf)(16 \ ft)\left(12 \ \frac{in}{ft}\right)^2}{47.2 \times 10^6 \ in^4}$$

$$= 1618 \ psi$$

$$0.2f'_c = (0.2)\left(3000 \ \frac{lbf}{in^2}\right) = 600 \ psi \quad [f_c > 0.2f'_c]$$

Thus, boundary elements are required.

Try adding 24 in × 24 in elements at the ends of the wall.

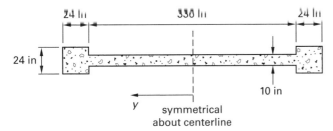

The revised index compressive stress corresponding to load combination 5 is

$$A_g = bl_w - (b - t_w)l_{w2}$$
$$= (24 \text{ in})(384 \text{ in}) - (24 \text{ in} - 10 \text{ in})(336 \text{ in})$$
$$= 4512 \text{ in}^2$$

$$I_g = \frac{bl_w^3 - (b - t_w)l_{w2}^3}{12}$$
$$= \frac{(24 \text{ in})(384 \text{ in})^3 - (24 \text{ in} - 10 \text{ in})(336 \text{ in})^3}{12}$$
$$= 69{,}000{,}000 \text{ in}^4$$

$$f_c = \frac{P_{u,5}}{A_{cv}} + \frac{M_u c}{I_g}$$
$$= \frac{917{,}000 \text{ lbf}}{4512 \text{ in}^2}$$
$$+ \frac{(28{,}270{,}000 \text{ ft-lbf})(16 \text{ ft})\left(12 \frac{\text{in}}{\text{ft}}\right)^2}{69{,}000{,}000 \text{ in}^4}$$
$$= 1150 \text{ psi}$$

$$0.2f_c' = (0.2)\left(3000 \frac{\text{lbf}}{\text{in}^2}\right)$$
$$= 600 \text{ psi} \quad [f_c > f_c']$$

Compute the longitudinal steel required at the boundaries based on the more severe overturning moment, loading case 7. Assume that all vertical web reinforcement resists overturning (two no. 4 bars at 16 in on centers gives $A_{sw} = 0.30 \text{ in}^2/\text{ft}$).

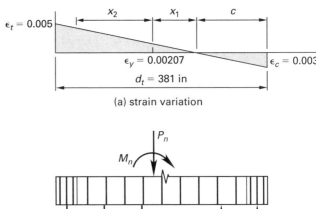

$P_{u,7} = 538{,}000 \text{ lbf}$

$M_u = 28{,}270 \text{ ft-lbf}$

T_w

lever arm $\approx$ 32 ft $-$ 2 ft $=$ 30 ft

T_u C_u

elevation

$$T_w = A_{sw}(l_w - 2h)f_y$$
$$= \left(0.30 \frac{\text{in}^2}{\text{ft}}\right)(32 \text{ ft} - (2)(2 \text{ ft}))\left(60{,}000 \frac{\text{lbf}}{\text{in}^2}\right)$$
$$= 504{,}000 \text{ lbf}$$

$$\sum M_{\text{right}} = T_u l_{\text{lever arm}} + (P_u + T_w)\left(\frac{l_{\text{lever arm}}}{2}\right) - M_u$$
$$= 0$$

$$T_u = \frac{M_u}{l_{\text{lever arm}}} - \frac{P_u + T_w}{2}$$
$$= \frac{28{,}270{,}000 \text{ ft-lbf}}{30 \text{ ft}} - \frac{538{,}000 \text{ lbf} + 504{,}000 \text{ lbf}}{2}$$
$$= 421{,}300 \text{ lbf}$$

$$T_u = \phi A_{st} f_y$$
$$A_{st} = \frac{T_u}{\phi f_y}$$
$$= \frac{421{,}300 \text{ lbf}}{(0.9)\left(60{,}000 \frac{\text{lbf}}{\text{in}^2}\right)}$$
$$= 7.80 \text{ in}^2$$

Try eight no. 9 bars ($A_{st} = (8)(1.00 \text{ in}^2) = 8.00 \text{ in}^2$). Per ACI 318 Sec. 21.9.6.4a, the boundary element must extend horizontally from the edges for a distance not less than $c - 0.1l_w$ or $c/2$, whichever is greater (c is the largest neutral axis depth), which corresponds to loading case 5. Compute the interaction values for the wall when the neutral axis depth is at the extreme position for a tension-controlled failure (that is, strain in the extreme tension side steel equals 0.005 when concrete strain reaches 0.003).

$\epsilon_t = 0.005$ x_2 x_1 c

$\epsilon_y = 0.00207$ $\epsilon_c = 0.003$

$d_t = 381 \text{ in}$

(a) strain variation

P_n

M_n

T_f T_{w2} T_{w1} C_w C_f

(b) elevation

From similar triangles,

$$\frac{c}{\epsilon_c} = \frac{d_t}{\epsilon_c + \epsilon_t}$$

$$c = \frac{d_t\epsilon_c}{\epsilon_c + \epsilon_t}$$

$$= \frac{(381 \text{ in})(0.003)}{0.003 + 0.005}$$

$$= 143 \text{ in}$$

$$a = \beta_1 c = (0.85)(143 \text{ in}) = 121.6 \text{ in}$$

Tension strains increase linearly from zero to the yield strain, 0.00207, over a distance x_1.

$$\frac{x_1}{\epsilon_y} = \frac{c}{0.003}$$

$$x_1 = \frac{(143 \text{ in})(0.00207)}{0.003} = 98.7 \text{ in}$$

$$x_2 = l_w - c - x_1 - h$$

$$= 384 \text{ in} - 143 \text{ in} - 98.7 \text{ in} - 24 \text{ in}$$

$$= 118.3 \text{ in}$$

The forces in the segments of the wall are

$$T_f = A_{st}f_y = (8.00 \text{ in}^2)\left(60,000 \frac{\text{lbf}}{\text{in}^2}\right) = 480,000 \text{ lbf}$$

$$T_{w1} = A_{sw}x_1(0.5f_y)$$

$$= \frac{\left(0.30 \frac{\text{in}^2}{\text{ft}}\right)(98.7 \text{ in})(0.5)\left(60,000 \frac{\text{lbf}}{\text{in}^2}\right)}{12 \frac{\text{in}}{\text{ft}}}$$

$$= 74,000 \text{ lbf}$$

$$T_{w2} = A_{sw}x_2f_y$$

$$= \frac{\left(0.30 \frac{\text{in}^2}{\text{ft}}\right)(118.3 \text{ in})\left(60,000 \frac{\text{lbf}}{\text{in}^2}\right)}{12 \frac{\text{in}}{\text{ft}}}$$

$$= 177,500 \text{ lbf}$$

$$C_f = 0.85f'_c(h^2 - A_{st}) + f_yA_{st}$$

$$= (0.85)\left(3000 \frac{\text{lbf}}{\text{in}^2}\right)\left((24 \text{ in})^2 - 8.00 \text{ in}^2\right)$$

$$\quad + \left(60,000 \frac{\text{lbf}}{\text{in}^2}\right)(8.00 \text{ in}^2)$$

$$= 1,928,000 \text{ lbf}$$

$$C_w = 0.85f'_c b(a-h) + (f_y - 0.85f'_c)A_{sw}(a-h)$$

$$= (0.85)\left(3000 \frac{\text{lbf}}{\text{in}^2}\right)(10 \text{ in})(121.6 \text{ in} - 24.0 \text{ in})$$

$$\quad + \left(60,000 \frac{\text{lbf}}{\text{in}^2} - (0.85)\left(3000 \frac{\text{lbf}}{\text{in}^2}\right)\right)$$

$$\quad \times \left(0.30 \frac{\text{in}^2}{\text{ft}}\right)\left(\frac{121.6 \text{ in} - 24.0 \text{ in}}{12 \frac{\text{in}}{\text{ft}}}\right)$$

$$= 2,629,000 \text{ lbf}$$

$$\sum F_{\text{vert}} = C_f + C_w - T_{w1} - T_{w2} - T_f - P_n = 0$$

$$P_n = C_f + C_w - T_{w1} - T_{w2} - T_f$$

$$= 1,928,000 \text{ lbf} + 2,629,000 \text{ lbf} - 74,000 \text{ lbf}$$

$$\quad - 177,500 \text{ lbf} - 480,000 \text{ lbf}$$

$$= 3,825,500 \text{ lbf}$$

The plastic centroid (pc) of the cross section is at midlength. Taking moments about this point,

$$\sum M_{\text{pc}} = 0$$

$$= C_f\left(\frac{l_w - h}{2}\right) + C_w\left(\frac{l_w - a}{2} - h\right)$$

$$\quad + T_f\left(\frac{l_w - h}{2}\right) + T_{w1}\left(\frac{l_w}{2} - h - x_2 - \frac{x_1}{3}\right)$$

$$\quad + T_{w2}\left(\frac{l_w}{2} - h - \frac{x_2}{2}\right) - M_n$$

$$M_n = C_f\left(\frac{l_w - h}{2}\right) + C_w\left(\frac{l_w - a}{2} - h\right) + T_f\left(\frac{l_w - h}{2}\right)$$

$$\quad + T_{w1}\left(\frac{l_w}{2} - h - x_2 - \frac{x_1}{3}\right) + T_{w2}\left(\frac{l_w}{2} - h - \frac{x_2}{2}\right)$$

$$= \frac{\begin{aligned} &(1,928,000 \text{ lbf})\left(\frac{384 \text{ in} - 24 \text{ in}}{2}\right) \\ &+ (2,629,000 \text{ lbf})\left(\frac{384 \text{ in} - 121.6 \text{ in}}{2} - 24 \text{ in}\right) \\ &+ (480,000 \text{ lbf})\left(\frac{384 \text{ in} - 24 \text{ in}}{2}\right) + (74,000 \text{ lbf}) \\ &\times \left(\frac{384 \text{ in}}{2} - 24 \text{ in} - 118.3 \text{ in} - \frac{98.7 \text{ in}}{3}\right) \\ &+ (177,500 \text{ lbf})\left(\frac{384 \text{ in}}{2} - 24 \text{ in} - \frac{118.3 \text{ in}}{2}\right) \end{aligned}}{12 \frac{\text{in}}{\text{ft}}}$$

$$= 61,322,000 \text{ ft-lbf}$$

The section is tension controlled, so

$$\phi P_n = (0.9)(3,825,000 \text{ lbf})$$

$$= 3,443,000 \text{ lbf} \quad [>P_u]$$

$$\phi M_n = (0.9)(61,322,000 \text{ ft-lbf})$$

$$= 55,190,000 \text{ ft-lbf} \quad [>M_u]$$

Thus, providing a boundary zone based on $c = 143$ in is conservative for all loading conditions. Let x be the horizontal extent of the boundary element.

$$x \geq \begin{cases} c - 0.5l_w = 143 \text{ in} - (0.1)(384 \text{ in}) \\ \qquad = 105 \text{ in} \quad [\text{controls}] \\ 0.5c = (0.5)(143 \text{ in}) = 72 \text{ in} \end{cases}$$

The vertical extent of the boundary element can terminate when the index stress is less than $0.15f'_c$. Design the confinement steel using ACI 318 Sec. 21.9.6.4. Arrange longitudinal bars such that $h_x = 10$ in. ACI 318 Eq. 21-2 gives

$$s_o = 4 \text{ in} + \frac{14 \text{ in} - h_x}{3}$$

$$= 4 \text{ in} + \frac{14 \text{ in} - 10 \text{ in}}{3}$$

$$= 5.3 \text{ in}$$

The maximum spacing of confinement steel is

$$s \leq \begin{cases} \dfrac{h}{4} = \dfrac{24 \text{ in}}{4} = 6 \text{ in} \\ 6d_{bl} = (6)(1.128 \text{ in}) = 6.7 \text{ in} \\ s_o = 5.3 \text{ in} \quad [\text{controls}] \end{cases}$$

Provide closed hoops at $5^{1}/_{4}$ in on centers. The area of confinement steel is the larger of the two values computed using ACI 318 Eq. 21-4 and Eq. 21-5.

$$A_{sh} \geq \begin{cases} 0.3sb_c\left(\dfrac{f'_c}{f_y}\right)\left(\left(\dfrac{A_g}{A_{ch}}\right) - 1\right) \\ \quad = (0.3)(5.25 \text{ in})(21 \text{ in})\left(\dfrac{3000 \frac{\text{lbf}}{\text{in}^2}}{60,000 \frac{\text{lbf}}{\text{in}^2}}\right) \\ \qquad \times \left(\dfrac{(24 \text{ in})^2}{(21 \text{ in})^2} - 1\right) \\ \quad = 0.51 \text{ in}^2 \quad [\text{controls}] \\ 0.09sb_c\left(\dfrac{f'_c}{f_y}\right) \\ \quad = (0.09)(5.25 \text{ in})(21 \text{ in})\left(\dfrac{3000 \frac{\text{lbf}}{\text{in}^2}}{60,000 \frac{\text{lbf}}{\text{in}^2}}\right) \\ \quad = 0.50 \text{ in}^2 \end{cases}$$

Use no. 4 hoops and crossties as shown to provide a minimum $A_{sh} = 3A_b = (3)(0.40 \text{ in}^2) = 60 \text{ in}^2$ on faces with three bars. Extend transverse reinforcement into the wall web over a distance of 105 in at each end. Extend the special transverse reinforcement into the footing at least 12 in.

14.3. From ACI 318 Sec. 21.7.5, the anchorage of a no. 9 hooked tension bar in footing is

$$l_{dh} \geq \begin{cases} 8d_b = (8)(1.128 \text{ in}) = 9.0 \text{ in} \\ 6 \text{ in} \\ \dfrac{f_y d_b}{65\sqrt{f'_c}} = \dfrac{(60,000 \text{ lbf})(1.128 \text{ in})}{65\sqrt{3000 \frac{\text{lbf}}{\text{in}^2}}} \\ \qquad = 19.0 \text{ in} \quad [\text{controls}] \end{cases}$$

The anchorage length for the boundary reinforcement is 19.0 in.

14.4. A detail of the wall reinforcement is shown.

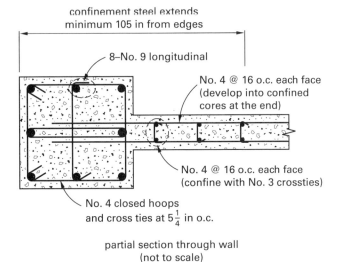

confinement steel extends
minimum 105 in from edges

8–No. 9 longitudinal

No. 4 @ 16 o.c. each face
(develop into confined
cores at the end)

No. 4 @ 16 o.c. each face
(confine with No. 3 crossties)

No. 4 closed hoops
and cross ties at $5\frac{1}{4}$ in o.c.

partial section through wall
(not to scale)

SOLUTION 15

15.1. Calculate the effect that the actual material strength will have on the shear requirements. For the original design, $f'_c = 3000$ psi, $f_y = 4000$ psi, and $f_{y,pr} = 1.25f_y = 50,000$ psi. The test values indicate actual strengths of $f'_c = 3400$ psi and $f_y = 72,000$ psi.

Assume that top and bottom steel is fully developed at faces of supports; include the contribution of compression steel to moment strength. For negative bending at the face of columns,

$$d^- = h - 2.5 \text{ in} = 24.5 \text{ in} - 2.5 \text{ in} = 22 \text{ in}$$

$$d' = 2.5 \text{ in}$$

$$b = b_w = 14 \text{ in}$$

$$A_s^- = n_{\text{bar}} A_b = (4)(1.00 \text{ in}^2) = 4.00 \text{ in}^2$$

$$A_s' = n_{\text{bar}} A_b = (3)(1.00 \text{ in}^2) = 3.00 \text{ in}^2$$

As an initial trial, neglect compression steel.

$$a = \frac{A_{\text{st}} f_y}{0.85 f_c' b_w}$$

$$= \frac{(4.00 \text{ in}^2)\left(72{,}000 \, \dfrac{\text{lbf}}{\text{in}^2}\right)}{(0.85)\left(3400 \, \dfrac{\text{lbf}}{\text{in}^2}\right)(14 \text{ in})}$$

$$= 7.12 \text{ in}$$

$$c = \frac{a}{\beta_1} = \frac{7.12 \text{ in}}{0.85}$$

$$= 8.38 \text{ in}$$

$$\frac{\epsilon_s'}{c - d'} = \frac{\epsilon_c}{d - c}$$

$$\epsilon_s' = \frac{\epsilon_c(c - d')}{d - c}$$

$$= \frac{(0.003)(8.38 \text{ in} - 2.5 \text{ in})}{22 \text{ in} - 8.38 \text{ in}}$$

$$= 0.0013$$

$$\epsilon_y = \frac{f_y}{E_s} = \frac{72{,}000 \, \dfrac{\text{lbf}}{\text{in}^2}}{29{,}000{,}000 \, \dfrac{\text{lbf}}{\text{in}^2}}$$

$$= 0.0025 \quad [\epsilon_s' < \epsilon_y]$$

Recomputing the depth of neutral axis by including the compression steel will result in a smaller value of c and hence a smaller value of ϵ_s', giving a smaller f_s'. Assume for a second trial a value of f_s' midway between zero and yield, $f_s' = 36{,}000$ psi.

$$a = \frac{A_s f_y - (f_s' - 0.85 f_c') A_s'}{0.85 f_c' b_w}$$

$$= \frac{(4.00 \text{ in}^2)\left(72{,}000 \, \dfrac{\text{lbf}}{\text{in}^2}\right) - \left(36{,}000 \, \dfrac{\text{lbf}}{\text{in}^2} - (0.85)\left(3400 \, \dfrac{\text{lbf}}{\text{in}^2}\right)\right)(3.00 \text{ in}^2)}{(0.85)\left(3400 \, \dfrac{\text{lbf}}{\text{in}^2}\right)(14 \text{ in})}$$

$$= 4.66 \text{ in}$$

$$c = \frac{a}{\beta_1} = \frac{4.66 \text{ in}}{0.85}$$

$$= 5.48 \text{ in}$$

$$\frac{\epsilon_s'}{c - d'} = \frac{\epsilon_c}{d - c}$$

$$\epsilon_s' = \frac{\epsilon_c(c - d')}{d - c}$$

$$= \frac{(0.003)(5.48 \text{ in} - 2.5 \text{ in})}{22 \text{ in} - 5.48 \text{ in}}$$

$$= 0.00054 \quad [< \epsilon_y = 0.0025]$$

$$f_s' = E_s \epsilon_s'$$

$$= \left(29{,}000{,}000 \, \frac{\text{lbf}}{\text{in}^2}\right)(0.00054)$$

$$= 15{,}700 \text{ psi}$$

Starting with a value of 36,000 psi resulted in a final value of 15,700 psi. Try again with an assumed value midway between these, such as $f_s' = 25{,}000$ psi.

$$a = \frac{A_s f_y - (f_s' - 0.85 f_c') A_s'}{0.85 f_c' b_w}$$

$$= \frac{\begin{aligned} &(4.00 \text{ in}^2)\left(72{,}000 \, \tfrac{\text{lbf}}{\text{in}^2}\right) \\ &- \left(25{,}000 \, \tfrac{\text{lbf}}{\text{in}^2} - (0.85)\left(3400 \, \tfrac{\text{lbf}}{\text{in}^2}\right)\right)(3.00 \text{ in}^2) \end{aligned}}{(0.85)\left(3400 \, \dfrac{\text{lbf}}{\text{in}^2}\right)(14 \text{ in})}$$

$$= 5.48 \text{ in}$$

$$c = \frac{a}{\beta_1} = \frac{5.48 \text{ in}}{0.85}$$

$$= 6.45 \text{ in}$$

$$\epsilon_s' = \frac{\epsilon_c(c - d')}{d - c}$$

$$= \frac{(0.003)(6.45 \text{ in} - 2.5 \text{ in})}{22 \text{ in} - 6.45 \text{ in}}$$

$$= 0.00076 \quad [< \epsilon_y = 0.0025]$$

$$f_s' = E_s \epsilon_s'$$

$$= \left(29{,}000{,}000 \, \frac{\text{lbf}}{\text{in}^2}\right)(0.00076)$$

$$= 22{,}000 \text{ psi}$$

This is close enough to the assumed value of 25,000 psi. Therefore, base the probable moment strength on a stress in compression steel equal to 25,000 psi.

$$M_{\text{pr,left}} = C_c\left(d - \frac{a}{2}\right) + C_s'(d - d')$$

$$= 0.85f_c'ba\left(d - \frac{a}{2}\right) + (f_s' - 0.85f_c')A_s'(d - d')$$

$$= \frac{\begin{aligned}(0.85)\left(3400\ \frac{\text{lbf}}{\text{in}^2}\right)(14\ \text{in}) \\ \times (5.48\ \text{in})\left(22.0\ \text{in} - \frac{5.48\ \text{in}}{2}\right) \\ + \left(25,000\ \frac{\text{lbf}}{\text{in}^2} - (0.85)\left(3400\ \frac{\text{lbf}}{\text{in}^2}\right)\right) \\ \times (3.00\ \text{in}^2)(22.0\ \text{in} - 2.5\ \text{in})\end{aligned}}{12\ \dfrac{\text{in}}{\text{ft}}}$$

$$= 464,000\ \text{ft-lbf}$$

For positive bending at the face of the opposite support,

$$d^+ = h - 2.5\ \text{in} = 24.5\ \text{in} - 2.5\ \text{in} = 22.0\ \text{in}$$

For positive bending, the slab contributes compression resistance over a width b_e. The neutral axis will likely be above the top steel, so treat this as a singly reinforced beam.

$$b_e \leq \begin{cases} \dfrac{L}{4} = \dfrac{(24\ \text{ft})\left(12\ \frac{\text{in}}{\text{ft}}\right)}{4} \\ \qquad = 72\ \text{in} \quad [\text{controls}] \\ \text{beam spacing (not applicable)} \\ b_w + 16h_s = 14\ \text{in} + (16)(4.5\ \text{in}) \\ \qquad = 86\ \text{in} \end{cases}$$

$$a = \frac{A_s f_y}{0.85 f_c' b} = \frac{(3.00\ \text{in}^2)\left(72,000\ \frac{\text{lbf}}{\text{in}^2}\right)}{(0.85)\left(3400\ \frac{\text{lbf}}{\text{in}^2}\right)(72\ \text{in})}$$

$$= 1.04\ \text{in}$$

$$c = \frac{a}{\beta_1} = \frac{1.04\ \text{in}}{0.85} = 1.22\ \text{in}$$

Thus, top steel is below the neutral axis as assumed.

$$M_{\text{pr,right}} = A_s f_y\left(d - \frac{a}{2}\right)$$

$$= (3.00\ \text{in}^2)\left(72,000\ \frac{\text{lbf}}{\text{in}^2}\right)\left(\frac{22.0\ \text{in} - \frac{1.04\ \text{in}}{2}}{12\ \frac{\text{in}}{\text{ft}}}\right)$$

$$= 387,000\ \text{ft-lbf}$$

From ACI 318 Sec. 21.5.4.1, the seismic shear is computed based on the probable moments acting at faces of supports in combination with the factored tributary gravity loads.

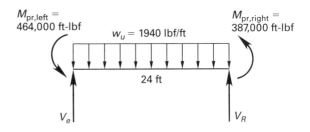

The gravity load acting in combination with earthquake load is

$$w_u = 1.2w_D + 0.5w_L$$

$$= (1.2)\left(1200\ \frac{\text{lbf}}{\text{ft}}\right) + (0.5)\left(1000\ \frac{\text{lbf}}{\text{ft}}\right)$$

$$= 1940\ \text{lbf/ft}$$

Maximum shear occurs at the left end when hinges form in the pattern shown.

$$V_e = \frac{w_u l_n}{2} + \frac{M_{\text{pr,left}} + M_{\text{pr,right}}}{l_n}$$

$$= \frac{\left(1940\ \frac{\text{lbf}}{\text{ft}}\right)(24\ \text{ft})}{2} + \frac{464,000\ \text{ft-lbf} + 387,000\ \text{ft-lbf}}{24\ \text{ft}}$$

$$= 58,800\ \text{lbf}$$

Thus, the development of hinges in the beam at the faces of the columns contributes the majority of the shear (about 60%). Given that the member was designed based on the specified material properties, it should have been designed for probable moment strengths that were based on $1.25f_y = (1.25)(40,000\ \text{psi}) = 50,000\ \text{psi}$ and $f_c' = 3000\ \text{psi}$. Considering that the f_c' has practically no effect on moment strength, the probable moment strengths can be estimated in direct proportion to the yield stresses. Thus,

$$V_{e,\text{original}} = \frac{w_u l_n}{2} + \left(\frac{f_y}{f_{y,\text{test}}}\right)\left(\frac{M_{\text{pr,left}} + M_{\text{pr,right}}}{l_n}\right)$$

$$= \frac{\left(1940\ \frac{\text{lbf}}{\text{ft}}\right)(24\ \text{ft})}{2} + \left(\frac{40,000\ \frac{\text{lbf}}{\text{in}^2}}{72,000\ \frac{\text{lbf}}{\text{in}^2}}\right)$$

$$\times \left(\frac{464,000\ \text{ft-lbf} + 387,000\ \text{ft-lbf}}{24\ \text{ft}}\right)$$

$$= 43,000\ \text{lbf}$$

Thus, the effect of the higher material strength is to increase the design shear by approximately 35% (from 43,000 lbf to 58,800 lbf).

15.2. Calculate the spacing of no. 3 stirrups to resist the shear. $V_e = 58{,}900$ lbf with $A_v = 0.22$ in², $f'_c = 3400$ psi, $f_y = 40{,}000$ psi for the stirrups, $b_w = 14$ in, and $d = 22$ in. Per ACI 318 Sec. 21.5.4, because the seismic contribution to the design shear exceeds one-half the required shear and the factored axial force is less than $A_g f'_c / 20$, the contribution of V_c to shear resistance is to be neglected. Thus,

$$V_s = \frac{V_e}{\phi}$$

$$= \frac{58{,}800 \text{ lbf}}{0.75}$$

$$= 78{,}400 \text{ lbf}$$

$$4\sqrt{f'_c}\, b_w d = 4\sqrt{3400 \ \frac{\text{lbf}}{\text{in}^2}}\,(14 \text{ in})(22 \text{ in})$$

$$= 71{,}800 \text{ lbf} \quad \left[V_s > 4\sqrt{f'_c}\, b_w d\right]$$

$$0.75\sqrt{f'_c} = 0.75\sqrt{3400 \ \frac{\text{lbf}}{\text{in}^2}}$$

$$= 44 \text{ psi} \quad [\text{use } 50 \text{ psi}]$$

$$s \le \begin{cases} \dfrac{d}{4} = \dfrac{22.0 \text{ in}}{4} = 5.5 \text{ in} \\[3mm] \dfrac{A_v f_y}{50 b_w} = \dfrac{(0.22 \text{ in}^2)\left(40{,}000 \ \frac{\text{lbf}}{\text{in}^2}\right)}{\left(50 \ \frac{\text{lbf}}{\text{in}^2}\right)(14 \text{ in})} \\[3mm] \qquad\qquad = 12.5 \text{ in} \\[3mm] \dfrac{A_v f_y d}{V_s} = \dfrac{(0.22 \text{ in}^2)\left(40{,}000 \ \frac{\text{lbf}}{\text{in}^2}\right)(22.0 \text{ in})}{78{,}400 \text{ lbf}} \\[3mm] \qquad\qquad = 2.5 \text{ in} \quad [\text{controls}] \end{cases}$$

In the hinge regions, closed hoops are required at minimum spacing (ACI 318 Sec. 21.5.3.2).

$$s \le \begin{cases} \dfrac{d}{4} = \dfrac{22 \text{ in}}{4} = 5.5 \text{ in} \\[2mm] 8 d_{b,l} = (8)(1.125 \text{ in}) = 9 \text{ in} \\[2mm] 24 d_{b,t} = (24)(0.375 \text{ in}) = 9 \text{ in} \\[2mm] 12 \text{ in} \end{cases}$$

Thus, shear requirement controls. Use no. 3 stirrups at $\boxed{2^{1}/_2 \text{ in o.c.}}$

SOLUTION 16

Assume that the roof system is simply supported by the walls. Use the ACI 318 strength design method, with values of $f'_c = 3000$ psi and $f_y = 40{,}000$ psi. For the 8 in slab with the given superimposed loads,

$$w_u = 1.2 w_D + 1.6 w_L = 1.2(w_{\text{sup}} + w_{\text{slab}}) + 1.6 w_L$$

$$= 1.2(w_{\text{sup}} + w_c h) + 1.6 w_L$$

$$= (1.2)\left(20 \ \frac{\text{lbf}}{\text{ft}^2} + \left(150 \ \frac{\text{lbf}}{\text{ft}^3}\right)\left(\frac{8 \text{ in}}{12 \ \frac{\text{in}}{\text{ft}}}\right)\right)$$

$$\quad + (1.6)\left(50 \ \frac{\text{lbf}}{\text{ft}^2}\right)$$

$$= 224 \text{ lbf/ft}^2$$

16.1. For the case with a 30 in deep girder, compare the stiffness in each direction. For this analysis, use an effective flange width of

$$b_e = 16 h_f + b_w = (16)(8 \text{ in}) + 36 \text{ in}$$

$$= 164 \text{ in}$$

$$A = b_e h_s + b(h - h_s)$$

$$= (164 \text{ in})(8 \text{ in}) + (36 \text{ in})(30 \text{ in} - 8 \text{ in})$$

$$= 2104 \text{ in}^2$$

$$\bar{y} = \frac{\sum A_i \bar{y}_i}{A} = \frac{b_e h_s \left(h - \dfrac{h_s}{2}\right) + \dfrac{b(h - h_s)^2}{2}}{A}$$

$$= \frac{(164 \text{ in})(8 \text{ in})\left(30 \text{ in} - \dfrac{8 \text{ in}}{2}\right) + \dfrac{(36 \text{ in})(30 \text{ in} - 8 \text{ in})^2}{2}}{2104 \text{ in}^2}$$

$$= 20.35 \text{ in}$$

$$I_b = \frac{b_e h_s^3}{12} + b_e h_s \left(h - \frac{h_s}{2} - \bar{y}\right)^2 + \frac{b_w (h - h_s)^3}{12}$$

$$\quad + b_w (h - h_s)\left(\frac{h - h_s}{2} - \bar{y}\right)^2$$

$$= \frac{(164 \text{ in})(8 \text{ in})^3}{12} + (164 \text{ in})(8 \text{ in})$$

$$\quad \times \left(30 \text{ in} - \frac{8 \text{ in}}{2} - 20.35 \text{ in}\right)^2$$

$$\quad + \frac{(36 \text{ in})(30 \text{ in} - 8 \text{ in})^3}{12} + (36 \text{ in})$$

$$\quad \times (30 \text{ in} - 8 \text{ in})\left(\frac{30 \text{ in} - 8 \text{ in}}{2} - 20.35 \text{ in}\right)^2$$

$$= 150{,}000 \text{ in}^4$$

Compare the moment of inertia of the beam to that of an 8 in thick slab of the same width spanning the perpendicular direction. Let r be the fraction of the force (applied at the intersection) carried by the beam.

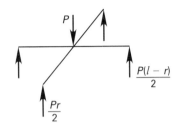

$$I_s = \frac{b_e h_s^3}{12} = \frac{(164 \text{ in})(8 \text{ in})^3}{12} = 6997 \text{ in}^4$$

$$\frac{rPL_1^3}{48EI_b} = \frac{(1-r)PL_2^3}{48EI_s}$$

$$r = \frac{L_2^3 I_b}{L_1^3 I_s + L_2^3 I_b}$$

$$= \frac{(30 \text{ ft})^3 (150{,}000 \text{ in}^4)}{(29.33 \text{ ft})^3 (6997 \text{ in}^4) + (30 \text{ ft})^3 (150{,}000 \text{ in}^4)}$$

$$= 0.96$$

Thus, the beam is rigid relative to the slab. Design the slab as a one-way slab spanning from wall to beam. Per the problem statement, use the moments at support centerline and neglect alternate span loading conditions. Consider a unit width of slab with values of $b = 12$ in and

$$d = h_s - \text{cover} - \frac{d_b}{2} = 8 \text{ in} - 1.0 \text{ in} - \frac{1.0 \text{ in}}{2} = 6.5 \text{ in}$$

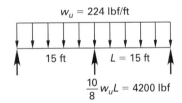

$w_u = 224$ lbf/ft

15 ft $L = 15$ ft

$\frac{10}{8} w_u L = 4200$ lbf

Critical shear occurs d-distance from the face of the interior support.

$$V_u = \frac{5 w_u L}{8} - \frac{w_u b_w}{2}$$

$$= \frac{(5)\left(224 \frac{\text{lbf}}{\text{ft}}\right)(15 \text{ ft})}{8} - \frac{\left(224 \frac{\text{lbf}}{\text{ft}}\right)(36 \text{ in})}{(2)\left(12 \frac{\text{in}}{\text{ft}}\right)}$$

$$= 1770 \text{ lbf}$$

$$\phi V_c = 2\phi\lambda\sqrt{f_c'} b_w d$$

$$= (2)(0.75)(1.0)\sqrt{3000 \frac{\text{lbf}}{\text{in}^2}}(12 \text{ in})(6.5 \text{ in})$$

$$= 6400 \text{ lbf}$$

Slab has adequate shear strength. Check the serviceability requirements of ACI 318 Sec. 9.5. For the 8 in slab with one end continuous, $f_y = 40{,}000$ psi.

$$h_s \geq \left(\frac{l_n}{24}\right)\left(0.4 + \frac{f_y}{100{,}000}\right)$$

$$\geq \left(\frac{(15 \text{ ft})\left(12 \frac{\text{in}}{\text{ft}}\right)}{24}\right)\left(0.4 + \frac{40{,}000 \frac{\text{lbf}}{\text{in}^2}}{100{,}000 \frac{\text{lbf}}{\text{in}^2}}\right)$$

$$\geq 6 \text{ in}$$

The 8 in slab is OK. Check the 30 in deep simple span beam (ACI 318 Sec. 9.5).

$$h_s \geq \left(\frac{l_n}{16}\right)\left(0.4 + \frac{f_y}{100{,}000}\right)$$

$$\geq \left(\frac{(29.33 \text{ ft})\left(12 \frac{\text{in}}{\text{ft}}\right)}{16}\right)\left(0.4 + \frac{40{,}000 \frac{\text{lbf}}{\text{in}^2}}{100{,}000 \frac{\text{lbf}}{\text{in}^2}}\right)$$

$$\geq 18 \text{ in}$$

The 30 in depth is adequate and deflection calculations are not required. Calculate flexural steel for the slab.

$$M_u^+ = \frac{w_u l_n^2}{8} = \frac{\left(224 \frac{\text{lbf}}{\text{ft}}\right)(15 \text{ ft})^2}{8}$$

$$= 6300 \text{ ft-lbf}$$

$$M_u = \phi M_n = \phi\rho b d^2 f_y\left(1 - 0.59\rho\left(\frac{f_y}{f_c'}\right)\right)$$

$$(6300 \text{ ft-lbf})\left(12 \frac{\text{in}}{\text{ft}}\right)$$

$$= 0.9\rho(12 \text{ in})(6.5 \text{ in})^2\left(40{,}000 \frac{\text{lbf}}{\text{in}^2}\right)$$

$$\times \left(1 - 0.59\rho\left(\frac{40{,}000 \frac{\text{lbf}}{\text{in}^2}}{3000 \frac{\text{lbf}}{\text{in}^2}}\right)\right)$$

$$\rho = 0.00430$$

$$A_s = \rho b d = (0.00430)(12 \text{ in})(6.5 \text{ in}) = 0.34 \text{ in}^2$$

Minimum slab reinforcement (ACI 318 Sec. 7.12) is

$$A_{s,\text{min}} = 0.002 b h_s$$

$$= (0.002)(12 \text{ in})(8 \text{ in})$$

$$= 0.19 \text{ in}^2 \quad [\text{does not control main steel}]$$

Use no. 5 bars at 11 in o.c. for top main steel.

For bottom main steel,

$$M_u^+ = \frac{9 w_u l_n^2}{128} = \frac{(9)\left(224 \frac{\text{lbf}}{\text{ft}}\right)(15 \text{ ft})^2}{128} = 3540 \text{ ft-lbf}$$

$$M_u = \phi M_n = \phi \rho b d^2 f_y \left(1 - 0.59\rho\left(\frac{f_y}{f'_c}\right)\right)$$

$$(3540 \text{ ft-lbf})\left(12 \frac{\text{in}}{\text{ft}}\right)$$

$$= 0.9\rho(12 \text{ in})(6.5 \text{ in})^2 \left(40,000 \frac{\text{lbf}}{\text{in}^2}\right)$$

$$\times \left(1 - 0.59\rho\left(\frac{40,000 \frac{\text{lbf}}{\text{in}^2}}{3000 \frac{\text{lbf}}{\text{in}^2}}\right)\right)$$

$$\rho = 0.00237$$

$$A_s = \rho b d = (0.00237)(12 \text{ in})(6.5 \text{ in}) = 0.19 \text{ in}^2$$

Use no. 4 bars at 12 in o.c. for bottom main steel. Use no. 4 bars at 12 in o.c. for temperature and shrinkage.

For the beam,

$$w_{u,b} = \frac{10\left(\frac{w_u}{1 \text{ ft}}\right)l_n}{8} + 1.2w_b = \frac{10\left(\frac{w_u}{1 \text{ ft}}\right)l_n}{8}$$
$$+ 1.2w_c b_w(h - h_s)$$

$$= \frac{(10)\left(\frac{224 \frac{\text{lbf}}{\text{ft}}}{1 \text{ ft}}\right)(15 \text{ ft})}{8} + (1.2)\left(150 \frac{\text{lbf}}{\text{ft}^3}\right)}{\left(12 \frac{\text{in}}{\text{ft}}\right)^2}$$

$$= 5190 \text{ lbf/ft}$$

$$M_u = \frac{w_{u,b}L^2}{8} = \frac{\left(5190 \frac{\text{lbf}}{\text{ft}}\right)(29.33 \text{ ft})^2}{8} = 558,000 \text{ ft-lbf}$$

$$d = h - \text{cover} - \frac{d_b}{2} = 30 \text{ in} - 2 \text{ in} - \frac{1 \text{ in}}{2} = 27.5 \text{ in}$$

$$b_e \leq \begin{cases} \dfrac{L}{4} = \dfrac{(29.33 \text{ ft})\left(12 \frac{\text{in}}{\text{ft}}\right)}{4} \\ \quad = 88 \text{ in} \quad [\text{controls}] \\ s = (15 \text{ ft})\left(12 \frac{\text{in}}{\text{ft}}\right) = 180 \text{ in} \\ b_w + 16h_s = 36 \text{ in} + (16)(8 \text{ in}) \\ \quad = 164 \text{ in} \end{cases}$$

$$M_u = \phi M_n = \phi \rho b d^2 f_y \left(1 - 0.59\rho\left(\frac{f_y}{f'_c}\right)\right)$$

$$(558,000 \text{ ft-lbf})\left(12 \frac{\text{in}}{\text{ft}}\right)$$

$$= 0.9\rho(88 \text{ in})(27.5 \text{ in})^2 \left(40,000 \frac{\text{lbf}}{\text{in}^2}\right)$$

$$\times \left(1 - 0.59\rho\left(\frac{40,000 \frac{\text{lbf}}{\text{in}^2}}{3000 \frac{\text{lbf}}{\text{in}^2}}\right)\right)$$

$$\rho = 0.00286$$

$$A_s = \rho b d = (0.00286)(88 \text{ in})(27.5 \text{ in}) = 6.92 \text{ in}^2$$

Check the neutral axis position at ultimate.

$$a = \frac{A_s f_y}{0.85 f'_c b} = \frac{(6.92 \text{ in}^2)\left(40,000 \frac{\text{lbf}}{\text{in}^2}\right)}{(0.85)\left(3000 \frac{\text{lbf}}{\text{in}^2}\right)(88 \text{ in})} = 1.23 \text{ in}$$

$$c = \frac{a}{\beta_1} = \frac{1.23 \text{ in}}{0.85} = 1.44 \text{ in} \quad [< h_s < 0.375d]$$

Therefore, beam behaves as rectangular and is a tension-controlled section.

$$A_{s,\min} \geq \begin{cases} \dfrac{200 b_w d}{f_y} = \dfrac{\left(200 \frac{\text{lbf}}{\text{in}^2}\right)(36 \text{ in})(27.5 \text{ in})}{40,000 \frac{\text{lbf}}{\text{in}^2}} \\ \quad = 4.95 \text{ in}^2 \\ \dfrac{3\sqrt{f'_c}b_w d}{f_y} = \dfrac{3\sqrt{3000 \frac{\text{lbf}}{\text{in}^2}}(36 \text{ in})(27.5 \text{ in})}{40,000 \frac{\text{lbf}}{\text{in}^2}} \\ \quad = 4.06 \text{ in}^2 \end{cases}$$

The required A_s exceeds the minimum required; therefore, use seven no. 9 bottom bars.

Check shear at d-distance from face of support.

$$V_c = 2\lambda\sqrt{f'_c}b_w d$$

$$= (2)(1.0)\sqrt{3000 \frac{\text{lbf}}{\text{in}^2}}(36 \text{ in})(27.5 \text{ in})$$

$$= 108,500 \text{ lbf}$$

$$V_u = \frac{w_u L}{2} - w_u d$$

$$= \frac{\left(5190 \frac{\text{lbf}}{\text{ft}}\right)(29.33 \text{ ft})}{2} - \left(5190 \frac{\text{lbf}}{\text{ft}}\right)\left(\frac{27.5 \text{ in}}{12 \frac{\text{in}}{\text{ft}}}\right)$$

$$= 64,200 \text{ lbf}$$

Minimum stirrup spacing controls. Use no. 4 stirrups at $d/2 = 27.5 \text{ in}/2 = 13 \text{ in o.c.}$ Discontinue stirrups when $V_u \le \phi V_c/2$.

$$V_u = \frac{w_u L}{2} - w_u x = \frac{\phi V_c}{2}$$

$$x = \frac{w_u L - \phi V_c}{2 w_u}$$

$$= \frac{\left(5190 \; \frac{\text{lbf}}{\text{ft}}\right)(29.33 \text{ ft}) - (0.75)(108{,}500 \text{ lbf})}{(2)\left(5190 \; \frac{\text{lbf}}{\text{ft}}\right)}$$

$$= 6.8 \text{ ft} \quad [\text{from each end}]$$

Therefore, the section may be drawn as shown.

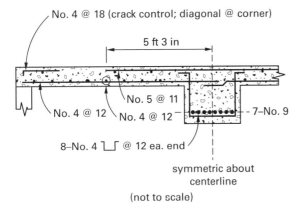

No. 4 @ 18 (crack control; diagonal @ corner)

5 ft 3 in

No. 5 @ 11

No. 4 @ 12   No. 4 @ 12   7–No. 9

8–No. 4 ⌐ @ 12 ea. end

symmetric about
centerline

(not to scale)

16.2. Redesign for the case where the beam is 36 in wide × 15 in deep. For the beam strip,

$$A = b_e h_s + b(h - h_s)$$

$$= (164 \text{ in})(8 \text{ in}) + (36 \text{ in})(15 \text{ in} - 8 \text{ in})$$

$$= 1564 \text{ in}^2$$

$$\bar{y} = \frac{\sum A_i \bar{y}_i}{A} = \frac{b_e h_s \left(h - \frac{h_s}{2}\right) + \frac{b(h - h_s)^2}{2}}{A}$$

$$= \frac{(164 \text{ in})(8 \text{ in})\left(15 \text{ in} - \frac{8 \text{ in}}{2}\right) + \frac{(36 \text{ in})(15 \text{ in} - 8 \text{ in})^2}{2}}{1564 \text{ in}^2}$$

$$= 9.78 \text{ in}$$

$$I_b = \frac{b_e h_s^3}{12} + b_e h_s \left(h - \frac{h_s}{2} - \bar{y}\right)^2 + \frac{b_w(h - h_s)^3}{12}$$

$$\qquad + b_w(h - h_s)\left(\frac{h - h_s}{2} - \bar{y}\right)^2$$

$$= \frac{(164 \text{ in})(8 \text{ in})^3}{12} + (164 \text{ in})(8 \text{ in})$$

$$\qquad \times \left(15 \text{ in} - \frac{8 \text{ in}}{2} - 9.78 \text{ in}\right)^2$$

$$\qquad + \frac{(36 \text{ in})(15 \text{ in} - 8 \text{ in})^3}{12}$$

$$\qquad + (36 \text{ in})(15 \text{ in} - 8 \text{ in})$$

$$\qquad \times \left(\frac{15 \text{ in} - 8 \text{ in}}{2} - 9.78 \text{ in}\right)^2$$

$$= 19{,}900 \text{ in}^4$$

$$\frac{r P L_1^3}{48 E I_b} = \frac{(1 - r) P L_2^3}{48 E I_s}$$

$$r = \frac{L_2^3 I_b}{L_1^3 I_s + L_2^3 I_b}$$

$$= \frac{(30 \text{ ft})^3 (19{,}900 \text{ in}^4)}{(29.33 \text{ ft})^3 (6997 \text{ in}^4) + (30 \text{ ft})^3 (19{,}900 \text{ in}^4)}$$

$$= 0.75$$

In this case, the system has similar stiffness in the two directions; therefore, design it as a two-way beam-slab. Use the distribution criteria of ACI 318 Sec. 13.6.

$$\alpha_{f1} = \frac{E I_b}{E I_s} = \frac{I_b}{I_s}$$

$$= \frac{19{,}900 \text{ in}^4}{6997 \text{ in}^4}$$

$$= 2.84$$

$$\alpha_{f2} = \frac{E I_s}{E I_s}$$

$$= 1.0$$

$$\frac{\alpha_{f1} L_2}{\alpha_{f2} L_1} = \frac{(2.84)(30 \text{ ft})}{(1.0)(29.33 \text{ ft})}$$

$$\approx 2.90$$

The stiffer direction will be designed to carry 65% of the slab loading. Thus, for the 29.33 ft span,

$$w_{u,b} = 0.65 w_u L_2 + 1.2 w_b$$

$$= 0.65 w_u L_2 + 1.2 w_c b_w (h - h_s)$$

$$= (0.65)\left(224 \ \frac{\text{lbf}}{\text{ft}^2}\right)(30 \ \text{ft}) + (1.2)\left(150 \ \frac{\text{lbf}}{\text{ft}^3}\right)$$

$$\times \frac{(36 \ \text{in})(15 \ \text{in} - 8 \ \text{in})}{\left(12 \ \frac{\text{in}}{\text{ft}}\right)^2}$$

$$= 4685 \ \text{lbf/ft}$$

$$M_0 = \frac{w_{u,b} L^2}{8}$$

$$= \frac{\left(4685 \ \frac{\text{lbf}}{\text{ft}}\right)(29.33 \ \text{ft})^2}{8}$$

$$= 503{,}800 \ \text{ft-lbf}$$

For the positive moment in an exterior span, ACI 318 Sec. 13.6.3.3 allocates 63% to the central strip and ACI 318 Sec. 13.6.5.1 assigns 85% of this value to the beam. The slab spanning parallel to the beam resists the complementary portion.

$$M_u = (0.63)(0.85) M_0$$

$$= (0.63)(0.85)(503{,}800 \ \text{ft-lbf})$$

$$= 269{,}800 \ \text{ft-lbf}$$

For this system, the effective width of the slab is given by ACI 318 Sec. 13.2.4.

$$b_e \leq \begin{cases} b_w + 2h_b = 36 \ \text{in} + (2)(15 \ \text{in}) = 66 \ \text{in} \quad \text{[controls]} \\ b_w + 8h_s = 36 \ \text{in} + (8)(8 \ \text{in}) = 100 \ \text{in} \end{cases}$$

$$d = h - \text{cover} - \frac{d_b}{2} = 15 \ \text{in} - 2.0 \ \text{in} - \frac{1 \ \text{in}}{2}$$

$$= 12.5 \ \text{in}$$

$$M_u = \phi M_n = \phi \rho b d^2 f_y \left(1 - 0.59 \rho \left(\frac{f_y}{f_c'}\right)\right)$$

$$(269{,}800 \ \text{ft-lbf})\left(12 \ \frac{\text{in}}{\text{ft}}\right)$$

$$= 0.9 \rho (66 \ \text{in})(12.5 \ \text{in})^2 \left(40{,}000 \ \frac{\text{lbf}}{\text{in}^2}\right)$$

$$\times \left(1 - 0.59\rho \left(\frac{40{,}000 \ \frac{\text{lbf}}{\text{in}^2}}{3000 \ \frac{\text{lbf}}{\text{in}^2}}\right)\right)$$

$$\rho = 0.00834$$

$$A_s = \rho b d = (0.00834)(66 \ \text{in}^2)(12.5 \ \text{in}) = 6.88 \ \text{in}^2$$

Compute the neutral axis at ultimate.

$$a = \frac{A_s f_y}{0.85 f_c' b} = \frac{(6.88 \ \text{in}^2)\left(40{,}000 \ \frac{\text{lbf}}{\text{in}^2}\right)}{(0.85)\left(3000 \ \frac{\text{lbf}}{\text{in}^2}\right)(66 \ \text{in})}$$

$$= 1.63 \ \text{in}$$

$$c = \frac{a}{\beta_1} = \frac{1.63 \ \text{in}}{0.85}$$

$$= 1.92 \ \text{in} \quad [<h_s < 0.375d]$$

Therefore, the beam behaves as rectangular and is a tension-controlled section. Use eight no. 9 bottom bars.

Check the shear at distance d from the face of support.

$$V_u = \frac{w_u L}{2} - w_u d$$

$$= \frac{\left(4820 \ \frac{\text{lbf}}{\text{ft}}\right)(29.33 \ \text{ft})}{2} - \left(4820 \ \frac{\text{lbf}}{\text{ft}}\right)\left(\frac{12.5 \ \text{in}}{12 \ \frac{\text{in}}{\text{ft}}}\right)$$

$$= 65{,}700 \ \text{lbf}$$

$$V_c = 2\lambda\sqrt{f_c'} b_w d$$

$$= (2)(1.0)\sqrt{3000 \ \frac{\text{lbf}}{\text{in}^2}}(36 \ \text{in})(12.5 \ \text{in})$$

$$= 49{,}300 \ \text{lbf}$$

Stirrups are required. Use no. 4 U-stirrups with $A_v = 2A_b = (2)(0.20 \ \text{in}^2) = 0.40 \ \text{in}^2$.

$$V_s = \frac{V_u}{\phi} - V_c$$

$$= \frac{65{,}700 \ \text{lbf}}{0.75} - 49{,}300 \ \text{lbf}$$

$$= 38{,}300 \ \text{lbf}$$

$$4\sqrt{f_c'} b_w d = 2V_c = (2)(49{,}300 \ \text{lbf})$$

$$= 98{,}600 \ \text{lbf} \quad [>V_s]$$

$$0.75\sqrt{f_c'} = 0.75\sqrt{3000 \ \frac{\text{lbf}}{\text{in}^2}}$$

$$= 41 \ \text{psi} \quad [\text{use 50 psi}]$$

$$s \leq \begin{cases} \dfrac{d}{2} = \dfrac{12.5 \text{ in}}{2} = 6.3 \text{ in} \\[2mm] \dfrac{A_v f_y}{50 b_w} = \dfrac{(0.40 \text{ in}^2)\left(40{,}000 \,\frac{\text{lbf}}{\text{in}^2}\right)}{\left(50 \,\frac{\text{lbf}}{\text{in}^2}\right)(36 \text{ in})} \\[4mm] \qquad = 8.9 \text{ in} \\[2mm] \dfrac{A_v f_y d}{V_s} = \dfrac{(0.40 \text{ in}^2)\left(40{,}000 \,\frac{\text{lbf}}{\text{in}^2}\right)(12.5 \text{ in})}{38{,}300 \text{ lbf}} \\[4mm] \qquad = 5.2 \text{ in} \quad [\text{controls}] \end{cases}$$

Use no. 4 U-stirrups at 5 in o.c. at the critical location. Terminate where $V_u < \phi V_c/2$.

$$V_u = \frac{w_u L}{2} - w_u x = \frac{\phi V_c}{2}$$

$$x = \frac{w_u L - \phi V_c}{2 w_u}$$

$$= \frac{\left(4820 \,\frac{\text{lbf}}{\text{ft}}\right)(29.33 \text{ ft}) - (0.75)(49{,}300 \text{ lbf})}{(2)\left(4820 \,\frac{\text{lbf}}{\text{ft}}\right)}$$

$$= 10.8 \text{ ft} \quad [\text{from each end}]$$

The slab must resist the remainder of the moment, M_0.

$$m_1 = \frac{(0.15)(0.63 M_0)}{B_1}$$

$$= \frac{(0.15)(0.63)(518{,}300 \text{ ft-lbf})}{15 \text{ ft}}$$

$$= 3300 \text{ ft-lbf/ft}$$

$$m_2 = \frac{0.37 M_0}{B_2} = \frac{(0.37)(518{,}300 \text{ ft-lbf})}{15 \text{ ft}}$$

$$= 12{,}800 \text{ ft-lbf/ft}$$

For the central region,

$$M_u = \phi M_n = \phi \rho b d^2 f_y \left(1 - 0.59 \rho \left(\frac{f_y}{f'_c}\right)\right)$$

$$(12{,}800 \text{ ft-lbf})\left(12 \,\frac{\text{in}}{\text{ft}}\right)$$

$$= 0.9 \rho (12 \text{ in})(6.5 \text{ in})^2 \left(40{,}000 \,\frac{\text{lbf}}{\text{in}^2}\right)$$

$$\times \left(1 - 0.59 \rho \left(\frac{40{,}000 \,\frac{\text{lbf}}{\text{in}^2}}{3000 \,\frac{\text{lbf}}{\text{in}^2}}\right)\right)$$

$$\rho = 0.00906$$

$$A_s = \rho b d = (0.00906)(12 \text{ in})(6.5 \text{ in})$$

$$= 0.71 \text{ in}^2 \quad [\text{per foot width}]$$

Minimum slab reinforcement is

$$A_{s,\min} = 0.002 b h_s$$

$$= (0.002)(12 \text{ in})(8 \text{ in})$$

$$= 0.19 \text{ in}^2 \quad [\text{does not control main steel}]$$

Use no. 7 bars at 10 in o.c. for bottom steel parallel to beam.

For the perpendicular direction,

$$M_0 = \frac{0.35 w_u L_1^2}{8}$$

$$= \frac{(0.35)\left(224 \,\frac{\text{lbf}}{\text{ft}}\right)(30 \text{ ft})^2}{8}$$

$$= 8800 \text{ ft-lbf}$$

$$M_u = \phi M_n = \phi \rho b d^2 f_y \left(1 - 0.59 \rho \left(\frac{f_y}{f'_c}\right)\right)$$

$$(8800 \text{ ft-lbf})\left(12 \,\frac{\text{in}}{\text{ft}}\right)$$

$$= 0.9 \rho (12 \text{ in})(6.5 \text{ in})^2 \left(40{,}000 \,\frac{\text{lbf}}{\text{in}^2}\right)$$

$$\times \left(1 - 0.59 \rho \left(\frac{40{,}000 \,\frac{\text{lbf}}{\text{in}^2}}{3000 \,\frac{\text{lbf}}{\text{in}^2}}\right)\right)$$

$$\rho = 0.00607$$

$$A_s = \rho b d = (0.00607)(12 \text{ in})(6.5 \text{ in})$$

$$= 0.47 \text{ in}^2 \quad [\text{per foot width}]$$

Use no. 6 bars at 10 in o.c.

Even though simple support conditions were assumed, provide minimum top steel (no. 4 at 12 in o.c. for crack) control extending a nominal distance (for example, 0.2L) into the span. Also, diagonal top and bottom steel is required at the corners (ACI 318 Sec. 13.3.6). Use no. 6 bars at 9 in o.c. The section may be drawn as shown.

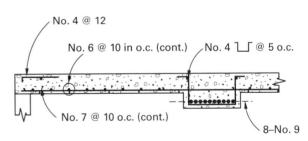

section
(wall reinforcement omitted for clarity)
(not to scale)

SOLUTION 17

17.1. Design a composite beam for the specified conditions. Check the adequacy of the given longitudinal steel, assuming that composite action can be achieved and that the shear strength of the section is adequate. Use the ACI 318 strength design method, with $f'_c = 3000$ psi and $f_y = 40,000$ psi. Allow 20 lbf/ft² for superimposed dead load. For office occupancy, ASCE/SEI7 Table 4-1 requires $w_o = 50$ lbf/ft². Reduce the live loads per ASCE/SEI7 Sec. 4.7.

$$L = L_o \left(0.25 + \frac{15}{\sqrt{K_{LL} A_t}} \right) \geq 0.5 L_o$$

$$= \left(50 \frac{\text{lbf}}{\text{ft}^2} \right) \left(0.25 + \frac{15}{\sqrt{(2)(12 \text{ ft})(35 \text{ ft})}} \right)$$

$$= 38 \text{ lbf/ft}^2$$

$$w_u = 1.2 w_D + 1.6 w_L$$

$$= 1.2 \big((w_{\text{sup}} + w_s) s + w_b \big) + 1.6 w_L s$$

$$= 1.2 \big((w_{\text{sup}} + w_c h_s) s + w_c b h_b \big) + 1.6 w_L s$$

$$= (1.2) \left(\begin{array}{c} \left(\left(20 \frac{\text{lbf}}{\text{ft}^2} + \left(150 \frac{\text{lbf}}{\text{ft}^3} \right) \left(\frac{6 \text{ in}}{12 \frac{\text{in}}{\text{ft}}} \right) \right) (12 \text{ ft}) \right. \\ \\ \left. + \left(150 \frac{\text{lbf}}{\text{ft}^3} \right) \left(\frac{16 \text{ in}}{12 \frac{\text{in}}{\text{ft}}} \right) \left(\frac{20 \text{ in}}{12 \frac{\text{in}}{\text{ft}}} \right) \right) \end{array} \right)$$

$$\quad + (1.6) \left(38 \frac{\text{lbf}}{\text{ft}^2} \right) (12 \text{ ft})$$

$$= 2500 \text{ lbf/ft}$$

The critical section is at the midspan where the opening destroys the T-flange on one side.

$$b_e \leq \begin{cases} b_w + \dfrac{L}{12} = 16 \text{ in} + \dfrac{(35 \text{ ft}) \left(12 \frac{\text{in}}{\text{ft}} \right)}{12} \\ \qquad = 51 \text{ in} \quad [\text{controls}] \\ \\ \text{spacing plus edge} = 72 \text{ in} + 8 \text{ in} \\ \qquad = 80 \text{ in} \\ \\ b_w + 6 h_s = 16 \text{ in} + (6)(6 \text{ in}) \\ \qquad = 52 \text{ in} \end{cases}$$

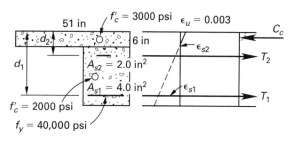

$$d_1 = h_b + h_s - \text{cover} - 0.5 d_b$$

$$= 20 \text{ in} + 6 \text{ in} - (2.5 \text{ in} + 0.375 \text{ in}) - (0.5)(1 \text{ in})$$

$$= 22.6 \text{ in}$$

$$d_2 = h_s + \text{cover} + 0.5 d_b$$

$$= 6 \text{ in} + (2.5 \text{ in} + 0.375 \text{ in}) + (0.5)(1 \text{ in})$$

$$= 9.4 \text{ in}$$

Assume that both layers of steel yield and that the beam behaves as a rectangular beam of width $b = b_e$.

$$a = \frac{(A_{s1} + A_{s2}) f_y}{0.85 f'_c b} \sqrt{f'_c}$$

$$= \frac{(4.00 \text{ in}^2 + 2.00 \text{ in}^2) \left(40,000 \frac{\text{lbf}}{\text{in}^2} \right)}{(0.85) \left(3000 \frac{\text{lbf}}{\text{in}^2} \right) (51 \text{ in})}$$

$$= 1.85 \text{ in}$$

$$c = \frac{a}{\beta_1} = \frac{1.85 \text{ in}}{0.85} = 2.2 \text{ in} \quad [< h_s = 6 \text{ in}]$$

The neutral axis is within the flange; therefore, the flexural behavior is the same as for a rectangular beam. Check strains in the two layers of longitudinal steel. By similar triangles,

$$\frac{\epsilon_{s1}}{c - d_1} = \frac{\epsilon_c}{c}$$

$$\epsilon_{s1} = \frac{(c - d_1) \epsilon_c}{c} = \frac{(22.6 \text{ in} - 2.2 \text{ in})(0.003)}{2.2 \text{ in}}$$

$$= 0.028 \quad [> 0.005 \text{ so tension controlled; } \phi = 0.9]$$

$$\frac{\epsilon_{s2}}{c - d_2} = \frac{\epsilon_c}{c}$$

$$\epsilon_{s2} = \frac{(c - d_2) \epsilon_c}{c}$$

$$= \frac{(9.4 \text{ in} - 2.2 \text{ in})(0.003)}{2.2 \text{ in}}$$

$$= 0.0098$$

$$\epsilon_y = \frac{f_y}{E_s} = \frac{40,000 \frac{\text{lbf}}{\text{in}^2}}{29,000,000 \frac{\text{lbf}}{\text{in}^2}}$$

$$= 0.00138 \quad [\epsilon_y < \epsilon_{s2}]$$

Therefore, both steel layers yield at ultimate. The flexural strength of the composite section at the critical location is

$$
\phi M_n = \phi\left(A_{s1}f_y\left(d_1 - \frac{a}{2}\right) + A_{s2}f_y\left(d_2 - \frac{a}{2}\right)\right)
$$

$$
= \frac{(0.9)\left(\begin{array}{c}(4.00 \text{ in}^2)\left(40{,}000 \dfrac{\text{lbf}}{\text{in}^2}\right) \\[4pt] \times \left(22.6 \text{ in} - \dfrac{1.85 \text{ in}}{2}\right) \\[4pt] + (2.00 \text{ in}^2)\left(40{,}000 \dfrac{\text{lbf}}{\text{in}^2}\right) \\[4pt] \times \left(9.4 \text{ in} - \dfrac{1.85 \text{ in}}{2}\right)\end{array}\right)}{\left(12 \dfrac{\text{in}}{\text{ft}}\right)\left(1000 \dfrac{\text{lbf}}{\text{kip}}\right)}
$$

$$
= 311 \text{ ft-kips}
$$

The given 35 ft span is to the outside of the 4 in thick walls, and the beam spans from center to center of the 10 in wide columns.

$$
\begin{aligned}
l_n &= L - 2t_w - h_c \\
&= 35 \text{ ft} - \frac{(2)(4 \text{ in})}{12 \dfrac{\text{in}}{\text{ft}}} - \frac{10 \text{ in}}{12 \dfrac{\text{in}}{\text{ft}}} \\
&= 33.5 \text{ ft}
\end{aligned}
$$

Top bars in the existing beam fully develop into a relatively rigid column-wall at both ends; therefore, both ends develop negative bending moments. Calculate the end moment based on $b = 16$ in, $d = d_1 - h = 22.6$ in $-$ 6 in $= 16.6$ in, and $f'_c = 2000$ psi. (The given compressive strength of the existing beam does not meet the 2500 psi minimum of ACI 318 Sec. 5.1.1; however, calculations will be based on the existing condition.)

$$
a = \frac{A_s f_y}{0.85 f'_c b} = \frac{(2.00 \text{ in}^2)\left(40{,}000 \dfrac{\text{lbf}}{\text{in}^2}\right)}{(0.85)\left(2000 \dfrac{\text{lbf}}{\text{in}^2}\right)(16 \text{ in})} = 2.94 \text{ in}
$$

$$
c = \frac{a}{\beta_1} = \frac{2.94 \text{ in}}{0.85} = 3.46 \text{ in} \quad [<0.375d, \text{ so } \phi = 0.9]
$$

$$
\begin{aligned}
\phi M_n^- &= \phi A_s f_y\left(d - \frac{a}{2}\right) \\
&= \frac{(0.9)(2.00 \text{ in}^2)\left(40{,}000 \dfrac{\text{lbf}}{\text{in}^2}\right)\left(16.6 \text{ in} - \dfrac{2.94 \text{ in}}{2}\right)}{12 \dfrac{\text{in}}{\text{ft}}} \\
&= 90{,}800 \text{ ft-lbf} \quad (90.8 \text{ ft-kips})
\end{aligned}
$$

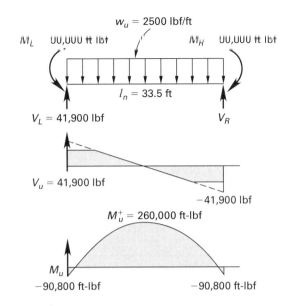

$w_u = 2500$ lbf/ft

M_L 90,000 ft-lbf M_H 90,000 ft-lbf

$l_n = 33.5$ ft

$V_L = 41{,}900$ lbf V_R

$V_u = 41{,}900$ lbf

$-41{,}900$ lbf

$M_u^+ = 260{,}000$ ft-lbf

M_u

$-90{,}800$ ft-lbf $-90{,}800$ ft-lbf

17.2. Calculate the maximum positive bending moment to determine if the composite section has adequate flexural strength.

$$
\begin{aligned}
M_u^+ &= \frac{w_u l_n^2}{8} - \phi M_n^- \\
&= \frac{\left(2500 \dfrac{\text{lbf}}{\text{ft}}\right)(33.5 \text{ ft})^2}{8} - 90{,}800 \text{ ft-lbf} \\
&= 260{,}000 \text{ ft-lbf} \quad [< \phi M_n = 311{,}000 \text{ ft-lbf}]
\end{aligned}
$$

Thus, the composite section has adequate flexural strength. Check shear capacity of the composite section. The existing smooth no. 3 bars are not acceptable under current codes, so disregard them. The specified cover of $2\frac{1}{2}$ in clear to existing stirrups will permit slots 2 in deep to be cut into the existing beam and new no. 3 stirrups installed. The new stirrups will perform multiple functions: resist vertical shear, tie the precast and cast-in-place sections, and increase horizontal shear transfer between the two parts. For the vertical shear, the critical shear occurs at distance d from support face.

$$
\begin{aligned}
V_u &= V_{\text{left}} - w_u d \\
&= 41{,}900 \text{ lbf} - \left(2500 \dfrac{\text{lbf}}{\text{ft}}\right)\left(\dfrac{22.6 \text{ in}}{12 \dfrac{\text{in}}{\text{ft}}}\right) \\
&= 37{,}200 \text{ lbf}
\end{aligned}
$$

$$
\begin{aligned}
V_c &= 2\lambda\sqrt{f'_c}\,b_w d = (2)(1.0)\sqrt{2000 \dfrac{\text{lbf}}{\text{in}^2}}(16 \text{ in})(22.6 \text{ in}) \\
&= 32{,}300 \text{ lbf}
\end{aligned}
$$

Try no. 3 stirrups ($A_v = 2A_b = (2)(0.11 \text{ in}^2) = 0.22 \text{ in}^2$).

$$
V_s = \frac{V_u}{\phi} - V_c = \frac{37{,}200 \text{ lbf}}{0.75} - 32{,}300 \text{ lbf} = 17{,}300 \text{ lbf}
$$

$$0.75\sqrt{f'_c} = 0.75\sqrt{2000 \ \frac{\text{lbf}}{\text{in}^2}} = 33.5 \text{ psi} \quad [\text{use 50 psi}]$$

$$s \leq \begin{cases} \dfrac{d}{2} = \dfrac{22.6 \text{ in}}{2} \\ \quad = 11.3 \text{ in} \\[4pt] \dfrac{A_v f_y}{50 b_w} = \dfrac{(0.22 \text{ in}^2)\left(40{,}000 \ \dfrac{\text{lbf}}{\text{in}^2}\right)}{\left(50 \ \dfrac{\text{lbf}}{\text{in}^2}\right)(16 \text{ in})} \\ \quad = 11 \text{ in} \quad [\text{controls}] \\[4pt] \dfrac{A_v f_y d}{V_s} = \dfrac{(0.22 \text{ in}^2)\left(40{,}000 \ \dfrac{\text{lbf}}{\text{in}^2}\right)(22.6 \text{ in})}{17{,}300 \text{ lbf}} \\ \quad = 11.5 \text{ in} \end{cases}$$

Use no. 3 U-stirrups at 11 in o.c.

Stirrups are not required where $V_u < \phi V_c/2$.

$$V_u = V_{\text{left}} - w_u x = \frac{\phi V_c}{2}$$

$$x = \frac{V_{\text{left}} - \dfrac{\phi V_c}{2}}{w_u} = \frac{41{,}900 \text{ lbf} - \dfrac{(0.75)(32{,}300 \text{ lbf})}{2}}{2500 \ \dfrac{\text{lbf}}{\text{ft}}}$$

$$= 12 \text{ ft}$$

> Stirrups are required nearly to midspan; therefore, use 11 in spacing throughout.

17.3. The required elevation is

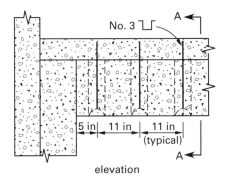

elevation

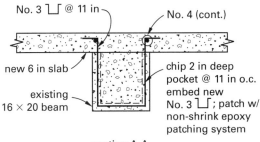

section A-A
(slab reinforcement and
existing beam reinforcement
omitted for clarity)

17.4. An additional required check is for horizontal shear transfer (ACI 318 Sec. 17.5). Let v_{nh} be the required horizontal shear strength required (psi).

$$V_u \leq \phi V_{nh} = \phi v_{nh} b_w d$$

$$41{,}900 \text{ lbf} \leq 0.75 v_{nh} (16 \text{ in})(22.6 \text{ in})$$

$$v_{nh} = 155 \text{ psi}$$

The required shear transfer can be accomplished by deliberately roughening the contact surface and using the minimum ties, which are provided by the no. 3 stirrups at 11 in o.c.

SOLUTION 18

18.1. Evaluate the proposed alteration to the structure. The column sizes are not given, and as the roof may be assumed to be simply supported, moment transfer from the central beam to supports is negligible. For roof live load, use ASCE/SEI7 Table 4-1, which gives $L_o = 20 \text{ lbf/ft}^2$. Base live load reduction on the tributary area for the shorter cantilever.

$$A_t = Ba = (13 \text{ ft})(25 \text{ ft})$$
$$= 325 \text{ ft}^2$$

$200 \text{ ft} \leq A_t \leq 600 \text{ ft}$, so,

$$R_1 = 1.2 - 0.001 A_t$$
$$= 1.2 - (0.001 \text{ ft}^{-2})(325 \text{ ft}^2)$$
$$= 0.875$$

$$F = \frac{\text{rise}}{\text{run}} = \frac{30 \text{ in}}{13 \text{ ft}} = 2.3 \text{ in/ft}$$

$$w_r = R_1 R_2 w_o = (0.875)(1.0)\left(20 \ \frac{\text{lbf}}{\text{ft}^2}\right)$$
$$= 17.5 \text{ lbf/ft}^2$$

$F < 4 \text{ in/ft}$, so,

$$R_2 = 1.0$$

$$w_r = R_1 R_2 w_o = (0.875)(1.0)\left(20 \ \frac{\text{lbf}}{\text{ft}^2}\right) = 17.5 \text{ lbf/ft}^2$$

Per footnote n to ASCE/SEI7 Table 4-1, analyze for the critical pattern of live loading on the structure. Critical loading occurs with live load plus dead load on the 43.5 ft span and with only dead load on the cantilevers. The problem statement indicates uniformly distributed loading; therefore, possible snow drift accumulation need not be considered. The roof slopes 3 ft in 22.5 ft.

$$L = \sqrt{l_{\text{horiz}}^2 + l_{\text{vert}}^2}$$

$$= \sqrt{(22.5 \text{ ft})^2 + (3 \text{ ft})^2}$$

$$= 22.7 \text{ ft}$$

$$w_s = w_c h \left(\frac{L}{l_{\text{horiz}}} \right) B$$

$$= \left(115 \ \frac{\text{lbf}}{\text{ft}^3} \right) \left(\frac{6 \text{ in}}{12 \ \frac{\text{in}}{\text{ft}}} \right) \left(\frac{22.7 \text{ ft}}{22.5 \text{ ft}} \right) (25 \text{ ft})$$

$$= 1450 \text{ lbf/ft}$$

Additional weight for the 24 in × 36 in beam on the central span is

$$w_b = w_c b (h_b - h_s)$$

$$= \left(115 \ \frac{\text{lbf}}{\text{ft}^3} \right) \left(\frac{24 \text{ in}}{12 \ \frac{\text{in}}{\text{ft}}} \right) \left(\frac{36 \text{ in} - 6 \text{ in}}{12 \ \frac{\text{in}}{\text{ft}}} \right)$$

$$= 575 \text{ lbf/ft}$$

For the cantilevers, w_b varies linearly from 575 lbf/ft at the supports to zero at the ends. Use the ACI 318 strength design method with values $f_c' = 3000$ psi and $f_y = 40{,}000$ psi. For the 23.5 ft cantilever,

$$w_{u,s} = 1.2 w_s = (1.2) \left(1450 \ \frac{\text{lbf}}{\text{ft}} \right)$$

$$= 1740 \text{ lbf/ft}$$

$$w_{u,b} = 1.2 w_b = (1.2) \left(575 \ \frac{\text{lbf}}{\text{ft}} \right)$$

$$= 690 \text{ lbf/ft}$$

$$M^- = \frac{w_{u,s} a^2}{2} + \frac{w_{u,b} a^2}{6}$$

$$= \frac{\left(1740 \ \frac{\text{lbf}}{\text{ft}} \right) (23.5 \text{ ft})^2}{2} + \frac{\left(690 \ \frac{\text{lbf}}{\text{ft}} \right) (23.5 \text{ ft})^2}{6}$$

$$= 544{,}000 \text{ ft-lbf}$$

For the 13 ft cantilever,

$$M^- = \frac{w_{u,s} a^2}{2} + \frac{w_{u,b} a^2}{6}$$

$$= \frac{\left(1740 \ \frac{\text{lbf}}{\text{ft}} \right) (13 \text{ ft})^2}{2} + \frac{\left(690 \ \frac{\text{lbf}}{\text{ft}} \right) (13 \text{ ft})^2}{6}$$

$$= 166{,}500 \text{ ft-lbf}$$

Critical loading on the 43.5 ft span is

$$w_u = 1.2 w_D + 1.6 w_L$$

$$= (1.2) \left(1450 \ \frac{\text{lbf}}{\text{ft}} + 575 \ \frac{\text{lbf}}{\text{ft}} \right) + (1.6) \left(438 \ \frac{\text{lbf}}{\text{ft}} \right)$$

$$= 3130 \text{ lbf/ft}$$

For the 43.5 ft span as originally built,

$$V_{\text{left}} = \frac{w_u L}{2} + \frac{M_{\text{left}}^- + M_{\text{right}}^-}{L}$$

$$= \frac{\left(3130 \ \frac{\text{lbf}}{\text{ft}} \right) (43.5 \text{ ft})}{2}$$

$$+ \frac{544{,}000 \text{ ft-lbf} + (-166{,}500 \text{ ft-lbf})}{43.5 \text{ ft}}$$

$$= 76{,}760 \text{ lbf}$$

The maximum positive bending moment occurs where $V = 0$.

$$V = V_{\text{left}} - w_u x = 0$$

$$x = \frac{V_{\text{left}}}{w_u} = \frac{76{,}760 \text{ lbf}}{3130 \ \frac{\text{lbf}}{\text{ft}}} = 24.5 \text{ ft}$$

$$M_u^+ = V_{\text{left}} x - \frac{w_u x^2}{2} - M_{\text{left}}^-$$

$$= (76{,}760 \text{ lbf})(24.5 \text{ ft})$$

$$- \frac{\left(3130 \ \frac{\text{lbf}}{\text{ft}} \right) (24.5 \text{ ft})^2}{2} - 544{,}000 \text{ ft-lbf}$$

$$= 397{,}200 \text{ ft-lbf}$$

Compute the maximum positive moment with the 23.5 ft cantilever removed (that is, $M_{\text{left}}^- = 0$).

$$V_{\text{left}} = \frac{w_u L}{2} + \frac{M_{\text{left}}^- + M_{\text{right}}^-}{L}$$

$$= \frac{\left(3130 \ \dfrac{\text{lbf}}{\text{ft}}\right)(43.5 \ \text{ft})}{2} + \frac{0 \ \text{ft-lbf} + (-166{,}500 \ \text{ft-lbf})}{43.5 \ \text{ft}}$$

$$= 64{,}250 \ \text{lbf}$$

$$V = V_{\text{left}} - w_u x = 0$$

$$x = \frac{V_{\text{left}}}{w_u} = \frac{64{,}250 \ \text{lbf}}{3130 \ \dfrac{\text{lbf}}{\text{ft}}} = 20.53 \ \text{ft}$$

$$M_u^+ = V_{\text{left}} x - \frac{w_u x^2}{2} - M_L^-$$

$$= (64{,}250 \ \text{lbf})(20.53 \ \text{ft})$$

$$- \frac{\left(3130 \ \dfrac{\text{lbf}}{\text{ft}}\right)(20.53 \ \text{ft})^2}{2} - 0 \ \text{ft-lbf}$$

$$= 659{,}400 \ \text{ft-lbf}$$

Thus, removing the cantilever causes a significant increase in bending moment (66%).

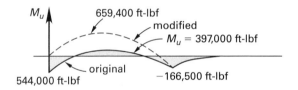

Deflection in the 43.5 ft span will also increase significantly. The reaction on the right support will increase about 10% from its original value, which might be a concern. Check the capacity of the member to handle the increased M_u.

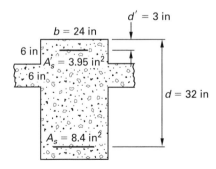

The beam is doubly reinforced at the critical section with five no. 8 bars in the compression region and 14 no. 7 bars in two layers in the tension region. Neglect compression in the slab and solve by trial and error. For trial 1, assume $f_s' = 30{,}000 \ \text{psi}$.

$$A_s' = n_{\text{bar}} A_b = (5)(0.79 \ \text{in}^2) = 3.95 \ \text{in}^2$$

$$A_s = n_{\text{bar}} A_b = (14)(0.60 \ \text{in}^2) = 8.40 \ \text{in}^2$$

$$C_c = T - C_s'$$

$$0.85 f_c' b a = A_s f_y - A_s'(f_s' - 0.85 f_c')$$

$$a = \frac{A_s f_y - A_s'(f_s' - 0.85 f_c')}{0.85 f_c' b}$$

$$= \frac{\begin{aligned}&(8.40 \ \text{in}^2)\left(40{,}000 \ \dfrac{\text{lbf}}{\text{in}^2}\right) - (3.95 \ \text{in}^2)\\[4pt] &\times \left(30{,}000 \ \dfrac{\text{lbf}}{\text{in}^2} - (0.85)\left(3000 \ \dfrac{\text{lbf}}{\text{in}^2}\right)\right)\end{aligned}}{(0.85)\left(3000 \ \dfrac{\text{lbf}}{\text{in}^2}\right)(24 \ \text{in})}$$

$$= 3.72 \ \text{in}$$

$$c = \frac{a}{\beta_1} = \frac{3.72 \ \text{in}}{0.85} = 4.38 \ \text{in}$$

From similar triangles in the strain diagram,

$$\frac{\epsilon_s'}{c - d'} = \frac{\epsilon_c}{c}$$

$$\epsilon_s' = \frac{(c - d')\epsilon_c}{c}$$

$$= \frac{(4.38 \ \text{in} - 3.0 \ \text{in})(0.003)}{4.38 \ \text{in}}$$

$$= 0.000945$$

$$\epsilon_y = \frac{f_y}{E_s} = \frac{40{,}000 \ \dfrac{\text{lbf}}{\text{in}^2}}{29{,}000{,}000 \ \dfrac{\text{lbf}}{\text{in}^2}}$$

$$= 0.00138 \quad [> \epsilon_s']$$

$$f_s' = E_s \epsilon_s' = \left(29{,}000{,}000 \ \dfrac{\text{lbf}}{\text{in}^2}\right)(0.000945)$$

$$= 27{,}400 \ \text{psi}$$

The difference between assumed and calculated compression stress is about 10%. For trial 2, try a value of f_s' about midway between the assumed and calculated values, such as $f_s' = 28{,}500 \ \text{psi}$.

$$C_c = T - C_s'$$

$$0.85 f_c' b a = A_s f_y - A_s'(f_s' - 0.85 f_c')$$

$$a = \frac{A_s f_y - A'_s(f'_s - 0.85f'_c)}{0.85f'_c b}$$

$$= \frac{\begin{array}{c}(8.40 \text{ in}^2)\left(40{,}000 \; \frac{\text{lbf}}{\text{in}^2}\right) - (3.95 \text{ in}^2) \\[2mm] \times \left(28{,}500 \; \frac{\text{lbf}}{\text{in}^2} - (0.85)\left(3000 \; \frac{\text{lbf}}{\text{in}^2}\right)\right)\end{array}}{(0.85)\left(3000 \; \frac{\text{lbf}}{\text{in}^2}\right)(24 \text{ in})}$$

$$= 3.82 \text{ in}$$

$$c = \frac{a}{\beta_1} = \frac{3.82 \text{ in}}{0.85} = 4.49 \text{ in}$$

$$\epsilon'_s = \frac{(c - d')\epsilon_c}{c}$$

$$= \frac{(4.49 \text{ in} - 3.0 \text{ in})(0.003)}{4.38 \text{ in}}$$

$$= 0.000996 \quad [< \epsilon_y]$$

$$f'_s = E_s \epsilon'_s = \left(29{,}000{,}000 \; \frac{\text{lbf}}{\text{in}^2}\right)(0.000996)$$

$$= 28{,}900 \text{ psi}$$

The calculated compression stress is close enough to the assumed value. Therefore, compute the moment capacity based on compression stress of 43,000 psi.

$$M_n = C_c\left(d - \frac{a}{2}\right) + C'_s(d - d')$$

$$= 0.85f'_c ba\left(d - \frac{a}{2}\right) + (f'_s - 0.85f'_c)A'_s(d - d')$$

$$= \frac{\begin{array}{c}(0.85)\left(3000 \; \frac{\text{lbf}}{\text{in}^2}\right)(24 \text{ in}) \\[2mm] \times (3.82 \text{ in})\left(32 \text{ in} - \dfrac{3.82 \text{ in}}{2}\right) \\[2mm] + \left(28{,}800 \; \frac{\text{lbf}}{\text{in}^2} - (0.85)\left(3000 \; \frac{\text{lbf}}{\text{in}^2}\right)\right) \\[2mm] \times (3.95 \text{ in}^2)(32.0 \text{ in} - 3.0 \text{ in})\end{array}}{\left(12 \; \frac{\text{in}}{\text{ft}}\right)\left(1000 \; \frac{\text{lbf}}{\text{kip}}\right)}$$

$$= 837 \text{ ft-kips}$$

Check strain in the extreme tension steel.

$$d_t = 36 \text{ in} - 3 \text{ in} = 33 \text{ in}$$

$$\frac{\epsilon_t}{d_t} - \frac{\epsilon_c}{c}$$

$$\epsilon_t = \frac{(d_t - c)\epsilon_c}{c}$$

$$= \frac{(33 \text{ in} - 4.49 \text{ in})(0.003)}{4.49 \text{ in}}$$

$$= 0.0190 \quad [> \epsilon_y, \text{ tension controlled}]$$

$$\phi = 0.9$$

$$\phi M_n = (0.9)(837 \text{ ft-kips})$$

$$= 754 \text{ ft-kips} \quad [\phi M_n > M_u]$$

Thus, the section has adequate flexural strength to resist the increased moment.

18.2. If it is necessary either to strengthen the beam or to compensate for the increased deflection, it should prove feasible to externally prestress the central span between lines 3 and 4. Anchors could be positioned so that negative eccentricity occurs at line 3 with a straight tendon sloped to zero eccentricity at line 4. The prestress force could be computed to induce an end moment at line 3 equivalent to the moment produced by the 23.5 ft cantilever.

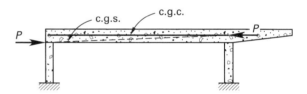

18.3. Installation would require the design of special anchorage blocks at lines 3 and 4. These could be installed on both sides of the girder at line 4 using drilled inserts. Slots would have to be cut through the slab to permit the prestressing tendon to pass. Special closure details would be needed to patch the slots after the tendons were installed. There are several post-tensioning systems that could be adapted for this type of application (for example, the Dywidag "threadbar" system).

SOLUTION 19

This problem relates to basic behavior of reinforced concrete members; the stress-strain curves for the steel and concrete are idealized and in some cases unrealistic (for example, the modulus of elasticity of steel corresponding to the given curve is 40,000 ksi).

19.1. Compute the safe working load on the beam. The effective depth is given as $d = h - 3 \text{ in} = 18 \text{ in} - 3 \text{ in} = 15 \text{ in}$. (The given stirrup spacing of 12 in o.c. does not conform to the ACI 318 minimum of $d/2 = 7.5$ in o.c. In a strict sense, the stirrups might not function as shear reinforcement because a diagonal crack could propagate between them. However, because this problem relates to theory rather than code criteria, stirrup capacity is included in the analysis.)

$$V_c = 2\lambda\sqrt{f_c'}b_w d$$

$$= (2)(1.0)\sqrt{3000\ \frac{\text{lbf}}{\text{in}^2}}(10\ \text{in})(15\ \text{in})$$

$$= 16{,}400\ \text{lbf}$$

$$V_s = \frac{A_v f_y d}{s}$$

$$= \frac{(0.40\ \text{in}^2)\left(40{,}000\ \frac{\text{lbf}}{\text{in}^2}\right)(15\ \text{in})}{12\ \text{in}}$$

$$= 20{,}000\ \text{lbf}$$

$$V_u = \phi(V_c + V_s)$$

$$= (0.75)(16{,}400\ \text{lbf} + 20{,}000\ \text{lbf})$$

$$= 27{,}300\ \text{lbf}$$

For a live load factor of 1.6, the safe load based on shear is

$$P = \frac{V_u}{1.6} = \frac{27{,}300\ \text{lbf}}{1.6} = 17{,}000\ \text{lbf}$$

To determine the working load limit based on flexure, use working stress design with an allowable steel stress of 20,000 psi and concrete compressive strength of $0.45f_c'$.

$$n = \frac{E_s}{E_c} = \frac{40{,}000\ \frac{\text{kips}}{\text{in}^2}}{3000\ \frac{\text{kips}}{\text{in}^2}} = 13$$

$$\rho n = \left(\frac{A_s}{bd}\right)n = \left(\frac{(2)(0.79\ \text{in}^2)}{(10\ \text{in})(15\ \text{in})}\right)(13) = 0.137$$

$$k = \sqrt{(\rho n)^2 + 2\rho n} - \rho n$$

$$= \sqrt{(0.137)^2 + (2)(0.137)} - 0.137$$

$$= 0.404$$

$$kd = (0.404)(15\ \text{in}) = 6.06\ \text{in}$$

$$d - kd = 15\ \text{in} - 6.06\ \text{in} = 8.94\ \text{in}$$

$$I_{cr} = \frac{b(kd)^3}{3} + nA_s(d - kd)^2$$

$$= \frac{(10\ \text{in})(6.06\ \text{in})^3}{3} + (13)(2)(0.79\ \text{in}^2)(8.94\ \text{in})^2$$

$$= 2380\ \text{in}^4$$

Check the moment capacity based on an allowable steel stress of 20,000 psi.

$$f_s = \frac{nM(d - kd)}{I_{cr}} \le 20{,}000\ \frac{\text{lbf}}{\text{in}^2}$$

$$M \le \frac{\left(20{,}000\ \frac{\text{lbf}}{\text{in}^2}\right)I_{cr}}{n(d - kd)}$$

$$\le \frac{\left(20{,}000\ \frac{\text{lbf}}{\text{in}^2}\right)(2380\ \text{in}^4)}{(13)(8.94\ \text{in})\left(12\ \frac{\text{in}}{\text{ft}}\right)\left(1000\ \frac{\text{lbf}}{\text{kip}}\right)}$$

$$\le 34.2\ \text{ft-kips}$$

Check the moment capacity based on an allowable concrete stress of $0.45f_c'$.

$$\frac{M(kd)}{I_{cr}} \le 0.45f_c'$$

$$M \le \frac{0.45f_c' I_{cr}}{kd}$$

$$\le \frac{(0.45)\left(3000\ \frac{\text{lbf}}{\text{in}^2}\right)(2380\ \text{in}^4)}{(6.06\ \text{in})\left(12\ \frac{\text{in}}{\text{ft}}\right)\left(1000\ \frac{\text{lbf}}{\text{kip}}\right)}$$

$$\le 44.2\ \text{ft-kips}$$

Steel stress governs: $M = 34.2$ ft-kips. The safe working load is

$$P = \frac{M}{a} = \frac{34.2\ \text{ft-kips}}{6\ \text{ft} - \dfrac{5\ \text{in}}{12\ \dfrac{\text{in}}{\text{ft}}}}$$

$$= \boxed{6.12\ \text{kips} \quad [\text{controls}]}$$

19.2. The stress at yield is 40,000 psi. Since behavior up to the yield stress is linear, the preceding result will lead directly to

$$\frac{P_y}{40{,}000\ \frac{\text{lbf}}{\text{in}^2}} = \frac{P}{20{,}000\ \frac{\text{lbf}}{\text{in}^2}}$$

$$P_y = \frac{P\left(40{,}000\ \frac{\text{lbf}}{\text{in}^2}\right)}{20{,}000\ \frac{\text{lbf}}{\text{in}^2}}$$

$$= \frac{(6.12\ \text{kips})\left(40{,}000\ \frac{\text{lbf}}{\text{in}^2}\right)}{20{,}000\ \frac{\text{lbf}}{\text{in}^2}}$$

$$= \boxed{12.2\ \text{kips}}$$

19.3. The given concrete stress-strain relationship applies to the linear strain distribution that exists in the compression zone. The usual assumptions of flexure theory apply (no tension in concrete, complete bond between steel and concrete, steel yields, and so on). Thus, for the trapezoidal stress distribution,

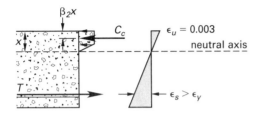

$$C_c = \beta_1 \beta_3 f'_c bx$$

$$\beta_3 = 1.0$$

$$\beta_1 f'_c bx = \text{area of compression stress distribution}$$

$$= \frac{2 f'_c bx}{3} + 0.5\left(\frac{f'_c bx}{3}\right)$$

$$= 0.83 f'_c bx$$

$$\beta_1 = \frac{0.83 f'_c bx}{f'_c bx} = 0.83$$

The position of the resultant with respect to the top fiber is

$$\beta_2 x = \frac{\sum (\text{force})(\text{lever arm})}{\text{compression force}}$$

$$= \frac{\left(\frac{x}{3}\right)\left(\frac{2 f'_c bx}{3}\right) + \left(\frac{f'_c bx}{6}\right)\left(\frac{2x}{3} + \frac{1x}{6}\right)}{0.83 f'_c bx}$$

$$\beta_2 = \frac{\dfrac{13 f'_c bx^2}{36}}{0.83 f'_c bx^2}$$

$$= 0.44$$

Thus,

$$\sum F_{\text{horiz}} = 0$$

$$C_c = T$$

$$\beta_1 f'_c bx = A_s f_y$$

$$x = \frac{A_s f_y}{\beta_1 f'_c b}$$

$$= \frac{(2)(0.79 \text{ in}^2)\left(40{,}000 \dfrac{\text{lbf}}{\text{in}^2}\right)}{(0.83)\left(3000 \dfrac{\text{lbf}}{\text{in}^2}\right)(10 \text{ in})}$$

$$= 2.54 \text{ in}$$

$$M_n = A_s f_y (d - \beta_2 x)$$

$$= \frac{(2)(0.79 \text{ in}^2)\left(40{,}000 \dfrac{\text{lbf}}{\text{in}^2}\right) \times (15.0 \text{ in} - (0.44)(2.54 \text{ in}))}{\left(12 \dfrac{\text{in}}{\text{ft}}\right)\left(1000 \dfrac{\text{lbf}}{\text{kip}}\right)}$$

$$= 73.1 \text{ ft-kips}$$

The ultimate load for the beam is

$$P_u = \frac{M_n}{a} = \frac{73.1 \text{ ft-kips}}{6 \text{ ft} - \dfrac{5 \text{ in}}{12 \dfrac{\text{in}}{\text{ft}}}}$$

$$= \boxed{13.1 \text{ kips}}$$

19.4. The ductility ratio, μ, is the ultimate curvature, ϕ_u, divided by the curvature at yield, ϕ_y. (ϵ_u is a dimensionless quantity, and μ is in rad/in; the formula is not dimensionally consistent.)

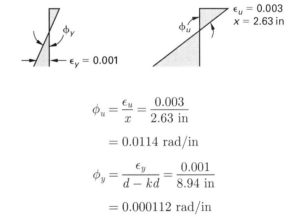

$$\phi_u = \frac{\epsilon_u}{x} = \frac{0.003}{2.63 \text{ in}}$$

$$= 0.0114 \text{ rad/in}$$

$$\phi_y = \frac{\epsilon_y}{d - kd} = \frac{0.001}{8.94 \text{ in}}$$

$$= 0.000112 \text{ rad/in}$$

Therefore, the ductility ratio is

$$\mu = \frac{\phi_u}{\phi_y} = \frac{0.0114 \dfrac{\text{rad}}{\text{in}}}{0.000112 \dfrac{\text{rad}}{\text{in}}}$$

$$= \boxed{10}$$

19.5. The failure mode would be described as a $\boxed{\text{ductile flexural failure.}}$

19.6. To determine adequacy of the column, apply the computed maximum working load, $P = 6.12$ kips, with a load factor of 1.6.

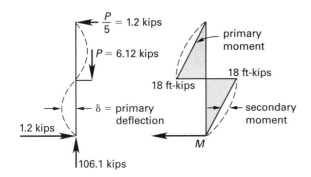

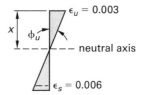

The column is subjected to critical combined axial compression plus bending.

$$P_u = 1.6P_{total}$$

$$= (1.6)(100 \text{ kips} + 6.12 \text{ kips})$$

$$= 170 \text{ kips}$$

$$M_u = (1.6)(18 \text{ ft-kips})$$

$$= 28.8 \text{ ft-kips}$$

Check the factored loading against the given interaction diagram. The point (28.8 ft-kips, 170 kips) is well inside the curve; therefore, the column strength is sufficient.

Secondary moments, $P\delta$, will increase along the column (see preceding diagram), but not at the point of maximum moment. Thus, it is not likely that secondary, or slenderness, effects will become critical for this loading. (This assumes that the column is braced against buckling, as stated in the problem.)

19.7. If A_s is increased to two no. 11 bars, then (as in Sol. 19.3),

$$x = \frac{A_s f_y}{\beta_1 f'_c b}$$

$$= \frac{(2)(1.56 \text{ in}^2)\left(40,000 \frac{\text{lbf}}{\text{in}^2}\right)}{(0.83)\left(3000 \frac{\text{lbf}}{\text{in}^2}\right)(10 \text{ in})}$$

$$= 5.01 \text{ in}$$

From similar triangles,

$$\frac{\epsilon_s}{d - x} = \frac{\epsilon_u}{x}$$

$$\epsilon_s = \frac{(d - x)\epsilon_u}{x}$$

$$= \frac{(15.0 \text{ in} - 5.01 \text{ in})(0.003)}{5.01 \text{ in}}$$

$$= 0.006 \quad [\epsilon_s > \epsilon_y]$$

Therefore, the beam will still show ductile behavior ($\phi_u \approx 5\phi_y$).

$$M_n = A_s f_y (d - \beta_2 x)$$

$$= \frac{(2)(1.56 \text{ in}^2)\left(40,000 \frac{\text{lbf}}{\text{in}^2}\right)}{\left(12 \frac{\text{in}}{\text{ft}}\right)\left(1000 \frac{\text{lbf}}{\text{kip}}\right)}$$

$$= 133 \text{ ft-kips}$$

$$P_u = \frac{M_n}{a} = \frac{133 \text{ ft-kips}}{6 \text{ ft} - \frac{5 \text{ in}}{12 \frac{\text{in}}{\text{ft}}}} = 23.8 \text{ kips}$$

The ultimate load computed on the basis of flexural strength exceeds the shear strength. The failure mode will be brittle shear failure.

SOLUTION 20

Analyze the thin shell structure for gravity dead plus live load using membrane theory.

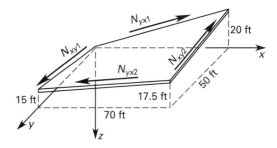

$$C_{xy1} = z_{50} - z_0 = 15 \text{ ft} - 0 \text{ ft} = 15 \text{ ft}$$

$$L_{y1} = \sqrt{C_{xy1}^2 + y^2} = \sqrt{(15 \text{ ft})^2 + (50 \text{ ft})^2} = 52.2 \text{ ft}$$

$$C_{yx1} = z_{70} - z_0 = 20 \text{ ft} - 0 \text{ ft} = 20 \text{ ft}$$

$$L_{x1} = \sqrt{C_{yx1}^2 + x^2} = \sqrt{(20 \text{ ft})^2 + (70 \text{ ft})^2} = 72.8 \text{ ft}$$

$C_{yx2} = z_0 - z_{70} = 15 \text{ ft} - 17.5 \text{ ft} = -2.5 \text{ ft}$

$L_{x2} = \sqrt{C_{yx2}^2 + y'} = \sqrt{(-2.5 \text{ ft})^2 + (70 \text{ ft})^2} = 70.04 \text{ ft}$

$C_{xy2} = z_0 - z_{50} = 20 \text{ ft} - 17.5 \text{ ft} = 2.5 \text{ ft}$

$L_{y2} = \sqrt{C_{xy2}^2 + y^2} = \sqrt{(2.5 \text{ ft})^2 + (50 \text{ ft})^2} = 50.06 \text{ ft}$

Follow the conventional practice of expressing the gravity load as an equivalent uniformly distributed load in lbf/ft^2 of horizontal projection. Include the extra weight of ribs over and above the 3 in slab (that is, calculate the rib weight based on 20 in − 3 in = 17 in of additional depth × 20 in of width).

$$w_{\text{rib}} = w_c bh$$

$$= \left(110 \, \frac{\text{lbf}}{\text{ft}^3}\right) \left(\frac{20 \text{ in}}{12 \, \frac{\text{in}}{\text{ft}}}\right) \left(\frac{17 \text{ in}}{12 \, \frac{\text{in}}{\text{ft}}}\right)$$

$$= 260 \text{ lbf/ft}$$

The service dead plus live load over the horizontal projection is

$$p_z = \frac{w_d(L_{x1} + L_{x2})(L_{y1} + L_{y2})}{BW} + w_l$$

$$= \frac{\left(47.2 \, \frac{\text{lbf}}{\text{ft}^2}\right)(72.8 \text{ ft} + 70.04 \text{ ft})}{\times \, (52.2 \text{ ft} + 50.06 \text{ ft})}{(100 \text{ ft})(140 \text{ ft})} + 12 \, \frac{\text{lbf}}{\text{ft}^2}$$

$$= 61 \text{ lbf/ft}^2$$

The dead load includes the concrete, ceiling, roof material, and ribs.

$$w_d = w_c h_s + w_{\text{ceiling}} + w_{\text{roofing}}$$
$$+ \frac{w_{\text{rib}}(2L_{x2} + 2L_{y2} + 4L_{x1} + 4L_{y1})}{BW}$$

$$= \left(110 \, \frac{\text{lbf}}{\text{ft}^3}\right)\left(\frac{3 \text{ in}}{12 \, \frac{\text{in}}{\text{ft}}}\right) + 3 \, \frac{\text{lbf}}{\text{ft}^2} + 3 \, \frac{\text{lbf}}{\text{ft}^2} + \left(260 \, \frac{\text{lbf}}{\text{ft}}\right)$$

$$\times \left(\frac{\begin{array}{c}(2)(70.04 \text{ ft}) + (2)(50.06 \text{ ft}) \\ + (4)(72.8 \text{ ft}) + (4)(52.2 \text{ ft})\end{array}}{(100 \text{ ft})(140 \text{ ft})}\right)$$

$$= 47.2 \text{ lbf/ft}^2$$

Equilibrium of the shell membrane requires stress resultants (force per unit length) uniformly distributed along the edges of each quadrant of the shell. Ribs are axially loaded by the equal and opposite stress resultants.

$$\sum F_x = N_{yx1} L_{x1} \cos \alpha_{x1} - N_{yx2} L_{x2} \cos \alpha_{x2} = 0$$

$$= N_{yx1}(72.8 \text{ ft})\left(\frac{70 \text{ ft}}{72.8 \text{ ft}}\right)$$

$$- N_{yx2}(70.04 \text{ ft})\left(\frac{70 \text{ ft}}{70.04 \text{ ft}}\right)$$

$$N_{yx1} = N_{yx2}$$

$$\sum F_y = N_{xy1} L_{y1} \cos \alpha_{y1} - N_{xy2} L_{y2} \cos \alpha_{y2} = 0$$

$$= N_{xy1}(52.2 \text{ ft})\left(\frac{50 \text{ ft}}{52.2 \text{ ft}}\right)$$

$$- N_{xy2}(50.06 \text{ ft})\left(\frac{50 \text{ ft}}{50.06 \text{ ft}}\right)$$

$$N_{xy1} = N_{xy2}$$

The sum of the moments about the z-axis equals zero, so

$$\sum M_{z\text{-axis}} = N_{xy2} L_{y2}(\cos \alpha_{y2})x - N_{yx2} L_{x2}(\cos x_2)y = 0$$

$$= N_{xy2}(50.06 \text{ ft})\left(\frac{50 \text{ ft}}{50.06 \text{ ft}}\right)(70 \text{ ft})$$

$$- N_{yx2}(72.8 \text{ ft})\left(\frac{70 \text{ ft}}{72.8 \text{ ft}}\right)(50 \text{ ft})$$

$$N_{xy2} = N_{yx2} = N_{xy}$$

Therefore,

$$\sum F_z = -p_z xy + N_{xy} L_{y1} \sin \alpha_{y1} + N_{xy} L_{x1} \sin \alpha_{x1}$$
$$+ N_{xy} L_{y2} \sin \alpha_{y2} - N_{xy} L_{x2} \sin \alpha_{x2} = 0$$

$$= \left(-61 \, \frac{\text{lbf}}{\text{ft}^2}\right)(50 \text{ ft})(70 \text{ ft}) + N_{xy}$$

$$\times \left(\begin{array}{c}(72.8 \text{ ft})\left(\dfrac{20 \text{ ft}}{72.8 \text{ ft}}\right) + (50.06 \text{ ft}) \\ \times \left(\dfrac{2.5 \text{ ft}}{50.06 \text{ ft}}\right) + (52.2 \text{ ft})\left(\dfrac{15 \text{ ft}}{52.2 \text{ ft}}\right) \\ - (70.04 \text{ ft})\left(\dfrac{2.5 \text{ ft}}{70.04 \text{ ft}}\right)\end{array}\right)$$

$$N_{xy} = 6100 \text{ lbf/ft}$$

20.1. The force per foot along the edges of the membrane reacts equal and opposite on the perimeter and interior ribs. Consider a free-body diagram of rib A.

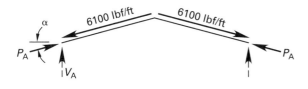

$$P_{\rm A} = N_{xy}L_{y1} = \left(6100\ \frac{\rm lbf}{\rm ft}\right)(52.2\ {\rm ft})$$

$$= \boxed{318{,}420\ {\rm lbf}}$$

$$V_{\rm A} = P_{\rm A}\sin\alpha_{y1} = (318{,}420\ {\rm lbf})\left(\frac{15\ {\rm ft}}{52.2\ {\rm ft}}\right)$$

$$= \boxed{91{,}500\ {\rm lbf}}$$

20.2. From the free-body diagram of rib A,

$$P_{\rm B} = N_{xy}L_{x1} = \left(6100\ \frac{\rm lbf}{\rm ft}\right)(72.8\ {\rm ft})$$

$$= \boxed{444{,}080\ {\rm lbf}}$$

$$V_{\rm B} = P_{\rm B}\sin\alpha_{x1} = (444{,}080\ {\rm lbf})\left(\frac{20\ {\rm ft}}{72.8\ {\rm ft}}\right)$$

$$= \boxed{122{,}000\ {\rm lbf}}$$

20.3. For rib C,

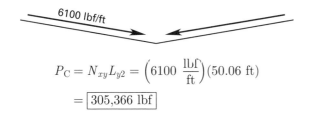

$$P_{\rm C} = N_{xy}L_{y2} = \left(6100\ \frac{\rm lbf}{\rm ft}\right)(50.06\ {\rm ft})$$

$$= \boxed{305{,}366\ {\rm lbf}}$$

20.4. For rib D,

$$P_{\rm D} = N_{xy}L_{x2} = \left(6100\ \frac{\rm lbf}{\rm ft}\right)(70.04\ {\rm ft})$$

$$= \boxed{427{,}244\ {\rm lbf}}$$

20.5. Design membrane reinforcement using ACI 318 Sec. 19.4. Principal stresses have the same magnitude as N_{xy} and occur on sections at 45° to the x- and y-axes. Place the main reinforcement in a single layer paralleling the x- and y-axes and design this reinforcement to satisfy shear friction requirements of ACI 318 Sec. 19.4.2.

Lightweight concrete weighing 110 lbf/ft³ is likely sand-lightweight; therefore, $\mu = \lambda = 0.85$.

$$\beta_l = \frac{w_l}{p_z} = \frac{12\ \dfrac{\rm lbf}{\rm ft^2}}{61\ \dfrac{\rm lbf}{\rm ft^2}} = 0.2$$

$$\beta_d = 1 - \beta_l = 1 - 0.2 = 0.8$$

$$V_u = 1.2\beta_d N_{xy} + 1.6\beta_l N_{xy}$$

$$= (1.2)(0.8)\left(6100\ \frac{\rm lbf}{\rm ft}\right) + (1.6)(0.2)\left(6100\ \frac{\rm lbf}{\rm ft}\right)$$

$$= 7810\ {\rm lbf/ft}$$

$$\phi V_f = \phi\mu A_{vf}f_y = V_u$$

$$A_{vf} = \frac{V_u}{\phi\mu f_y}$$

$$= \frac{7810\ \dfrac{\rm lbf}{\rm ft}}{(0.75)(0.85)\left(60{,}000\ \dfrac{\rm lbf}{\rm in^2}\right)}$$

$$= 0.20\ {\rm in^2/ft}$$

ACI 318 Sec. 19.4.3 requires

$$A_{s,\min} = 0.0018bh$$

$$= (0.0018)\left(12\ \frac{\rm in}{\rm ft}\right)(3\ {\rm in})$$

$$= 0.06\ {\rm in^2/ft}\quad[\text{does not control}]$$

Therefore, the shear friction reinforcement controls.

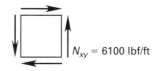

Check ductility requirements of ACI 318 Sec. 19.4.

$$N_{\max,c} = \phi P_n = \phi(0.4f'_c)bh$$

$$= (0.65)(0.4)\left(3000\ \frac{\rm lbf}{\rm in^2}\right)\left(12\ \frac{\rm in}{\rm ft}\right)(3\ {\rm in})$$

$$= 28{,}000\ {\rm lbf/ft}$$

$$N_{\max,t} = \phi(0.707A_{vf})f_y$$

$$= (0.9)(0.707)\left(0.20\ \frac{\rm in^2}{\rm ft}\right)\left(60{,}000\ \frac{\rm lbf}{\rm in^2}\right)$$

$$= 7600\ {\rm lbf/ft}\quad[\text{controls}]$$

The tension resistance is well below the compression strength; therefore, the steel will yield before the concrete crushes. The tensile strength, 7600 lbf/linear ft, is close enough to the required strength; therefore, use no. 4 bars at 12 in o.c. (this is close enough).

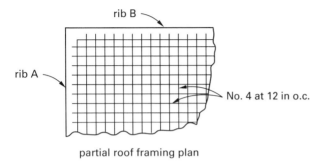

partial roof framing plan

20.6. Check vertical equilibrium at the intersection of ribs C and D.

$$\sum F_z = -2P_D \sin \alpha_{x2} + 2P_C \sin \alpha_{y2}$$
$$= (-2)(427{,}244 \text{ lbf}) \left(\frac{2.5 \text{ ft}}{70.04 \text{ ft}} \right)$$
$$\qquad + (2)(305{,}366 \text{ lbf}) \left(\frac{2.5 \text{ ft}}{50.06 \text{ ft}} \right)$$
$$= 0$$

Therefore, equilibrium is satisfied; the roof is stable.

The total vertical reaction at the corner is

$$V_{\text{total}} = V_A + V_B = 91{,}500 \text{ lbf} + 122{,}000 \text{ lbf}$$
$$= 213{,}500 \text{ lbf}$$

Check vertical equilibrium of the entire roof.

$$\sum F_z = -p_z BW + 4V_{\text{total}}$$
$$= \left(-61 \ \frac{\text{lbf}}{\text{ft}^2} \right)(100 \text{ ft})(140 \text{ ft}) + (4)(213{,}500 \text{ lbf})$$
$$= 0$$

Thus, the vertical forces are transferred to the corners; there is no need for a center column.

3 Structural Steel Design

PROBLEM 1

1.1. When and why would the use of high-strength friction bolts be specified?

1.2. How is the installation of high-strength bolts in a friction-type connection different from their installation in a bearing-type connection?

1.3. There are three methods for installing high-strength steel bolts such as ASTM A325-SC. Name and describe them.

1.4. Identify three possible corrective measures for the oversized holes, as shown. Steel is ASTM A36.

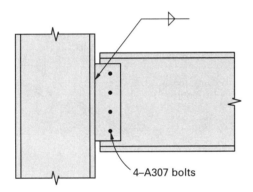

4–A307 bolts

1.5. A contractor was to install ASTM A325 high-strength bolts as required by the drawings. The heads of the bolts already installed have marks as shown. Are these the required high-strength bolts? Explain.

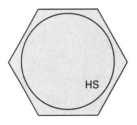

HS

1.6. A rectangular steel tube spans between two rigid supports as shown. A load of 12 kips is supported by an outrigger at midspan. Assume that steel supports are used and that support arms and connections are adequate. What are the relative merits of each of the following kinds of steel sections with respect to torsional resistance? (No calculations are required.)

(a) rectangular tube sections

(b) pipe sections

(c) wide flange sections

(d) channel sections

(e) angle sections

(f) square structural tube sections

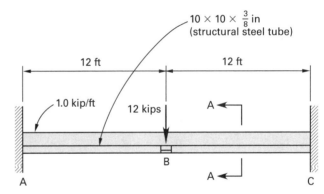

beam elevation

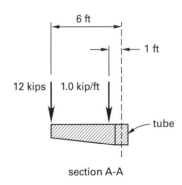

section A-A

1.7. What is the P-Δ effect in high-rise design?

1.8. What is lamellar tearing and where might it occur in the joint shown?

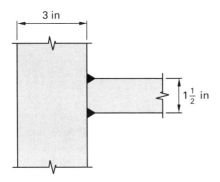

1.9. Moment-resisting steel frames are often used in high-rise buildings to resist lateral forces. Should columns or girders be designed to sustain inelastic action, particularly plastic hinging, when under severe earthquake motions? Explain.

1.10. What is a prequalified welded joint?

1.11. The W12 × 50 beams shown are loaded in such a way that a moment connection equal in capacity to that of the W12 × 50 beams has to be provided at the built-up girder. Draw a moment connection. Describe briefly the advantages of your solution. No calculations are required.

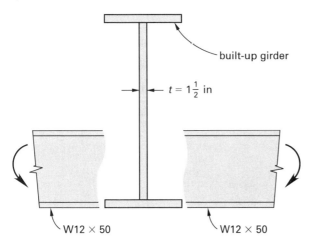

1.12. A beam-to-column connection is shown. Assume the tab plate and welds are adequate. Steel is ASTM A36. The web thickness of the supported beam is $^7/_{16}$ in. Determine if the design is adequate for a beam reaction of 82 kips dead load plus live load. Use ASD.

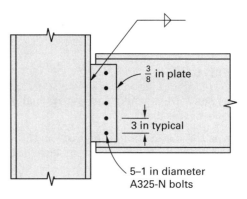

1.13. Compute the longitudinal shear capacity of a $^5/_{16}$ in fillet weld made with E80XX series electrodes. Show all calculations.

1.14. Using the *AISC Manual* and LRFD methods, compute the ultimate capacity of each of the two double-angle connections shown. Which would be expected to fail first under inelastic seismic overloads? Why?

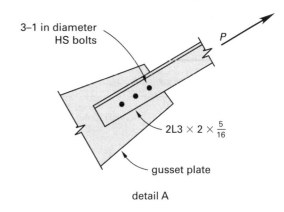

detail A

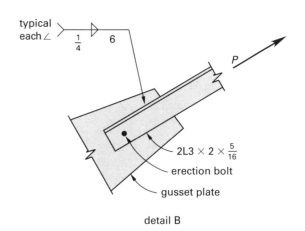

detail B

PROBLEM 2

One story of a column tier for a moment-resisting frame system is shown. The columns in the moment-resisting frame system support one-third of the gravity loads for the controlling load case, with the remainder supported by columns that are not part of the lateral force resisting system. Columns are fully braced in the out-of-plane direction. The given girder and column sizes were determined during the preliminary design analysis. The axial load and moments were determined by a first-order computer analysis satisfying all requirements of AISC Specification Sec. C2.1(2). They are the result of service-level dead, live, and wind loads. The moments given are center-to-center of members and are entirely due to wind load. The axial force is solely due to dead and live loads.

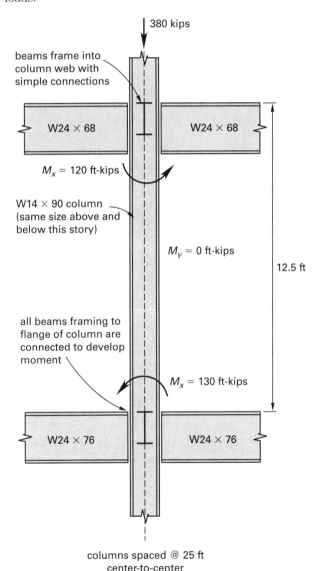

Design Criteria

- ASTM A992 steel

- do not investigate beam-to-column connections

- use the effective length method of AISC App. 7

- use the approximate second-order analysis of AISC App. 8

2.1. Determine the adequacy of the column due to the forces given. Use ASD. Show all calculations.

PROBLEM 3

The steel fabricator has submitted the following detail for a steel beam-to-column connection. Calculations show the beam to have a reaction of 16.0 kips.

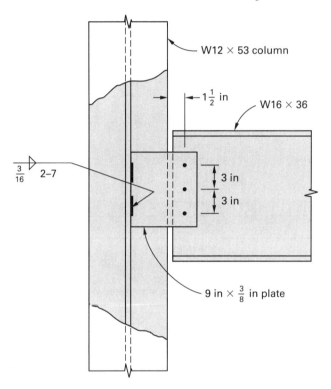

Design Criteria

- ASTM A36 steel

- E70XX welding electrodes

- $^7/_8$ in diameter A325-X bolts

3.1. Evaluate the connection and determine if the detail is acceptable. Use ASD. Show all calculations.

PROBLEM 4

The partial framing plan shown indicates a typical bay in an existing floor system. The W18×35 beams were designed to support a uniform dead and live load of 1 kip/ft including the beam weight. 18 studs, $3/4$ in in diameter by 3 in and spaced at 20 in on center, were welded to the top flange to accommodate a future additional load. It is now proposed to place two concentrated loads of 3000 lbf each at the beam's one-third points.

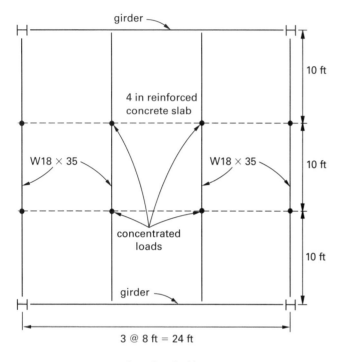

plan of typical bay
showing locations of concentrated loads

Design Criteria

- $f'_c = 3$ ksi hardrock concrete
- ASTM A36 steel

4.1. Investigate the adequacy of the W18×35 beams to support the original design load and the proposed concentrated loads. Use LRFD. Neglect deflection and assume the pin connection at the girder is adequate for shear. Show all calculations.

PROBLEM 5

A bracket for a beam-to-column connection is shown.

Design Criteria

- $P = 20$ kips
- ASTM A36 steel plate and column
- E70XX welding electrodes
- tensile strength range = 70 ksi minimum

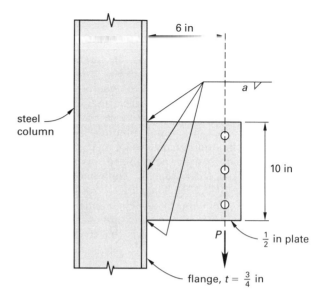

5.1. Determine the size of fillet welds for static loads. (See a on figure.) Use ASD.

5.2. Determine the size of fillet welds for 100% reversible stresses for over 2,000,000 cycles. Show all calculations.

PROBLEM 6

The detail shows the construction stage at which a local job was stopped. At this point, it was determined that the steel beam will be 35% overstressed in bending when fully loaded. Remedial action is needed immediately. The beam cannot be replaced, and new columns may not be installed. The beam spans 30 ft and was designed for a uniformly distributed load.

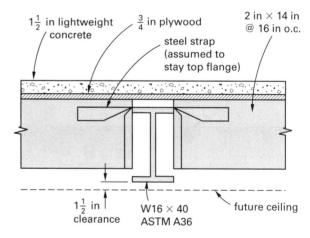

6.1. Recommend a remedial solution and draw a complete detail. Show all calculations. Calculate the maximum load (in kips per foot) that the reinforced beam can safely support. Use LRFD.

6.2. Describe briefly what instructions should be made to the contractor regarding any action he or she must undertake before proceeding with the remedial solution.

PROBLEM 7

The figure shown represents a column splice detail. The wide flange sections that are to be spliced are shown in their relative positions.

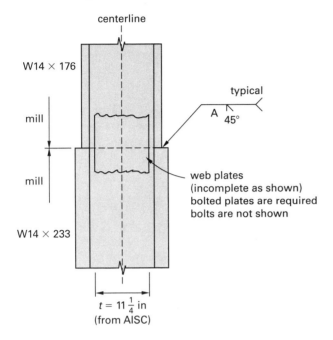

Design Criteria

- E70XX welding electrodes

- A325-SC (1 in maximum) bolts

- ASTM A36 steel ($F_b = 24$ ksi)

- column splice to be developed for 50% of moment capacity and 50% of shear capacity

7.1. What should the size be for partial penetration bevel weld shown on the figure as weld A?

7.2. Determine the web plates (sizes and thickness) and the bolts to complete the detail. Bolts act in double shear. Draw a detail showing the placement of bolts, including spacing and edge distances.

PROBLEM 8

A partial floor framing plan of a typical office building is shown. Girder G1 passes through an elevator core as shown. Elevator operation requirements allow only a 7 in wide girder flange at the elevator core.

Design Criteria

- G1 is simple span girder

- girder self-weight = 100 lbf/ft

- maximum depth of girder = 25 in

- built-up girder not acceptable

- no lateral ties within elevator core

- allowable live load reduction per ASCE/SEI7

- ASTM A36 steel

- dead loads as follows:
  - $3^1/4$ in lightweight fill on 3 in steel deck = 48 lbf/ft^2
  - partitions = 20 lbf/ft^2 (including wall weight around elevator)
  - ceiling, mechanical, and miscellaneous = 10 lbf/ft^2
  - beam framing = 5 lbf/ft^2

- live loads = 80 lbf/ft^2

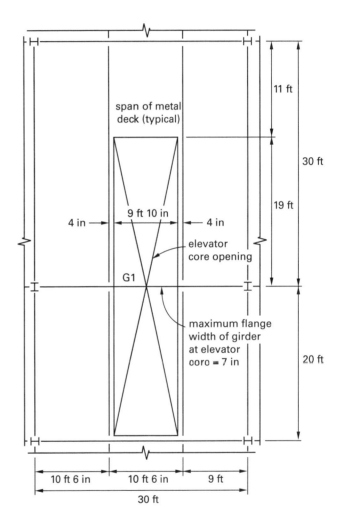

8.1. Design girder G1. Use rolled section or modified rolled section to meet the design requirements. Use LRFD.

8.2. Explain how a rolled section could be modified to satisfy the requirements.

PROBLEM 9

A large clear floor area must be provided as shown. (See *Illustration for Prob. 9.*) Column 2 is to be supported on a beam that spans columns 1 and 3.

Design Criteria

- ASTM A992 steel for W shapes

- ASTM A36 steel for angles and plates

- ASTM A307 bolts

- sidesway prevented in both directions at the roof

- struts shown on section X-X provide lateral support normal to the beam

- ratio of dead load to total load $= 0.5$

9.1. What is the lightest W27 member that may be used for beam A? Assume lateral support is provided at midspan only. Use LRFD.

9.2. Draw a detail of an all-bolted unstiffened beam seat connection at B. Show supporting data.

9.3. What is the lightest W8 member that may be used for the column on line 1?

Illustration for Prob. 9

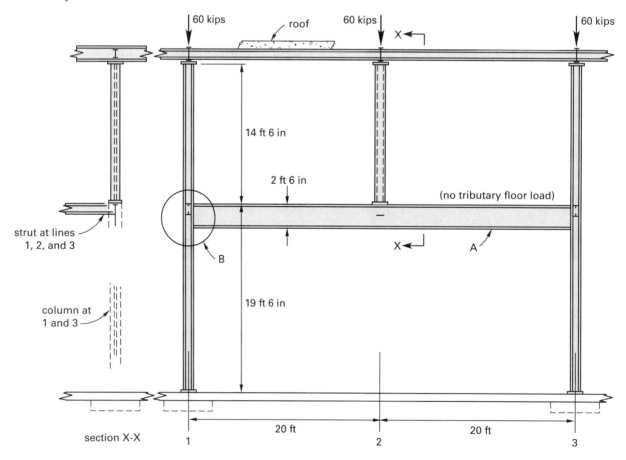

PROBLEM 10

A 12-story office building 90 ft × 150 ft in plan dimension is to be designed using a structural steel space frame to resist 100% of the lateral forces. A seismic analysis has been made and the F_x forces tributary to the typical interior frame are as shown. The given forces are the strength level values.

Design Criteria

- ASTM A992 steel
- seismic design category D
- $S_{DS} = 1.0$
- special steel moment frame using beam sections
- use IBC and AISC
- $I_e = 1.0$
- diaphragms are rigid
- field welding permitted
- ASTM A325 bolts
- top flanges of beams are continuously supported

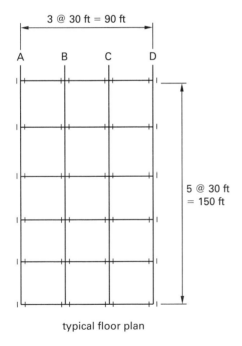

typical floor plan

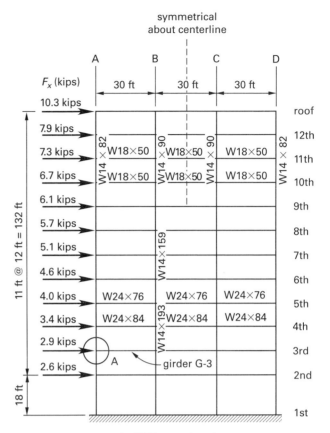

typical W-frame elevation

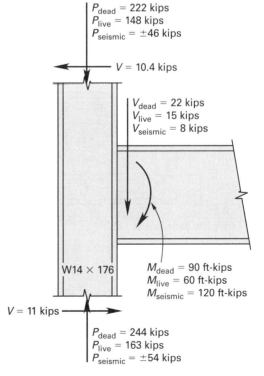

section A

10.1. Using an appropriate approximate method, compute

(a) preliminary beam moments and shears at the tenth floor

(b) column axial loads, shears, and moments at the top of the ninth story

(c) column axial loads, shears, and moments at the bottom of the tenth story

Present answers in the form of free-body and moment diagrams. Show all calculations.

10.2. Using the preliminary beam and column sizes shown, estimate the story drift at the tenth story.

10.3. Design and draw a detail of the connection of girder G3 at the third floor to the exterior column. Use LRFD. Assume a dead-to-total gravity load ratio of 0.6.

PROBLEM 11

An existing steel beam must be modified to provide an opening as shown. (See *Illustration for Prob. 11*.) The beam is a W24 × 84 with a 30 kip load at midspan.

Design Criteria

* assume compression flange is stayed against buckling

* neglect beam weight

* ASTM A36 steel

11.1. Certain reinforcement plates are shown. Verify the adequacy of the plates. Use LRFD.

11.2. Determine the length of welds required.

11.3. Show the welding required (E70 electrodes).

11.4. Describe a construction sequence to accomplish the desired changes.

Illustration for Prob. 11

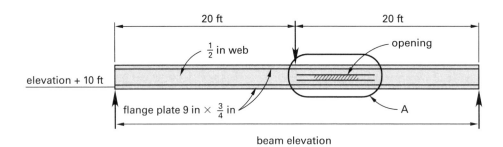

beam elevation

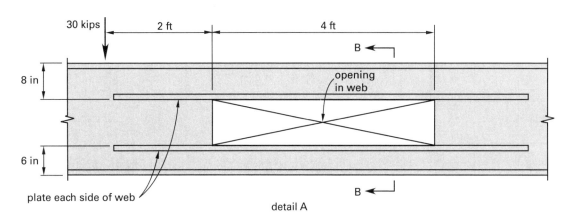

detail A

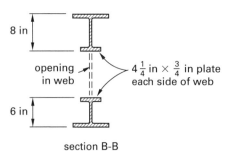

section B-B

PROBLEM 12

The following drawing shows a bridge supported by a simply supported runway girder that spans 22 ft between columns.

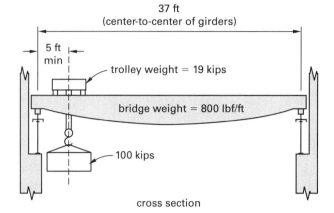

cross section

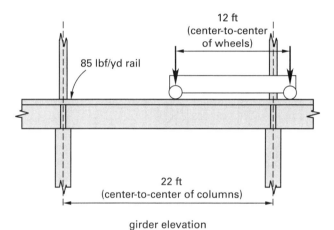

girder elevation

Design Criteria

- girder is rolled W shape with standard channel cap

- cab-operated trolley crane rolls on 85 lbf/yd rail fastened to top flange

- ASTM A36 steel

- impact and lateral loads to be considered

- maximum stresses occur infrequently, reduction in stress range for fatigue not necessary

12.1. Design the runway girder with the most economical proportions, taking into account both vertical and horizontal loading. Only the top flange of the W shape may be combined with the channel to resist horizontal loading. Use ASD.

12.2. Design welds to connect the channel to the W shape. Give type and length.

PROBLEM 13

An existing simple $W18 \times 46$ beam is shown. The top flange of the beam is laterally supported, and the support is not accessible. The existing load is 3150 lbf/ft. A tension cover plate is to be added to the bottom flange so that a new concentrated load of 5150 lbf may be added at midspan. Before the plate is added, the beam will be shored and the existing load deflection essentially eliminated.

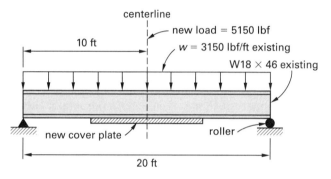

Design Criteria

- cover plate to be $7^{1}/_{2}$ in wide

- cover plate to be fastened with $^{5}/_{8}$ in diameter high-tension (A325-SC) bolts in two rows, $3^{1}/_{2}$ in gage

- cover plate and fasteners will add 350 lbf to new load at midspan

- for beam and cover plate, $F_y = 36$ ksi

- all loads are at service level

13.1. Determine the minimum thickness (to the nearest $^{1}/_{8}$ in) and length of the cover plate needed to resist the new load. (Specify the cover plate in this form: thickness $\times 7^{1}/_{2}$ in $\times$ length.)

13.2. Find the spacing for pairs of $^{5}/_{8}$ in diameter A325-SC tension bolts.

PROBLEM 14

An existing steel beam, which is continuous over one support (at B as shown), supports a uniform load along its full length. The load and spans are shown.

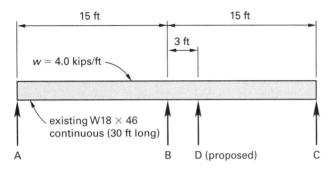

Certain modifications require the support of B to be relocated to D. Two procedures are proposed.

- Procedure 1: Snug the new support at D tight to the existing beam soffit, and then remove the support at B.

- Procedure 2: At point B, release restraints by support B against upward movement of the beam. Jack up point D to the same elevation as A and C. Install the support at D, and then remove the support at B.

Design Criteria

- ASTM A36 steel

- top flange of beam continuously supported

14.1. Determine the final reaction at D for each of the two proposed procedures. Use ASD.

14.2. Verify that the existing beam section is adequate for the modification, using the second proposal.

14.3. Determine the minimum column size for a new support at D. The unsupported column height is 10 ft 0 in.

14.4. Design and draw a detail for a welded beam-to-column connection at D.

PROBLEM 15

A simply supported beam is shown.

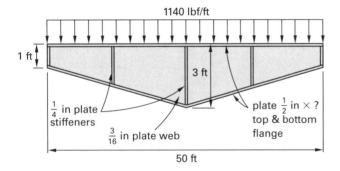

Design Criteria

- uniform load = 1140 lbf/ft (includes beam weight)

- top flange of beam is fully stayed

- allowable bending stress = 22.0 ksi

15.1. Approximately where is the critical depth?

15.2. What is the minimum flange width (to the nearest $1/4$ in) at the critical depth?

15.3. What is the deflection of the beam at midspan due to total load?

PROBLEM 16

The mechanical equipment shown has moving parts that could subject the structure to an annoying vibration.

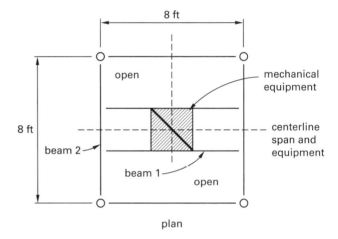

plan

Design Criteria

- equipment weight = 300 lbf

- 1725 rpm motor

- ASTM A992 steel

- 10 in wide flange beams

16.1. Select the lightest W10 beams for beams 1 and 2 that will support the gravity load. Use LRFD.

16.2. In what ways can the beams be modified or changed to eliminate the vibration problem?

16.3. What change could be recommended to minimize or solve the vibration problem? Show calculations.

PROBLEM 17

A high-rise steel frame will be erected on top of a reinforced concrete deck. All work is new. Full continuity is desired from the steel column through the column base to the concrete below. The design loads are shown.

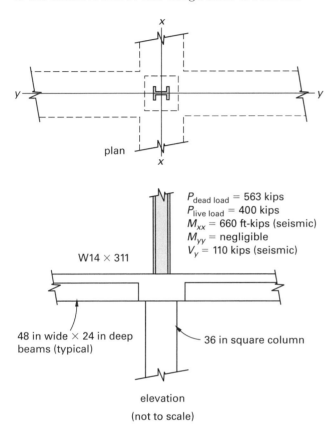

plan

W14 × 311

$P_{dead\ load}$ = 563 kips
$P_{live\ load}$ = 400 kips
M_{xx} = 660 ft-kips (seismic)
M_{yy} = negligible
V_y = 110 kips (seismic)

48 in wide × 24 in deep beams (typical)

36 in square column

elevation
(not to scale)

Design Criteria

- $f'_c - 3$ ksi
- ASTM A36 steel
- seismic performance category D
- $S_{DS} = 0.5$
- redundancy factor = 1.0
- given loads include amplification as required by ANSI/AISC 341 Sec. D2.6

17.1. Design a square steel base plate and a base connection to satisfy the design criteria. Use LRFD. Show the plate thickness, stiffeners (if any), required welding or other connection method, and size and placement of anchor bolts. Draw a detail of your base connection showing all the necessary information to complete the installation.

PROBLEM 18

The cross section of a long covered walk between two hotel units is shown. The bent shown is made of steel tubing.

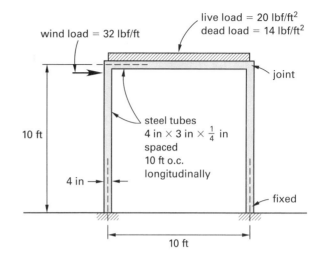

wind load = 32 lbf/ft

live load = 20 lbf/ft²
dead load = 14 lbf/ft²

joint

steel tubes
4 in × 3 in × $\frac{1}{4}$ in spaced
10 ft o.c.
longitudinally

10 ft

4 in

fixed

10 ft

Design Criteria

- use ASCE/SEI7 load combinations
- bents spaced 10 ft on center longitudinally
- column legs fixed at base
- neglect tube weight
- ASTM A500 group B ($F_y = 46$ ksi)

18.1. Verify that the given tube sizes are adequate. Use ASD. Consider both the vertical and lateral loads. Show calculations for maximum moments, shears, and any other critical design values, for both the beam member and the columns. Use the effective length method of AISC App. 7 and the approximate second-order analysis of AISC App. 8.

18.2. Draw a detail of the joint at the upper right corner of the bent. Show the necessary welding and any other considerations to provide a completely fabricated joint.

PROBLEM 19

One bay of an existing building is shown. Modifications to the building require the frames to be altered. The existing columns and horizontal top beam must remain in the frame to minimize any disturbance to the structure and function of the building.

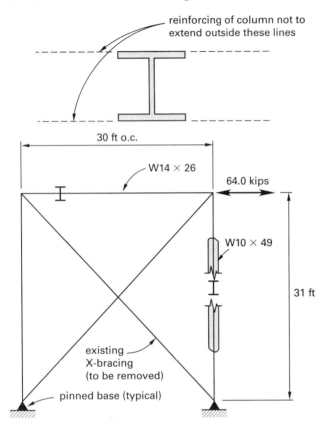

existing frames

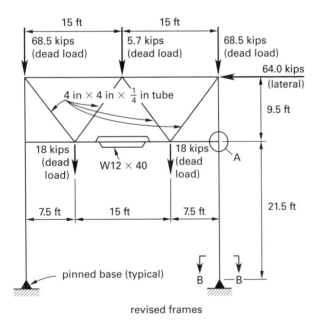

revised frames

Design Criteria

- column bases and joints are pinned
- continuous full-height columns
- web members pinned at column connection
- dimensions to centerlines of members
- columns braced perpendicularly to frame between floor and new lower beam
- frame weight to be ignored
- lateral force from wind
- ASTM A992 steel

19.1. Determine the loads in all members of the modified frames due to the dead loads plus lateral loads.

19.2. Reinforce the existing columns as required, and draw a detail of the proposed modification. Use LRFD.

19.3. Draw details of joint A and section B-B.

PROBLEM 20

A plan of the first floor framing designed for a live load of 50 lbf/ft^2 is shown. (See *Illustration for Prob. 20.*) The office function of a local printing plant has been occupying the first floor area shown. The building owner has requested the following changes.

- Convert the office area to a computer room with a uniformly distributed live load of 100 lbf/ft^2.
- Remove the existing wood construction from the floor and replace with metal deck and nonstructural concrete fill to provide the same floor elevation as at present.
- Strengthen the existing steel framing members and connections, if necessary, to support the revised design.

Design Criteria

- ASTM A36 steel
- ASTM A307 bolts
- E70XX welding electrodes
- floor dead load = 50 lbf/ft^2 (partitions included)

20.1. Review the existing steel floor system and connections. Use LRFD. Identify inadequate members and connections, if any. Neglect column stresses.

20.2. Provide new details for the strengthening of inadequate members. Provide calculations to support your work and conclusions. Demonstrate adequacy of members for deflection and for bending and shear stresses. Neglect composite action between metal deck and concrete floor.

Illustration for Prob. 20

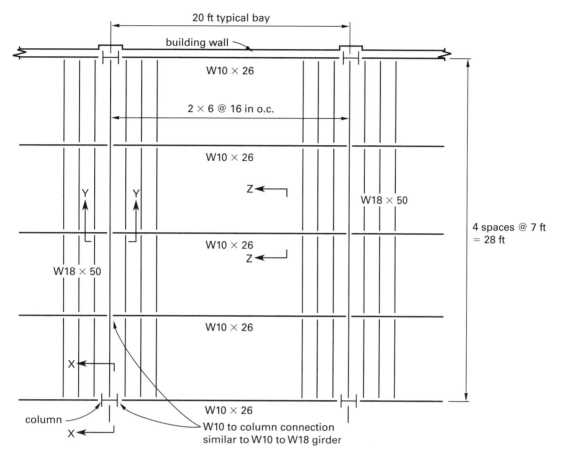

first floor plan

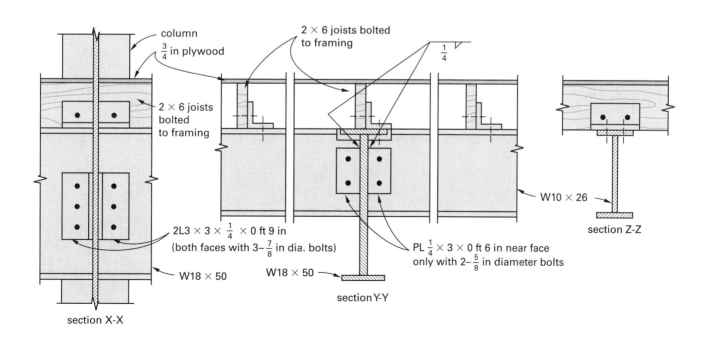

PROBLEM 21

A portion of a roof plan is shown. This is a part of a large single-story industrial plant. Steel joists and steel girders consisting of rolled steel shapes support an incombustible roof and a hung ceiling.

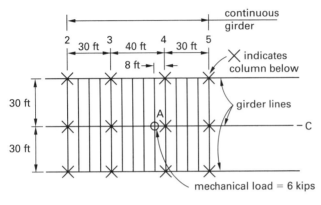

partial roof plan

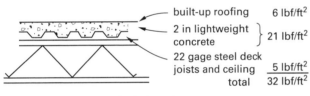

typical section

Design Criteria

- continuous roof girders over supports on lines 3 and 4

- simply supported roof girders at lines 2 and 5

- ASTM A992 steel

- roof live load = 12 lbf/ft^2

- estimated beam weight = 50 lbf/ft

21.1. Determine the lightest rolled shape steel girder section along line C that will extend between column lines 2 and 5. Use LRFD. Show calculations for the maximum moment and buckling, and explain how lateral support will be provided. Use inelastic analysis per AISC App. 1.

PROBLEM 22

A truss joint is shown.

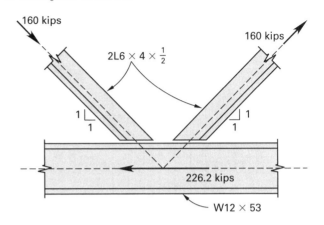

Design Criteria

- ASTM A36 steel

- ASTM A325 slip-critical bolts

22.1. Design and detail the shown connection using bolts to connect the angles to the gusset and welds to connect the gusset to the bottom chord. Use ASD. Show through calculations that the transfer of all the forces has been accounted for, and that the stresses in the critical sections are within the allowable limits.

PROBLEM 23

An aircraft hangar with a cantilevered roof is shown. (See *Illustration for Prob. 23.*) The main roof girder supports open web steel joists at 8 ft 4 in on center. The indicated uniform dead load includes an allowance for the weight of the main roof girder ABF.

Design Criteria

- ASTM A992 steel

- all joists are pinned

- loads are as follows

	dead load	live load
P	9.5 kips	–
w	0.62 kips/ft	0.63 kips/ft

23.1. Design the main roof girder using the most economical W section. Use LRFD. Consider strength only (i.e., do not check deflections and ponding).

23.2. Complete the details of section A-A. Add any necessary items and state your reason for each addition or modification.

Illustration for Prob. 23

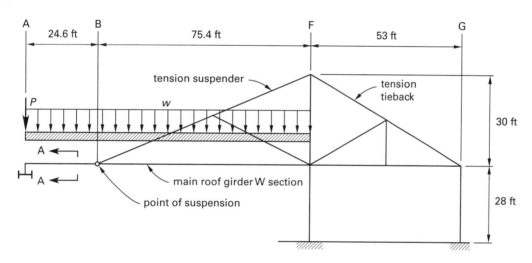

elevation—typical cantilever bent

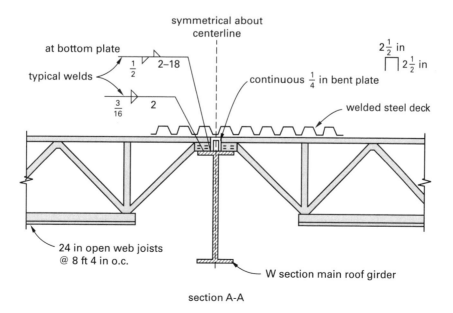

section A-A

PROBLEM 24

The end of one wing of a two story concrete hospital is shown. (See *Illustration for Prob. 24*.) The column shown on line 2 must be removed to install a new operating room. The existing ceiling line must be maintained, and the ceiling space cannot be used for the new structure.

The 8 in exterior wall on line 1 and the columns on line 3 can support new loads. The roof deflection cannot be increased from its current condition. A new steel beam will be added above the roof line as indicated.

Design Criteria

* roof live load $= 20$ lbf/ft^2 (not reducible)

* $f'_c = 3$ ksi stone concrete

* ASTM A992 steel for W shapes

* ASTM A36 steel for other shapes

24.1. Design a new steel beam to support the existing roof in the region where the existing column will be removed. Use LRFD. Show calculations for moment, shear, and buckling. Determine an appropriate beam size.

24.2. Design a suitable connection to support the existing roof at line 2 after the new beam is installed and the column removed. Draw a detail of the design showing all the necessary structural elements.

24.3. Describe steps to be followed during construction to install and complete the structural requirements of the beam and connection. Do not include finish work on the roofing and waterproofing of the steel.

Illustration for Prob. 24

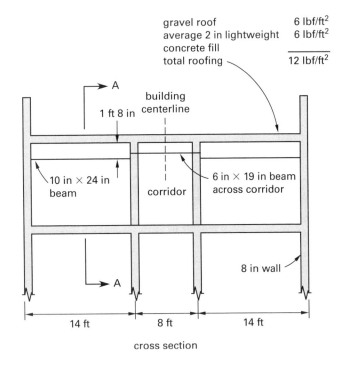

cross section

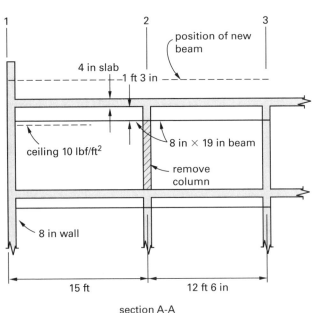

section A-A

SOLUTION 1

1.1. Friction type, or slip-critical, fasteners are used whenever slippage of the joint cannot be tolerated. Uses include applications where joints will be subjected to stress reversals, where bolts share load with weld groups in a connection, where oversized holes are used to facilitate erection, and where slippage would impair serviceability (such as splices in long-span trusses).

1.2. For friction-type connections, the condition of the faying surfaces governs the coefficient of friction and so has a significant effect on capacity. For most applications, faying surfaces are cleaned of mill scale and left unpainted.

Proper bolt tension is also important in a friction-type connection. For bearing-type fasteners, bolts need only be tightened to snug-tight condition.

1.3. There are three methods of installing A325-SC bolts.

- *Turn-of-nut tightening.* All bolts in the group are brought to snug-tight condition, then each bolt is tightened an additional turn (ranging from $^1/_3$ to 1 full turn depending on the length of the bolt).

- *Calibrated wrench tightening.* A torque wrench is calibrated using a device that measures bolt tension. Calibration is performed using the specific bolt and surface conditions used on the job.

- *Direct tension indicator tightening.* A load indicator washer or other device is used to gauge the tension.

1.4. Possible corrective measures for the oversized holes include

- replacing the A307 bolts with appropriate friction-type fasteners with hardened washers

- reaming the holes and using larger diameter A307 bolts

- field welding the shear tab-to-web connection to carry the total reaction

1.5. These are not the required bolts. The markings on the hexagonal head of an A325 are shown.

1.6. Closed cross sections are the most efficient in resisting torsion. Of the given sections, pipe resists torsion best, then square structural tube, and then rectangular tube. The closed sections offer relatively large resistance to torsional deformations. For the pipe, an applied torque induces only shear stress. The square and rectangular tubes can experience some warping stresses, but these are usually not significant.

Open cross sections (for example, W, C, and L sections) are relatively flexible and weak in torsional resistance. Warping stresses may become significant for these sections and are additive to the normal stresses caused by the bending moment. Of the given sections, the W shape (wide flange section) offers the advantage that its shear center coincides with its geometric centroid, which means that for most applications, loading will not induce torsion. The channel, on the other hand, has its shear center outside the section, which makes it difficult to apply transverse loading without causing torsion (twisting). The shear center of the angle is at the centerline junction of the two legs, and it is also difficult to load without some twisting.

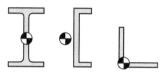

1.7. The P-Δ effect is the bending moment and deformation (so-called secondary bending) that is induced in a frame by the gravity forces (P-force) acting through the joint translations (Δ) caused by the externally applied forces.

1.8. Lamellar tearing is a separation in the through-thickness direction of thick steel plates caused by stresses induced by restraint to weld shrinkage. In the detail shown, the potential for lamellar tearing exists in both plates.

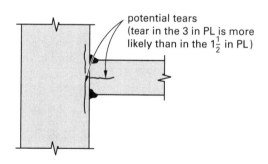

potential tears
(tear in the 3 in PL is more likely than in the $1\frac{1}{2}$ in PL)

1.9. Plastic hinges should form first in the girders of a rigid frame, rather than in the columns. This permits energy imparted by ground motion to be dissipated by inelastic deformation without collapse of the structure.

1.10. According to the *AISC Manual*, prequalified welds are commonly used joint details that are exempt from qualification testing. However, prescribed form, geometry, and procedures must be followed to deposit sound weld metal.

1.11. The best solution is to connect the flanges as shown. A slot large enough to pass the connection splice plate could probably be made without reinforcing the web opening. This connection also has the advantage of using field bolting rather than field welding.

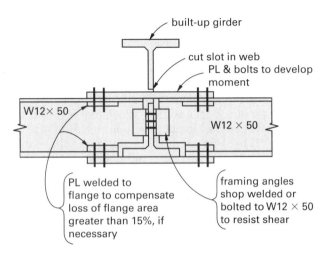

The alternatives to a plate through the web involve either stiff tension hangers (structural tees) bolted through the web, or welded splice plates on opposite sides of the web. The hanger type is inefficient and would interfere with the shear splice in such a shallow member. Welding splice plates to the relatively thick (1.5 in) web would be a bad design choice because lamellar tearing could occur.

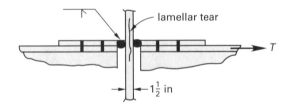

1.12. Using ASD, for A325-N bolts with a 1 in diameter and in single shear, AISC Table 7-1 gives the available shear strength per bolt as

$$\frac{r_n}{\Omega_v} = 21.2 \text{ kips/bolt}$$

For A36 steel ($F_u = 58$ ksi) and standard size holes with 3 in spacing, AISC Table 7-4 gives

$$\frac{r_n}{\Omega_v} = 67.4 \text{ kips/in-bolt}$$

Since the web thickness of the beam is greater than $^3/_8$ in, the bearing capacity of the bolts is controlled by the $^3/_8$ in plate thickness.

$$\left(\frac{r_n}{\Omega_v}\right)t = \left(67.4 \, \frac{\text{kips}}{\text{in-bolt}}\right)(0.375 \text{ in})$$
$$= 25.3 \text{ kips/bolt}$$

The strength of 21.2 kips/bolt controls. There are five bolts, so the total available shear strength is

$$R = n_{\text{bolt}}\left(\frac{r_n}{\Omega_v}\right) = (5 \text{ bolts})\left(21.2 \, \frac{\text{kips}}{\text{bolt}}\right) = 106 \text{ kips}$$

Given that the plate and weld are adequate, the capacity of the bolts is more than enough, and it is not necessary to check the bolts for eccentric shear.

The connection is adequate for 82 kips.

1.13. Use AISC Table J2.5. For a $^5/_{16}$ in E80XX fillet weld, the strength per inch of weld based on the effective weld throat is

$$R_n = 0.707\omega(0.6F_{\text{EXX}})L$$
$$= (0.707)\left(\tfrac{5}{16}\text{ in}\right)(0.6)\left(80 \, \frac{\text{kips}}{\text{in}^2}\right)\left(1 \, \frac{\text{in}}{\text{in}}\right)$$
$$= 10.6 \text{ kips/in}$$

By LRFD ($\phi = 0.75$),

$$\phi R_n = (0.75)\left(10.6 \, \frac{\text{kips}}{\text{in}}\right) = 7.97 \text{ kips/in}$$

By ASD ($\Omega = 2.00$),

$$\frac{R_n}{\Omega} = \frac{10.6 \, \dfrac{\text{kips}}{\text{in}}}{2.00} = 5.3 \text{ kips/in}$$

1.14. Use LRFD to calculate the ultimate capacity of the details. The gusset plate is not specified; therefore, assume it is adequate. For the $2L3 \times 2 \times {}^5/_{16}$, with a single gage of three 1 in diameter A325-SC standard size punched holes, assume the bolts are at usual pitch 3 in on centers and that the 2L shapes have short legs back to back.

$$A_n = A_g - 2(d_b + 0.125 \text{ in})t$$
$$= 2.92 \text{ in}^2 - (2)(1.00 \text{ in} + 0.125 \text{ in})(0.3125 \text{ in})$$
$$= 2.22 \text{ in}^2$$

The connection eccentricity, $\overline{x}$, is found in AISC Table 1-7. The shear lag factor, U, is calculated as in AISC Table D3.1.

$$U = 1 - \frac{\overline{x}}{L} = 1 - \frac{1.01 \text{ in}}{6 \text{ in}} = 0.83$$
$$A_e = U A_n = (0.83)(2.22 \text{ in}^2) = 1.84 \text{ in}^2$$

From AISC Chap. D,

$$\phi P_n \le \begin{cases} \phi F_y A_g = (0.9)\left(36 \, \dfrac{\text{kips}}{\text{in}^2}\right)(2.92 \text{ in}^2) \\ \qquad = 95 \text{ kips} \\ \phi F_u A_e = (0.75)\left(58 \, \dfrac{\text{kips}}{\text{in}^2}\right)(1.84 \text{ in}^2) \\ \qquad = 80 \text{ kips} \quad \text{[controls]} \end{cases}$$

For the bolts, use the design strengths from AISC Table 7-3 and Table 7-4. For 1 in diameter A325-SC bolts in double shear, AISC Table 7-3 gives

$$\phi r_v = 34.6 \text{ kips/bolt}$$

For two $^5/_{16}$ angles, the equivalent thickness is $^5/_8$ in. AISC Table 7-4 gives

$$\phi r_v = \left(101 \frac{\text{kips}}{\text{in-bolt}}\right) t = \left(101 \frac{\text{kips}}{\text{in-bolt}}\right)(0.625 \text{ in})$$

$$= 63.1 \text{ kips/bolt}$$

The double shear capacity is lower and controls bolt strength.

$$\phi P_n = n_{\text{bolt}} \phi r_v = (3 \text{ bolts})\left(34.6 \frac{\text{kips}}{\text{bolt}}\right) = 104 \text{ kips}$$

Therefore, tensile fracture governs: $\phi P_n = 80$ kips.

For the $2L3 \times 2 \times {}^5/_{16}$ attached with a single leg, using a $^1/_4$ in E70XX fillet weld both sides, assume short legs back to back and no end returns. (The capacity of the erection bolt cannot be included per AISC Sec. J1.8.) The length of the weld and section modulus of the weld are

$$L_w = 2L = (2)(6 \text{ in}) = 12 \text{ in}$$

$$S_w = \frac{2L^2}{6} = \frac{(2)(6 \text{ in})^2}{6} = 12 \text{ in}^2$$

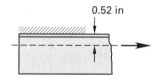

0.52 in

Calculate the shear flow from the force, the bending moment, and their resultant shear flow.

$$f_v = \frac{P_u}{L_w} = \frac{P_u}{12 \text{ in}} = \frac{0.083 P_u}{\text{in}}$$

$$f_b = \frac{M}{S_w} = \frac{P_u e}{S_w} = \frac{P_u(0.52 \text{ in})}{12 \text{ in}^2} = \frac{0.043 P_u}{\text{in}}$$

$$f = \sqrt{f_v^2 + f_b^2} = \sqrt{\left(\frac{0.083 P_u}{\text{in}}\right)^2 + \left(\frac{0.043 P_u}{\text{in}}\right)^2}$$

$$= \frac{0.0936 P_u}{\text{in}}$$

For a minimum field weld of $^1/_4$ in, the weld capacity is found by multiplying the number of sixteenths of an inch in the weld size, D, by the weld capacity per sixteenth of an inch. For an E70 fillet weld, the strength per inch per $^1/_{16}$ in weld size is 1.39 kips/in per $^1/_{16}$ in. (See AISC Part 8.)

$$q_w = Dq = (4)\left(1.39 \frac{\text{kips}}{\text{in}}\right)$$

$$= 5.56 \text{ kips/in} \quad [\text{each side}]$$

Set the shear strength equal to the weld capacity.

$$f = q_w$$

$$\frac{0.0936 P_u}{\text{in}} = 5.56 \text{ kips/in}$$

$$P_u = 59 \text{ kips}$$

Detail B would be expected to fail before detail A. The weld would probably fail first, followed by shearing of the erection bolt. The capacity of the eccentrically loaded weld is significantly lower than the fracture strength that governs the bolted connection of detail A. Also, the eccentricity in the welded connection concentrates the inelastic demand near the ends of the weld, which lowers the number of cycles that can be tolerated before failure.

SOLUTION 2

2.1. Check the adequacy of the $W14 \times 90$ column. Use ASD, and assume that the given loads represent the controlling combination of ASCE/SEI7 Sec. 2.4.1.

For the $W14 \times 90$ column: $A_g = 26.5$ in^2, $b/2t_f = 10.2$, $h/t_w = 25.9$, $r_x = 6.14$ in, $r_y = 3.70$ in, $Z_x = 157$ in^3, $S_x = 143$ in^3, $L_p = 15.2$ ft, $L_r = 42.6$ ft, and $I_x = 999$ in^4.

For the $W24 \times 68$: $I_x = 1830$ in^4.

For the $W24 \times 76$: $I_x = 2100$ in^4.

Per AISC Comm. Sec. 7.2, for a frame with sidesway uninhibited and ignoring inelastic effects,

$$G_B = \frac{\sum \dfrac{I_c}{L_c}}{\sum \dfrac{I_g}{I_{tg}}} = \frac{(2)\left(\dfrac{999 \text{ in}^4}{12.5 \text{ ft}}\right)}{(2)\left(\dfrac{1830 \text{ in}^4}{25 \text{ ft}}\right)} = 1.09$$

$$G_A = \frac{\sum \dfrac{I_c}{L_c}}{\sum \dfrac{I_g}{L_g}} = \frac{(2)\left(\dfrac{999 \text{ in}^4}{12.5 \text{ ft}}\right)}{(2)\left(\dfrac{2100 \text{ in}^4}{25 \text{ ft}}\right)} = 0.95$$

From AISC Fig. C-A-7.2, these values of G_A and G_B give $K_x = 1.3$. The weak axis is braced at the floor levels, so $K_y = 1.0$.

$$\left(\frac{KL}{r}\right)_x = \frac{(1.3)(12.5 \text{ ft})\left(12 \frac{\text{in}}{\text{ft}}\right)}{6.14 \text{ in}}$$

$$= 32$$

$$\left(\frac{KL}{r}\right)_y = \frac{(1.0)(12.5 \text{ ft})\left(12 \frac{\text{in}}{\text{ft}}\right)}{3.70 \text{ in}}$$

$$= 41 \quad [\text{controls axial compression capacity}]$$

For A992 steel ($F_y = 50$ ksi), compute the strength of the column in the absence of bending. (See AISC Sec. E3.)

$$\sqrt{\frac{E}{F_y}} = \sqrt{\frac{29,000 \ \frac{\text{kips}}{\text{in}^2}}{50 \ \frac{\text{kips}}{\text{in}^2}}} = 24.08$$

$$4.71 \sqrt{\frac{E}{F_y}} = (4.71)(24.08) = 113$$

$KL/r < 4.71\sqrt{E/F_y}$, so use AISC Eq. E3-2.

$$F_e = \frac{\pi^2 E}{\left(\dfrac{KL}{r}\right)_y^2} = \frac{\pi^2 \left(29,000 \ \frac{\text{kips}}{\text{in}^2}\right)}{(41)^2}$$

$$= 170 \ \text{kips/in}^2$$

$$F_{cr} = 0.658^{F_y/F_e} F_y = (0.658)^{50 \ \frac{\text{kips}}{\text{in}^2} / 170 \ \frac{\text{kips}}{\text{in}^2}} \left(50 \ \frac{\text{kips}}{\text{in}^2}\right)$$

$$= 44.2 \ \text{kips/in}^2$$

$$P_c = \frac{F_{cr} A_g}{\Omega_c} = \frac{\left(44.2 \ \frac{\text{kips}}{\text{in}^2}\right)(26.5 \ \text{in}^2)}{1.67} = 701 \ \text{kips}$$

For bending about the x-axis, use the approximate second-order analysis of AISC App. 8.

$$P_{e2} = \frac{\pi^2 E A_g}{\left(\dfrac{KL}{r}\right)_x^2} = \frac{\pi^2 \left(29,000 \ \frac{\text{kips}}{\text{in}^2}\right)(26.5 \ \text{in}^2)}{(32)^2}$$

$$= 7400 \ \text{kips}$$

Since columns in moment-resisting frames support one-third of the gravity loads,

$$\frac{P_{\text{story}}}{P_{e,\text{story}}} = \frac{3P_{\text{nt}}}{P_{e2}} = \frac{(3)(380 \ \text{kips})}{7400 \ \text{kips}} = 0.154$$

$$B_2 = \frac{1}{1 - \dfrac{\alpha P_{\text{story}}}{P_{e,\text{story}}}} = \frac{1}{1 - (1.6)(0.154)} = 1.33$$

$$M_r = B_1 M_{\text{nt}} + B_2 M_{\text{lt}}$$

$$= B_1(0 \ \text{ft-kips}) + (1.33)(130 \ \text{ft-kips})$$

$$= 173 \ \text{ft-kips}$$

$$P_r = P_{\text{nt}} + B_2 P_{\text{lt}} = 380 \ \text{kips} + (1.33)(0 \ \text{kips})$$

$$= 380 \ \text{kips}$$

Check local buckling using AISC Table B.4. For the flange,

$$\lambda_p = 0.38 \sqrt{\frac{E}{F_y}} = (0.38)(24.08) = 9.15$$

$$\frac{b_f}{2t_f} = 10.2 \quad [> \lambda_p, \text{ so not compact}]$$

For the web,

$$\lambda_p = 3.76 \sqrt{\frac{E}{F_y}} = (3.76)(24.08) = 90.5$$

$$\frac{h}{t_w} = 25.9 \quad [< \lambda_p]$$

The section is not compact, based on local buckling in the flange. The flexural strength is governed either by the flexural capacity in lateral-torsional buckling (AISC Sec. F2.2) or by local flange buckling (AISC Sec. F3.2), whichever is smaller. The unsupported length is

$$L_b = 12.5 \ \text{ft} \quad [< L_p = 15.2 \ \text{ft}, \text{ so use AISC Eq. F2-1}]$$

From AISC Sec. F2.2, the plastic bending moment for the unsupported length is

$$M_n = M_p = F_y Z_x = \left(50 \ \frac{\text{kips}}{\text{in}^2}\right)(157 \ \text{in}^3)$$

$$= 7850 \ \text{in-kips}$$

Checking AISC Sec. F3.2 for a rolled W shape with noncompact flanges,

$$\lambda = \frac{b_f}{2t_f} = 10.2$$

$$\lambda_{pf} = \lambda_p = 9.15$$

$$\lambda_{rf} = \lambda_r = 1.0 \sqrt{\frac{E}{F_y}} = 24.08$$

$$M_n = M_p - (M_p - 0.7 F_y S_x)\left(\frac{\lambda - \lambda_{pf}}{\lambda_{rf} - \lambda_{pf}}\right)$$

$$= 7850 \ \text{in-kips} - \left(\begin{array}{c} 7850 \ \text{in-kips} - (0.7) \\ \times \left(50 \ \frac{\text{kips}}{\text{in}^2}\right)(143 \ \text{in}^3) \end{array} \right)$$

$$\times \left(\frac{10.2 - 9.15}{24.1 - 9.15}\right)$$

$$= 7650 \ \text{in-kips} \quad [\text{controls}]$$

$$M_c = \frac{M_n}{\Omega_c} = \frac{7650 \ \text{in-kips}}{1.67} = 4580 \ \text{in-kips}$$

Check combined axial compression and bending. (See AISC Sec. H1.)

$$\frac{P_r}{P_c} = \frac{380 \ \text{kips}}{701 \ \text{kips}} = 0.54 \quad [> 0.2, \text{ so use AISC Eq. H1-1a}]$$

$$\frac{P_r}{P_c} + \frac{8}{9}\left(\frac{M_{rx}}{M_{cx}}\right) = 0.54 + \left(\tfrac{8}{9}\right)\left(\frac{(173 \text{ ft-kips})\left(12 \frac{\text{in}}{\text{ft}}\right)}{4580 \text{ in-kips}}\right)$$

$$= 0.94 \quad [\leq 1.0]$$

The W14 × 90 is adequate in combined axial plus bending.

SOLUTION 3

3.1. Check the given $^3/_6$ in E70XX intermittent fillet welds using the elastic method. Use ASD.

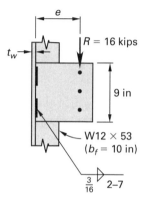

For the W12 × 53, $b_f = 10.0$ in and $t_w = 0.34$ in.

$$e = 0.5(b_f - t_w) + 1.5 \text{ in}$$

$$= (0.5)(10.0 \text{ in} - 0.34 \text{ in}) + 1.5 \text{ in}$$

$$= 6.3 \text{ in}$$

$$M = Re = (16 \text{ kips})(6.3 \text{ in})$$

$$= 101 \text{ in-kips}$$

$$L_w = 2(L_{\text{top}} + L_{\text{bottom}}) = (2)(2 \text{ in} + 2 \text{ in}) = 8 \text{ in}$$

$$I_w = (2)(2)\left(\frac{L_{\text{top}}^3}{12} + L_{\text{top}}d^2\right)$$

$$= (2)(2)\left(\frac{(2 \text{ in})^3}{12} + (2 \text{ in})(3.5 \text{ in})^2\right)$$

$$= 100.6 \text{ in}^3$$

$$S_w = \frac{I_w}{c} = \frac{100.6 \text{ in}^3}{4.5 \text{ in}} = 22.4 \text{ in}^2$$

$$f_v = \frac{R}{L_w} = \frac{16 \text{ kips}}{8 \text{ in}} = 2 \text{ kips/in}$$

$$f_b = \frac{M}{S_w} = \frac{101 \text{ in-kip}}{22.4 \text{ in}^2} = 4.5 \text{ kips/in}$$

$$f_r = \sqrt{f_v^2 + f_b^2} = \sqrt{\left(2.0 \frac{\text{kips}}{\text{in}}\right)^2 + \left(4.5 \frac{\text{kips}}{\text{in}}\right)^2}$$

$$= 4.9 \text{ kips/in}$$

For A36 steel, shear yielding controls the base material strength. (See AISC Sec. J4.2.) Since welds are on both sides of the $^3/_8$ in plate, only half the plate thickness is available to resist stresses from one $^3/_{16}$ in weld. From AISC Table J2.5, the strength of the fillet weld on one side is

$$q \leq \begin{cases} D\left(\dfrac{f_w}{\Omega}\right) = (3)\left(0.928 \dfrac{\text{kips}}{\text{in}}\right) \\ \quad = 2.78 \text{ kips/in} \quad [\text{controls}] \\ \left(\dfrac{0.6F_u}{\Omega}\right)\left(\dfrac{t}{2}\right) = \left(\dfrac{(0.6)\left(58 \dfrac{\text{kips}}{\text{in}^2}\right)}{2.0}\right)\left(\dfrac{0.375 \text{ in}}{2}\right) \\ \quad = 3.26 \text{ kips/in} \end{cases}$$

D is the number of sixteenths of an inch in the weld length. The weld strength, q, is less than the required shear strength, f_r, so the weld is overstressed.

The connection is no good.

Even if the weld strength had been adequate, the detail shows a poor design. Using the 9 in by $^3/_8$ in plate would create high punching shear stress in the relatively thin web of the column. A better detail would show a seated connection, which would reduce the eccentricity and transfer the shear closer to the column flanges. If, for some reason, the connection must be made with the given eccentricity, horizontal plates should be welded across between the flanges to provide resistance to the bending in the plate.

SOLUTION 4

4.1. Check the beam for the proposed loading using LRFD, and take the load factor, U, as 1.6 (conservative). Given: A36 steel ($F_y = 36$ ksi), 4 in thick concrete slab, normal weight concrete, $f'_c = 3$ ksi, and eighteen $^3/_4$ in diameter headed studs uniformly spaced. From AISC Table 3-21, the nominal horizontal shear, Q_n, is 21.0 kips per stud.

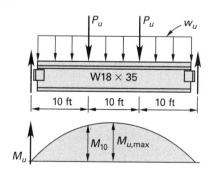

Per AISC Sec. I-3.1a, the effective flange width is

$$b_e \leq \begin{cases} 0.25L = (0.25)(30 \text{ ft})\left(12 \dfrac{\text{in}}{\text{ft}}\right) \\ \quad = 90 \text{ in} \quad [\text{controls}] \\ \text{beam spacing} = (8 \text{ ft})\left(12 \dfrac{\text{in}}{\text{ft}}\right) \\ \quad = 96 \text{ in} \end{cases}$$

Shear and deflections are assumed to be adequate, so only the flexural strength needs to be checked. For the W18 × 35,

$$\frac{h}{t_w} = 53.5$$

$$3.76\sqrt{\frac{E}{F_y}} = 3.76\sqrt{\frac{29{,}000 \dfrac{\text{kips}}{\text{in}^2}}{36 \dfrac{\text{kips}}{\text{in}^2}}}$$

$$= 107 \quad \left[h/t_w \leq 3.76\sqrt{E/F_y}\right]$$

Per AISC Sec. I1.2a, use the plastic stress distribution with $\phi = 0.9$.

$$w_u = 1.6w = (1.6)\left(1 \frac{\text{kip}}{\text{ft}}\right) = 1.6 \text{ kips/ft}$$

$$P_u = 1.6P = (1.6)(3 \text{ kips}) = 4.8 \text{ kips}$$

$$M_{10} = P_u a + (0.5)\big(0.5 w_u L + w_u(5 \text{ ft})\big)a$$

$$= (4.8 \text{ kips})(10 \text{ ft})$$

$$+ (0.5)\left(\begin{array}{c} (0.5)\left(1.6 \dfrac{\text{kips}}{\text{ft}}\right)(30 \text{ ft}) \\ + \left(1.6 \dfrac{\text{kips}}{\text{ft}}\right)(5 \text{ ft}) \end{array} \right)(10 \text{ ft})$$

$$= 208 \text{ ft-kips}$$

$$M_{u,\text{max}} = P_u a + 0.125 w_u L^2$$

$$= (4.8 \text{ kips})(10 \text{ ft}) + (0.125)\left(1.6 \frac{\text{kips}}{\text{ft}}\right)(30 \text{ ft})^2$$

$$= 228 \text{ ft-kips}$$

Uniformly spaced studs are acceptable per AISC; however, AISC Sec. I8.2c requires that the number of studs between a concentrated load and a point of zero moment must develop the moment at the load point. Therefore, check the design strength, ϕM_n, for the W18 × 35 required to develop the moment at the load point, $M_{10} = 208$ ft-kips. Over this length, there are six uniformly spaced studs.

$$\sum Q_n = (6 \text{ studs})\left(21.0 \frac{\text{kips}}{\text{stud}}\right) = 126 \text{ kips}$$

$$a = \frac{\sum Q_n}{0.85 f'_c b_e} = \frac{126 \text{ kips}}{(0.85)\left(3 \dfrac{\text{kips}}{\text{in}^2}\right)(90 \text{ in})} = 0.55 \text{ in}$$

$$Y2 = h - 0.5a = 4 \text{ in} - (0.5)(0.55 \text{ in}) = 3.73 \text{ in}$$

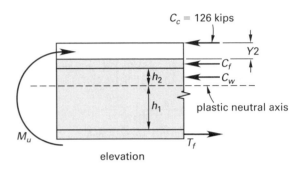

elevation

The *AISC Manual* does not tabulate moment capacity for composite beams of A36 steel. However, the strength can be found from equilibrium of the plastic stress distribution. For the W18 × 35: $A_g = 10.3$ in^2, $d = 17.7$ in, $t_w = 0.30$ in, $b_f = 6.0$ in, and $t_f = 0.425$ in. Neglect the fillets and assume the cross section to consist of two flange plates and a web plate of uniform thickness (conservative). For horizontal equilibrium,

$$\sum F_{\text{horiz}} = T_f + T_w - C_w - C_f - C_c - 0$$

$$= F_y(b_f t_f + t_w h_1 - t_w h_2 - b_f t_f) - 126 \text{ kips}$$

$$t_w(h_1 - h_2) = \frac{126 \text{ kips}}{F_y} = \frac{126 \text{ kips}}{36 \dfrac{\text{kips}}{\text{in}^2}} = 3.50 \text{ in}^2$$

$$t_w(h_1 - h_2) = t_w\big((h_1 + h_2 - h_2) - h_2\big)$$

$$= t_w\big((h_1 + h_2) - 2h_2\big)$$

$$= t_w\big((d - 2t_f) - 2h_2\big)$$

$$h_2 = \frac{d}{2} - t_f - \frac{t_w(h_1 - h_2)}{2t_w}$$

$$= \frac{17.7 \text{ in}}{2} - 0.425 \text{ in} - \frac{3.50 \text{ in}^2}{(2)(0.30 \text{ in})}$$

$$= 2.59 \text{ in}$$

$$h_1 = \frac{3.50 \text{ in}^2}{t_w} + h_2$$

$$= \frac{3.50 \text{ in}^2}{0.30 \text{ in}} + 2.59 \text{ in}$$

$$= 14.26 \text{ in}$$

Taking moments about the force C_c,

$$\phi M_n = \phi \begin{pmatrix} F_y b_f t_f (d - t_f) + F_y t_w h_1 \\ \times \left(Y2 + d - t_f - \dfrac{h_1}{2} \right) \\ - F_y t_w h_2 \left(Y2 + t_f + \dfrac{h_1}{2} \right) \end{pmatrix}$$

$$= \frac{(0.9) \begin{pmatrix} \left(36\ \dfrac{\text{kips}}{\text{in}^2} \right)(6.0\ \text{in})(0.425\ \text{in}) \\ \times (17.7\ \text{in} - 0.425\ \text{in}) \\ + \left(36\ \dfrac{\text{kips}}{\text{in}^2} \right)(0.3\ \text{in})(14.26\ \text{in}) \\ \times \begin{pmatrix} 3.73\ \text{in} + 17.7\ \text{in} \\ - 0.425\ \text{in} - \dfrac{14.26\ \text{in}}{2} \end{pmatrix} \\ - \left(36\ \dfrac{\text{kips}}{\text{in}^2} \right)(0.3\ \text{in})(2.59\ \text{in}) \\ \times \left(3.73\ \text{in} + 0.425\ \text{in} + \dfrac{2.59\ \text{in}}{2} \right) \end{pmatrix}}{12\ \dfrac{\text{in}}{\text{ft}}}$$

$$= 268\ \text{ft-kips} \quad [> M_{10}]$$

The capacity at the concentrated load point is larger than the $M_{u,\text{max}}$ at midspan, and the capacity at midspan will be greater than at $x = 10$ ft, where nine studs are effective, so there is no need to recompute the capacity at midspan.

> The W18 × 35 composite beam is adequate in flexure.

SOLUTION 5

5.1. Determine the fillet weld size for static and fatigue loading. Given: ASTM A36 steel ($F_y = 36$ ksi) and E70XX electrodes. Assume given loads are service values, and use ASD.

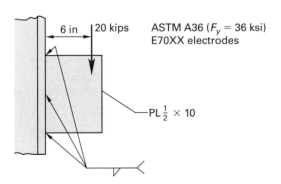

The one-sided fillet weld is not recommended. (See AISC Fig. 8-8.) However, assume that lateral deformation cannot occur and that the detail is acceptable for static loading. The following calculations are made using the elastic method, based on the given weld pattern. For the 20 kip static load, from AISC Table J2.4, the minimum weld size is $w_{\min} = {}^3/_{16}$ in based on the thinner $1/2$ in plate. Assume that only the vertical weld resists shear and that the vertical plus two returns resist the bending.

$$L_w = 10\ \text{in}$$

$$I_w = \frac{L_w^3}{12} + 2t \left(\frac{L_w}{2} \right)^2$$

$$= \frac{(10\ \text{in})^3}{12} + (2)(0.5\ \text{in}) \left(\frac{10\ \text{in}}{2} \right)^2$$

$$= 108.3\ \text{in}^3$$

$$S_w = \frac{I_w}{0.5 L_w} = \frac{108.3\ \text{in}^3}{(0.5)(10\ \text{in})} = 21.7\ \text{in}^2$$

$$f_v = \frac{P}{L_w} = \frac{20\ \text{kips}}{10\ \text{in}} = 2.0\ \text{kips/in}$$

$$f_b = \frac{M}{S_w} = \frac{Pe}{S_w} = \frac{(20\ \text{kips})(6\ \text{in})}{21.7\ \text{in}^2}$$

$$= 5.5\ \text{kips/in}$$

$$f_r = \sqrt{f_v^2 + f_b^2}$$

$$= \sqrt{\left(2.0\ \frac{\text{kips}}{\text{in}} \right)^2 + \left(5.5\ \frac{\text{kips}}{\text{in}} \right)^2}$$

$$= 5.9\ \text{kips/in}$$

The weld size, in sixteenths of an inch, is

$$w \geq \frac{f_r}{q_w} = \frac{5.9\ \dfrac{\text{kips}}{\text{in}}}{0.928\ \dfrac{\text{kip}}{\text{in}}} = 6.3 \quad \left[\text{say}\ {}^7/_{16}\ \text{in} \right]$$

For A36 steel ($F_y = 36$ ksi and $F_u = 58$ ksi), the base material strength is the lower of the values computed for shear yield and shear rupture.

$$q_{\text{beam}} \leq \begin{cases} \dfrac{0.6 F_y t}{\Omega} = \dfrac{(0.6)\left(36\ \dfrac{\text{kips}}{\text{in}^2} \right)(0.5\ \text{in})}{1.5} \\ \qquad = 7.2\ \text{kips/in} \quad [\text{controls}] \\[4pt] \dfrac{0.6 F_u t}{\Omega} = \dfrac{(0.6)\left(58\ \dfrac{\text{kips}}{\text{in}^2} \right)(0.5\ \text{in})}{2.0} \\ \qquad = 8.7\ \text{kips/in} \end{cases}$$

> Use $^7/_{16}$ in E70 fillet welds for the static condition.

5.2. For 2,000,000 cycles of fully reversible stress on the throat of a transverse fillet weld, AISC Table A-3.1 and Sec. 8.2 categorize the detail as stress category F and give the fatigue parameters as $C_f = 150 \times 10^{10}$ and $F_{TH} = 8$ ksi. For stress category F, AISC Eq. A-3-2 gives the design stress range as

$$F_{SR} \geq \begin{cases} \left(\dfrac{C_f}{N}\right)^{0.167} = \left(\dfrac{150 \times 10^{10}}{2 \times 10^6}\right)^{0.167} \\ \qquad = 9.6 \text{ ksi} \quad [\text{controls}] \\ F_{TH} = 8 \text{ ksi} \end{cases}$$

The actual stress range on the effective throat of the weld is

$$f_{SR} = \frac{2f_r}{0.707w} \leq F_{SR}$$

The needed weld size, then, is

$$w \geq \frac{2f_r}{0.707 F_{SR}} = \frac{(2)\left(5.9 \; \dfrac{\text{kips}}{\text{in}}\right)}{(0.707)\left(9.6 \; \dfrac{\text{kips}}{\text{in}^2}\right)}$$

$$= \boxed{1.73 \text{ in} \quad [\text{impractical}]}$$

The detail would have to be revised to accommodate the fatigue loading.

SOLUTION 6

6.1. Prepare a remedial solution for the overstressed beam. Use LRFD. The flexural strength at midspan must be 35% greater than the nominal moment strength, M_n, of the W16 × 40 of A36 steel: $A_s = 11.8$ in^2, $I_x = 518$ in^4, $Z_x = 73.0$ in^3, $b_f/2t_f = 6.93$, $h/t_w = 46.5$, $r_y = 1.57$ in, $b_f = 7.00$ in, $t_f = 0.505$ in, $d = 16.0$ in, $t_w = 0.305$ in, and $k_{des} = 0.907$ in. From AISC Table B4.1,

$$\lambda_{pf} = 0.38\sqrt{\frac{E}{F_y}} = 0.38\sqrt{\frac{29{,}000 \; \dfrac{\text{kips}}{\text{in}^2}}{36 \; \dfrac{\text{kips}}{\text{in}^2}}} = 10.8$$

$$\frac{b_f}{2t_f} = 6.93 \quad [< \lambda_{pf}]$$

$$\lambda_{pw} = 3.76\sqrt{\frac{E}{F_y}} = 3.76\sqrt{\frac{29{,}000 \; \dfrac{\text{kips}}{\text{in}^2}}{36 \; \dfrac{\text{kips}}{\text{in}^2}}} = 106.7$$

$$\frac{h}{t_w} = 46.5 \quad [< \lambda_{pw}, \text{ compact}]$$

The section is compact and laterally braced by the joists spaced 16 in on center.

$$L_p = 1.76 r_y \sqrt{\frac{E}{F_y}}$$

$$= (1.76)(1.57 \text{ in})\sqrt{\frac{29{,}000 \; \dfrac{\text{kips}}{\text{in}^2}}{36 \; \dfrac{\text{kips}}{\text{in}^2}}}$$

$$= 78.5 \text{ in} \quad [> L_b = 16 \text{ in}]$$

Thus, the existing beam can develop its full plastic flexural strength.

$$M_p = F_y Z_x = \frac{\left(36 \; \dfrac{\text{kips}}{\text{in}^2}\right)(73.0 \text{ in}^3)}{12 \; \dfrac{\text{in}}{\text{ft}}} = 219 \text{ ft-kips}$$

The existing beam is 35% overloaded; therefore, the beam must be modified to increase its flexural capacity to at least $(1.35)(219 \text{ ft-kips}) = 296$ ft-kips. One possibility is to create partial composite action with the 1.5 in concrete slab, but the separation between the slab and top of flange of the steel girder would make it difficult to create reliable composite action. Another possibility is to add a cover plate that fits within the 1.5 in clearance between the existing beam and ceiling. Adding a plate of A36 steel that has the same area as the W16 × 40 (11.8 in^2) would move the plastic neutral axis to the bottom of the flange. Try a PL1$\frac{1}{2}$ × 8 (slightly larger in area by 0.2 in^2, which can be neglected when computing M_p).

$$M_p = F_y\big(A_s(0.5d) + bt(0.5t)\big)$$

$$= \frac{\left(36 \; \dfrac{\text{kips}}{\text{in}^2}\right)\left(\begin{array}{c}(11.8 \text{ in}^2)(0.5)(16.0 \text{ in}) + (8 \text{ in}) \\ \times (1.5 \text{ in})(0.5)(1.5 \text{ in})\end{array}\right)}{12 \; \dfrac{\text{in}}{\text{ft}}}$$

$$= 310 \text{ ft-kips} \quad [> M_n \text{ required}]$$

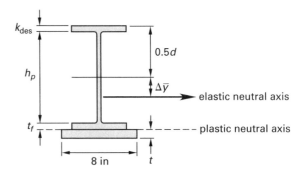

Check local buckling for the singly symmetrical section. (See AISC Table B4.1, case 11.)

$$A = A_{PL} + A_{W16 \times 40} = 12.0 \text{ in}^2 + 11.8 \text{ in}^2 = 23.8 \text{ in}^2$$

$$\Delta\overline{y} = \frac{A_{\text{PL}}\big(0.5(d+t)\big)}{A}$$

$$= \frac{(12.0\text{ in}^2)(0.5)(16.0\text{ in}+1.5\text{ in})}{23.8\text{ in}^2}$$

$$= 4.41\text{ in}$$

$$I = (I_x + A\overline{y}^2)_{\text{W16}\times40} + \frac{bt^3}{12} + A_{\text{PL}}\big(0.5(d+t)-\Delta\overline{y}\big)^2$$

$$= 518\text{ in}^4 + (11.8\text{ in}^2)(4.41\text{ in})^2 + \frac{(8\text{ in})(1.5\text{ in})^3}{12}$$

$$\qquad + (12.0\text{ in}^2)\big((0.5)(16.0\text{ in}+1.5\text{ in})-4.41\text{ in}\big)^2$$

$$= 976\text{ in}^4$$

$$M_{yc} = \frac{F_y I}{0.5d + \Delta\overline{y}} = \frac{\left(36\ \dfrac{\text{kips}}{\text{in}^2}\right)(976\text{ in}^4)}{\big((0.5)(16.0\text{ in})+4.41\text{ in}\big)\left(12\ \dfrac{\text{in}}{\text{ft}}\right)}$$

$$= 236\text{ ft-kips}$$

$$h_c = 2\big(\Delta\overline{y}+0.5(d-k_{\text{des}})\big)$$

$$= (2)\big(4.41\text{ in}+(0.5)(16.0\text{ in}-0.91\text{ in})\big)$$

$$= 23\text{ in}$$

$$h_p = 2(d-k_{\text{des}}) = (2)(16.0\text{ in}-0.91\text{ in})$$

$$= 30.1\text{ in}$$

$$\frac{h_c}{t_w} = \frac{23\text{ in}}{0.305\text{ in}} = 75.4$$

$$\lambda_{pw} = \frac{\dfrac{h_c}{h_p}\sqrt{\dfrac{E}{F_y}}}{\left(0.54\left(\dfrac{M_p}{M_{yc}}\right)-0.09\right)^2}$$

$$= \frac{\dfrac{23\text{ in}}{30.1\text{ in}}\sqrt{\dfrac{29{,}000\ \dfrac{\text{kips}}{\text{in}^2}}{36\ \dfrac{\text{kips}}{\text{in}^2}}}}{\left((0.54)\left(\dfrac{310\text{ ft-kips}}{226\text{ ft-kips}}\right)-0.09\right)^2}$$

$$= 57$$

$$\lambda_{rw} = 5.7\sqrt{\frac{E}{F_y}} = 5.7\sqrt{\frac{29{,}000\ \dfrac{\text{kips}}{\text{in}^2}}{36\ \dfrac{\text{kips}}{\text{in}^2}}}$$

$$= 161 \quad [\lambda_{pw}<\lambda<\lambda_{rw}]$$

The section is noncompact. Check AISC Eq. F4-9b.

$$R_{pc} \leq \begin{cases} \dfrac{M_p}{M_{yc}} - \left(\dfrac{M_p}{M_{yc}}-1\right)\left(\dfrac{\lambda-\lambda_{pw}}{\lambda_{rw}-\lambda_{pw}}\right) \\[2mm] \quad = \dfrac{310\text{ ft-kips}}{236\text{ ft-kips}} - \left(\dfrac{310\text{ ft-kips}}{236\text{ ft-kips}}-1\right) \\[2mm] \qquad \times \left(\dfrac{75.4-57}{161-57}\right) \\[2mm] \quad = 1.26 \quad [\text{controls}] \\[2mm] \dfrac{M_p}{M_{yc}} = 1.31 \end{cases}$$

$$M_n = R_{pc}M_{yc} = (1.26)(236\text{ ft-kips}) = 297\text{ ft-kips}$$

The modified section has the required flexural strength. Use PL$1^1/_2 \times 8$.

Compute the uniform load capacity of the strengthened member.

$$M_u = 0.125w_u L^2 = \phi_b M_n$$

$$= (0.9)(297\text{ ft-kips})$$

$$= 267\text{ ft-kips}$$

$$w_u = \frac{\phi_b M_n}{0.125L^2}$$

$$= \frac{267\text{ ft-kips}}{(0.125)(30\text{ ft})^2}$$

$$= 2.37\text{ kips/ft}$$

No connection or support information is given in the problem statement, so assume these items do not need to be checked. Theoretically, the cover plate can be terminated where the bending moment is less than

$$\phi M_p = (0.9)(267\text{ ft-kips}) = 197\text{ ft-kips}$$

$$R = 0.5w_u L$$

$$= (0.5)\left(2.37\ \frac{\text{kips}}{\text{ft}}\right)(30\text{ ft})$$

$$= 35.6\text{ kips}$$

$$M_{ux} = Rx - 0.5w_u x^2$$

$$= (35.6\text{ kips})x - (0.5)\left(2.37\ \frac{\text{kips}}{\text{ft}}\right)x^2$$

$$= 197\text{ ft-kips}$$

$$x = (7.4\text{ ft},\ 22.6\text{ ft})$$

Extend the cover plate far enough beyond the theoretical cutoff points to develop its share of flexural strength at the cutoff location. The stresses at top and bottom of the cover plate at the cutoff points are

$$f_{b1} = \frac{M(0.5d - \Delta\overline{y})}{I}$$

$$= \frac{(197 \text{ ft-kips})\left(12 \frac{\text{in}}{\text{ft}}\right)\left((0.5)(16.0 \text{ in}) - 4.41 \text{ in}\right)}{976 \text{ in}^4}$$

$$= 8.7 \text{ kips/in}^2$$

$$f_{b2} = \frac{M(0.5d - \Delta\overline{y} + t)}{I}$$

$$= \frac{(197 \text{ ft-kips})\left(12 \frac{\text{in}}{\text{ft}}\right)\left(\begin{array}{c}(0.5 \text{ in})(16.0 \text{ in}) \\ -4.41 \text{ in} + 1.5 \text{ in}\end{array}\right)}{976 \text{ in}^4}$$

$$= 12.3 \text{ kips/in}^2$$

$$T_{\text{PL}} = 0.5(f_{b1} + f_{b2})A_{\text{PL}}$$

$$= (0.5)\left(8.7 \frac{\text{kips}}{\text{in}^2} + 12.3 \frac{\text{kips}}{\text{in}^2}\right)(12.0 \text{ in}^2)$$

$$= 126 \text{ kips}$$

For the thinner plate joined ($t = t_f = 0.505$ in), AISC Table J2.4 requires $^1/_4$ in fillet welds. For an E70 fillet weld, the strength is 1.39 kips/in per $^1/_{16}$ in of weld size. (See AISC Part 8.)

$$q_w = \left(\text{no. of } \frac{1}{16} \text{ in}\right)\left(\text{strength per } \frac{1}{16} \text{ in}\right)$$

$$= (4)\left(1.39 \frac{\text{kips}}{\text{in}}\right)$$

$$= 5.56 \text{ kips/in}$$

$$L_w = \frac{T_{\text{PL}}}{q_w} = \frac{126 \text{ kips}}{\left(5.56 \frac{\text{kips}}{\text{in}}\right)\left(12 \frac{\text{in}}{\text{ft}}\right)}$$

$$= 1.9 \text{ ft} \quad [\text{total, half each side}]$$

However, AISC Eq. F13-7 requires that the welds extend a minimum of two times the plate width beyond the theoretical cutoff point; therefore, extend by a length of $(2)(8 \text{ in}) = 16 \text{ in} = 1.3 \text{ ft}$. The actual cutoff points are

$$x_1 - 0.5L_w = 7.4 \text{ ft} - 1.3 \text{ ft}$$

$$= 6.1 \text{ ft}$$

$$x_2 + 0.5L_w = 22.6 \text{ ft} + 1.3 \text{ ft}$$

$$= 23.9 \text{ ft}$$

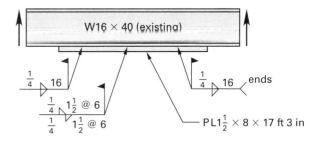

6.2. The weld heat may be sufficient to lower the yield point significantly, so shore the floor during installation. Precautions must also be taken to protect the wood floor joists during welding. The construction sequence should be as follows.

First, shore the joists on both sides adjacent to the W16 × 40.

Second, clamp the PL1$^1/_2$ × 8 to the lower flange.

Third, weld the plate.

Fourth, remove the clamps and shores.

SOLUTION 7

7.1. Per the problem statement, design the connection to transfer 50% of the shear and moment strength of the smaller (W14 × 176) section. The axial force is not specified and will not be included in the design. Use LRFD. For the W14 × 176: $Z_x = 320$ in^3, $d = 15.2$ in, $b_f = 15.7$ in, $t_w = 0.83$ in, and $t_f = 1.31$ in. For A36 steel, $F_y = 36$ ksi. Calculate the required flexural strength of the connection.

$$M_u = 0.5M_p = (0.5)\left(36 \frac{\text{kips}}{\text{in}^2}\right)(320 \text{ in}^3) = 5760 \text{ in-kips}$$

From AISC Table J2.1, for a partial penetration weld with 45° bevel by the shielded manual arc process, the effective throat is $^1/_8$ in smaller than the weld size. From AISC Table J2.5, for tension normal to the weld axis,

$$\phi f_w \leq \begin{cases} \phi F_u t = (0.75)\left(58 \frac{\text{kips}}{\text{in}^2}\right)D \\ \\ \quad = \left(43.5 \frac{\text{kips}}{\text{in}^2}\right)D \\ \\ \phi(0.6F_{\text{EXX}})t = (0.8)(0.6)\left(70 \frac{\text{kips}}{\text{in}^2}\right)D \\ \\ \quad = \left(33.6 \frac{\text{kips}}{\text{in}^2}\right)D \quad [\text{controls}] \end{cases}$$

D is the required throat thickness.

$$0.5M_p = \phi f_w b_f D(d - D)$$

$$5760 \text{ in-kip} = \left(33.6 \ \frac{\text{kips}}{\text{in}^2}\right)(15.7 \text{ in})$$
$$\times D(15.2 \text{ in} - D)$$

$$D^2 - (15.2 \text{ in})D + 10.92 \text{ in}^2 = 0 \text{ in}^2$$

$$D = 0.756 \text{ in}$$

The required weld size is

$$s = D + 0.125 \text{ in} = 0.756 \text{ in} + 0.125 \text{ in} = 0.881 \text{ in}$$

Use a weld $^{15}/_{16}$ in long (BTC-P4).

The size computed for strength exceeds the minimum weld size required by AISC Table J2.3.

7.2. Determine the number of bolts required to transfer 50% of the shear capacity of the W14 × 176. The bolts are required to be slip-critical. The factored shear force to be transferred across the web splice is

$$V_u = 0.5\big(\phi(0.6F_y)dt_w\big)$$

$$= (0.5)(0.6)(0.9)\left(36 \ \frac{\text{kips}}{\text{in}^2}\right)(15.2 \text{ in})(0.83 \text{ in})$$

$$= 122 \text{ kips}$$

Try 1 in diameter A325-SC bolts arranged as close together as the code permits to alleviate the stresses caused by eccentricity of the connection. Per AISC Sec. J3.10, bearing should not be critical on the $t_w =$ 0.83 in web material. Therefore, space the bolts at $2.67d$, say 2.75 in o.c. Use 11 in wide plates with 4 bolts across.

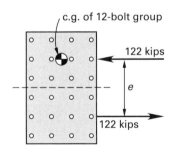

$$\text{edge distance} = 0.5(w - 3s)$$
$$= (0.5)\big(11 \text{ in} - (3)(2.75 \text{ in})\big)$$
$$= 1.375 \text{ in}$$

The edge distance cannot be provided with a sheared plate. (See AISC Table J3.4.) Therefore, specify that plates be gas cut. Try a 12-bolt group on each side of the splice. Use the elastic method of analysis for the eccentrically loaded fastener group. (See AISC Part 7.)

$$e = 3s = (3)(2.75 \text{ in}) = 8.25 \text{ in}$$

$$M = V_u e = (122 \text{ kips})(8.25 \text{ in}) = 1000 \text{ in-kips}$$

$$r_v = \frac{V_u}{n_{\text{bolt}}} = \frac{122 \text{ kips}}{12} = 10.1 \text{ kips}$$

$$\sum(x_i^2 + y_i^2) = (8)(2.75 \text{ in})^2 + (6)(4.125 \text{ in})^2$$
$$+ (6)(1.375 \text{ in})^2$$
$$= 174 \text{ in}^2$$

$$r_{mx} = \frac{My}{\sum(x_i^2 + y_i^2)} = \frac{(1000 \text{ in-kips})(2.75 \text{ in})}{174 \text{ in}^2}$$
$$= 15.8 \text{ kips}$$

$$r_{my} = \frac{Mx}{\sum(x_i^2 + y_i^2)} = \frac{(1000 \text{ in-kips})(4.125 \text{ in})}{174 \text{ in}^2}$$
$$= 23.7 \text{ kips}$$

$$r_u = \sqrt{(r_{mx} + r_v)^2 + r_{my}^2}$$
$$= \sqrt{(15.8 \text{ kips} + 10.1 \text{ kips})^2 + (23.7 \text{ kips})^2}$$
$$= 35.1 \text{ kips/bolt}$$

From AISC Table 7-3, for a 1 in diameter slip-critical A325 bolt in double shear, the tabulated capacity is 34.6 kips per bolt, which is close enough to be adequate. Use twelve 1 in diameter A325-SC bolts on each side.

Check the ultimate shear in the plates.

$$V_{u, \text{PL}} = 0.5V_u = (0.5)(122 \text{ kips})$$
$$= 61 \text{ kips} \quad [\text{per plate}]$$
$$A_{nv} = (w - 4d_{\text{hole}})l$$
$$= \big(11 \text{ in} - (4)(1.125 \text{ in})\big)t$$
$$= (6.5 \text{ in})t$$

Check AISC Sec. J4.2. For shear fracture (from AISC Eq. J4-4),

$$V_u \le \phi(0.6F_u)A_{nv}$$
$$\le \phi(0.6F_u)(6.5 \text{ in})t$$
$$t \ge \frac{V_u}{\phi(0.6F_u)(6.5 \text{ in})}$$
$$\ge \frac{61 \text{ kips}}{(0.75)(0.6)\left(58 \ \dfrac{\text{kips}}{\text{in}^2}\right)(6.5 \text{ in})}$$
$$\ge 0.36 \text{ in}$$

For shear yield (from AISC Eq. J4-3),

$$A_{gv} = wt = (11.0 \text{ in})t$$

$$
\begin{aligned}
V_u &\le \phi(0.6F_y)A_{gv} \\
&\le \phi(0.6F_y)(11.0 \text{ in})t \\
t &\ge \frac{V_u}{\phi(0.6F_y)(11.0 \text{ in})} \\
&\ge \frac{61 \text{ kips}}{(1.00)(0.6)\left(36 \dfrac{\text{kips}}{\text{in}^2}\right)(11.0 \text{ in})} \\
&\ge 0.26 \text{ in}
\end{aligned}
$$

The larger value for fracture governs.

Use two $\text{PL}\frac{3}{8} \times 11 \times 1$ ft 5 in.

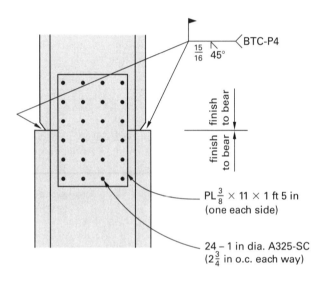

BTC-P4

$\frac{15}{16}$ $45°$

finish to bear

finish to bear

$\text{PL}\frac{3}{8} \times 11 \times 1$ ft 5 in
(one each side)

$24 - 1$ in dia. A325-SC
($2\frac{3}{4}$ in o.c. each way)

SOLUTION 8

8.1. Design girder G1 using either standard or modified wide flange sections. Use LRFD. According to the design criteria, the given 20 lbf/ft² partition load is dead load. The dead loads are

$$
\begin{aligned}
w_D &= w_{\text{slab}} + w_{\text{partitions}} + w_{\text{ceiling/mechanical}} + w_{\text{beam}} \\
&= 48 \frac{\text{lbf}}{\text{ft}^2} + 20 \frac{\text{lbf}}{\text{ft}^2} + 10 \frac{\text{lbf}}{\text{ft}^2} + 5 \frac{\text{lbf}}{\text{ft}^2} \\
&= 83 \text{ lbf/ft}^2
\end{aligned}
$$

The live loads are

$$w_L = 80 \text{ lbf/ft}$$

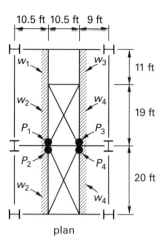

plan

For the girder, the tributary area is

$$
\begin{aligned}
A_T &= (0.5L_1 + 0.5L_2)(0.5s_1 + 0.5s_2) \\
&= \big((0.5)(30 \text{ ft}) + (0.5)(20 \text{ ft})\big) \\
&\quad \times \big((0.5)(10.5 \text{ ft}) + (0.5)(9 \text{ ft})\big) \\
&= 245 \text{ ft}^2
\end{aligned}
$$

Per ASCE/SEI7 Sec. 4.7.2, for A_T less than 400 ft², no reduction in live load is permitted, so use $w_L = 80$ lbf/ft².

$$
\begin{aligned}
w_{u,1} &= 1.2w_D s + 1.6w_L s \\
&= \frac{(1.2)\left(83 \dfrac{\text{lbf}}{\text{ft}^2}\right)(10.5 \text{ ft}) + (1.6)\left(80 \dfrac{\text{lbf}}{\text{ft}^2}\right)(10.5 \text{ ft})}{1000 \dfrac{\text{lbf}}{\text{kip}}} \\
&= 2.39 \text{ kips/ft}
\end{aligned}
$$

$$w_{u,2} = 0.5w_{u,1} = (0.5)\left(2.39 \frac{\text{kips}}{\text{ft}}\right) = 1.20 \text{ kips/ft}$$

$$w_{u,3} = \left(\frac{s_1}{s_2}\right)w_{u,1} = \left(\frac{9.75 \text{ ft}}{10.5 \text{ ft}}\right)\left(2.39 \frac{\text{kips}}{\text{ft}}\right) = 2.22 \text{ kips/ft}$$

$$w_{u,4} = \left(\frac{s_3}{s_2}\right)w_{u,1} = \left(\frac{4.5 \text{ ft}}{10.5 \text{ ft}}\right)\left(2.39 \frac{\text{kips}}{\text{ft}}\right) = 1.02 \text{ kips/ft}$$

$$
\begin{aligned}
P_1 &= \frac{0.5w_{u,1}a^2 + 0.5w_{u,2}(L_1 - a)(L_1 + a)}{L_1} \\
&= \frac{(0.5)\left(2.39 \dfrac{\text{kips}}{\text{ft}}\right)(11 \text{ ft})^2 + (0.5)\left(1.20 \dfrac{\text{kips}}{\text{ft}}\right)}{\times (30 \text{ ft} - 11 \text{ ft})(30 \text{ ft} + 11 \text{ ft})}{30 \text{ ft}} \\
&= 20.4 \text{ kips}
\end{aligned}
$$

$$P_2 = \frac{0.5 w_{u,2} L_2^2}{L_2}$$

$$= \frac{(0.5)\left(1.20 \ \frac{\text{kips}}{\text{ft}}\right)(20 \ \text{ft})^2}{20 \ \text{ft}}$$

$$= 12.0 \ \text{kips}$$

$$P_3 = \frac{0.5 w_{u,3} a^2 + 0.5 w_{u,4}(L_1 - a)(L_1 + a)}{L_1}$$

$$= \frac{\begin{aligned}(0.5)\left(2.22 \ \frac{\text{kips}}{\text{ft}}\right)(11 \ \text{ft})^2 + (0.5)\left(1.02 \ \frac{\text{kips}}{\text{ft}}\right) \\ \times (30 \ \text{ft} - 11 \ \text{ft})(30 \ \text{ft} + 11 \ \text{ft})\end{aligned}}{30 \ \text{ft}}$$

$$= 17.7 \ \text{kips}$$

$$P_4 = \left(\frac{w_{u,4}}{w_{u,2}}\right) P_2 = \left(\frac{1.02 \ \frac{\text{kips}}{\text{ft}}}{1.20 \ \frac{\text{kips}}{\text{ft}}}\right)(12.0 \ \text{kips}) = 10.2 \ \text{kips}$$

$$w_{ug} = 1.2 w_g = (1.2)\left(0.1 \ \frac{\text{kip}}{\text{ft}}\right) = 0.12 \ \text{kip/ft}$$

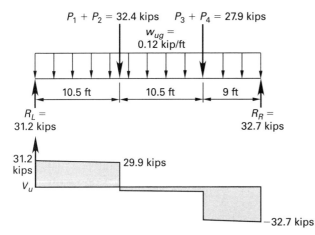

$$\sum M_L = (P_1 + P_2)a + (P_3 + P_4)(2a) + 0.5 w_{ug} L^2 - R_R L$$
$$= 0$$

$$R_R = \frac{(P_1 + P_2)a + (P_3 + P_4)(2a) + 0.5 w_{ug} L^2}{L}$$

$$= \frac{\begin{aligned}(20.4 \ \text{kips} + 12.0 \ \text{kips})(10.5 \ \text{ft}) \\ + (17.7 \ \text{kips} + 10.2 \ \text{kips})(2)(10.5 \ \text{ft}) \\ + (0.5)\left(0.12 \ \frac{\text{kip}}{\text{ft}}\right)(30 \ \text{ft})^2\end{aligned}}{30 \ \text{ft}}$$

$$= 32.7 \ \text{kips}$$

$$\sum F_y = R_L - (P_1 + P_2) - (P_3 + P_4) - w_{ug} L + R_R = 0$$
$$R_L = P_1 + P_2 + P_3 + P_4 + w_{ug} L - R_R$$
$$= 20.4 \ \text{kips} + 12.0 \ \text{kips} + 17.7 \ \text{kips} + 10.2 \ \text{kips}$$
$$+ \left(0.12 \ \frac{\text{kip}}{\text{ft}}\right)(30 \ \text{ft}) - 32.7 \ \text{kips}$$
$$= 31.2 \ \text{kips}$$

Maximum moment at the left end occurs at $x = 10.5$ ft where the shear line passes through zero. For the simple beam, the maximum moment is equivalent to the area of the shear diagram that is under the shear line from the point of zero moment to the point of maximum moment.

$$M_u = 0.5(R_L + V_{10.5})a$$
$$= (0.5)(31.2 \ \text{kips} + 29.9 \ \text{kips})(10.5 \ \text{ft})$$
$$= 321 \ \text{ft-kips}$$

From the problem statement, $L_b = 10.5$ ft, $C_b = 1.0$, and $F_y = 36$ ksi. Try a W24 × 62, with $d = 23.7$ in, $b_f = 7.0$ in, $k_{\text{des}} = 1.09$ in, $b_f/2t_f = 5.97$, $h/t_w = 50.1$, $Z_x = 153$ in^3, $S_x = 131$ in^3, $I_y = 34.5$ in^4, $r_y = 1.38$ in, and $J = 1.71$ in^4.

$$\sqrt{\frac{E}{F_y}} = \sqrt{\frac{29{,}000 \ \frac{\text{kips}}{\text{in}^2}}{36 \ \frac{\text{kips}}{\text{in}^2}}} = 28.4$$

Check local buckling using AISC Table B4.1.

$$\lambda_{pf} = 0.38 \sqrt{\frac{E}{F_y}} = 0.38 \sqrt{\frac{29{,}000 \ \frac{\text{kips}}{\text{in}^2}}{36 \ \frac{\text{kips}}{\text{in}^2}}}$$

$$= 10.8 \quad [b_f/2t_f < \lambda_{pf}]$$

$$\lambda_{pw} = 3.76 \sqrt{\frac{E}{F_y}} = 3.76 \sqrt{\frac{29{,}000 \ \frac{\text{kips}}{\text{in}^2}}{36 \ \frac{\text{kips}}{\text{in}^2}}}$$

$$= 107 \quad [h/t_w < \lambda_{pw}, \text{ section is compact}]$$

$$L_p = 1.76 r_y \sqrt{\frac{E}{F_y}}$$

$$= \frac{(1.76)(1.38 \ \text{in}) \sqrt{\frac{29{,}000 \ \frac{\text{kips}}{\text{in}^2}}{36 \ \frac{\text{kips}}{\text{in}^2}}}}{12 \ \frac{\text{in}}{\text{ft}}}$$

$$= 5.7 \ \text{ft}$$

$$h_o = d - 2k_{\text{des}} = 23.7 \ \text{in} - (2)(1.09 \ \text{in}) = 21.5 \ \text{in}$$

$$r_{ts} = \sqrt{\frac{I_y h_o}{2 S_x}} = \sqrt{\frac{(34.5 \ \text{in}^4)(21.5 \ \text{in})}{(2)(131 \ \text{in}^3)}} = 1.68 \ \text{in}$$

From AISC Eq. F2-6,

$$L_r = 1.95 r_{ts} \frac{E}{0.7F_y} \sqrt{\frac{Jc}{S_x h_0} + \sqrt{\left(\frac{Jc}{S_x h_0}\right)^2 + 6.76 \left(\frac{0.7F_y}{E}\right)^2}}$$

$$= \frac{(1.95)(1.68 \text{ in}) \left(\dfrac{29{,}000 \frac{\text{kips}}{\text{in}^2}}{(0.7)\left(36 \frac{\text{kips}}{\text{in}^2}\right)}\right)}{12 \frac{\text{in}}{\text{ft}}} \times \sqrt{\frac{(1.71 \text{ in}^4)(1.0)}{(131 \text{ in}^3)(21.5 \text{ in})} + \sqrt{\left(\frac{(1.71 \text{ in}^4)(1.0)}{(131 \text{ in}^3)(21.5 \text{ in})}\right)^2 + (6.76)\left(\frac{(0.7)\left(36 \frac{\text{kips}}{\text{in}^2}\right)}{29{,}000 \frac{\text{kips}}{\text{in}^2}}\right)^2}}$$

$$= 17.0 \text{ ft}$$

$$M_p = F_y Z_x = \left(36 \frac{\text{kips}}{\text{in}^2}\right)(153 \text{ in}^3) = 5500 \text{ in-kips}$$

$$\phi M_n = \phi C_b \left(M_p - (M_p - 0.7F_y S_x)\left(\frac{L_b - L_p}{L_r - L_p}\right)\right)$$

$$= \frac{(0.9)(1.0) \times \left(5500 \text{ in-kips} - \left(5500 \text{ in-kips} - (0.7)\left(36 \frac{\text{kips}}{\text{in}^2}\right) \times (131 \text{ in}^3)\right) \times \left(\frac{10.5 \text{ ft} - 5.7 \text{ ft}}{17.0 \text{ ft} - 5.7 \text{ ft}}\right)\right)}{12 \frac{\text{in}}{\text{ft}}}$$

$$= 342 \text{ ft-kips} \quad [> M_u]$$

Check the live load deflection, Δ_L. Approximate the concentrated service live load as a percentage of the factored loads previously computed.

$$\Delta_L = \left(\frac{P_{L,1} b_1}{12EI}\right)(0.75L^2 - b_1^2) + \left(\frac{P_{L,2} a_2}{12EI}\right)(0.75L^2 - a_2^2)$$

$$P_{L,1} = \left(\frac{w_L}{1.2w_D + 1.6w_L}\right)(P_1 + P_2)$$

$$= \left(\frac{80 \frac{\text{lbf}}{\text{ft}^2}}{(1.2)\left(83 \frac{\text{lbf}}{\text{ft}^2}\right) + (1.6)\left(80 \frac{\text{lbf}}{\text{ft}^2}\right)}\right) \times (20.4 \text{ kips} + 12.0 \text{ kips})$$

$$= 11.4 \text{ kips}$$

$$P_{L,2} = \left(\frac{w_L}{1.2w_D + 1.6w_L}\right)(P_3 + P_4)$$

$$= \left(\frac{80 \frac{\text{lbf}}{\text{ft}^2}}{(1.2)\left(83 \frac{\text{lbf}}{\text{ft}^2}\right) + (1.6)\left(80 \frac{\text{lbf}}{\text{ft}^2}\right)}\right) \times (17.7 \text{ kips} + 10.2 \text{ kips})$$

$$= 9.8 \text{ kips}$$

$$b_1 = L - a_1 = 30 \text{ ft} - 10.5 \text{ ft} = 19.5 \text{ ft}$$

$$a_2 = (2)(10.5 \text{ ft}) = 21 \text{ ft}$$

Therefore, the live load deflection ($I_x = 1550 \text{ in}^4$) is

$$\Delta_L = \left(\frac{P_{L,1} b_1}{12EI_x}\right)(0.75L^2 - b_1^2) + \left(\frac{P_{L,2} a_2}{12EI_x}\right)(0.75L^2 - a_2^2)$$

$$= \left(\begin{array}{l}\left(\dfrac{(11.4 \text{ kips})(19.5 \text{ ft})}{(12)\left(29{,}000 \frac{\text{kips}}{\text{in}^2}\right)(1550 \text{ in}^4)}\right) \\ \times ((0.75)(30 \text{ ft})^2 - (19.5 \text{ ft})^2) \\ + \left(\dfrac{(9.8 \text{ kips})(21.0 \text{ ft})}{(12)\left(29{,}000 \frac{\text{kips}}{\text{in}^2}\right)(1550 \text{ in}^4)}\right) \\ \times ((0.75)(30 \text{ ft})^2 - (21.0 \text{ ft})^2)\end{array}\right) \times \left(12 \frac{\text{in}}{\text{ft}}\right)^3$$

$$= 0.38 \text{ in}$$

$$\Delta_L < \frac{L}{360} = \frac{(30 \text{ ft})\left(12 \frac{\text{in}}{\text{ft}}\right)}{360} = 1 \text{ in}$$

Use a W24 × 62 without modification.

8.2. In this case, a standard shape with 7.0 in flange width is available. If a standard section were not available, it would be necessary to select a heavier shape that would furnish the required moment strength when trimmed to a 7 in width. The flange tips could be gas-cut by the same amount on each side to maintain a doubly symmetrical cross section. The reduced section would taper back to full width to alleviate stress concentrations as shown in the following partial plan view.

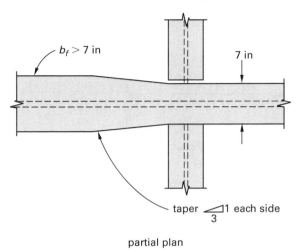

partial plan

SOLUTION 9

9.1. Design the lightest W27 member for the transfer girder. Use LRFD. Lateral support is provided at the supports and at midspan, so the laterally unbraced length of compression flange is $L_b = 20$ ft. From AISC Table 3-1, for a concentrated load and lateral brace at midspan, $C_b = 1.67$. Steel is ASTM A992 ($F_y = 50$ ksi). Given that the ratio of dead load to total load is 0.5, the controlling gravity load factor is

$$U = 1.2D + 1.6L = (1.2)(0.5) + (1.6)(0.5) = 1.4$$

$$P_u = UP = (1.4)(60 \text{ kips}) = 84 \text{ kips}$$

Approximate the girder weight as 0.1 kip/ft. Then,

$$w_u = 1.2 w_D = (1.2)\left(0.1 \frac{\text{kip}}{\text{ft}}\right) = 0.12 \text{ kip/ft}$$

$$M_u = 0.25 P_u L + 0.125 w_u L^2$$

$$= (0.25)(84 \text{ kips})(40 \text{ ft}) + (0.125)$$

$$\times \left(0.12 \frac{\text{kip}}{\text{ft}}\right)(40 \text{ ft})^2$$

$$= 864 \text{ ft-kips}$$

From AISC Table 3-2, select a trial section based on the fully plastic capacity, then verify the trial using the values of C_b and L_b.

Try W27 × 84: $\phi_b M_p = 915$ ft-kips, $L_p = 7.3$ ft, $L_r = 20.8$ ft, $d = 26.7$ in, $t_w = 0.46$ in, $b_f = 10.0$ in, $t_f = 0.64$ in, $k_{\text{des}} = 1.24$ in, $b_f/2t_f = 7.78$, $h/t_w = 52.7$, $Z_x = 244$ in^3, $I_x = 2850$ in^4, $S_x = 213$ in^3, $I_y = 106$ in^4, and $r_y = 2.07$ in.

The actual unbraced length of compression flange falls within the range $L_p < L_b < L_r$; therefore, calculate the moment strength using AISC Eq. F2-2.

$$M_n \leq \begin{cases} M_p = F_y Z_x = \left(50 \dfrac{\text{kips}}{\text{in}^2}\right)(244 \text{ in}^3) \\ \quad = 12{,}200 \text{ in-kips} \quad [\text{controls}] \\ C_b\left(M_p - (M_p - 0.7 F_y S_x)\left(\dfrac{L_b - L_p}{L_r - L_p}\right)\right) \\ \quad = \dfrac{(0.9)(1.67)\begin{pmatrix} 12{,}200 \text{ in-kips} \\ -\begin{pmatrix} 12{,}200 \text{ in-kip} \\ -(0.7)\left(50 \dfrac{\text{kips}}{\text{in}^2}\right) \\ \times (213 \text{ in}^3) \end{pmatrix} \\ \times \left(\dfrac{20 \text{ ft} - 7.3 \text{ ft}}{20.8 \text{ ft} - 7.3 \text{ ft}}\right) \end{pmatrix}}{12 \dfrac{\text{in}}{\text{ft}}} \\ \quad = 12{,}900 \text{ in-kips} \end{cases}$$

$$\phi M_p = \frac{(0.9)(12{,}200 \text{ in-kips})}{12 \dfrac{\text{in}}{\text{ft}}}$$

$$= 915 \text{ ft-kips} \quad [> M_u, \text{ so OK}]$$

Check local buckling using AISC Table B-4.

$$\lambda_{pf} = 0.38\sqrt{\frac{E}{F_y}} = 0.38\sqrt{\frac{29{,}000 \dfrac{\text{kips}}{\text{in}^2}}{50 \dfrac{\text{kips}}{\text{in}^2}}} = 9.15$$

$$\frac{b_f}{2t_f} = 7.78 \quad [< \lambda_{pf}]$$

$$\lambda_{pw} = 3.76\sqrt{\frac{E}{F_y}} = 3.76\sqrt{\frac{29{,}000 \dfrac{\text{kips}}{\text{in}^2}}{50 \dfrac{\text{kips}}{\text{in}^2}}} = 90.6$$

$$\frac{h}{t_w} = 52.7 \quad [< \lambda_{pw}]$$

The section is compact for flexure (this can also be verified by AISC Table 3-2). Check live load deflection ($I_x = 2850$ in^4).

$$\Delta_L = \frac{P_L L^3}{48 E I_x}$$

$$= \frac{(30 \text{ kips})\left((40 \text{ ft})\left(12 \dfrac{\text{in}}{\text{ft}}\right)\right)^3}{(48)\left(29{,}000 \dfrac{\text{kips}}{\text{in}^2}\right)(2850 \text{ in}^4)}$$

$$= 0.84 \text{ in}$$

$$\frac{L}{360} = \frac{(40 \text{ ft})\left(12 \dfrac{\text{in}}{\text{ft}}\right)}{360} = 1.33 \text{ in} \quad [\Delta_L < L/360]$$

Serviceability is OK. Check the shear ($d = 29.8$ in, $t_w = 0.55$ in, $h/t_w < 260$).

$$\phi V_n = \phi(0.6F_y)dt_w$$

$$= (1.0)(0.6)\left(50 \ \frac{\text{kips}}{\text{in}^2}\right)(26.7 \text{ in})(0.46 \text{ in})$$

$$= 369 \text{ kips}$$

$$V_u = 0.5P_u + 0.5w_u L$$

$$= (0.5)(84 \text{ kips}) + (0.5)\left(0.12 \ \frac{\text{kips}}{\text{ft}}\right)(40 \text{ ft})$$

$$= 44.4 \text{ kips} \quad [< \phi V_n]$$

Use W27 × 84.

9.2. Select a bolted seated connection. The required bearing length for a W27 × 84 end reaction with $V_u = 44.4$ kips and $N/d < 0.2$ is found from AISC Eq. J10-5a.

$$\phi R_n = V_u = 0.40\phi t_w^2 \left(1 + 3\left(\frac{N_{\text{req}}}{d}\right)\left(\frac{t_w}{t_f}\right)^{1.5}\right)\sqrt{\frac{EF_{yw}t_f}{t_w}}$$

$$N_{\text{req}} = \left(\frac{d}{3}\right)\left(\frac{V_u\sqrt{\dfrac{t_w}{EF_y t_f}}}{0.4\phi t_w^2} - 1.0\right)\left(\frac{t_f}{t_w}\right)^{1.5}$$

$$= \left(\frac{26.7 \text{ in}}{3}\right)$$

$$\times \left(\frac{(44.4 \text{ kips})\sqrt{\dfrac{\dfrac{0.46 \text{ in}}{\left(29,000 \ \dfrac{\text{kips}}{\text{in}^2}\right)} \times \left(50 \ \dfrac{\text{kips}}{\text{in}^2}\right) \times (0.64 \text{ in})}{(0.4)(0.75)(0.46 \text{ in})^2}} - 1.0\right)$$

$$\times \left(\frac{0.64 \text{ in}}{0.46 \text{ in}}\right)^{1.5}$$

$$= -7.4 \text{ in}$$

The negative result implies that web crippling does not control; however, AISC Sec. J10.2 requires a minimum bearing length of at least $k_{\text{des}} = 1.24$ in.

Select an unstiffened seated connection from AISC Table 10-5 based on a bearing length of 1.24 in and a required strength of 44.4 kips. This requires an angle thickness of $^5/_8$ in and a length of 6 in. For $^3/_4$ in A307 bolts in single shear,

$$\phi_v r_n = 8.97 \text{ kips/bolt}$$

$$n > \frac{V_u}{\phi_v r_n} = \frac{44.4 \text{ kips}}{8.97 \ \dfrac{\text{kips}}{\text{bolt}}}$$

$$= 4.9 \text{ bolts} \quad \left[\begin{array}{l}\text{use six bolts in type C} \\ \text{pattern of AISC Fig. 10-7}\end{array}\right]$$

Use an L8 × 4 × $^5/_8$ seat with six $^3/_4$ in A307 bolts. Provide L3 × 3 × $^1/_4$ top angle to provide stability about the longitudinal axis of the W27 × 84, and attach with $^3/_4$ in A307 bolts.

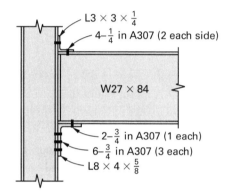

9.3. Design the lightest W8 column for column line 1. Assume an eccentricity of load from the new seated connection to centerline of the column equal to 6 in, which causes bending about the strong axis of the column.

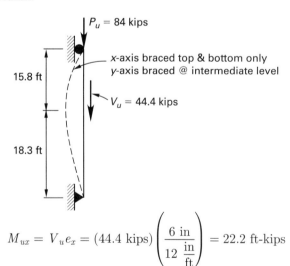

$$M_{ux} = V_u e_x = (44.4 \text{ kips})\left(\frac{6 \text{ in}}{12 \ \dfrac{\text{in}}{\text{ft}}}\right) = 22.2 \text{ ft-kips}$$

Distribute the moment to the top and bottom column segments.

$$K_{\text{bottom}} = \frac{3EI}{h_{\text{bottom}}} = \frac{3EI}{18.3\text{ ft}}$$

$$K_{\text{top}} = \frac{3EI}{h_{\text{top}}} = \frac{3EI}{15.8\text{ ft}}$$

$$M_{\text{bottom}} = M_{ux}\left(\frac{K_{\text{bottom}}}{K_{\text{bottom}} + K_{\text{top}}}\right)$$

$$= (22.2\text{ ft-kips})\left(\frac{\dfrac{3EI}{18.3\text{ ft}}}{\dfrac{3EI}{18.3\text{ ft}} + \dfrac{3EI}{15.8\text{ ft}}}\right)$$

$$= 10.3\text{ ft-kips}$$

The required axial compression strength is

$$P_u = UP + V_u$$
$$= (1.4)(60\text{ kips}) + 44.4\text{ kips}$$
$$= 128\text{ kips}$$

The column is braced at the second floor about its weak axis, but is braced only at the ground and roof about the strong axis.

$$(KL)_x = h_{\text{top}} + h_{\text{bottom}} = 15.8\text{ ft} + 18.3\text{ ft} = 34.1\text{ ft}$$
$$(KL)_y = h_{\text{bottom}} = 18.3\text{ ft}$$
$$L_b = 18.3\text{ ft}$$

Select a trial W8 column that has in excess of $\phi_c P_n > P_u$ for the weak axis case. Then check and refine the trial design using the combined axial compression and bending requirements. From AISC Table 4-1, a section larger than a W8×31 is required (a W8×31 was tried and found to be unsuitable due to flange local buckling limits). Try a W8×35: $A_g = 10.3$ in^2, $b_f/2t_f = 8.1$, $h/t_w = 20.5$, $r_x = 3.51$ in, $r_y = 2.03$ in, $I_x = 127$ in^4, $S_x = 31.2$ in^3, $Z_x = 34.7$ in^3, $L_p = 7.17$ ft, and $L_r = 27.0$ ft.

$$\left(\frac{KL}{r}\right)_x = \frac{(34.1\text{ ft})\left(12\frac{\text{in}}{\text{ft}}\right)}{3.51\text{ in}}$$
$$= 117\quad[\text{controls axial compression capacity}]$$

$$\left(\frac{KL}{r}\right)_y = \frac{(18.3\text{ ft})\left(12\frac{\text{in}}{\text{ft}}\right)}{2.03\text{ in}}$$
$$= 108$$

$$4.71\sqrt{\frac{E}{F_y}} = 4.71\sqrt{\frac{29{,}000\frac{\text{kips}}{\text{in}^2}}{50\frac{\text{kips}}{\text{in}^2}}}$$
$$= 113\quad[< KL/r = 117,\text{ so use AISC Eq. E3-3}]$$

$$F_e = \frac{\pi^2 E}{\left(\frac{KL}{r}\right)_x^2} = \frac{\pi^2\left(29{,}000\frac{\text{kips}}{\text{in}^2}\right)}{(117)^2} = 20.9\text{ ksi}$$

$$F_{cr} = 0.877F_e = (0.877)\left(20.9\frac{\text{kips}}{\text{in}^2}\right) = 18.3\text{ ksi}$$

$$P_c = \phi F_{cr} A_g = (0.9)\left(18.3\frac{\text{kips}}{\text{in}^2}\right)(10.3\text{ in}^2)$$
$$= 170\text{ kips}$$

For bending without axial load, $L_p = 7.17$ ft $< L_b = 18.3$ ft $< L_r = 27.0$ ft; therefore, use AISC Eq. F2-2. From Table 3-1, $C_b = 1.67$. Check local buckling (AISC Table B4.1).

$$\lambda_{pf} = 0.38\sqrt{\frac{E}{F_y}} = 0.38\sqrt{\frac{29{,}000\frac{\text{kips}}{\text{in}^2}}{50\frac{\text{kips}}{\text{in}^2}}} = 9.15$$

$$\frac{b_f}{2t_f} = 8.1\quad[<\lambda_{pf}]$$

$$\lambda_{pw} = 3.76\sqrt{\frac{E}{F_y}} = 3.76\sqrt{\frac{29{,}000\frac{\text{kips}}{\text{in}^2}}{50\frac{\text{kips}}{\text{in}^2}}} = 90.5$$

$$\frac{h}{t_w} = 20.5\quad[<\lambda_{pw}]$$

Therefore, the section is compact.

$$M_n \leq \begin{cases} C_b\left(M_p - (M_p - 0.7F_y S_x)\left(\dfrac{L_b - L_p}{L_r - L_p}\right)\right) \\[4pt] \quad = (1.67)\left(\begin{array}{l}1735\text{ in-kips} \\ -\left(\begin{array}{l}1735\text{ in-kips} - (0.7) \\ \times\left(50\frac{\text{kips}}{\text{in}^2}\right) \\ \times(31.2\text{ in}^3)\end{array}\right) \\ \times\left(\dfrac{18.3\text{ ft} - 7.17\text{ ft}}{27.0\text{ ft} - 7.17\text{ ft}}\right)\end{array}\right) \\[4pt] \quad = 2295\text{ in-kips} \\[4pt] M_p = F_y Z_x = \left(50\frac{\text{kips}}{\text{in}^2}\right)(34.7\text{ in}^3) \\[4pt] \quad = 1735\text{ in-kips}\quad[\text{controls}] \end{cases}$$

$$\phi M_n = \phi M_p = \frac{(0.9)(1735\text{ in-kips})}{12\frac{\text{in}}{\text{ft}}} = 130\text{ ft-kips}$$

The column can sway at the point of maximum bending moment; therefore, compute a moment amplification per AISC App. 8.1.

For a column with an intermediate applied moment, the conservative value of C_m is 1.0. P_r is the axial force at the base. α is 1.0 for LRFD.

$$P_{e1} = \frac{\pi^2 EI^*}{(K_1 L)^2}$$

$$= \frac{\pi^2 \left(29{,}000 \ \frac{\text{kips}}{\text{in}^2}\right)(127 \ \text{in}^4)}{\left((1.0)(34.1 \ \text{ft})\left(12 \ \frac{\text{in}}{\text{ft}}\right)\right)^2}$$

$$= 217 \ \text{kips}$$

$$B_1 = \frac{C_m}{1 - \dfrac{\alpha P_r}{P_{e1}}}$$

$$= \frac{1.0}{1.0 - \dfrac{(1.0)(128 \ \text{kips})}{217 \ \text{kips}}}$$

$$= 2.44$$

$$M_{rx} = B_1 M_{\text{bottom}} = (2.44)(10.3 \ \text{ft-kips})$$

$$= 25.1 \ \text{ft-kips}$$

$$P_r = P_u = 128 \ \text{kips}$$

$$\frac{P_r}{P_c} = \frac{128 \ \text{kips}}{170 \ \text{kips}} = 0.75 \quad [> 0.2, \text{ so use AISC Eq. H1-1a}]$$

$$\frac{P_r}{P_c} + \frac{8}{9}\left(\frac{M_{rx}}{M_{cx}}\right) = 0.75 + \left(\tfrac{8}{9}\right)\left(\frac{25.1 \ \text{ft-kips}}{130 \ \text{ft-kips}}\right)$$

$$= 0.92 \quad [< 1.0, \text{ so OK}]$$

Therefore, $\boxed{\text{use a W8} \times 35.}$

SOLUTION 10

10.1. Use the portal method to compute the preliminary seismic moments and shears at the tenth floor.

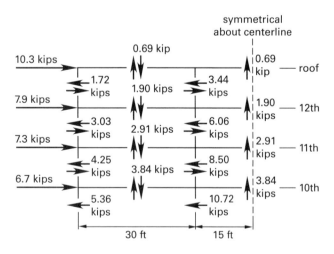

The distribution of shears shown is based on the portal assumptions that exterior columns resist one-half as much shear as interior columns and that points of inflection occur at midspan of columns and girders. Therefore, the distribution of shears at the tenth floor is as follows.

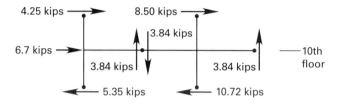

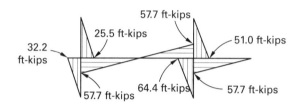

portal method analysis bending moments at 10th floor
left half of symmetrical structure

10.2. Estimate the tenth-story drift. For a frame of this proportion, axial, shear, and panel zone deformations are relatively small. The drift can be estimated based on member centerline dimensions and flexural deformations in the columns and girders (see Becker, Naeim, and Teal).

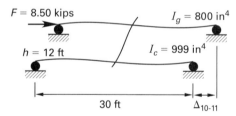

For the girder: W18 $\times$ 50 ($I_x = 800 \ \text{in}^4$). For the column: W14 $\times$ 90 ($I_x = 999 \ \text{in}^4$).

$$\Delta = \frac{Fh^3}{12 EI_{\text{col}}} + \frac{Fh^2 L}{12 EI_{\text{girder}}}$$

$$\Delta_{10-11} = \frac{(8.5 \ \text{kips})\left((12 \ \text{ft})\left(12 \ \frac{\text{in}}{\text{ft}}\right)\right)^3}{(12)\left(29{,}000 \ \frac{\text{kips}}{\text{in}^2}\right)(999 \ \text{in}^4)}$$

$$+ \frac{(8.5 \ \text{kips})\left((12 \ \text{ft})\left(12 \ \frac{\text{in}}{\text{ft}}\right)\right)^2 \left((30 \ \text{ft})\left(12 \ \frac{\text{in}}{\text{ft}}\right)\right)}{(12)\left(29{,}000 \ \frac{\text{kips}}{\text{in}^2}\right)(800 \ \text{in}^4)}$$

$$= 0.30 \ \text{in}$$

This is an estimate of the elastic drift between floors 10 and 11 caused by the strength-level lateral forces. The corresponding maximum inelastic displacement is estimated using the coefficient C_d from ASCE/SEI7 Table 12.2-1. For a special steel moment frame, $C_d = 5.5$ and the maximum inelastic displacement is

$$\Delta_{M,10-11} = C_d\Delta_{10-11} = (5.5)(0.30 \text{ in}) = \boxed{1.65 \text{ in}}$$

If the girder cross sections are reduced to assure that plastic hinges occur away from the face of columns (as in Sol. 11.3), the reduction in girder rigidity increases the drift on the order of 5% (see Iwankin).

$$\Delta_{M,10-11} = (1.05)(1.65 \text{ in}) = 1.75 \text{ in}$$

10.3. Design the connection for G3 at the third floor level. Use ANSI/AISC 341 and 358. Of the nonproprietary connections that have been tested, the reduced beam section offers one of the more reliable approaches to assuring that plastic hinges will form in the beams away from the column faces, and is the choice for this solution.

For the non-reduced W24 × 84 girder: $d = 24.10$ in, $t_w = 0.47$ in, $b_f = 9.02$ in, $t_f = 0.77$ in, $Z_x = 224$ in^3, and $S_x = 196$ in^3.

For the W14 × 176 column: $A_g = 51.8$ in^2, $d = 15.22$ in, $t_w = 0.83$ in, $b_f = 15.65$ in, $t_f = 1.31$ in, $k = 1.91$ in, and $Z_x = 320$ in^3.

Try reducing the flexural strength 40% using an 18 in long circular "dogbone," starting 6 in from the face of the column (so that the probable plastic hinge centers at 15 in = 1.25 ft from the column face). The required reduction in flange width, b_R, is

$$Z_{RBS} - 0.6Z_x = Z_x - b_R t_f(d - t_f)$$

$$b_R = \frac{Z_x - 0.6Z_x}{t_f(d - t_f)}$$

$$= \frac{(0.4)(224 \text{ in}^3)}{(0.77 \text{ in})(24.10 \text{ in} - 0.77 \text{ in})}$$

$$= 5.00 \text{ in}$$

This results in a flange width at the plastic hinge location of $b_f - b_R = 9.02$ in $- 5.0$ in $= 4.02$ in. For yielding to occur in the girders, the strength of the columns must satisfy ANSI/AISC 341 Eq. E3-1.

$$\frac{\sum M_{pc}^*}{\sum M_{pb}^*} > 1.0 \quad \left[M_{pc}^* = Z_c\left(F_{yc} - \frac{P_u}{A_g}\right)\right]$$

Seismic loads are reversible and the combination that is critical for this check occurs when seismic loads are additive to gravity loads.

$$P_{u,3-4} = 1.2P_D + 0.5P_L + 0.2S_{DS}D + P_E$$

$$= (1.2)(222 \text{ kips}) + (0.5)(148 \text{ kips})$$

$$+ (0.2)(1.0)(222 \text{ kips}) + 46 \text{ kips}$$

$$= 431 \text{ kips}$$

Below the third floor, additional loads increase the factored load in the column.

$$P_{u,2-3} = 1.2P_D + 0.5P_L + 0.2S_{DS}D + P_E$$

$$= (1.2)(244 \text{ kips}) + (0.5)(163 \text{ kips})$$

$$+ (0.2)(1.0)(244 \text{ kips}) + 54 \text{ kips}$$

$$= 477 \text{ kips}$$

$$\sum M_{pc}^* = Z_c\left(F_{yc} - \frac{P_{u,3-4}}{A_g}\right) + Z_c\left(F_{yc} - \frac{P_{u,2-3}}{A_g}\right)$$

$$= \frac{\begin{aligned}(320 \text{ in}^3)\left(50 \frac{\text{kips}}{\text{in}^2} - \frac{431 \text{ kips}}{51.8 \text{ in}^2}\right) \\ + (320 \text{ in}^3)\left(50 \frac{\text{kips}}{\text{in}^2} - \frac{477 \text{ kips}}{51.8 \text{ in}^2}\right)\end{aligned}}{12 \frac{\text{in}}{\text{ft}}}$$

$$= 2197 \text{ ft-kips}$$

To determine $\sum M_{pb}^*$, the probable moment strength of the girders at the reduced section must be computed.

$$\sum M_{pb}^* = \sum(M_{pr} + M_v)$$

Using ANSI/AISC 358 Eq. 2.4.3-2,

$$C_{pr} \leq \begin{cases} \dfrac{F_y + F_u}{2F_y} = \dfrac{50 \frac{\text{kips}}{\text{in}^2} + 65 \frac{\text{kips}}{\text{in}^2}}{(2)\left(50 \frac{\text{kips}}{\text{in}^2}\right)} \\ \qquad = 1.15 \quad \text{[controls]} \\ 1.2 \end{cases}$$

From ANSI/AISC 341 Table A3.1, $R_y = 1.1$ for ASTM A992 steel. Given that the reduced beam section has a plastic moment capacity equal to $0.6M_{pb}$ of the full section,

$$M_{pr} = C_{pr}R_yF_yZ_e$$

$$= \frac{(1.15)(1.1)\left(50 \frac{\text{kips}}{\text{in}^2}\right)(0.6)(224 \text{ in}^3)}{12 \frac{\text{in}}{\text{ft}}}$$

$$= 708 \text{ ft-kips}$$

The span between plastic hinges is

$$L' = L - 2(0.5d_c + a + 0.5b)$$

$$= 30 \text{ ft} - \frac{(2)\big((0.5)(15.2 \text{ in}) + 6 \text{ in} + (0.5)(18 \text{ in})\big)}{12 \dfrac{\text{in}}{\text{ft}}}$$

$$= 26.2 \text{ ft}$$

The factored gravity load in combination with seismic is

$$w_u = \frac{(1.2 + 0.2 S_{DS}) V_D + 0.5 V_L}{L}$$

$$= \frac{(1.2 + (0.2)(1.0))(22 \text{ kips}) + (0.5)(15 \text{ kips})}{30 \text{ ft}}$$

$$= 1.28 \text{ kips/ft}$$

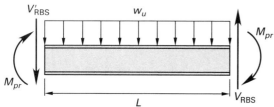

$$V_{\text{RBS}} = \frac{2M_{pr}}{L'} + 0.5 w_u L'$$

$$= \frac{(2)(708 \text{ ft-kips})}{26.2 \text{ ft}} + (0.5)\left(1.28 \frac{\text{kips}}{\text{ft}}\right)(26.2 \text{ ft})$$

$$= 70.8 \text{ kips} \quad [\text{up}]$$

$$V'_{\text{RBS}} = \frac{-2M_{pr}}{L'} + 0.5 w_u L'$$

$$= \frac{(-2)(708 \text{ ft-kips})}{26.2 \text{ ft}} + (0.5)\left(1.28 \frac{\text{kips}}{\text{ft}}\right)(26.2 \text{ ft})$$

$$= -37.3 \text{ kips} \quad [\text{down}]$$

For an exterior column, only M_{pr} and V_{RBS} contribute to the bending moment at the column centerline.

$$\sum M^*_{pb} = M_{pr} + V_{\text{RBS}}(0.5d_c + a + 0.5b)$$

$$= 708 \text{ ft-kips} + (70.8 \text{ kips})$$

$$\times \left(\frac{(0.5)(15.22 \text{ in}) + 6 \text{ in} + (0.5)(18 \text{ in})}{12 \dfrac{\text{in}}{\text{ft}}} \right)$$

$$= 842 \text{ ft-kips}$$

$$\frac{\sum M^*_{pc}}{\sum M^*_{pb}} = \frac{2197 \text{ ft-kips}}{842 \text{ ft-kips}} = 2.6 \quad [>1.0]$$

Therefore, the weak beam, strong column criterion is satisfied. Yielding at the reduced beam section is reasonably assured.

Check panel zone strength. At the face of the column,

$$M_{uf} \geq 1.2M_D + 0.5M_L + 0.2S_{DS}M_D + M_E$$

$$\geq (1.2)(90 \text{ ft-kips}) + (0.5)(60 \text{ ft-kips})$$

$$+ (0.2)(1.0)(90 \text{ ft-kips}) + 120 \text{ ft-kips}$$

$$= 267 \text{ ft-kips}$$

ANSI/AISC 341 requires that, as a minimum, the panel zone shear must be computed by projecting the probable moment at the plastic hinge to the face of the column.

$$M_{uf} \geq M_{pr} + V_{\text{RBS}}(a + 0.5b)$$

$$= (708 \text{ ft-kips}) + (70.8 \text{ kips})\left(\frac{6 \text{ in} + (0.5)(18 \text{ in})}{12 \dfrac{\text{in}}{\text{ft}}} \right)$$

$$= 797 \text{ ft-kips} \quad [\text{controls}]$$

$$V = \frac{M_{uf}}{d_b - t_f}$$

$$= \frac{(797 \text{ ft-kips})\left(12 \dfrac{\text{in}}{\text{ft}}\right)}{24.10 \text{ in} - 0.77 \text{ in}}$$

$$= 410 \text{ kips}$$

$$\phi V_n = \phi 0.6 F_y d_c t_w \left(1 + \frac{3b_{cf} t^2_{cf}}{d_b d_c t_w}\right)$$

$$= (1.0)(0.6)\left(50 \frac{\text{kips}}{\text{in}^2}\right)(15.22 \text{ in})(0.83 \text{ in})$$

$$\times \left(1 + \frac{(3)(15.65 \text{ in})(1.31 \text{ in})^2}{(24.10 \text{ in})(15.22 \text{ in})(0.83 \text{ in})}\right)$$

$$= 480 \text{ kips}$$

From ANSI/AISC 341 Eq. E3-7,

$$t_z \geq \frac{d_z + w_z}{90} = \frac{(d_b - 2t_{bf}) + (d_c - 2t_{cf})}{90}$$

$$= \frac{(24.10 \text{ in} - (2)(0.77 \text{ in})) + (15.22 \text{ in} - (2)(1.31 \text{ in}))}{90}$$

$$= 0.39 \text{ in} \quad [< t_{cw} = 0.83 \text{ in}]$$

The panel zone shear strength is adequate without doubler plates. Check the need for continuity plates (ANSI/AISC 341 Sec. E3.6f).

$$
t_{cf} \geq \begin{cases} 0.4\sqrt{1.8 b_{bf} t_{bf}\left(\dfrac{F_{yb}R_{yb}}{F_{yc}R_{yc}}\right)} \\[4ex] \quad = 0.4\sqrt{\begin{array}{c}(1.8)(9.02\text{ in})(0.77\text{ in}) \\ \times \left(\dfrac{\left(50\,\dfrac{\text{kips}}{\text{in}^2}\right)(1.1)}{\left(50\,\dfrac{\text{kips}}{\text{in}^2}\right)(1.1)}\right)\end{array}} \\[6ex] \quad = 1.41\text{ in} \\[2ex] \dfrac{b_{bf}}{6} = \dfrac{9.02\text{ in}}{6} \\[2ex] \quad = 1.5\text{ in} \quad \begin{bmatrix} t_{cf} = 1.31\text{ in, so requires} \\ \text{continuity plates}\end{bmatrix} \end{cases}
$$

ANSI/AISC 341 Sec. E3.6f(2) requires plates extending to mid-depth of columns and with thickness equal to at least one-half the beam flange thickness. For this case, $t_{bf} = 0.77$ in, so use PL$^1/_2 \times 4 \times 0$ ft 6 in (ASTM A36). Use E70 full penetration welds to develop strength along edges, and full penetration groove welds to connect plates to column flanges. (See ANSI/AISC 341 Sec. E3.6f(3).)

Design the web shear connection to transfer the projected probable shear at the face of column.

$$
\begin{aligned}
V_u &= \frac{2M_{pr}}{L'} + 0.5 w_u L \\
&= \frac{(2)(708\text{ ft-kips})}{26.2\text{ ft}} + (0.5)\left(1.28\,\frac{\text{kips}}{\text{ft}}\right)(30\text{ ft}) \\
&= 73.2\text{ kips}
\end{aligned}
$$

Use a single plate connection with $^7/_8$ in diameter A325-SC bolts. Getting the value of ϕr_v from AISC Table 7-3,

$$
n \geq \frac{V_u}{\phi r_v} = \frac{73.2\text{ kips}}{13.2\,\dfrac{\text{kips}}{\text{bolt}}} = 6\text{ bolts}
$$

Six bolts at 3 in o.c. will fit between weld access holes in the web of the W24 × 84 (say, an 18 in long plate). Assuming A36 material for the connection plate ($F_y = 36$ ksi and $F_u = 58$ ksi), try a $^3/_8$ in plate. Check shear yield in the plate.

$$
\begin{aligned}
\phi V_n &= \phi 0.6 F_y L t \\
&= (1.0)(0.6)\left(36\,\frac{\text{kips}}{\text{in}^2}\right)(18\text{ in})(0.375\text{ in}) \\
&= 146\text{ kips}
\end{aligned}
$$

Check shear rupture.

$$
\begin{aligned}
\phi V_n &= \phi 0.6 F_u L_n t \\
&= (0.75)(0.6)\left(58\,\frac{\text{kips}}{\text{in}^2}\right)(18\text{ in} - (6)(1\text{ in})) \\
&\quad \times (0.375\text{ in}) \\
&= 117\text{ kips}
\end{aligned}
$$

Use PL$^3/_8 \times 4\frac{1}{2} \times 1$ ft 6 in with six $\frac{7}{8}$ in A325-SC bolts.

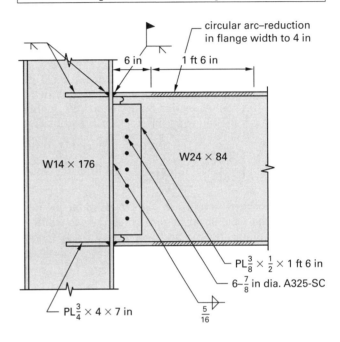

SOLUTION 11

11.1. Verify adequacy of the reinforcement across the 48 in wide by 10 in deep web opening. Use LRFD. The design criteria give that the compression flange is fully braced, that beam weight is negligible, and that ASTM A36 steel is to be used ($F_y = 36$ ksi). Assuming 50% dead load and 50% live load,

$$
\begin{aligned}
U &= 1.2D + 1.6L \\
&= (1.2)(0.5) + (1.6)(0.5) \\
&= 1.4
\end{aligned}
$$

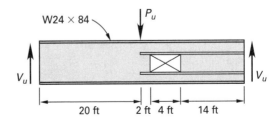

$$
P_u = UP = (1.4)(30\text{ kips}) = 42\text{ kips}
$$
$$
V_u = 0.5 P_u = (0.5)(42\text{ kips}) = 21\text{ kips}
$$

Calculate the bending moment at the centerline of the web opening.

$$M_u = V_u x = (21 \text{ kips})(16 \text{ ft}) = 336 \text{ ft-kips}$$

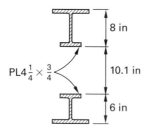

For the W24 × 84: $A_g = 24.7 \text{ in}^2$, $d = 24.1 \text{ in}$, $t_w = 0.47 \text{ in}$, $b_f = 9.0 \text{ in}$, $t_f = 0.77 \text{ in}$, $S_t = 196.0 \text{ in}^3$, and $Z_x = 224 \text{ in}^3$. As the depth of the beam is 24 in, the height of the spacing is 10.1 in. A section with reinforced opening must satisfy the following interaction equation (see Darwin).

$$\left(\frac{\phi M_n}{\phi M_m}\right)^3 + \left(\frac{\phi V_n}{\phi V_m}\right)^3 \leq 1.0$$

The ϕ-factor here is 0.9. M_m is the nominal moment strength when the shear is zero (must be less than or equal to the plastic moment capacity of the section without an opening).

$$M_p = F_y Z_x = \frac{\left(36 \dfrac{\text{kips}}{\text{in}^2}\right)(224 \text{ in}^3)}{12 \dfrac{\text{in}}{\text{ft}}}$$
$$= 672 \text{ ft-kips}$$

Locate the plastic neutral axis (PNA) of the reinforced section.

$$A_{\text{ten}} = A_{\text{comp}} = (0.5)(A_g - t_w h_{\text{hole}} + 4 b_p t_p)$$
$$= (0.5)\left(\begin{array}{c} 24.7 \text{ in}^2 - (0.47 \text{ in})(10 \text{ in}) \\ + (4)(4.25 \text{ in})(0.75 \text{ in}) \end{array}\right)$$
$$= 16.35 \text{ in}^2$$

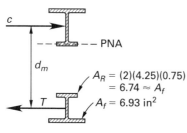

$$d_m = d - 0.5(d_{\text{comp}} + d_{\text{ten}})$$
$$= 24.1 \text{ in} - (0.5)(8 \text{ in} + 6 \text{ in})$$
$$= 17.1 \text{ in}$$

$$M_m \leq \begin{cases} F_y A_{\text{ten}} d_m - \dfrac{\left(36 \dfrac{\text{kips}}{\text{in}^2}\right)(16.35 \text{ in}^2)(17.1 \text{ in})}{12 \dfrac{\text{in}}{\text{ft}}} \\ \qquad = 839 \text{ ft-kips} \\ M_p = 672 \text{ ft-kips} \quad \text{[controls]} \end{cases}$$

Compute V_m, which is the sum of the shear capacities of the T sections above and below the opening in the absence of a bending moment. Use the AISC Design Guide 2 Sec. 3.6 (see Darwin).

$$V_m = V_{mt} + V_{mb}$$

V_{mt} and V_{mb} represent the shear capacity of the top and bottom T sections, respectively. For the top region,

$$V_{mt} \leq \begin{cases} \left(\dfrac{\sqrt{6} + \mu}{\nu + \sqrt{3}}\right) V_{pt} \\ V_{pt} \end{cases}$$

$$V_{pt} = \frac{F_y t_w s_t}{\sqrt{3}}$$
$$= \frac{\left(36 \dfrac{\text{kips}}{\text{in}^2}\right)(0.47 \text{ in})(8 \text{ in})}{\sqrt{3}}$$
$$= 78 \text{ kips}$$

The area of reinforcement is

$$A_r = 2 b_p t_p = (2)(4.25 \text{ in})(0.75 \text{ in}) = 6.38 \text{ in}^2$$

From the problem statement, the depth of the top T section is $s_t = 8 \text{ in}$. The shear parameter when reinforcement is used is

$$\bar{s}_t = s_t - \frac{A_r}{2 b_f} = 8 \text{ in} - \frac{6.38 \text{ in}^2}{(2)(9.0 \text{ in})} = 7.64 \text{ in}$$

The aspect ratio of the T section is

$$\nu = \frac{a_o}{\bar{s}_t} = \frac{48 \text{ in}}{7.64 \text{ in}} = 6.3$$

$$P_r \leq \begin{cases} F_y A_r = \left(36 \dfrac{\text{kips}}{\text{in}^2}\right)(6.38 \text{ in}^2) \\ \qquad = 230 \text{ kips} \quad \text{[controls]} \\ \dfrac{F_y t_w a_o}{2\sqrt{3}} = \dfrac{\left(36 \dfrac{\text{kips}}{\text{in}^2}\right)(0.47 \text{ in})(48 \text{ in})}{2\sqrt{3}} \\ \qquad = 234 \text{ kips} \end{cases}$$

The distance from the outside edge of the flange to the centroid of reinforcement is

$$d_r = s_t - 0.5t_p = 8 \text{ in} - (0.5)(0.75 \text{ in}) = 7.63 \text{ in}$$

$$\mu = \frac{2P_r d_r}{V_{pt}s_t} = \frac{(2)(230 \text{ kips})(7.63 \text{ in})}{(78 \text{ kips})(8 \text{ in})} = 5.62$$

Therefore,

$$V_{mt} \leq \begin{cases} \left(\dfrac{\sqrt{6}+\mu}{\nu+\sqrt{3}}\right)V_{pt} = \left(\dfrac{\sqrt{6}+5.62}{6.3+\sqrt{3}}\right)(78 \text{ kips}) \\ \qquad = 78.4 \text{ kips} \\ V_{pt} = 78 \text{ kips} \quad \text{[controls]} \end{cases}$$

Similarly, for the T section below the opening,

$$V_{pb} = \frac{F_y t_w s_b}{\sqrt{3}} = \frac{\left(36 \, \dfrac{\text{kips}}{\text{in}^2}\right)(0.47 \text{ in})(6 \text{ in})}{\sqrt{3}} = 58.6 \text{ kips}$$

The depth of the top T section (from the problem statement) is $s_b = 6$ in. The shear parameter when reinforcement is used is

$$\bar{s}_b = s_b - \frac{A_r}{2b_f} = 6 \text{ in} - \frac{6.38 \text{ in}^2}{(2)(9.0 \text{ in})} = 5.64 \text{ in}$$

The aspect ratio of the T section is

$$\nu = \frac{a_o}{\bar{s}_b} = \frac{48 \text{ in}}{5.64 \text{ in}} = 8.51$$

$$\mu = \frac{2P_r d_r}{V_{pt}s_b} = \frac{(2)(230 \text{ kips})(5.63 \text{ in})}{(78 \text{ kips})(6 \text{ in})} = 5.53$$

$$V_{mb} \leq \begin{cases} \left(\dfrac{\sqrt{6}+\mu}{\nu+\sqrt{3}}\right)V_{pb} = \left(\dfrac{\sqrt{6}+5.53}{8.51+\sqrt{3}}\right)(58.6 \text{ kips}) \\ \qquad = 45.7 \text{ kips} \quad \text{[controls]} \\ V_{pb} = 58.6 \text{ kips} \end{cases}$$

Thus, the moment-shear interaction equation, from Eq. 3-3 in the AISC Design Guide 2 (see Darwin), gives

$$V_m = V_{mt} + V_{mb} = 78 \text{ kips} + 45.7 \text{ kips} = 123.7 \text{ kips}$$

The points on the interaction diagram are

$$\frac{M_u}{\phi M_m} = \frac{336 \text{ ft-kips}}{(0.9)(672 \text{ ft-kips})} = 0.56$$

$$\frac{V_u}{\phi V_m} = \frac{21 \text{ kips}}{(0.9)(123.7 \text{ kips})} = 0.19$$

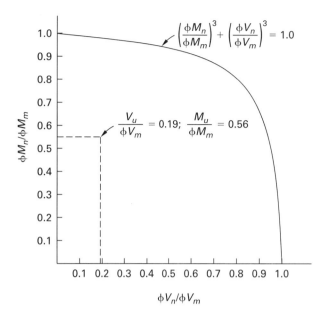

The points fall well within the boundary of the interaction diagram, so the proposed reinforced opening is
adequate.

Considering the low value of the interaction equation, it is likely that one or both of the reinforcing stiffeners could be eliminated. However, the problem statement requires only verification of the proposed opening detail and further refinement will not be pursued.

11.2. Design fillet welds to develop the full strength of the plates beyond the end of the opening.

$$P_r = 0.5A_r F_y$$
$$\quad = (0.5)(6.38 \text{ in}^2)\left(36 \, \frac{\text{kips}}{\text{in}^2}\right)$$
$$\quad = 115 \text{ kips}$$

Fillet welds are along a $^3/_4$ in thick plate onto a 0.47 in web. From AISC Table J2.4, use $^3/_{16}$ in welds on each side. One-half of t_w is the available base material thickness. For an E70 weld, the strength is 1.39 kips/in per $^1/_{16}$ in of weld size. (See AISC Part 8.)

$$\phi F_w \leq \begin{cases} \phi 0.6 F_y (0.5t_w) = (1.0)(0.6)\left(36 \, \dfrac{\text{kips}}{\text{in}^2}\right) \\ \qquad \times (0.5)(0.47 \text{ in}) \\ \qquad = 5.1 \text{ kips/in} \\ \phi 0.6 F_u (0.5t_w) = (0.75)(0.6)\left(58 \, \dfrac{\text{kips}}{\text{in}^2}\right) \\ \qquad \times (0.5)(0.47 \text{ in}) \\ \qquad = 6.1 \text{ kips/in} \\ n_{16\text{ths}}\phi f_w = (3)\left(1.39 \, \dfrac{\text{kips}}{\text{in}}\right) \\ \qquad = 4.2 \text{ kips/in} \quad \text{[controls]} \end{cases}$$

The required weld length is

$$L_w \geq \frac{P_f}{\phi F_w} = \frac{115 \text{ kips}}{4.2 \frac{\text{kips}}{\text{in}}} = \boxed{27 \text{ in} \quad [\text{say } 13.5 \text{ in each side}]}$$

11.3. The required weld details are as shown.

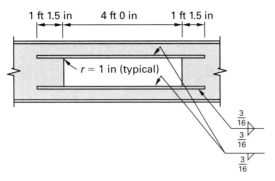

11.4. The construction sequence is to

- shore
- drill 2 in diameter holes at each corner of the opening
- weld stiffeners tangent to corner holes
- cut and remove the web portion

SOLUTION 12

12.1. Design the crane runway girder using ASD. Per ASCE/SEI7 Sec. 4.9, the impact factor for a cab-operated bridge crane is 25%, which applies to the crane bridge wheel load. The crane spans 37 ft simply supported and its maximum reactions occur when the trolley and lift are at the limit, 5 ft from the end.

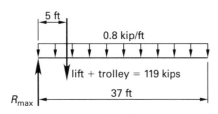

elevation of bridge girder

$$\sum M_R = P(L - 5 \text{ ft}) + (wL)(0.5L) - R_{\max}L = 0$$
$$R_{\max} = \frac{P(L - 5 \text{ ft}) + 0.5wL^2}{L}$$
$$= \frac{(119 \text{ kips})(37 \text{ ft} - 5 \text{ ft}) + (0.5)\left(0.8 \frac{\text{kip}}{\text{ft}}\right)(37 \text{ ft})^2}{37 \text{ ft}}$$
$$= 118 \text{ kips}$$

Increase $R_{\max}$ by 25% (impact factor) and distribute equally to two wheels spaced 12 ft on center.

$$R = 0.5(1 + I)R_{\max}$$
$$= (0.5)(1 + 0.25)(118 \text{ kips})$$
$$= 74 \text{ kips}$$

The resultant of the two moving forces is midway between them.

$$\overline{x} = (0.5)(12 \text{ ft}) = 6 \text{ ft}$$

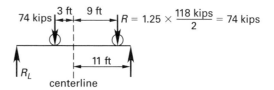

elevation of crane girder

The moving loads are supported by the crane girder, which spans 22 ft simply supported. The maximum bending moment occurs under the wheel closest to the centerline when the distance from girder centerline to the resultant of the two moving forces is equidistant from centerline to the nearest force, as shown. To find the reaction at the left end of girder for this load position,

$$\sum M_R = 2R(0.5L - 0.5\overline{x}) - R_LL = 0$$
$$R_L = \frac{2R(0.5L - 0.5\overline{x})}{L}$$
$$= \frac{(2)(74 \text{ kips})\big((0.5)(22 \text{ ft}) - (0.5)(6 \text{ ft})\big)}{22 \text{ ft}}$$
$$= 53.8 \text{ kips}$$

Maximum bending moment occurs under the wheel load nearest midspan and includes the moment due to live load plus impact, plus the dead load of the girder and crane rail.

$$M_{\max} = R_L(0.5L - 0.5\overline{x}) + M_d$$
$$= (53.8 \text{ kips})\big((0.5)(22 \text{ ft}) - (0.5)(6 \text{ ft})\big) + M_d$$
$$= 430 \text{ ft-kips} + M_d$$

Another possibility for maximum bending moment occurs when only one force is on the span, positioned at midspan, and the other force is off the span. For this case,

$$M_{\max} = 0.25RL = (0.25)(74 \text{ kips})(22 \text{ ft}) = 407 \text{ ft-kips}$$

The case with both wheels on the span is larger and controls. Maximum shear occurs when one wheel is adjacent to a support.

$$\sum M_R = V_{\max}L - RL - R(L - 12 \text{ ft}) = 0$$

$$V_{\max} = \frac{RL + R(L - 12 \text{ ft})}{L}$$

$$= \frac{(74 \text{ kips})(22 \text{ ft}) + (74 \text{ kips})(22 \text{ ft} - 12 \text{ ft})}{22 \text{ ft}}$$

$$= 108 \text{ kips}$$

Per ASCE/SEI7 Sec. 4.9.4, apply a lateral force equal to 20% of the trolley plus lift at the top of the rail. Distribute half of the lateral force to each parallel crane girder and half to each wheel.

$$H = (0.5)(0.5)\big(0.2(W_{\text{trolley}} + W_{\text{lift}})\big)$$

$$= (0.5)(0.5)(0.2)(19 \text{ kips} + 100 \text{ kips})$$

$$= 5.95 \text{ kips} \quad [\text{per wheel}]$$

The maximum bending moment about the weak axis occurs with wheel loads positioned for critical strong axis bending. The magnitude of this moment is in direct proportion to the strong axis moment.

$$M_{ry} = \left(\frac{H}{R}\right)M$$

$$= \left(\frac{5.95 \text{ kips}}{74 \text{ kips}}\right)(430 \text{ ft-kips})$$

$$= 34.6 \text{ ft-kips}$$

Follow the usual procedures for design of crane runway girders (see Fisher and Buettner). The class of service for the crane is not specified. Assume that maximum stresses occur infrequently so that no reduction in the stress range is required for fatigue. The recommended limits on live load deflections are $\Delta_y \leq L/1000$ and $\Delta_x = L/400$. Thus, with one wheel at midspan and the other off the girder,

$$\Delta_y = \frac{RL^3}{48EI_x} \leq 0.001L$$

$$I_x \geq \frac{RL^2}{0.048E}$$

$$\geq \frac{(74 \text{ kips})\Big((22 \text{ ft})\Big(12 \frac{\text{in}}{\text{ft}}\Big)\Big)^2}{(0.048)\Big(29,000 \frac{\text{kips}}{\text{in}^2}\Big)}$$

$$\geq 3705 \text{ in}^4$$

$$\Delta_x = \frac{HL^3}{48EI_y} \leq 0.0025L$$

$$I_y \geq \frac{HL^2}{0.12E}$$

$$\geq \frac{(5.95 \text{ kips})\Big((22 \text{ ft})\Big(12 \frac{\text{in}}{\text{ft}}\Big)\Big)^2}{(0.12)\Big(29,000 \frac{\text{kips}}{\text{in}^2}\Big)}$$

$$\geq 119 \text{ in}^4$$

From AISC Table 1-19, try a W27 × 84 in combination with a C15 × 33.9: $I_x = 4050$ in^4, $I_y = 420$ in^4, $S_{x1} = 237$ in^3, and $S_{x2} = 403$ in^3.

$$w_D = w_{W+C} + w_{\text{rail}}$$

$$= \frac{118 \frac{\text{lbf}}{\text{ft}} + 28 \frac{\text{lbf}}{\text{ft}}}{1000 \frac{\text{lbf}}{\text{kip}}}$$

$$= 0.146 \text{ kip/ft}$$

Calculate the dead load moment at the point of $M_{\max}$, 3 ft from the centerline.

$$V_D = 0.5wL = (0.5)\left(0.146 \frac{\text{kip}}{\text{ft}}\right)(22 \text{ ft})$$

$$= 1.6 \text{ kips}$$

$$M_D = 0.5(V_D + 0.5\bar{x}w_D)(0.5L - 0.5\bar{x})$$

$$= (0.5)\left(1.6 \text{ kips} + (0.5)(6 \text{ ft})\left(0.146 \frac{\text{kip}}{\text{ft}}\right)\right)$$

$$\times \big((0.5)(22 \text{ ft}) - (0.5)(6 \text{ ft})\big)$$

$$= 8 \text{ ft-kips}$$

$$M_{rx} = M + M_D = 430 \text{ ft-kips} + 8 \text{ ft-kips} = 438 \text{ ft-kips}$$

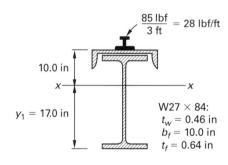

For a single W27 × 84: $A = 24.7$ in^2, $d = 26.7$ in, $t_w = 0.46$ in, $b_f = 10.0$ in, $t_f = 0.64$ in, $k_{\text{des}} = 1.24$ in, $J = 2.81$ in^4, $b_f/2t_f = 7.78$, $h/t_w = 52.7$, $Z_x = 244$ in^3, $S_x = 213$ in^3, $I_y = 106$ in^4, and $r_y = 2.07$ in.

For a single $C15 \times 33.9$: $A = 10.0$ in^2, $t_w = 0.40$ in, $\bar{x} = 0.79$ in, $I_x = 315$ in^4, $Z_x = 50.8$ in^3, and $I = 1.01$ in^4

For the $C15 \times 33.9$ in combination with the $W27 \times 84$ (AISC Table 1-19): $I_x = 4050$ in^4, $y_1 = 17.1$ in, $y_2 = 10.0$ in, $S_{xt} = I_x/y_1 = 237$ in^3, $S_{xc} = I_x/y_2 = 403$ in^3, $Z_x = 316$ in^3, and $y_p = 23.9$ in.

Check local buckling (AISC Table B4.1). The critical flange element is the stiffened web of the $C15 \times 33.9$ between the flange tips of the $W27 \times 84$.

$$\lambda_{pf} = 1.12\sqrt{\frac{E}{F_y}} = 1.12\sqrt{\frac{29{,}000 \frac{\text{kips}}{\text{in}^2}}{36 \frac{\text{kips}}{\text{in}^2}}} = 31.8$$

$$\frac{b_f}{t_w} = \frac{10.0 \text{ in}}{0.40 \text{ in}} = 25.0 \quad [<\lambda_{pf}]$$

For the web, case 16 in AISC Table B4.1 applies.

$$\lambda_{pw} = \frac{\frac{h_c}{h_p}\sqrt{\frac{E}{F_y}}}{\left(0.54\left(\frac{M_p}{M_y}\right) - 0.09\right)^2}$$

$$= \frac{\left(\frac{16.7 \text{ in}}{3.12 \text{ in}}\right)\sqrt{\frac{29{,}000 \frac{\text{kips}}{\text{in}^2}}{36 \frac{\text{kips}}{\text{in}^2}}}}{\left((0.54)\left(\frac{11{,}376 \text{ in-kips}}{8532 \text{ in-kips}}\right) - 0.09\right)^2} = 383$$

$$\lambda_r = 5.7\sqrt{\frac{E}{F_y}} = 5.7\sqrt{\frac{29{,}000 \frac{\text{kips}}{\text{in}^2}}{36 \frac{\text{kips}}{\text{in}^2}}}$$

$$= 162 \quad [\text{controls}]$$

$$h_c = 2(y_2 - t_{w,C} - k_{des,W})$$

$$= (2)(10.0 \text{ in} - 0.40 \text{ in} - 1.24 \text{ in})$$

$$= 16.7 \text{ in}$$

$$M_p = F_y Z_x = \left(36 \frac{\text{kips}}{\text{in}^2}\right)(316 \text{ in}^3)$$

$$= 11{,}376 \text{ in-kips}$$

$$M_y = F_y S_{xt} = \left(36 \frac{\text{kips}}{\text{in}^2}\right)(237 \text{ in}^3)$$

$$= 8532 \text{ in-kips}$$

$$h_p = 2(d - y_p - k_{des} - t_{w,C})$$

$$= (2)(27.1 \text{ in} - 23.9 \text{ in} - 1.24 \text{ in} - 0.40 \text{ in})$$

$$= 3.12 \text{ in}$$

$$\frac{h_c}{t_{w,W}} = \frac{16.7 \text{ in}}{0.46 \text{ in}} = 36.3 \quad [<\lambda_r]$$

The combination section is therefore compact, provided that the C section is continuously attached to the $W27 \times 84$. Compute the flexural strength for bending about the strong axis per AISC Sec. F4.2(ii).

$$r_t = \sqrt{\frac{I_{x,C} + \frac{t_f b_f^3}{12}}{A_C + (b_f t_f)_W + \frac{t_{w,W}(0.5h_c)}{6}}}$$

$$= \sqrt{\frac{315 \text{ in}^4 + \frac{(0.64 \text{ in})(10.0 \text{ in})^3}{12}}{10.0 \text{ in}^2 + (0.64 \text{ in})(10.0 \text{ in}) + \frac{(0.46 \text{ in})(0.5)(16.7 \text{ in})}{6}}}$$

$$= 4.65 \text{ in}$$

$$L_p = 1.1 r_t \sqrt{\frac{E}{F_y}}$$

$$= \frac{(1.1)(4.65 \text{ in})\sqrt{\frac{29{,}000 \frac{\text{kips}}{\text{in}^2}}{36 \frac{\text{kips}}{\text{in}^2}}}}{12 \frac{\text{in}}{\text{ft}}}$$

$$= 12.1 \text{ ft}$$

The unbraced length, $L_b = 22$ ft, is greater than the limiting length for full plastic capacity; therefore, calculate the limit for inelastic lateral-torsional buckling per AISC Eq. F4-8.

$$M_{yc} = F_y S_{xc} = \left(36 \frac{\text{kips}}{\text{in}^2}\right)(403 \text{ in}^3)$$

$$= 14{,}500 \text{ in-kips}$$

$$R_{pc} = \frac{M_p}{M_{yc}} = \frac{11{,}376 \text{ in-kips}}{14{,}500 \text{ in-kips}}$$

$$= 0.78$$

$$\frac{S_{xt}}{S_{xc}} = \frac{237 \text{ in}^3}{403 \text{ in}^3} = 0.59 \quad [<0.7, \text{ so use AISC Eq. F4-6b}]$$

$$F_L = F_y\left(\frac{S_{xt}}{S_{xc}}\right) = \left(36 \frac{\text{kips}}{\text{in}^2}\right)(0.59)$$

$$= 21.2 \text{ ksi}$$

$$h_o = (d - 2k_{des})_W$$

$$= 26.7 \text{ in} - (2)(1.24 \text{ in})$$

$$= 24.2 \text{ in}$$

$$J = J_C + J_W$$

$$= 1.01 \text{ in}^4 + 2.81 \text{ in}^4$$

$$= 3.82 \text{ in}^4$$

From AISC Eq. F4-8,

$$L_r = 1.95 r_t \frac{E}{F_L} \sqrt{\frac{J}{S_{xc}h_0} + \sqrt{\left(\frac{J}{S_{xc}h_0}\right)^2 + 6.76\left(\frac{F_L}{E}\right)^2}}$$

$$= \frac{(1.95)(4.65 \text{ in})\left(\dfrac{29{,}000 \frac{\text{kips}}{\text{in}^2}}{21.2 \frac{\text{kips}}{\text{in}^2}}\right)}{12 \frac{\text{in}}{\text{ft}}} \times \sqrt{\frac{3.82 \text{ in}^4}{(403 \text{ in}^3)(24.2 \text{ in})} + \sqrt{\left(\frac{3.82 \text{ in}^4}{(403 \text{ in}^3)(24.2 \text{ in})}\right)^2 + 6.76\left(\frac{21.2 \frac{\text{kips}}{\text{in}^2}}{29{,}000 \frac{\text{kips}}{\text{in}^2}}\right)^2}}$$

$$= 50.0 \text{ ft}$$

For the wheel loads positioned to produce $M_{max} = 430$ ft-kips,

$$M_A = R_L x = (53.8 \text{ kips})(5.5 \text{ ft})$$
$$= 296 \text{ ft-kips}$$

$$M_B = R_L x - P(0.5\overline{x})$$
$$= (53.8 \text{ kips})(11.0 \text{ ft}) - (74 \text{ kips})(3 \text{ ft})$$
$$= 370 \text{ ft-kips}$$

$$M_C = R_L x - P(x - 8 \text{ ft})$$
$$= (53.8 \text{ kips})(16.5 \text{ ft}) - (74 \text{ kips})(16.5 \text{ ft} - 8 \text{ ft})$$
$$= 259 \text{ ft-kips}$$

$$C_b = \frac{12.5 M_{max}}{2.5 M_{max} + 3 M_A + 4 M_B + 3 M_C}$$
$$= \frac{(12.5)(430 \text{ ft-kips})}{(2.5)(430 \text{ ft-kips}) + (3)(296 \text{ ft-kips}) + (4)(370 \text{ ft-kips}) + (3)(259 \text{ ft-kips})}$$
$$= 1.27$$

$$M_n \leq \begin{cases} C_b\left(R_{pc}M_{yc} - (R_{pc}M_{yc} - F_L S_{xc})\left(\dfrac{L_b - L_p}{L_r - L_p}\right)\right) \\ = (1.27)\left(\begin{array}{c}(0.78)(14{,}500 \text{ in-kips}) \\ - \left(\begin{array}{c}(0.78)(14{,}500 \text{ in-kips}) \\ -\left(21.2 \frac{\text{kips}}{\text{in}^2}\right)(403 \text{ in}^3)\end{array}\right) \\ \times \left(\dfrac{22 \text{ ft} - 12.1 \text{ ft}}{50.0 \text{ ft} - 12.1 \text{ ft}}\right)\end{array}\right) \\ = 13{,}350 \text{ in-kips} \\ R_{pc}M_{yc} = (0.78)(14{,}500 \text{ in-kips}) \\ \quad = 11{,}310 \text{ in-kips} \quad \text{[controls]} \end{cases}$$

$$M_{cx} = \frac{M_n}{\Omega_b} = \frac{11{,}310 \text{ in-kips}}{(1.67)\left(12 \frac{\text{in}}{\text{ft}}\right)} = 564 \text{ ft-kips}$$

Assume that the bending caused by the horizontal crane force is resisted by the channel and top flange of the W27 × 84.

$$\frac{M_n}{\Omega_b} = \frac{F_y\left(Z_{x,C} + \dfrac{t_f b_f^2}{4}\right)}{\Omega_b}$$
$$= \frac{\left(36 \frac{\text{kips}}{\text{in}^2}\right)\left(50.8 \text{ in}^3 + \dfrac{(0.64 \text{ in})(10.0 \text{ in})^2}{4}\right)}{(1.67)\left(12 \frac{\text{in}}{\text{ft}}\right)}$$
$$= 120 \text{ ft-kips}$$

Check combined bending about both axes, per AISC Eq. H1-1b.

$$\left|\frac{M_{rx}}{M_{cx}} + \frac{M_{ry}}{M_{cy}}\right| = \left|\frac{438 \text{ ft-kips}}{564 \text{ ft-kips}} + \frac{34.6 \text{ ft-kips}}{120 \text{ ft-kips}}\right|$$
$$= 1.06 \quad [> 1.0]$$

The overstress (6%) is acceptable, considering the small likelihood that absolute maximum bending moments will occur simultaneously about both principal axes. Therefore, flexural strength is adequate. Check shear.

$$\frac{V_n}{\Omega_v} = \frac{(0.6 F_y) d t_w}{\Omega_v}$$
$$= \frac{(0.6)\left(36 \frac{\text{kips}}{\text{in}^2}\right)(26.7 \text{ in})(0.46 \text{ in})}{1.5}$$
$$= 176 \text{ kips} \quad [> V_{max}, \text{ so OK}]$$

Use a W27 × 84 capped with a C15 × 33.9.

12.2. Design welds connecting the C15 × 33.9 to the W27 × 84. Use fillet welds, E70XX. Per AISC Table J2.4: $\omega_{min} = {}^{3}/_{10}$ in, and weld strength is governed by the horizontal shear stress transfer under vertical load.

$$\overline{y} = y_2 - \overline{x}_C = 10.0 \text{ in} - 0.79 \text{ in} = 9.21 \text{ in}$$

$$Q = A_c\overline{y} = (10.0 \text{ in}^2)(9.21 \text{ in}) = 92.1 \text{ in}^3$$

$$V = V_{max} + V_D = 108 \text{ kips} + 1.6 \text{ kips}$$
$$= 109.6 \text{ kips}$$

$$q = \frac{VQ}{I_x} = \frac{(109.6 \text{ kips})(92.1 \text{ in}^3)}{4050 \text{ in}^4}$$
$$= 2.5 \text{ kips/in}$$

The maximum shear flow, 2.5 kips/in, is shared by two welds. Thus, the minimum size weld, $^{3}/_{16}$ in, is adequate.

$$f_w = n_{16ths}q_w = (4)\left(0.928 \, \frac{\text{kips}}{\text{in}^2}\right) = 3.7 \text{ kips/in}$$

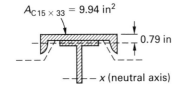

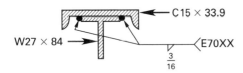

SOLUTION 13

13.1. Design a tension-side cover plate for the existing beam.

$w = 3.15 \text{ kips/ft} + 0.046 \text{ kips/ft}$
$= 3.20 \text{ kips/ft}$

$P = 5.15 \text{ kips} + 0.35 \text{ kips}$
$= 5.5 \text{ kips}$

10 ft   10 ft

$R = 2.75 \text{ kips} + (3.20 \text{ kips/ft})(10 \text{ ft})$
$= 34.65 \text{ kips}$

$$M_{max} = 0.125wL^2 + 0.25PL$$
$$= (0.125)\left(3.20 \, \frac{\text{kips}}{\text{ft}}\right)(20 \text{ ft})^2$$
$$+ (0.25)(5.5 \text{ kips})(20 \text{ ft})$$
$$= 187 \text{ ft-kips}$$

For the W18 × 46 of A36 steel: $A_g = 13.5$ in², $d = 18.1$ in, $t_w = 0.36$ in, $b_f = 6.06$ in, $t_f = 0.605$ in, $b_f/2t_f = 5.01$, $h/t_w = 44.6$, $I_x = 712$ in⁴, $S_x = 78.8$ in³, and $Z_x = 90.7$ in³. Check local buckling (AISC Table B4.1).

$$\lambda_{pf} = 0.38\sqrt{\frac{E}{F_y}} = 0.38\sqrt{\frac{29,000 \, \frac{\text{kips}}{\text{in}^2}}{36 \, \frac{\text{kips}}{\text{in}^2}}} = 10.8$$

$$\frac{b_f}{2t_f} = 5.01 \quad [< \lambda_{pf}]$$

$$\lambda_{pw} = 3.76\sqrt{\frac{E}{F_y}} = 3.76\sqrt{\frac{29,000 \, \frac{\text{kips}}{\text{in}^2}}{36 \, \frac{\text{kips}}{\text{in}^2}}} = 106.7$$

$$\frac{h}{t_w} = 44.6 \quad [< \lambda_{pw}, \text{ compact}]$$

The section is compact before adding a cover plate. (With a cover plate, a greater portion of the W18 × 46 web will be in compression and case 16 of AISC Table B4.1 must be checked.) The compression flange is fully braced (per the problem statement). Therefore,

$$\frac{M_n}{\Omega_b} = \frac{M_p}{\Omega_b} = \frac{F_yZ_x}{\Omega_b}$$
$$= \frac{\left(36 \, \frac{\text{kips}}{\text{in}^2}\right)(90.7 \text{ in}^3)}{(1.67)\left(12 \, \frac{\text{in}}{\text{ft}}\right)}$$
$$= 163 \text{ ft-kips}$$

Flexural strength therefore must be increased. The problem requires a 7.5 in wide cover plate. Try adding a plate that approximately matches the area of the W18 × 46 web (making the plastic neutral axis (PNA) at the bottom web-flange junction).

$$A_w = A_g - 2b_ft_f$$
$$= 13.5 \text{ in}^2 - (2)(6.06 \text{ in})(0.605 \text{ in})$$
$$= 6.17 \text{ in}^2$$
$$t \approx \frac{A_w}{b_{PL}} = \frac{6.17 \text{ in}^2}{7.5 \text{ in}} = 0.82 \text{ in}$$

Try a PL$^{3}/_4$ × 7.5. The area of the cover plate, $A_{PL} = (7.5 \text{ in})(0.75 \text{ in}) = 5.625$ in², is less than the web area, $A_w = 6.16$ in², so the PNA will be above the web-flange junction. Ignore the fillets and treat the W18 × 46 web as a uniform plate of equivalent width t_{we}.

$$t_{we} = \frac{A_w}{d - 2t_f}$$
$$= \frac{6.16 \text{ in}^2}{18.1 \text{ in} - (2)(0.605 \text{ in})}$$
$$= 0.365 \text{ in}$$

$$A_{\text{ten}} = A_{\text{comp}}$$

$$b_f t_f + t_{\text{we}} h_p = A_{\text{PL}} + b_f t_f + t_{\text{we}}(d - 2t_f - h_p)$$

$$h_p = \frac{A_{\text{PL}} + t_{\text{we}}(d - 2t_f)}{2t_{\text{we}}}$$

$$= \frac{\begin{array}{c}(5.625 \text{ in}^2) + (0.365 \text{ in}) \\ \times (18.1 \text{ in} - (2)(0.605 \text{ in}))\end{array}}{(2)(0.365 \text{ in})}$$

$$= 16.2 \text{ in}$$

$$h_2 = d - 2t_f - h_p$$

$$= 18.1 \text{ in} - (2)(0.605 \text{ in}) - 16.2 \text{ in}$$

$$= 0.8 \text{ in}$$

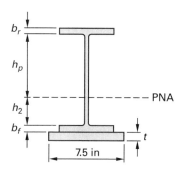

The PNA is 0.8 in above the bottom the web-flange junction. Take moments of area about the PNA.

$$Z_x = b_f t_f(d - t_f) + A_{\text{PL}}(0.5t + t_f + h_2)$$
$$\quad + 0.5 t_{\text{we}} h_2^2 + 0.5 t_{\text{we}} h_p^2$$

$$= (6.06 \text{ in})(0.605 \text{ in})(18.1 \text{ in} - 0.605 \text{ in})$$

$$\quad + (5.625 \text{ in}^2)\big((0.5)(0.75 \text{ in}) + 0.605 \text{ in} + 0.8 \text{ in}\big)$$

$$\quad + (0.5)(0.365 \text{ in})(0.8 \text{ in})^2$$

$$\quad + (0.5)(0.365 \text{ in})(16.2 \text{ in})^2$$

$$= 122 \text{ in}^3$$

$$\frac{M_n}{\Omega_b} = \frac{M_p}{\Omega_b} = \frac{F_y Z_x}{\Omega_b}$$

$$= \frac{\left(36 \dfrac{\text{kips}}{\text{in}^2}\right)(122 \text{ in}^3)}{(1.67)\left(12 \dfrac{\text{in}}{\text{ft}}\right)}$$

$$= 219 \text{ ft-kips}$$

This is well in excess of the required flexural strength, 187 kips, which suggests that a thinner plate would probably work. Repeat the calculations assuming a PL$^3/_8 \times 7.5$.

$$A_{\text{PL}} = (7.5 \text{ in})(0.375 \text{ in}) = 2.81 \text{ in}^2$$

$$A_{\text{ten}} = A_{\text{comp}}$$

$$b_f t_f + t_{\text{we}}(0.5 h_p) = A_{\text{PL}} + b_f t_f + t_{\text{we}}(d - 2t_f - 0.5 h_p)$$

$$h_p = \frac{A_{\text{PL}} + t_{\text{we}}(d - 2t_f)}{2t_{\text{we}}}$$

$$= \frac{2.81 \text{ in}^2 + (0.365 \text{ in})\big(18.1 \text{ in} - (2)(0.605 \text{ in})\big)}{(2)(0.365 \text{ in})}$$

$$= 12.3 \text{ in}$$

$$h_2 = d - 2t_f - h_p$$

$$= 18.1 \text{ in} - (2)(0.605 \text{ in}) - 12.3 \text{ in}$$

$$= 4.6 \text{ in}$$

$$Z_x = b_f t_f(d - t_f) + A_{\text{PL}}(0.5t + t_f + h_2)$$
$$\quad + 0.5 t_{\text{we}} h_2^2 + 0.5 t_{\text{we}} h_p^2$$

$$= (6.06 \text{ in})(0.605 \text{ in})(18.1 \text{ in} - 0.605 \text{ in})$$

$$\quad + (2.81 \text{ in}^2)\left(\begin{array}{c}(0.5)(0.375 \text{ in}) \\ + 0.605 \text{ in} + 4.6 \text{ in}\end{array}\right)$$

$$\quad + (0.5)(0.365 \text{ in})(4.6 \text{ in})^2$$

$$\quad + (0.5)(0.365 \text{ in})(12.3 \text{ in})^2$$

$$= 111 \text{ in}^3$$

$$\frac{M_n}{\Omega_b} = \frac{M_p}{\Omega_b} = \frac{F_y Z_x}{\Omega_b} = \frac{\left(36 \dfrac{\text{kips}}{\text{in}^2}\right)(111 \text{ in}^3)}{(1.67)\left(12 \dfrac{\text{in}}{\text{ft}}\right)}$$

$$= 199 \text{ ft-kips}$$

This is close enough. Recheck local buckling for the singly symmetrical section using AISC Table B4.1, case 16.

$$A = A_{\text{PL}} + A_{\text{W18}\times46} = 2.81 \text{ in}^2 + 13.5 \text{ in}^2 = 16.3 \text{ in}^2$$

$$\bar{y} = \frac{A_{\text{PL}}\big(0.5(d + t)\big)}{A}$$

$$= \frac{(2.81 \text{ in}^2)(0.5)(18.1 \text{ in} + 0.375 \text{ in})}{16.3 \text{ in}^2}$$

$$= 1.6 \text{ in}$$

$$I = (I_x + A\bar{y}^2)_{\text{W18}\times46} + A_{\text{PL}}\big(0.5(d + t) - \bar{y}\big)^2$$

$$= 712 \text{ in}^4 + (13.5 \text{ in}^2)(1.6 \text{ in})^2 + (2.81 \text{ in}^2)$$

$$\quad \times \big((0.5)(18.1 \text{ in} + 0.375 \text{ in}) - 1.6 \text{ in}\big)^2$$

$$= 910 \text{ in}^4$$

$$M_y = \dfrac{F_y I}{0.5d + \overline{y}}$$

$$= \dfrac{\left(36 \; \dfrac{\text{kips}}{\text{in}^2}\right)(910 \text{ in}^4)}{\left((0.5)(18.1 \text{ in}) + 1.6 \text{ in}\right)\left(12 \; \dfrac{\text{in}}{\text{ft}}\right)}$$

$$= 256 \text{ ft-kips}$$

$$M_p = F_y Z_x = \dfrac{\left(36 \; \dfrac{\text{kips}}{\text{in}^2}\right)(111 \text{ in}^3)}{12 \; \dfrac{\text{in}}{\text{ft}}} = 333 \text{ ft-kips}$$

$$h_c = 2\left(\overline{y} + 0.5(d - 2t_f)\right)$$

$$= (2)\left(1.6 \text{ in} + (0.5)\left(18.1 \text{ in} - (2)(0.605 \text{ in})\right)\right)$$

$$= 20.1 \text{ in}$$

$$\lambda_{pw} = \dfrac{\dfrac{h_c}{h_p}\sqrt{\dfrac{E}{F_y}}}{\left(0.54\left(\dfrac{M_p}{M_y}\right) - 0.09\right)^2}$$

$$= \dfrac{\dfrac{20.1 \text{ in}}{24.6 \text{ in}}\sqrt{\dfrac{29{,}000 \; \dfrac{\text{kips}}{\text{in}^2}}{36 \; \dfrac{\text{kips}}{\text{in}^2}}}}{\left((0.54)\left(\dfrac{333 \text{ ft-kips}}{256 \text{ ft-kips}}\right) - 0.09\right)^2}$$

$$= 61.8$$

$$\dfrac{h_c}{t_w} = \dfrac{20.1 \text{ in}}{0.36 \text{ in}} = 55.8 \quad [< \lambda_{pw}]$$

Therefore, the cover-plated section is compact and can develop the required moment capacity.

The plate is to be attached with $^5/_8$ in diameter A325-SC (slip-critical) bolts. The tensile rupture of the tension flange must also be checked, per AISC Sec. F13.1. Assume holes are drilled $^1/_{16}$ in larger than the bolt diameter.

$$A_{fn} = b_f t_f - 2(d_b + 0.125 \text{ in})t_f$$

$$= (6.06 \text{ in})(0.605 \text{ in}) - (2)$$

$$\times (0.625 \text{ in} + 0.125 \text{ in})(0.605 \text{ in})$$

$$= 2.76 \text{ in}^2$$

$$\dfrac{F_y}{F_u} = \dfrac{36 \; \dfrac{\text{kips}}{\text{in}^2}}{58 \; \dfrac{\text{kips}}{\text{in}^2}} = 0.62 \quad [< 0.8, \text{ so } Y_t = 1.0]$$

$$F_u A_{fn} = \left(58 \; \dfrac{\text{kips}}{\text{in}^2}\right)(2.76 \text{ in}^2) = 160 \text{ kips}$$

$$Y_t F_y A_{fg} = Y_t F_y b_f t_f$$

$$= (1.0)\left(36 \; \dfrac{\text{kips}}{\text{in}^2}\right)(6.06 \text{ in})(0.605 \text{ in})$$

$$= 132 \text{ kips}$$

Therefore, the holes in the flanges do not reduce flexural strength. For the $\text{PL}^3/_8 \times 7.5$,

$$A_n = \left(w - 2(d_b + 0.125 \text{ in})\right)t$$

$$= \left(7.5 \text{ in} - (2)(0.625 \text{ in} + 0.125 \text{ in})\right)(0.375 \text{ in})$$

$$= 2.25 \text{ in}^2$$

$$\dfrac{P_n}{\Omega_t} \leq \begin{cases} \dfrac{F_y A_g}{\Omega_t} = \dfrac{\left(36 \; \dfrac{\text{kips}}{\text{in}^2}\right)(2.81 \text{ in}^2)}{1.67} \\[6pt] \qquad = 61 \text{ kips} \quad [\text{controls}] \\[12pt] \dfrac{F_u U A_n}{\Omega_t} = \dfrac{\left(58 \; \dfrac{\text{kips}}{\text{in}^2}\right)(1.0)(2.25 \text{ in}^2)}{2.00} \\[6pt] \qquad = 65.3 \text{ kips} \end{cases}$$

Thus, the plate is limited by yielding of the gross area and the plastic capacity can be attained.

Calculate the length of the cover plate required. The theoretical cutoff points occur where the flexural strength of the bare W18 × 46 equals the bending moment.

$$M - \dfrac{M_n}{\Omega_b} - 163 \text{ ft-kips}$$

$$M = Rx - 0.5wx^2$$

$$163 \text{ ft-kips} = (34.65 \text{ kips})x - (0.5)\left(3.19 \; \dfrac{\text{kips}}{\text{ft}}\right)x^2$$

$$(163 \text{ ft-kips})\left(1 \; \dfrac{\text{ft}}{\text{kip}}\right) = \left((34.65 \text{ kips})\left(1 \; \dfrac{\text{ft}}{\text{kip}}\right)\right)x - (0.5)$$

$$\times \left(\left(3.19 \; \dfrac{\text{kips}}{\text{ft}}\right)\left(1 \; \dfrac{\text{ft}}{\text{kip}}\right)\right)x^2$$

$$163 \text{ ft}^2 = (34.65 \text{ ft})x - 1.595x^2$$

$$0 \text{ ft}^2 = 1.595x^2 - (34.65 \text{ ft})x + 163 \text{ ft}^2$$

$$= x^2 - (21.72 \text{ ft})x + 102.2 \text{ ft}^2$$

$$x = 6.9 \text{ ft} \quad [\text{from left support}]$$

From symmetry, the theoretical cutoff point on the right side is 6.9 ft from the right support. AISC Sec. F13.3, requires that the cover plate extend far enough beyond the theoretical cutoff point to develop the plate's share of flexural strength at the cutoff location. At the cutoff points, $M = 163$ ft-kips.

$$f_b = \frac{M\left(0.5(d+t) - \overline{y}\right)}{I}$$

$$= \frac{(163 \text{ ft-kips})\left(12 \frac{\text{in}}{\text{ft}}\right) \times \left((0.5)(18.1 \text{ in} + 0.375 \text{ in}) - 1.6 \text{ in}\right)}{910 \text{ in}^4}$$

$$= 16.4 \text{ ksi}$$

$$F_{\text{PL}} = f_b A_{\text{PL}} = \left(16.4 \frac{\text{kips}}{\text{in}^2}\right)(2.81 \text{ in}^2) = 46.1 \text{ kips}$$

For $\frac{5}{8}$ in A325-SC bolts in single shear, $r_v/\Omega_v = 4.29$ kips/bolt.

$$n \geq \frac{F_{\text{PL}}}{\frac{r_v}{\Omega_v}} = \frac{46.1 \text{ kips}}{4.29 \frac{\text{kips}}{\text{bolt}}}$$

$$= 10.7 \text{ bolts} \quad [\text{say 6 rows of 2 bolts each (12 total)}]$$

$$\text{extension} = (\text{no. of spaces})s + \text{end distance}$$

$$= (5)(3 \text{ in}) + 1.5 \text{ in}$$

$$= 16.5 \text{ in}$$

$$\text{total plate length} = (L - 2x) + 2(\text{extension})$$

$$= \left(20 \text{ ft} - (2)(6.9 \text{ ft})\right) + (2)\left(\frac{16.5 \text{ in}}{12 \frac{\text{in}}{\text{ft}}}\right)$$

$$= 8.95 \text{ ft}$$

> Use $\text{PL}\frac{3}{8} \times 7.5 \times 9$ ft 6 in.

13.2. The total number of bolts required between the point of maximum moment and the end of the cover plate must develop the yield strength of the cover plate.

$$\frac{P_n}{\Omega_t} \leq \frac{F_y A_{\text{PL}}}{\Omega_t} = \frac{\left(36 \frac{\text{kips}}{\text{in}^2}\right)(2.81 \text{ in}^2)}{1.67} = 61 \text{ kips}$$

$$n \geq \frac{\frac{P_n}{\Omega_t}}{\frac{r_v}{\Omega_v}} = \frac{61 \text{ kips}}{4.29 \frac{\text{kips}}{\text{bolt}}}$$

$$= 14 \text{ bolts} \quad [\text{say 7 rows of 2 bolts each (14 total)}]$$

The bolt spacing required between the cutoff points varies as the shear flow, VQ/I.

$$Q = A_{\text{PL}}(0.5d + 0.5t_{\text{PL}} - \overline{y})$$

$$= \left((7.5 \text{ in})(0.375 \text{ in})\right)\left(\begin{array}{c}(0.5)(18.1 \text{ in}) + (0.5) \\ \times (0.375 \text{ in}) - 1.6 \text{ in}\end{array}\right)$$

$$= 21.5 \text{ in}^3$$

At the cutoff point, $x = 6.5$ ft,

$$V = R - wx$$

$$= 34.65 \text{ kips} - \left(3.19 \frac{\text{kips}}{\text{ft}}\right)(6.5 \text{ ft})$$

$$= 13.9 \text{ kips}$$

$$q = \frac{VQ}{I} = \frac{(13.9 \text{ kips})(21.5 \text{ in}^3)}{910 \text{ in}^4} = 0.33 \text{ kip/in}$$

Two $\frac{5}{8}$ in A325-SC bolts will resist a lateral force of (2 bolts)(4.29 kips/bolt) = 8.6 kips. The bolts must be spaced no farther apart than

$$s \leq \frac{q_{\text{pair}}}{q} = \frac{8.6 \text{ kips}}{0.33 \frac{\text{kip}}{\text{in}}} = 26.1 \text{ in}$$

Per AISC Sec. J3.5a, the maximum spacing is

$$S_{\text{max}} \leq \begin{cases} 24t_{\text{thinner plate}} = (24)(0.375 \text{ in}) \\ \qquad\qquad = 9.0 \text{ in} \quad [\text{controls}] \\ 12 \text{ in} \end{cases}$$

> Use six rows of two $\frac{5}{8}$ in bolts each at 3 in on center at ends, and eight rows of two $\frac{5}{8}$ in bolds at 9 in on center over the middle 7 ft of plate.

SOLUTION 14

14.1. Determine the final reaction at D for each of the two proposed construction schemes. Use ASD.

Of the two proposals, the second one is simpler in that the final vertical displacements at supports A, D, and C are identical; thus, the beam behaves as a two-span continuous beam on unyielding simple supports. Any method of linear structural analysis provides the reactions and internal shears and moments. For this solution, analyze the two-span beam using moment distribution.

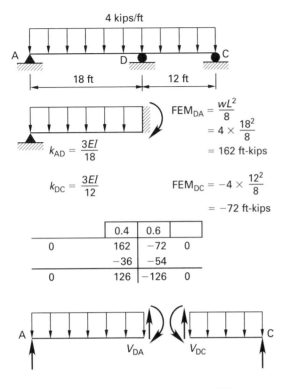

From the free-body diagram of member AD,

$$\sum M_A - 0.5wL_{AD}^2 + M_{DA} - V_{DA}L_{AD} = 0$$
$$= (0.5)\left(4\ \frac{\text{kips}}{\text{ft}}\right)(18\ \text{ft})^2 + 126\ \text{ft-kips}$$
$$- V_{DA}(18\ \text{ft})$$
$$V_{DA} = 43\ \text{kips}\quad[\text{upward}]$$

From the free-body diagram for member DC,

$$\sum M_C = -0.5wL_{DC}^2 - M_{DC} + V_{DC}L_{DC} = 0$$
$$= (-0.5)\left(4\ \frac{\text{kips}}{\text{ft}}\right)(12\ \text{ft})^2 - 126\ \text{ft-kips}$$
$$+ V_{DC}(12\ \text{ft})$$
$$V_{DC} = 34.5\ \text{kips}\quad[\text{upward}]$$

Thus, summing vertical forces at joint D gives the reaction at D for construction scheme 2.

$$\sum F_y = -V_{DA} - V_{DC} + R_D = 0$$
$$= -43\ \text{kips} - 34.5\ \text{kips} + R_{D2}$$
$$R_{D2} = 77.5\ \text{kips}\quad[\text{upward}]$$

The alternative construction sequence is to place the new column in a tight fit and then remove the existing column. Theoretically, this sequence will result in a final deflection at the support D that is not the same as at the end supports. Due to symmetry in spans and loading (assuming both spans simultaneously loaded with 4 kips/ft), the deflection can be calculated assuming a fixed-hinged uniformly loaded beam of span 15 ft (AISC Table 3-23, case 12). Solving for the deflection at point D in the existing condition gives

$$\Delta_x = \left(\frac{wx}{48EI}\right)(L^3 - 3Lx^2 + 2x^3)\quad\begin{bmatrix}L = 15\ \text{ft} = 180\ \text{in}\\x = 12\ \text{ft} = 144\ \text{in}\end{bmatrix}$$

$$= \left(\frac{\left(4\ \dfrac{\text{kips}}{\text{ft}}\right)(12\ \text{ft})}{(48)\left(29{,}000\ \dfrac{\text{kips}}{\text{in}^2}\right)(712\ \text{in}^4)}\right)$$

$$\times \left(\begin{array}{c}(180\ \text{in})^3 - (3)(180\ \text{in})(144\ \text{in})^2\\+ (2)(144\ \text{in})^3\end{array}\right)$$

$$= 0.029\ \text{in}\quad[\text{downward}]$$

The change in reaction at D due to this residual deflection equals the concentrated force that is applied at D in the released structure to produce the deflection. Thus,

$$\Delta_D = \frac{Pa^2b^2}{3EIL} = 0.029\ \text{in}$$

L is the full length of 30 ft = 360 in, a = 216 in, and b = 144 in.

$$P = \left(\frac{3EIL}{a^2b^2}\right)(0.029\ \text{in})$$

$$= \left(\frac{(3)\left(29{,}000\ \dfrac{\text{kips}}{\text{in}^2}\right)(712\ \text{in}^4)(360\ \text{in})}{(216\ \text{in})^2(144\ \text{in})^2}\right)(0.029\ \text{in})$$

$$= 0.7\ \text{kip}$$

Superposition gives the final reaction at D for construction scheme 1.

$$R_{D1} = R_{D2} - P = 77.5\ \text{kips} - 0.7\ \text{kip} = \boxed{76.8\ \text{kips}}$$

14.2. Verify the W18 × 46 for the modified structure: $d = 18.1$ in, $b_f = 6.06$ in, $k_{des} = 1.01$ in, $b_f/2t_f = 5.01$, $h/t_w = 44.6$, $Z_x = 90.7$ in³, $S_x = 78.8$ in³, $I_y = 22.5$ in⁴, $r_y = 1.29$ in, and $J = 1.22$ in⁴.

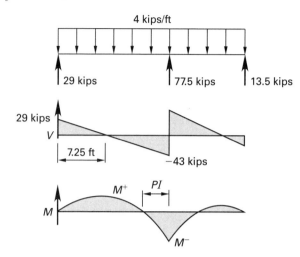

Maximum positive moment occurs in span AD where shear equals zero.

$$V = V_{AD} - wx = 0 \text{ kips}$$

$$x = \frac{V_{AD}}{w} = \frac{29 \text{ kips}}{4 \dfrac{\text{kips}}{\text{ft}}} = 7.25 \text{ ft}$$

$$M^+ = 0.5 V_{AD} x = (0.5)(29 \text{ kips})(7.25 \text{ ft})$$
$$= 105 \text{ ft-kips} \quad [\text{with } L_b = 0 \text{ ft}]$$

$$M^- = 126 \text{ ft-kips} \quad [\text{with } L_b = 0 \text{ ft}]$$

M^- controls flexural strength. Compute C_b for the segment AD.

$$M = V_{AD} x - 0.5 wx^2$$

$$M_A = \left| (29 \text{ kips})(4.5 \text{ ft}) - (0.5)\left(4 \frac{\text{kips}}{\text{ft}}\right)(4.5 \text{ ft})^2 \right|$$
$$= 90 \text{ ft-kips}$$

$$M_B = \left| (29 \text{ kips})(9.0 \text{ ft}) - (0.5)\left(4 \frac{\text{kips}}{\text{ft}}\right)(9.0 \text{ ft})^2 \right|$$
$$= 99 \text{ ft-kips}$$

$$M_C = \left| (29 \text{ kips})(13.5 \text{ ft}) - (0.5)\left(4 \frac{\text{kips}}{\text{ft}}\right)(13.5 \text{ ft})^2 \right|$$
$$= 27 \text{ ft-kips}$$

$$M_{max} = |M_u^-| = |-126 \text{ ft-kips}| = 126 \text{ ft-kips}$$

$$C_b = \frac{12.5 M_{max}}{2.5 M_{max} + 3 M_A + 4 M_B + 3 M_C}$$
$$= \frac{(12.5)(126 \text{ ft-kips})}{(2.5)(126 \text{ ft-kips}) + (3)(90 \text{ ft-kips}) + (4)(99 \text{ ft-kips}) + (3)(27 \text{ ft-kips})} = 1.48$$

$$\sqrt{\frac{E}{F_y}} = \sqrt{\frac{29{,}000 \dfrac{\text{kips}}{\text{in}^2}}{36 \dfrac{\text{kips}}{\text{in}^2}}} = 28.4$$

$$L_p = 1.76 r_y \sqrt{\frac{E}{F_y}} = \frac{(1.76)(1.29 \text{ in})(28.4)}{12 \dfrac{\text{in}}{\text{ft}}} = 5.4 \text{ ft}$$

Check local buckling.

$$\frac{b_f}{2t_f} = 5.01 < 0.38 \sqrt{\frac{E}{F_y}} = (0.38)(28.4) = 10.8 = \lambda_{pf}$$

$$\frac{h}{t_w} = 44.6 < 3.76 \sqrt{\frac{E}{F_y}} = (3.76)(28.4) = 107 = \lambda_{pw}$$

Therefore, the section is compact.

$$h_o = d - 2k_{des} = 18.1 \text{ in} - (2)(1.01 \text{ in}) = 16.1 \text{ in}$$

$$r_{ts} = \sqrt{\frac{I_y h_o}{2 S_x}} = \sqrt{\frac{(22.5 \text{ in}^4)(16.1 \text{ in})}{(2)(78.8 \text{ in}^3)}} = 1.52 \text{ in}$$

From AISC Eq. F2-6,

$$L_r = 1.95 r_{ts} \frac{E}{0.7 F_y} \sqrt{\frac{Jc}{S_x h_0} + \sqrt{\left(\frac{Jc}{S_x h_0}\right)^2 + 6.76 \left(\frac{0.7 F_y}{E}\right)^2}}$$

$$= \frac{(1.95)(1.52 \text{ in})\left(\dfrac{29{,}000 \dfrac{\text{kips}}{\text{in}^2}}{(0.7)\left(36 \dfrac{\text{kips}}{\text{in}^2}\right)}\right)}{12 \dfrac{\text{in}}{\text{ft}}} \times \sqrt{\frac{(1.22 \text{ in}^4)(1.0)}{(78.8 \text{ in}^3)(16.1 \text{ in})} + \sqrt{\left(\frac{(1.22 \text{ in}^4)(1.0)}{(78.8 \text{ in}^3)(16.1 \text{ in})}\right)^2 + (6.76)\left(\frac{(0.7)\left(36 \dfrac{\text{kips}}{\text{in}^2}\right)}{29{,}000 \dfrac{\text{kips}}{\text{in}^2}}\right)^2}}$$

$$= 16.6 \text{ ft}$$

Since $L_b = 18$ ft $> L_r = 16.6$ ft, compute M_n using AISC Eq. F2-4.

$$\frac{L_b}{r_{ts}} = \frac{(18 \text{ ft})\left(12 \frac{\text{in}}{\text{ft}}\right)}{1.52 \text{ in}} = 142$$

$$M_n = F_{cr} S_x = \left(\frac{C_b \pi^2 E}{\left(\frac{L_b}{r_{ts}}\right)^2}\sqrt{1 + 0.078\left(\frac{Jc}{S_x h_o}\right)\left(\frac{L_b}{r_{ts}}\right)^2}\right) S_x$$

$$= \left(\frac{\dfrac{1.48\pi^2\left(29{,}000 \frac{\text{kips}}{\text{in}^2}\right)}{(142)^2}}{\sqrt{1 + (0.078)\left(\dfrac{(1.22 \text{ in}^4)(1.0)}{(78.8 \text{ in}^3)(16.1 \text{ in})}\right)(142)^2}}\right)$$

$$\times (78.8 \text{ in}^3)$$

$$= 2621 \text{ in-kips}$$

$$M_p = F_y Z_x = \left(36 \frac{\text{kips}}{\text{in}^2}\right)(90.7 \text{ in}^3)$$

$$= 3265 \text{ in-kips} \quad [M_n < M_p]$$

Therefore,

$$\frac{M_n}{\Omega_b} = \frac{\dfrac{2621 \text{ in-kips}}{1.67}}{12 \frac{\text{in}}{\text{ft}}} = 130 \text{ ft-kips} > M^- = 126 \text{ ft-kips}$$

The W18 × 46 has adequate flexural strength. Check the shear strength. (See AISC Sec. G2.)

$$V_n = 0.6 F_y d t_w C_v$$

$$= (0.6)\left(36 \frac{\text{kips}}{\text{in}^2}\right)(18.1 \text{ in})(0.36 \text{ in})(1.0)$$

$$= 141 \text{ kips}$$

$$\frac{V_n}{\Omega_v} = \frac{141 \text{ kips}}{1.5} = 94 \text{ kips} \quad [> V_{max} = 43 \text{ kips}]$$

Therefore, the existing W18 × 46 is $\boxed{\text{OK}}$ for the proposed modification.

14.3. Design a new column at D. $P = R_D = 77.5$ kips with $KL_x = KL_y = 10$ ft. Since the problem specifies ASTM A36 steel and the *AISC Manual* does not contain tabulated capacities of sections with $F_y = 36$ ksi, use a trial-and-error approach to find an acceptable solution. Try the following.

$$F_{cr} = 0.75 F_y = (0.75)\left(36 \frac{\text{kips}}{\text{in}^2}\right) = 27.0 \text{ ksi}$$

$$\frac{P_n}{\Omega_c} = \frac{F_{cr} A_g}{\Omega_c} = \frac{\left(27.0 \frac{\text{kips}}{\text{in}^2}\right) A_g}{1.67} \geq P = 77.5 \text{ kips}$$

$$A_g \geq 4.80 \text{ in}^2$$

Try a W6 × 20. $A_g = 5.87$ in^2, $r_x = 2.66$ in, and $r_y = 1.50$ in (which controls).

$$\frac{KL_y}{r_y} = \frac{(1.0)(10.0 \text{ ft})\left(12 \frac{\text{in}}{\text{ft}}\right)}{1.50 \text{ in}} = 80$$

$$\sqrt{\frac{E}{F_y}} = \sqrt{\frac{29{,}000 \frac{\text{kips}}{\text{in}^2}}{36 \frac{\text{kips}}{\text{in}^2}}} = 28.4$$

$$\frac{KL}{r} = \frac{KL_y}{r_y} = 80$$

$$4.71\sqrt{\frac{E}{F_y}} = (4.71)(28.4)$$

$$= 134 \quad \left[KL/r < 4.71\sqrt{E/F_y}\right]$$

Use AISC Eq. E3-2.

$$F_e = \frac{\pi^2 E}{\left(\frac{KL}{r}\right)^2} = \frac{\pi^2\left(29{,}000 \frac{\text{kips}}{\text{in}^2}\right)}{(80)^2} = 44.7 \text{ ksi}$$

$$F_{cr} = 0.658^{F_y/F_e} F_y = \left(0.658^{36 \frac{\text{kips}}{\text{in}^2}/44.7 \frac{\text{kips}}{\text{in}^2}}\right)\left(36 \frac{\text{kips}}{\text{in}^2}\right)$$

$$= 25.7 \text{ ksi}$$

$$\frac{P_n}{\Omega_c} = \frac{F_{cr} A_g}{\Omega_c} = \frac{\left(25.7 \frac{\text{kips}}{\text{in}^2}\right)(5.87 \text{ in}^2)}{1.67}$$

$$= 90 \text{ kips} \quad [> R_{D1} = 77.5 \text{ kips}]$$

$\boxed{\text{Select a W6 × 20.}}$

14.4. The column-to-girder connection is to be made by field welding. The top flange is braced; in order for the W18 × 46 to be laterally braced on the bottom flange at the support, provide fitted stiffeners to the lower flange and web directly over the column. For simplicity, use two sets of stiffeners directly in line with the flanges of the W6 × 20 and furnishing roughly the same area.

$$A_{st} = 0.5 b_f t_f = (0.5)(6.02 \text{ in})(0.365 \text{ in})$$

$$= 1.1 \text{ in}^2 \quad [\text{say PL}^3/_8 \times 3]$$

Since bearing stiffeners are used, local web crippling cannot occur and does not need to be checked. Stop stiffeners short of the top flange of the W18 × 46.

$$L = d - 2k_{\text{det}} = 18.1 \text{ in} - (2)(1.25 \text{ in})$$
$$= 15.6 \text{ in} \quad [\text{say } 15 \text{ in}]$$

Use minimum size fillet welds of $^1/_4$ in. To cap the W6 × 20, use a plate that is wider than the flange of the W18 × 46 ($b_f = 6.06$ in). Thus, the proposed connection is

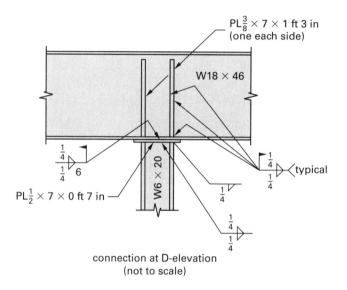

PL$\frac{3}{8}$ × 7 × 1 ft 3 in (one each side)

W18 × 46

$\frac{1}{4}$ 6

PL$\frac{1}{2}$ × 7 × 0 ft 7 in

W6 × 20

$\frac{1}{4}$ typical

connection at D-elevation
(not to scale)

SOLUTION 15

15.1. Compute the minimum flange width at the critical section. Base the design on an elastic analysis with maximum computed bending stress equal to $f_b = Mc/I$. Because F_b is already specified, local buckling of flange and web need not be checked.

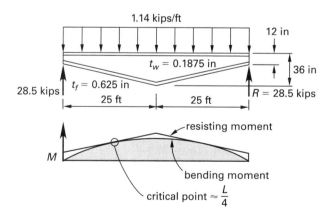

1.14 kips/ft

12 in

$t_w = 0.1875$ in

36 in

28.5 kips

$t_f = 0.625$ in

$R = 28.5$ kips

25 ft | 25 ft

resisting moment

M

bending moment

critical point $\approx \dfrac{L}{4}$

The depth, d, varies linearly from 12 in at the ends to 36 in at midspan; $t_f = ^5/_8$ in; $t_w = ^3/_{16}$ in; $F_b = 22$ ksi. As an approximation,

$$I_x = 2A_f\left(\frac{d}{2}\right)^2 + \frac{t_w d^3}{12}$$

$$S_x = \frac{2I_x}{d} = \frac{2\left(2A_f\left(\dfrac{d}{2}\right)^2 + \dfrac{t_w d^3}{12}\right)}{d}$$

$$= A_f d + \frac{t_w d^2}{6}$$

$$M_x = 0.5wLx - 0.5wx^2$$

$$= (0.5)\left(1.14 \, \frac{\text{kips}}{\text{ft}}\right)(50 \text{ ft})x - (0.5)\left(1.14 \, \frac{\text{kips}}{\text{ft}}\right)x^2$$

$$= (28.5 \text{ kips})x - \left(0.57 \, \frac{\text{kip}}{\text{ft}}\right)x^2$$

To get a trial value of A_f, assume that the critical location for bending stress is at $0.25L$ from supports where $d = 24$ in.

$$x = 0.25L = (0.25)(50 \text{ ft}) = 12.5 \text{ ft}$$

$$M_x = (28.5 \text{ kips})x - \left(0.57 \, \frac{\text{kip}}{\text{ft}}\right)x^2$$

$$M_{x=12.5 \text{ ft}} = (28.5 \text{ kips})(12.5 \text{ ft}) - \left(0.57 \, \frac{\text{kip}}{\text{ft}}\right)(12.5 \text{ ft})^2$$

$$= 267 \text{ ft-kips}$$

$$M_{x=12.5 \text{ ft}} = M_R = F_b S_x$$

$$267 \text{ ft-kips} = \left(22 \, \frac{\text{kips}}{\text{in}^2}\right)\left(\begin{array}{c} A_f(24 \text{ in}) \\ + \dfrac{(0.1875 \text{ in})(24 \text{ in})^2}{6} \end{array}\right)$$

$$A_f = 5.31 \text{ in}^2$$

For the $^5/_8$ in thick plate,

$$b_f = \frac{A_f}{t_f} = \frac{5.31 \text{ in}^2}{0.625 \text{ in}} = 8.5 \text{ in}$$

Compute the actual section properties based on the trial dimensions. To confirm the location of the critical bending stress, check the actual bending stress in the vicinity of $0.25L = 12.5$ ft; say at $x = 10$ ft, 12.5 ft, and 15 ft. At $x = 10$ ft,

$$d = d_e + mx$$

$$= 12 \text{ in} + \left(\frac{36 \text{ in} - 12 \text{ in}}{25 \text{ ft}}\right)(10 \text{ ft})$$

$$= 21.6 \text{ in}$$

$$M_x = (28.5 \text{ kips})x - \left(0.57 \frac{\text{kip}}{\text{lt}}\right)x^2$$

$$M_{x=10 \text{ ft}} = (28.5 \text{ kips})(10.0 \text{ ft}) - \left(0.57 \frac{\text{kip}}{\text{ft}}\right)(10.0 \text{ ft})^2$$

$$= 228 \text{ ft-kips}$$

$$I_x = 2A_f\left(0.5(d - t_f)^2\right) + \frac{t_w(d - 2t_f)^3}{12}$$

$$= (2)\left((8.5 \text{ in})(0.625 \text{ in})\right)(0.5)(21.6 \text{ in} - 0.625 \text{ in})^2$$

$$+ \frac{(0.1875 \text{ in})\left(21.6 \text{ in} - (2)(0.625 \text{ in})\right)^3}{12}$$

$$= 1300 \text{ in}^4$$

$$S_x = \frac{2I_x}{d} = \frac{(2)(1300 \text{ in}^4)}{21.6 \text{ in}} = 120 \text{ in}^3$$

$$f_b = \frac{M_{x=10 \text{ ft}}}{S_x} = \frac{(228 \text{ ft-kips})\left(12 \frac{\text{in}}{\text{ft}}\right)}{120 \text{ in}^3}$$

$$= 22.8 \text{ ksi} \quad [> F_b = 22 \text{ ksi, so OK}]$$

Similarly, using $A_f = 8.5 \times 0.625 = 5.31 \text{ in}^2$, stresses at other locations are computed and tabulated as follows.

x (ft)	d (in)	M_x (ft-kips)	I_x (in^4)	S_x (in^3)	f_b (ksi)
10.0	21.6	228	1300	120	22.8
12.5	24.0	267	1640	136	23.6 (max)
15.0	26.4	299	2010	152	23.5

This confirms that the critical point for bending stress is near the quarter point (close enough for all practical purposes).

Critical bending stress occurs at $\boxed{12.5 \text{ ft (and 37.5 ft)}.}$

15.2. The actual bending stress exceeds that allowed by roughly 7%; therefore, increase the flange width proportionally.

$$b_f = \left(\frac{f_{b,\text{max}}}{F_b}\right) b_{f,\text{trial}}$$

$$= \left(\frac{23.6 \frac{\text{kips}}{\text{in}^2}}{22.0 \frac{\text{kips}}{\text{in}^2}}\right)(8.5 \text{ in})$$

$$= 9 \text{ in} \quad [\text{to nearest } 1/4 \text{ in}]$$

Check the stress at $x = 12.5 \text{ ft}$ using $b_f = 9 \text{ in}$.

$$I_x = 2A_f\left(0.5(d - t_f)^2\right) + \frac{t_w(d - 2t_f)^3}{12}$$

$$= (2)\left((9.0 \text{ in})(0.625 \text{ in})\right)(0.5)(24.0 \text{ in} - 0.625 \text{ in})^2$$

$$+ \frac{(0.1875 \text{ in})\left(24.0 \text{ in} - (2)(0.625 \text{ in})\right)^3}{12}$$

$$= 1720 \text{ in}^4$$

$$S_x = \frac{2I_x}{d} = \frac{(2)(1720 \text{ in}^4)}{24.0 \text{ in}} = 143.3 \text{ in}^3$$

$$f_b = \frac{M_{x=12.5 \text{ ft}}}{S_x} = \frac{(267 \text{ ft-kips})\left(12 \frac{\text{in}}{\text{ft}}\right)}{143.3 \text{ in}^3}$$

$$= 22.3 \text{ ksi} \quad [\approx F_b = 22 \text{ ksi}]$$

The minimum width for the $5/8$ in thick plate is $\boxed{9 \text{ in.}}$

15.3. Maximum deflection occurs at midspan, where the slope is zero. Various methods of linear elastic analysis can be used to compute the maximum deflection (virtual work, moment-area, etc.). Given that EI varies with d^3, the simplest approach appears to be the conjugate beam method, using a numerical scheme to approximate areas and moments of area under the M/EI-diagrams. Therefore, break the beam into, say, 10 segments with M/EI approximated as constant over each segment. Using symmetry, factor E out to simplify calculations. Approximate the component areas of the conjugate beam loading as the area of a trapezoid with a base $0.1L = 60 \text{ in}$.

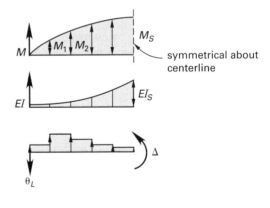

location (ft)	M (ft-kips)	I (in^4)	M/I (kips/in^3)	A_i (kips/in^2)	a_i (in)
0	0	383	0	–	–
1	128.3	795	1.94	58.2	270
2	228.0	1370	2.00	118.2	210
3	299.0	2120	1.69	110.7	150
4	342.0	3050	1.35	91.1	90
5	356.0	4180	1.02	84.9	30

$$A_i = 0.5\left(\left(\frac{M}{I}\right)_{i-1} + \left(\frac{M}{I}\right)_i\right)(0.1L)$$

$$A_1 = (0.5)\left(\frac{0 \text{ ft-kip}}{383 \text{ in}^4} + \frac{128.3 \text{ ft-kips}}{795 \text{ in}^4}\right)$$

$$\times (0.1)(50 \text{ ft})\left(12 \ \frac{\text{in}}{\text{ft}}\right)^2$$

$$= 58.2 \text{ ksi}$$

$$A_2 = (0.5)\left(\frac{128.3 \text{ ft-kips}}{795 \text{ in}^4} + \frac{228.0 \text{ ft-kips}}{1370 \text{ in}^4}\right)$$

$$\times (0.1)(50 \text{ ft})\left(12 \ \frac{\text{in}}{\text{ft}}\right)^2$$

$$= 118.2 \text{ ksi}$$

$$A_3 = (0.5)\left(\frac{228.0 \text{ ft-kips}}{1370 \text{ in}^4} + \frac{299.0 \text{ ft-kips}}{2120 \text{ in}^4}\right)$$

$$\times (0.1)(50 \text{ ft})\left(12 \ \frac{\text{in}}{\text{ft}}\right)^2$$

$$= 110.7 \text{ ksi}$$

$$A_4 = (0.5)\left(\frac{299.0 \text{ ft-kips}}{2120 \text{ in}^4} + \frac{342.0 \text{ ft-kips}}{3050 \text{ in}^4}\right)$$

$$\times (0.1)(50 \text{ ft})\left(12 \ \frac{\text{in}}{\text{ft}}\right)^2$$

$$= 91.1 \text{ ksi}$$

$$A_5 = (0.5)\left(\frac{342.0 \text{ ft-kips}}{3050 \text{ in}^4} + \frac{356.0 \text{ ft-kips}}{4180 \text{ in}^4}\right)$$

$$\times (0.1)(50 \text{ ft})\left(12 \ \frac{\text{in}}{\text{ft}}\right)^2$$

$$- 84.9 \text{ ksi}$$

$$\theta_L = \sum_{1=1}^{5} \frac{A_i}{E}$$

$$= \frac{\begin{array}{c}58.2 \ \frac{\text{kips}}{\text{in}^2} + 118.2 \ \frac{\text{kips}}{\text{in}^2} + 110.7 \ \frac{\text{kips}}{\text{in}^2} \\ + 91.1 \ \frac{\text{kips}}{\text{in}^2} + 84.9 \ \frac{\text{kips}}{\text{in}^2}\end{array}}{29{,}000 \ \frac{\text{kips}}{\text{in}^2}}$$

$$= 0.01596 \text{ rad}$$

$$\Delta_{\max} = 0.5\theta_L L - \sum_{i=1}^{5} \frac{A_i a_i}{E}$$

$$= (0.5)(0.01596 \text{ rad})\left((50 \text{ ft})\left(12 \ \frac{\text{in}}{\text{ft}}\right)\right)$$

$$\left(58.2 \ \frac{\text{kips}}{\text{in}^2}\right)(270 \text{ in}) + \left(118.2 \ \frac{\text{kips}}{\text{in}^2}\right)$$

$$\times (210 \text{ in}) + \left(110.7 \ \frac{\text{kips}}{\text{in}^2}\right)(150 \text{ in})$$

$$+ \left(91.1 \ \frac{\text{kips}}{\text{in}^2}\right)(90 \text{ in})$$

$$- \frac{+ \left(84.9 \ \frac{\text{kips}}{\text{in}^2}\right)(30 \text{ in})}{29{,}000 \ \frac{\text{kips}}{\text{in}^2}}$$

$$= \boxed{2.45 \text{ in} \quad [\text{downward}]}$$

SOLUTION 16

16.1. Select the lightest W10 section to support the gravity loads. Use LRFD. Per ASCE/SEI7 Sec. 4.6.3, apply an impact factor of 1.2 for light machinery that is driven by shaft or motor. For beam 1, estimate the factored beam weight as 0.02 kip/ft.

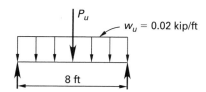

$$P_u = 1.6(\text{impact factor})0.5W$$

$$= (1.6)(1.2)(0.5)(0.3 \text{ kip})$$

$$= 0.3 \text{ kip}$$

$$M_u = 0.25P_u L + 0.125w_u L^2$$

$$= (0.25)(0.3 \text{ kip})(8 \text{ ft}) + (0.125)\left(0.02 \ \frac{\text{kip}}{\text{ft}}\right)(8 \text{ ft})^2$$

$$= 0.76 \text{ ft-kip}$$

$$L_b = 8 \text{ ft}, \ C_b = 1.0$$

The lightest W10 section of A992 steel is a W10 × 12, which has a capacity of $\phi M_n = 28$ ft-kips with $L_b = 8$ ft, more than adequate. $\boxed{\text{Use W10} \times 12 \text{ for all beams}}$ (lightly stressed in flexure).

16.2. Check the vibration potential with the machine operating at 1725 rev/min. (1 revolution = 1 cycle.)

$$\overline{f} = 1725 \ \frac{\text{rev}}{\text{min}} = \frac{1725 \ \dfrac{\text{rev}}{\text{min}}}{60 \ \dfrac{\text{sec}}{\text{min}}} = 29 \ \text{rev/sec}$$

Check the natural frequency of the beam-girder-motor system. Idealize the system as a single degree of freedom with its mass lumped at the load point. The weight of the beams is significant relative to the motor weight. Approximate the equivalent mass for vertical vibration as that of the machine plus two times beam 1, plus the central half of beam 2.

$$W = W_{\text{machine}} + 2w_b L + 2(0.5w_b L)$$

$$= 300 \ \text{lbf} + (2)\left(12 \ \frac{\text{lbf}}{\text{ft}}\right)(8 \ \text{ft})$$

$$+ (2)(0.5)\left(12 \ \frac{\text{lbf}}{\text{ft}}\right)(8 \ \text{ft})$$

$$= 588 \ \text{lbf}$$

$$m = \frac{W}{g} = \frac{588 \ \text{lbf}}{g\left(1000 \ \dfrac{\text{lbf}}{\text{kip}}\right)} = \frac{0.59 \ \text{kip}}{g}$$

To compute the equivalent stiffness, apply a concentrated force, k, at midspan (for the W10 × 12, $I_x = 53.8 \ \text{in}^4$).

$$\Delta = \Delta_b + \Delta_g$$

$$= \frac{kL_b^3}{48E(2I_b)} + \frac{0.5kL_g^3}{48EI_g}$$

$$= \frac{k\left((8 \ \text{ft})\left(12 \ \frac{\text{in}}{\text{ft}}\right)\right)^3}{(96)\left(29{,}000 \ \dfrac{\text{kips}}{\text{in}^2}\right)(53.8 \ \text{in}^4)}$$

$$+ \frac{0.5k\left((8 \ \text{ft})\left(12 \ \frac{\text{in}}{\text{ft}}\right)\right)^3}{(48)\left(29{,}000 \ \dfrac{\text{kips}}{\text{in}^2}\right)(53.8 \ \text{in}^4)}$$

$$= \left(0.01181 \ \frac{\text{in}}{\text{kip}}\right)k$$

Setting Δ equal to 1 in and solving for k gives

$$\Delta = \left(0.01181 \ \frac{\text{in}}{\text{kip}}\right)k = 1 \ \text{in}$$

$$k = 84.7 \ \text{kips/in}$$

For damping on the order of that in a steel frame (about 5% of critical damping), the natural frequency is practically unaffected by damping. Therefore,

$$f_1 = f$$

$$= \frac{1}{2\pi}\sqrt{\frac{k}{m}}$$

$$= \frac{1}{2\pi}\sqrt{\frac{\left(84.7 \ \dfrac{\text{kips}}{\text{in}}\right)\left(12 \ \dfrac{\text{in}}{\text{ft}}\right)}{\dfrac{0.59 \ \text{kip}}{32.2 \ \dfrac{\text{ft}}{\text{sec}^2}}}}$$

$$= 37.5 \ \text{Hz}$$

There is no lateral bracing in the system, so vibration can occur in the principal horizontal directions as well as vertically. In either direction, the stiffness is furnished by two W10 × 12 beams bending about their weak axis ($I_y = 2.18 \ \text{in}^4$).

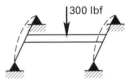

Calculate the natural frequency for a beam with a weight of w_b and a midspan weight of W vibrating laterally (see Young, Budynas, and Sadegh).

$$f = \frac{3.13}{\sqrt{\dfrac{(W + 0.486w_b L_b)L_b^3}{48EI}}}$$

$W = W_{\text{machine}} = 0.3 \ \text{kip}$, $w = 2w_{\text{girder}} = (2)(0.012 \ \text{kip/ft}) = 0.024 \ \text{kip/ft}$, $I = 2I_y = (2)(2.18 \ \text{in}^4) = 4.36 \ \text{in}^4$, and $L = 8 \ \text{ft} = 96 \ \text{in}$.

$$f = \frac{3.13 \ \dfrac{\text{rev-in}^{0.5}}{\text{sec}}}{\sqrt{\dfrac{\left(\begin{array}{c}0.3 \ \text{kip} + (0.486) \\ \times \left(0.024 \ \dfrac{\text{kip}}{\text{ft}}\right)(8 \ \text{ft})\end{array}\right)(96 \ \text{in})^3}{(48)\left(29{,}000 \ \dfrac{\text{kips}}{\text{in}^2}\right)(4.36 \ \text{in}^4)}}}$$

$$= 13.1 \ \text{rev/sec}$$

Thus, for vertical motion, $f = 37.5 \ \text{rev/sec}$; for horizontal motion, $f = 13.1 \ \text{rev/sec}$.

The dynamic amplification factor, defined as the ratio of deflection under dynamic loading to the deflection under the same force amplitude applied statically, is plotted as shown (see Paz and Leigh).

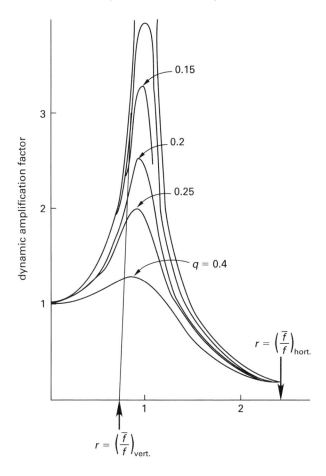

For the vertical vibration mode,

$$r = \frac{\bar{f}}{f} = \frac{29 \, \frac{\text{rev}}{\text{sec}}}{37.5 \, \frac{\text{rev}}{\text{sec}}}$$

$$= 0.8 \quad [\text{too close to unity, no good}]$$

For the horizontal vibration mode,

$$r = \frac{\bar{f}}{f} = \frac{29 \, \frac{\text{rev}}{\text{sec}}}{13.1 \, \frac{\text{rev}}{\text{sec}}} = 2.2 \quad [\text{OK}]$$

For this system, the dynamic magnification of vertical deflection will be large and likely objectionable. The practical solution is to use a lighter, more flexible beam that still has adequate strength and stiffness, but has a lower natural frequency of vibration.

16.3. The maximum moment under design loads is only 0.76 ft-kip, so a W6 × 9 has enough strength ($\phi M_n = 14$ ft-kips with $L_b = 8$ ft), but with $I_x = 16.4$ in⁴ (roughly one-third the stiffness of the W10 × 12) and with $I_y = 2.2$ in⁴ (practically the same as the W10 × 12). For this member, vibration in the vertical mode gives

$$\Delta = \Delta_{\text{beam}} + \Delta_{\text{girder}} = 1$$

$$= \frac{kL_b^3}{48E(2I_b)} + \frac{0.5kL_g^3}{48EI_g}$$

$$= \frac{k\left((8 \text{ ft})\left(12 \, \frac{\text{in}}{\text{ft}}\right)\right)^3}{(96)\left(29{,}000 \, \frac{\text{kips}}{\text{in}^2}\right)(16.4 \text{ in}^4)}$$

$$+ \frac{0.5k\left((8 \text{ ft})\left(12 \, \frac{\text{in}}{\text{ft}}\right)\right)^3}{(48)\left(29{,}000 \, \frac{\text{kips}}{\text{in}^2}\right)(16.4 \text{ in}^4)}$$

$$= \left(0.03845 \, \frac{\text{in}}{\text{kip}}\right)k = 1 \text{ in}$$

$$k = 26.0 \text{ kips/in}$$

$$W = W_{\text{machine}} + 2w_{\text{beam}}L + 2(0.5w_{\text{beam}}L)$$

$$= 300 \text{ lbf} + (2)\left(9 \, \frac{\text{lbf}}{\text{ft}}\right)(8 \text{ ft}) + (2)(0.5)\left(9 \, \frac{\text{lbf}}{\text{ft}}\right)(8 \text{ ft})$$

$$= 516 \text{ lbf}$$

$$m = \frac{W}{g} = \frac{516 \text{ lbf}}{g\left(1000 \, \frac{\text{lbf}}{\text{kip}}\right)} = \frac{0.52 \text{ kip}}{g}$$

$$f_1 = f$$

$$= \frac{1}{2\pi}\sqrt{\frac{k}{m}}$$

$$= \frac{1}{2\pi}\sqrt{\frac{\left(26.0 \, \frac{\text{kips}}{\text{in}}\right)\left(32.2 \, \frac{\text{ft}}{\text{sec}^2}\right)\left(12 \, \frac{\text{in}}{\text{ft}}\right)}{0.52 \text{ kip}}}$$

$$= 22 \text{ Hz}$$

$$r = \frac{\bar{f}}{f} = \frac{29}{22} = 1.32$$

The dynamic amplification factor plot graph shows that for this condition, the dynamic amplification is at a reasonable level.

SOLUTION 17

Design the steel-to-concrete column connection using LRFD.

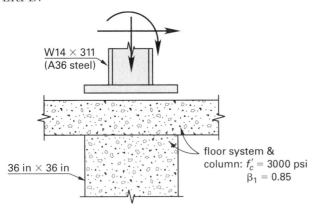

W14 × 311
(A36 steel)

36 in × 36 in

floor system & column: f'_c = 3000 psi
β_1 = 0.85

Consider load combinations 2, 5, and 7 from Sec. 12.4 and Sec. 2.3 of ASCE/SEI7.

$$P_{u,2} = 1.2P_D + 1.6P_L$$
$$= (1.2)(563 \text{ kips}) + (1.6)(400 \text{ kips})$$
$$= 1316 \text{ kips} \quad \text{[controls]}$$

$$M_{u,2} = 1.2M_D + 1.6M_L$$
$$= (1.2)(0 \text{ ft-kips}) + (1.6)(0 \text{ ft-kips})$$
$$= 0 \text{ ft-kips}$$

$$P_{u,5} = 1.2P_D + 0.5P_L + P_E + 0.2S_{DS}P_D$$
$$= (1.2)(563 \text{ kips}) + (0.5)(400 \text{ kips}) + 0 \text{ kips}$$
$$\quad + (0.2)(0.5)(563 \text{ kips})$$
$$= 932 \text{ kips}$$

$$M_{u,5} = 1.2M_D + 0.5M_L + M_E + 0.2S_SM_D$$
$$= (1.2)(0 \text{ ft-kips}) + (1.6)(0 \text{ ft-kips}) + 660 \text{ ft-kips}$$
$$\quad + (0.2)(0.5)(0 \text{ ft-kips})$$
$$= 660 \text{ ft-kips} \quad \text{[controls]}$$

$$P_{u,7} = 0.9P_D + P_E - 0.2S_{DS}P_D$$
$$= (0.9)(563 \text{ kips}) + 0 \text{ kips} - (0.2)(0.5)(563 \text{ kips})$$
$$= 450 \text{ kips}$$

$$M_{u,7} = 0.9M_D + M_E - 0.2S_SM_D$$
$$= (0.9)(0 \text{ ft-kips}) + 660 \text{ ft-kips}$$
$$\quad + (0.2)(0.5)(0 \text{ ft-kips})$$
$$= 660 \text{ ft-kips} \quad \text{[controls]}$$

Load combinations 2 and 7 are the controlling conditions. Check bearing on the concrete for loading combination 2 (uniform axial compression). Per AISC Sec. J8,

for bearing on less than the full area of concrete, taking $\sqrt{A_2/A_1}$ as its limiting value of 2.0,

$$\phi P_p = \phi 0.85 f'_c A_1 \sqrt{\frac{A_2}{A_1}}$$
$$= (0.65)(0.85)\left(3 \text{ } \frac{\text{kips}}{\text{in}^2}\right) A_1 (2.0)$$
$$= \left(3.06 \text{ } \frac{\text{kips}}{\text{in}^2}\right) A_1 \geq P_{u,2} = 1316 \text{ kips}$$
$$A_1 \geq 397 \text{ in}^2$$

For load combination 7, the moment is best resisted by a plate with a length as large as practicable. The anchor bolts will extend through the floor and embed within the confined core of the concrete column below. To maintain clearance within the column core, the maximum distance from column centerline to anchor bolt is

$$a = \frac{h}{2} - \text{cover} - d_s - d_l - d_a$$
$$\approx \frac{36 \text{ in}}{2} - 1.5 \text{ in} - 0.5 \text{ in} - \frac{14 \text{ in}}{8} - 1.5 \text{ in}$$
$$\approx 12 \text{ in}$$

d_s is the tie bar diameter, d_l is longitudinal bar diameter, and d_a is the anchor bolt diameter. Therefore, try a 20 in × 28 in base plate with the longer dimension resisting the moment.

$$A_1 = BN = (20 \text{ in})(28 \text{ in}) = 560 \text{ in}^2$$

This is greater than 430 in², the area required for uniform axial compression.

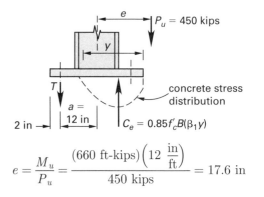

$$e = \frac{M_u}{P_u} = \frac{(660 \text{ ft-kips})\left(12 \text{ } \frac{\text{in}}{\text{ft}}\right)}{450 \text{ kips}} = 17.6 \text{ in}$$

y is the distance from the compression edge of base plate to the neutral axis under the design loading.

$$\sum F_{yu} = T + P - C_c = 0$$
$$= T + P_u - 0.85f'_c b(0.85y)$$

$$T = 0.85 f'_c b(0.85y) - P_u$$

$$= (0.85)\left(3 \ \frac{\text{kips}}{\text{in}^2}\right)(20 \ \text{in})(0.85y) - 450 \ \text{kips}$$

$$= \left(43.35 \ \frac{\text{kips}}{\text{in}}\right)y - 450 \ \text{kips}$$

Take moments about the line of action of P_u.

$$\sum M = T(a + e) - C_c(e - 0.5h + 0.5\beta_1 y) = 0$$

$$= \left(\left(43.35 \ \frac{\text{kips}}{\text{in}}\right)y - 450 \ \text{kips}\right)$$

$$\times (12 \ \text{in} + 17.6 \ \text{in}) - \left(\left(43.35 \ \frac{\text{kips}}{\text{in}}\right)y\right)$$

$$\times \left(17.6 \ \text{in} - 14 \ \text{in} - (0.5)(0.85)y\right)$$

$$y = 16.0 \ \text{in}$$

Therefore,

$$T = \left(43.35 \ \frac{\text{kips}}{\text{in}}\right)y - 450 \ \text{kips}$$

$$= \left(43.35 \ \frac{\text{kips}}{\text{in}}\right)(16.0 \ \text{in}) - 450 \ \text{kips}$$

$$= 244 \ \text{kips}$$

Try four anchor bolts on each side (eight total). Assume all bolts resist the seismic shear.

$$V_u = \frac{V_{ex}}{n} = \frac{110 \ \text{kips}}{8 \ \text{rods}} = 13.8 \ \text{kips/rod}$$

Try a 1.5 in diameter ASTM F1554 grade 55 anchor rod. From AISC Table 2-6, $F_u = 75$ ksi.

$$A_b = 0.25\pi d_b^2 = 0.25\pi(1.5 \ \text{in})^2 = 1.77 \ \text{in}^2$$

Per AISC Table J3.2, assume that the threads are included in the shear plane.

$$F_{nt} = 0.75 F_u = (0.75)\left(75 \ \frac{\text{kips}}{\text{in}^2}\right) = 56.3 \ \text{ksi}$$

$$F_{nv} = 0.45 F_u = (0.45)\left(75 \ \frac{\text{kips}}{\text{in}^2}\right) = 33.8 \ \text{ksi}$$

$$f_v = \frac{V}{A_b} = \frac{13.8 \ \text{kips}}{1.77 \ \text{in}^2} = 7.8 \ \text{ksi}$$

For combined shear and tension (per AISC Sec. J3.7),

$$F'_{nt} = 1.3 F_{nt} - \left(\frac{F_{nt}}{\phi F_{nv}}\right)f_v \le F_{nt}$$

$$= (1.3)\left(56.3 \ \frac{\text{kips}}{\text{in}^2}\right) - \left(\frac{56.3 \ \frac{\text{kips}}{\text{in}^2}}{(0.75)\left(33.8 \ \frac{\text{kips}}{\text{in}^2}\right)}\right)$$

$$\times \left(7.8 \ \frac{\text{kips}}{\text{in}^2}\right)$$

$$= 55.9 \ \text{ksi} \quad [< F_{nt} = 56.3 \ \text{ksi}]$$

$$\phi R_n = \phi F'_{nt} A_b$$

$$= (0.75)\left(55.9 \ \frac{\text{kips}}{\text{in}^2}\right)\left(1.77 \ \frac{\text{in}^2}{\text{bolt}}\right)$$

$$= 74.2 \ \text{kips/bolt}$$

$$n \ge \frac{T}{\phi R_n} = \frac{244 \ \text{kips}}{74.2 \ \dfrac{\text{kips}}{\text{bolt}}}$$

$$= 3.3 \ \text{bolts} \quad \left[\begin{array}{c}\text{use four 1.5 in diameter F1554 grade 55}\\ \text{anchor bolts each side (8 total)}\end{array}\right]$$

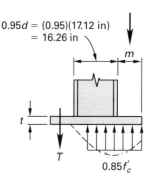

Check bearing pressure for load combination 2.

$$f_p = \frac{P_u}{A} = \frac{P_u}{BN} = \frac{1316 \ \text{kips}}{(20 \ \text{in})(28 \ \text{in})}$$

$$= 2.35 \ \text{ksi} \quad [< 0.85 f'_c, \ \text{so does not control}]$$

Therefore, plate bending stresses control. Determine the required plate thickness (for A36 steel) based on the compression side bending moment. The critical section occurs at a distance, m, from the longer edge of the base plate.

$$m = 0.5(h - 0.95d)$$

$$= (0.5)\big(28 \ \text{in} - (0.95)(17.12 \ \text{in})\big)$$

$$= 5.87 \ \text{in}$$

$$m_u = (0.5)(0.85 f'_c m^2)\left(1 \frac{\text{in}}{\text{in}}\right)$$

$$= 0.5(0.85)\left(3 \frac{\text{kips}}{\text{in}^2}\right)(5.87 \text{ in})^2\left(1 \frac{\text{in}}{\text{in}}\right)$$

$$= 44 \text{ in-kips/in}$$

On the tension side,

$$m_u = \frac{T\left(a - \frac{0.95d}{2}\right)}{B}$$

$$= \frac{(224 \text{ kips})\left(12 \text{ in} - \frac{(0.95)(17.1 \text{ in})}{2}\right)}{20 \text{ in}}$$

$$= 47.2 \text{ in-kips/in}$$

Therefore, the tension side governs.

$$\phi M_p = \frac{\phi F_y t^2}{4} = m_u$$

$$t = \sqrt{\frac{4 m_u}{\phi F_y}} = \sqrt{\frac{(4)(47.2 \text{ in-kips})}{(0.9)\left(36 \frac{\text{kips}}{\text{in}^2}\right)}} = 2.42 \text{ in}$$

Use PL$2^{1}/_2 \times 20 \times$ 2 ft 4 in.

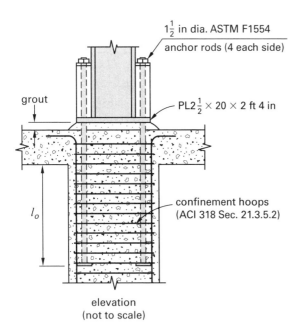

1$\frac{1}{2}$ in dia. ASTM F1554
anchor rods (4 each side)

grout

PL2$\frac{1}{2}$ × 20 × 2 ft 4 in

l_o

confinement hoops
(ACI 318 Sec. 21.3.5.2)

elevation
(not to scale)

Assume all shear transfers through the web. Per AISC Table J2.4, use $^5/_{16}$ in minimum size fillet welds. For E70XX electrodes, the strength is 1.39 kips/in per $^1/_{16}$ in of weld size. (See AISC Part 8.)

$$q \le D\phi F_w = (5)\left(1.39 \frac{\text{kips}}{\text{in}}\right) = 6.95 \text{ kips/in}$$

$$L_w \ge \frac{V_u}{q} = \frac{110 \text{ kips}}{6.95 \frac{\text{kips}}{\text{in}}}$$

$$= 15.8 \text{ in} \quad [< 2T = (2)(10.0 \text{ in}) = 20 \text{ in}]$$

Use $^5/_{16}$ in fillet welds each side of web. For the flange connection, mill the contact surfaces so that compression stress is not critical. Follow FEMA 267A recommendation and avoid partial penetration weld of flange-to-plate connection for seismic loading. Transfer the tension force, 244 kips, through side plates welded to the column flange. Assuming the three plates are fully effective,

$$V_u = \frac{T}{n} = \frac{244 \text{ kips}}{3 \text{ side plates}} = 81.3 \text{ kips/side plate}$$

$$M_u = V_u(a - 0.5d)$$

$$= (81.3 \text{ kips})\big(12 \text{ in} - (0.5)(17.12 \text{ in})\big)$$

$$= 280 \text{ in-kips}$$

Use plates thick enough to develop the strength of $^5/_{16}$ in fillet welds to both sides. From AISC Sec. J4.2, for A36 steel,

$$\phi f_{\text{BM}} \le \begin{cases} \phi(0.6F_y) = (1.0)(0.6)\left(36 \frac{\text{kips}}{\text{in}^2}\right) \\ \qquad = 21.6 \text{ ksi} \quad [\text{controls}] \\ \phi(0.6F_u) = (0.75)(0.6)\left(58 \frac{\text{kips}}{\text{in}^2}\right) \\ \qquad = 26.1 \text{ ksi} \end{cases}$$

$$t \ge \frac{2q}{\phi f_{\text{BM}}} = \frac{(2)\left(6.95 \frac{\text{kips}}{\text{in}}\right)}{21.6 \frac{\text{kips}}{\text{in}^2}}$$

$$= 0.64 \text{ in} \quad [\text{say } ^3/_4 \text{ in plate}]$$

Determine the length required to resist the combined shear and bending using the elastic method.

$$L_w \ge \frac{V_u}{2q} = \frac{81.3 \text{ kips}}{(2)\left(6.95 \frac{\text{kips}}{\text{in}}\right)}$$

$$\ge 5.8 \text{ in} \quad [\text{try 12 in long plates}]$$

$$f_v = \frac{V_u}{2L_w} = \frac{81.3 \text{ kips}}{(2)(12 \text{ in})} = 3.4 \text{ kips/in}$$

$$f_b = \frac{M_u}{S_w} = \frac{V_u e}{2\left(\dfrac{L_w^2}{6}\right)} = \frac{(81.3 \text{ kips})(2 \text{ in})}{(2)\left(\dfrac{(12 \text{ in})^2}{6}\right)}$$

$$= 3.39 \text{ kips/in}$$

$$f_r = \sqrt{f_v^2 + f_b^2}$$

$$= \sqrt{\left(3.4 \ \frac{\text{kips}}{\text{in}}\right)^2 + \left(3.39 \ \frac{\text{kips}}{\text{in}}\right)^2}$$

$$= 4.80 \text{ kips/in} \quad [\approx q, \text{ so OK}]$$

For a minimum field weld of $^1/_4$ in, the weld capacity is

$$q \le D\phi F_w = (4)\left(1.39 \ \frac{\text{kips}}{\text{in}}\right)$$

$$= 5.56 \text{ kips/in} \quad [\text{OK}]$$

> Use three $\text{PL}\frac{3}{4} \times 5 \times 1$ ft 0 in plates with the same size bearing plate, and $\frac{5}{16}$ in fillet welds.

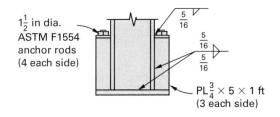

1$\frac{1}{2}$ in dia.
ASTM F1554
anchor rods
(4 each side)

$\text{PL}\frac{3}{4} \times 5 \times 1$ ft
(3 each side)

SOLUTION 18

18.1. Verify the tube size shown for the structure. Use ASD. For the given wind load,

$$P = p_W s = \left(\frac{32 \ \dfrac{\text{lbf}}{\text{ft}}}{1000 \ \dfrac{\text{lbf}}{\text{kip}}}\right)(10 \text{ ft}) = 0.32 \text{ kip}$$

Analyze the rigid frame for lateral load using the portal method.

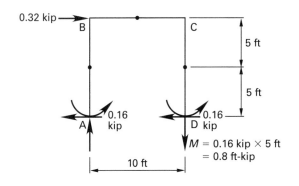

The columns resist an equal share of the lateral force, 0.32 kip (0.16 kip each). Assuming that the inflection point is at midheight, the moment at the base of the column is

$$M_w = V(0.5h) = (0.16 \text{ kip})(0.5)(10 \text{ ft}) = 0.8 \text{ ft-kip}$$

$$\sum M_A = Ph - P_W L - 2M = 0$$

$$P_W = \frac{Ph - 2M}{L}$$

$$= \frac{(0.32 \text{ kip})(10 \text{ ft}) - (2)(0.8 \text{ ft-kip})}{10 \text{ ft}}$$

$$= 0.16 \text{ kip}$$

Use the moment distribution method to analyze the gravity live plus dead loads (see Leet, Uang, and Gilbert).

$$w = (w_D + w_L)s = \frac{\left(14 \ \dfrac{\text{lbf}}{\text{ft}^2} + 20 \ \dfrac{\text{lbf}}{\text{ft}^2}\right)(10 \text{ ft})}{1000 \ \dfrac{\text{lbf}}{\text{kip}}}$$

$$= 0.34 \text{ kip/ft}$$

$$\text{FEM} = \frac{\pm wL^2}{12} = \frac{\pm\left(0.34 \ \dfrac{\text{kip}}{\text{ft}}\right)(10 \text{ ft})^2}{12}$$

$$= \pm 2.83 \text{ ft-kips}$$

Starting with the fixed end moments (FEMs) at joints B and C, an initial distribution at B shows that moments of +1.42 ft-kips at BA and BC are required to balance the joint (indicated by underlined values in the table). From the distributed moment at BC, one-half carries over to joint C, which results in an unbalanced moment of $(2.83 + 0.71)$ ft-kips at C. Joint C is then balanced by moments -1.77 ft-kips at CB and CD; with one half of 1.77 ft-kips carried over to joint B. Successive balancing and carrying over is performed between joints B and C until the final balanced moments are insignificant (-0.01 ft-kip). Summation of all moments (FEMs, distribution, and carryover moments) gives the final moments of 1.89 ft-kips (positive when clockwise on a joint).

joint	BA	BC	CB	CD
DF	0.5	0.5	0.5	0.5
FEM:	0	−2.83	2.83	0
Bal.B,CO	1.42	1.42	0.71	
Bal.C,CO		−0.88	−1.77	−1.77
Bal.B,CO	0.44	0.44	0.22	
Bal.C,CO		−0.06	−0.11	−0.11
Bal.B,CO	0.03	0.03	0.02	
Bal.C,Stop			−0.01	−0.01
$\Sigma =$	1.89	−1.89	1.89	−1.89

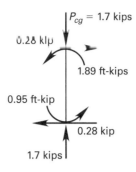

P_{cg} = 1.7 kips

0.28 klp

1.89 ft-kips

0.95 ft-kip

0.28 kip

1.7 kips

Check loading combinations 2, 5, and 6 from ASCE/SEI7 Sec. 2.4.1. The relative dead and wind load effects are such that combination 7 is not critical and need not be included in the combinations. For loading combination 2,

$$P_2 = P = 1.7 \text{ kips} \quad \text{[controls]}$$

$$M_2 = 1.89 \text{ ft-kips}$$

$$P_5 = \left(\frac{w_D}{w}\right)P + 0.6P_W$$

$$= \left(\frac{14 \frac{\text{lbf}}{\text{ft}^2}}{34 \frac{\text{lbf}}{\text{ft}^2}}\right)(1.7 \text{ kips}) + (0.6)(0.16 \text{ kip})$$

$$= 0.8 \text{ kip}$$

For loading combination 5,

$$M_5 = \left(\frac{w_D}{w}\right)M_2 + 0.6M_w$$

$$= \left(\frac{14 \frac{\text{lbf}}{\text{ft}^2}}{34 \frac{\text{lbf}}{\text{ft}^2}}\right)(1.89 \text{ ft-kips}) + (0.6)(0.8 \text{ ft-kip})$$

$$= 1.26 \text{ ft-kips}$$

For loading combination 6,

$$P_6 = \left(\frac{w_D}{w}\right)P + 0.75\left(\frac{w_L}{w}\right)P + (0.75)(0.6)P_W$$

$$= \left(\frac{14 \frac{\text{lbf}}{\text{ft}^2}}{34 \frac{\text{lbf}}{\text{ft}^2}}\right)(1.7 \text{ kips})$$

$$+ (0.75)\left(\frac{20 \frac{\text{lbf}}{\text{ft}^2}}{34 \frac{\text{lbf}}{\text{ft}^2}}\right)(1.7 \text{ kips})$$

$$+ (0.75)(0.6)(0.16 \text{ kip})$$

$$= 1.52 \text{ kips}$$

$$M_6 = \left(\frac{w_D}{w}\right)M_2 + 0.75\left(\frac{w_L}{w}\right)M_2 + (0.75)(0.6)M_w$$

$$= \left(\frac{14 \frac{\text{lbf}}{\text{ft}^2}}{34 \frac{\text{lbf}}{\text{ft}^2}}\right)(1.89 \text{ ft-kips})$$

$$+ (0.75)\left(\frac{20 \frac{\text{lbf}}{\text{ft}^2}}{34 \frac{\text{lbf}}{\text{ft}^2}}\right)(1.89 \text{ ft-kips})$$

$$+ (0.75)(0.6)(0.8 \text{ ft-kip})$$

$$= 2.0 \text{ ft-kips}$$

Loading combination 2 controls. Check the HSS4 × 3 × 0.250. Section properties are as follows: $A_g = 2.91 \text{ in}^2$, $b/t = 9.88$, $h/t = 14.2$, $I_x = 6.15 \text{ in}^4$, $I_y = 3.91 \text{ in}^4$, $r_x = 1.45 \text{ in}$, $r_y = 1.16 \text{ in}$, $Z_x = 3.12 \text{ in}^3$, and $F_y = 46 \text{ ksi}$. For the x-axis, using AISC Fig. C-A-7.2, $G_A = 1.0$ because the base is fixed.

$$G_B = \frac{\left(\frac{I}{L}\right)_c}{\left(\frac{I}{L}\right)_g} = \frac{\left(\frac{3.91 \text{ in}^4}{10 \text{ ft}}\right)_c}{\left(\frac{3.91 \text{ in}^4}{10 \text{ ft}}\right)_g} = 1.0$$

From AISC Fig. C-A-7.2, $K_x = 1.3$. The y-axis is braced out of plane; therefore, $K_y = 1.0$.

$$\frac{KL_x}{r_x} = \frac{(1.3)\left((10 \text{ ft})\left(12 \frac{\text{in}}{\text{ft}}\right)\right)}{1.45 \text{ in}} = 108 \quad \text{[controls]}$$

$$\frac{KL_y}{r_y} = \frac{(1.0)\left((10 \text{ ft})\left(12 \frac{\text{in}}{\text{ft}}\right)\right)}{1.16 \text{ in}} = 103$$

$$\sqrt{\frac{E}{F_y}} = \sqrt{\frac{29{,}000 \frac{\text{kips}}{\text{in}^2}}{46 \frac{\text{kips}}{\text{in}^2}}} = 25.1$$

Check the compactness per AISC Table B4.1.

$$\frac{h}{t} = 14.2 < 2.42\sqrt{\frac{E}{F_y}} = (2.42)(25.1) = 60.7$$

$$\frac{b}{t} = 9.88 < 1.12\sqrt{\frac{E}{F_y}} = (1.12)(25.1)$$

$$= 28.1 \quad \text{[section is compact]}$$

$$\frac{KL}{r} = \frac{KL_x}{r_x} = 108 < 4.71\sqrt{\frac{E}{F_y}} = (4.71)(25.1) = 118$$

$$F_e = \frac{\pi^2 E}{\left(\frac{KL}{r}\right)^2}$$

$$= \frac{\pi^2 \left(29{,}000 \ \frac{\text{kips}}{\text{in}^2}\right)}{(108)^2}$$

$$= 24.5 \ \text{ksi}$$

$$F_{\text{cr}} = 0.658^{F_y/F_e} F_y$$

$$= \left(0.658^{46 \frac{\text{kips}}{\text{in}^2} / 24.5 \frac{\text{kips}}{\text{in}^2}}\right) \left(46 \ \frac{\text{kips}}{\text{in}^2}\right)$$

$$= 21.0 \ \text{ksi}$$

$$P_c = \frac{F_{\text{cr}} A_g}{\Omega_c}$$

$$= \frac{\left(21.0 \ \frac{\text{kips}}{\text{in}^2}\right)(2.91 \ \text{in}^2)}{1.67}$$

$$= 36.6 \ \text{kips}$$

Calculate the amplified moment for the unbraced frame per AISC App. 8.

For loading combination 2, the moments associated with lateral translation are zero. However, AISC App. 7.2.2 requires that for moments caused by a notional load, N_i, AISC Eq. C2-1 must be considered.

$$N_i = 0.002\alpha Y_i$$

$$= (0.002)(1.6)\left(0.34 \ \frac{\text{kips}}{\text{ft}}\right)(10 \ \text{ft})$$

$$= 0.011 \ \text{kips}$$

This force is insignificant and can be neglected. Calculating the amplification factor using AISC Eq. A-8-3,

$$C_m = 0.6 - 0.4\left(\frac{M_1}{M_2}\right)$$

$$= 0.6 - (0.4)\left(\frac{0.95 \ \text{ft-kips}}{1.89 \ \text{ft-kips}}\right)$$

$$= 0.4$$

$$P_{e1} = \frac{\pi^2 EI^*}{(K_1 L)^2}$$

$$= \frac{\pi^2 \left(29{,}000 \ \frac{\text{kips}}{\text{in}^2}\right)(6.15 \ \text{in}^4)}{\left((1.3)(120 \ \text{in})\right)^2}$$

$$= 72.3 \ \text{kips}$$

$$B_1 = \frac{C_m}{1 - \dfrac{\alpha P_r}{P_{e1}}} \geq 1.0$$

$$= \frac{0.4}{1 - \dfrac{(1.6)(1.7 \ \text{kips})}{72.3 \ \text{kips}}}$$

$$= 0.42 \quad [\text{use } 1.0]$$

$$M_r = B_1 M = (1.0)(1.89 \ \text{ft-kips})$$

$$= 1.89 \ \text{ft-kips}$$

Calculate the flexural strength per AISC Sec. F7.

$$M_n = M_p = F_y Z_x$$

$$= \frac{\left(46 \ \frac{\text{kips}}{\text{in}^2}\right)(3.12 \ \text{in}^3)}{12 \ \frac{\text{in}}{\text{ft}}}$$

$$= 12.0 \ \text{ft-kips}$$

$$M_{cx} = \frac{M_n}{\Omega} = \frac{12.0 \ \text{ft-kips}}{1.67} = 7.2 \ \text{ft-kips}$$

Combine axial compression and bending per AISC Sec. H1.

$$\frac{P_r}{P_c} = \frac{1.7 \ \text{kips}}{36.6 \ \text{kips}}$$

$$= 0.05 \quad [< 0.2, \text{ therefore, use AISC Eq. H1-1b}]$$

$$\frac{P_r}{2P_c} + \frac{M_{rx}}{M_{cx}} = \frac{0.05}{2} + \frac{1.89 \ \text{ft-kips}}{7.2 \ \text{ft-kips}}$$

$$= 0.29 \quad [< 1.0, \text{ so OK}]$$

Therefore, the $\boxed{\text{HSS4} \times 3 \times 0.250 \text{ is OK.}}$

18.2. Detail of the beam-to-column connection.

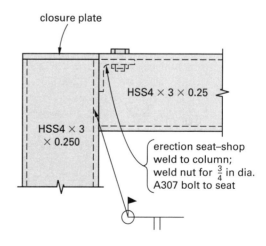

connection detail view
(not to scale)

SOLUTION 19

19.1. Analyze the modified frame for dead load and lateral forces. The W10 × 49 columns are oriented to bend about their weak axes. For dead load analysis, the column shear is practically zero under the symmetrical loading, because of the flexibility of the W10 × 49 about its weak axis.

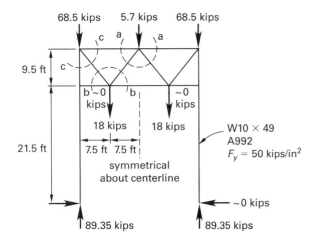

The free-body diagram for the cut a-a is

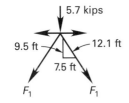

$$\sum F_y = -2F_1 \cos \alpha - 5.7 \text{ kips} = 0$$

$$F_1 = \frac{-5.7 \text{ kips}}{2 \cos \alpha}$$

$$= \frac{-5.7 \text{ kips}}{(2)\left(\dfrac{9.5 \text{ ft}}{12.1 \text{ ft}}\right)}$$

$$= -3.63 \text{ kips} \quad [\text{compression}]$$

For the cut b-b,

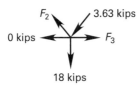

$$\sum F_y = -18 \text{ kips} - (3.63 \text{ kips})\left(\frac{9.5 \text{ ft}}{12.1 \text{ ft}}\right) + F_2\left(\frac{9.5 \text{ ft}}{12.1 \text{ ft}}\right)$$

$$= 0$$

$$F_2 = 26.6 \text{ kips} \quad [\text{tension}]$$

$$\sum F_x = -\left(\frac{7.5 \text{ ft}}{12.1 \text{ ft}}\right)(26.6 \text{ kips})$$

$$\qquad -\left(\frac{7.5 \text{ ft}}{12.1 \text{ ft}}\right)(3.63 \text{ kips}) + F_3 = 0$$

$$F_3 = 18.7 \text{ kips} \quad [\text{tension}]$$

For the cut c-c,

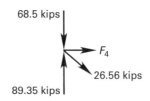

$$\sum F_x = F_4 + \left(\frac{7.5 \text{ ft}}{12.1 \text{ ft}}\right)(26.6 \text{ kips}) = 0$$

$$F_4 = -16.5 \text{ kips} \quad [\text{compression}]$$

For analysis of the lateral wind force, the distribution of lateral force is distributed equally with each column receiving $0.5V = (0.5)(64.0 \text{ kips}) = 32.0$ kips. Summing moments about the left support gives

$$\sum M_{lt} = Vh + P_{rt}L = 0$$

$$= (64.0 \text{ kips})(31 \text{ ft}) + P_{rt}(30 \text{ ft})$$

$$P_{rt} = -66.1 \text{ kips} \quad [66.1 \text{ kips downward}]$$

$$\sum F_y = P_{lt} - 66.1 \text{ kips} = 0$$

$$P_{lt} = 66.1 \text{ kips} \quad [\text{upward}]$$

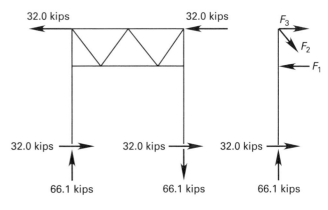

From a free-body diagram cut adjacent to the left column,

$$\sum M_{y=31 \text{ ft}} = (32.0 \text{ kips})(31 \text{ ft}) - (9.5 \text{ ft})F_1 = 0$$

$$F_1 = 104.4 \text{ kips} \quad [\text{compression}]$$

$$\sum F_y = -\left(\frac{9.5 \text{ ft}}{12.1 \text{ ft}}\right)F_2 + 66.1 \text{ kips} = 0$$

$$F_2 = 84.2 \text{ kips} \quad [\text{tension}]$$

$$\sum F_x = 32.0 \text{ kips} - 104.4 \text{ kips} + \left(\frac{7.5 \text{ ft}}{12.1 \text{ ft}}\right)$$
$$\times (84.2 \text{ kips}) + F_3 - 32.0 \text{ kips} = 0$$

$$F_3 = 52.2 \text{ kips} \quad \text{[tension]}$$

Summarizing,

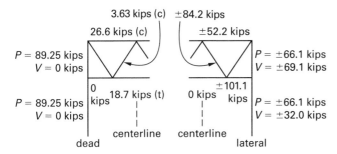

The lateral force is reversible; therefore, the axial force in each member due to lateral force can be either tension or compression.

19.2. Reinforce the existing column as required using LRFD. The critical load combination is axial compression plus bending.

AISC App. 7.2.2 requires that a notional load, N_i, be applied in the direction of the greatest destabilizing effect. From AISC Eq. C2-1,

$$N_i = 0.002\alpha Y_i = (0.002\alpha)(2)(1.2P_D)$$
$$= (0.002)(1.0)(2)(1.2)(89.35 \text{ kips})$$
$$= 0.43 \text{ kip}$$

This force is applied laterally and produces an axial force and a moment that are added to the wind effect.

$$P_{\text{not}} = \frac{Y_i}{V}P_W = \left(\frac{0.43 \text{ kip}}{64 \text{ kips}}\right)(62 \text{ kips}) = 0.42 \text{ kip}$$

$$M_{\text{not}} = \frac{Y_i}{2}h' = \left(\frac{0.43 \text{ kip}}{2}\right)(21.5 \text{ ft}) = 4.6 \text{ ft-kips}$$

$$M_W = Vh' = (32 \text{ kips})(21.5 \text{ ft})$$
$$= 688 \text{ ft-kips}$$

$$P_{nt} = 1.2P_D = (1.2)(89.35 \text{ kips})$$
$$= 107 \text{ kips}$$

$$P_{lt} = P_W + P_{\text{not}}$$
$$= 66.1 \text{ kips} + 0.42 \text{ kip}$$
$$= 66.5 \text{ kips}$$

$$M_{nt} = 0$$

$$M_{lt} = 1.2M_D + M_W + M_{\text{not}}$$
$$= (1.2)(0 \text{ ft-kips}) + 688 \text{ ft-kips} + 4.6 \text{ ft-kips}$$
$$= 693 \text{ ft-kips}$$

Check bending about the weak axis at elevation 21.5 ft. (W10 × 49 section properties: $Z_y = 28.3$ in^3, $S_x = 18.7$ in^3.)

$$M_{py} \leq \begin{cases} F_y Z_y = \dfrac{\left(50 \text{ }\dfrac{\text{kips}}{\text{in}^2}\right)(28.3 \text{ in}^3)}{12 \text{ }\dfrac{\text{in}}{\text{ft}}} \\ \qquad = 117 \text{ ft-kips} \quad \text{[controls]} \\ \\ 1.6F_yS_y = \dfrac{(1.6)\left(50 \text{ }\dfrac{\text{kips}}{\text{in}^2}\right)(18.7 \text{ in}^3)}{12 \text{ }\dfrac{\text{in}}{\text{ft}}} \\ \qquad = 125 \text{ ft-kips} \end{cases}$$

The modified frame requires a substantial increase in bending resistance at elevation 21.5 ft. The additional strength needed is

$$\phi M_n = \phi(M_{py,\text{W10}} + \Delta M_p) = 693 \text{ ft-kips}$$

$$\phi \Delta M_p = \phi M_n - \phi M_{py,\text{W10}}$$
$$= 693 \text{ ft-kips} - (0.9)(117 \text{ ft-kips})$$
$$= 588 \text{ ft-kips}$$

Assume that the compression flange can be laterally braced at $y = 21.5$ ft, then $L_b = 21.5$ ft and the equivalent of a W27 × 94 bending about its strong axis is needed to furnish this strength. Therefore, try two WT10.5 × 50.5 members welded to the web of the W10 × 49.

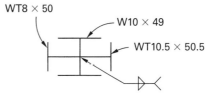

For the properties of the built-up section, use data for a W21 × 101 as being approximately and conservatively equivalent to the two WT10.5 × 50.5 members.

$$A_g = A_{g,\text{W10}} + A_{g,\text{W21}} = 14.4 \text{ in}^2 + 29.8 \text{ in}^2$$
$$= 44.2 \text{ in}^2$$

$$I_x = I_{x,\text{W21}} + I_{y,\text{W10}} = 2420 \text{ in}^4 + 93.4 \text{ in}^4$$
$$= 2513 \text{ in}^4$$

$$I_y = I_{y,\text{W21}} + I_{x,\text{W10}} = 248 \text{ in}^4 + 272 \text{ in}^4 = 520 \text{ in}^4$$

$$r_x = \sqrt{\frac{I_x}{A_g}} = \sqrt{\frac{2513 \text{ in}^4}{44.2 \text{ in}^2}} = 7.54 \text{ in}$$

$$r_y = \sqrt{\frac{I_y}{A_g}} = \sqrt{\frac{520 \text{ in}^4}{44.2 \text{ in}^2}} = 3.43 \text{ in}$$

Check axial compression without bending. Use the idealized case of a column pinned at base with sway rotational restraint at the top (from AISC Table C-A-7.1, case f, $K = 2.0$). The column is assumed to be braced at the base and at the elevation of 21.5 ft about its weak axis.

$$\frac{KL_x}{r_x} = \frac{(2.0)(21.5 \text{ ft})\left(12 \ \frac{\text{in}}{\text{ft}}\right)}{7.54 \text{ in}} = 68$$

$$\frac{KL_y}{r_y} = \frac{(1.0)(21.5 \text{ ft})\left(12 \ \frac{\text{in}}{\text{ft}}\right)}{3.43 \text{ in}}$$

$$= 75 \quad [\text{controls axial compression capacity}]$$

$$\sqrt{\frac{E}{F_y}} = \sqrt{\frac{29,000 \ \dfrac{\text{kips}}{\text{in}^2}}{50 \ \dfrac{\text{kips}}{\text{in}^2}}} = 24.08$$

A W21 × 101 is slender in axial compression; however, the built-up column effectively stiffens the web at the intersection, resulting in a stiffened compression element with b/t half the h/t_w for a W21 × 101. Thus, checking AISC Table B4.1, case 8, gives

$$\frac{b}{t} = 0.5\left(\frac{h}{t_w}\right) = (0.5)(37.5)$$

$$= 18.8 \quad [< 1.49\sqrt{E/F_y} = (1.49)(24.08) = 35.9]$$

The section is not slender. Compute the axial compression strength using AISC Sec. E3.

$$F_e = \frac{\pi^2 E}{\left(\dfrac{KL}{r}\right)^2} = \frac{\pi^2\left(29,000 \ \dfrac{\text{kips}}{\text{in}^2}\right)}{(75)^2} = 50.8 \text{ ksi}$$

$$F_{\text{cr}} = 0.658^{F_y/F_e} F_y = \left(0.658^{50 \frac{\text{kips}}{\text{in}^2}/50.8 \frac{\text{kips}}{\text{in}^2}}\right)\left(50 \ \frac{\text{kips}}{\text{in}^2}\right)$$

$$= 33.1 \text{ ksi}$$

$$P_c = \phi P_n = \phi F_{\text{cr}} A_g = (0.9)\left(33.1 \ \frac{\text{kips}}{\text{in}^2}\right)(44.2 \text{ in}^2)$$

$$= 1317 \text{ kips}$$

Check bending without axial compression. For the W21 × 101, $Z_x = 253$ in³, $S_x = 227$ in³, $L_p = 10.2$ ft, $L_r = 30.1$ ft, and $L_b = 21.5$ ft; the section is compact for $F_y = 50$ ksi. Per AISC Table 3-1, for moment varying linearly from zero to maximum at 21.5 ft, C_b equals 1.67.

$$M_p = F_y Z_x = \left(50 \ \frac{\text{kips}}{\text{in}^2}\right)(253 \text{ in}^3) = 12,650 \text{ in-kips}$$

$$\phi M_n = \phi C_b \left(M_p - (M_p - 0.7 F_y S_x)\left(\frac{L_b - L_p}{L_r - L_p}\right)\right) \leq \phi M_p$$

$$= \frac{(0.9)(1.67)\left(\begin{array}{c} 12,650 \text{ in-kips} \\ -\left(\begin{array}{c} 12,650 \text{ in-kips} \\ -(0.7)\left(50 \ \dfrac{\text{kips}}{\text{in}^2}\right) \\ \times (227 \text{ in}^3) \end{array}\right) \\ \times \left(\dfrac{21.5 \text{ ft} - 10.2 \text{ ft}}{30.1 \text{ ft} - 10.2 \text{ ft}}\right) \end{array}\right)}{12 \ \dfrac{\text{in}}{\text{ft}}}$$

$$= 1250 \text{ ft-kips}$$

$$\phi M_p = \frac{(0.9)(12,650 \text{ in-kips})}{12 \ \dfrac{\text{in}}{\text{ft}}}$$

$$= 949 \text{ ft-kips} \quad [\text{controls because } \phi M_n > \phi M_p]$$

$$M_c = \phi M_p = 949 \text{ ft-kips}$$

For the two columns, find the story drift under lateral loads by bending deformation. Use zero for the slope midway between the top and bottom chords.

$$\Delta_H = \frac{H h_e^3}{3E(2I_x)}$$

$$= \frac{(64.0 \text{ kips})\left(\left(21.5 \text{ ft} + \dfrac{9.5 \text{ ft}}{2}\right)\left(12 \ \dfrac{\text{in}}{\text{ft}}\right)\right)^3}{(3)\left(29,000 \ \dfrac{\text{kips}}{\text{in}^2}\right)(2)(2513 \text{ in}^4)}$$

$$= 4.57 \text{ in}$$

From AISC App. 8,

$$P_{e,\text{story}} = \frac{R_M H L}{\Delta_H}$$

$$= \frac{(0.85)(64.0 \text{ kips})(31.0 \text{ ft})\left(12 \ \dfrac{\text{in}}{\text{ft}}\right)}{4.57 \text{ in}}$$

$$= 4428 \text{ kips}$$

$$P_{\text{story}} = (2)(1.2)P_D = (2)(1.2)(89.35 \text{ kips})$$

$$= 214 \text{ kips}$$

$$B_2 = \frac{1}{1 - \dfrac{\alpha P_{\text{story}}}{P_{e,\text{story}}}} = \frac{1}{1 - \dfrac{(1)(214 \text{ kips})}{4428 \text{ kips}}}$$

$$= 1.05$$

$$P_r = P_{nt} + B_2 P_{lt}$$

$$= 107 \text{ kips} + (1.05)(66.5 \text{ kips})$$

$$= 177 \text{ kips}$$

$$M_r = B_1 M_{nt} + B_2 M_{lt}$$
$$= B_1(0 \text{ ft-kips}) + (1.05)(693 \text{ ft-kips})$$
$$= 728 \text{ ft-kips}$$
$$\frac{P_r}{P_c} = \frac{177 \text{ kips}}{1317 \text{ kips}}$$
$$= 0.13 \quad [\, < 0.2, \text{ so check AISC Eq. H1-1b}]$$
$$\frac{P_r}{2P_c} + \frac{M_r}{M_c} = \frac{0.13}{2} + \frac{728 \text{ ft-kips}}{949 \text{ ft-kips}} = 0.83 \quad [<1.0, \text{ so OK}]$$

> Use two WT10.5 × 50.5 members combined with the W10 × 49.

Size welds to attach each WT10.5 × 50.5 to the web of the W10 × 49. Use $^3/_{16}$ in E70XX (from AISC Table J2.4) to weld both sides of the web of the W10 × 49 ($t_w = 0.34$ in). For A992 steel, shear rupture controls base material strength. Try $^3/_{16}$ in fillet welds on both sides.

$$q \leq \begin{cases} D(\phi f_w) = (3)\left(1.39 \ \dfrac{\text{kips}}{\text{in}}\right) \\ \qquad = 4.17 \text{ kips/in} \quad [\text{controls}] \\ \phi(0.6F_u)\left(\dfrac{t_w}{2}\right) = (0.75)(0.6)\left(65 \ \dfrac{\text{kips}}{\text{in}^2}\right)\left(\dfrac{0.34 \text{ in}}{2}\right) \\ \qquad = 5.00 \text{ kips/in} \end{cases}$$

$$L_w = \frac{F_y A_{g,\text{WT}}}{2q}$$
$$= \frac{\left(50 \ \dfrac{\text{kips}}{\text{in}^2}\right)(14.9 \text{ in}^2)}{(2)\left(4.17 \ \dfrac{\text{kips}}{\text{in}}\right)}$$
$$= 89 \text{ in} \quad [\text{total required each side}]$$

Connection of the W12 × 40 lower chord ($t_w = {}^5/_{16}$ in) results in

$$P_u = 101 \text{ kips}$$

The member is subject to tension at one end, and compression at the other. For the compression loading, brace the weak axis at the panel points to provide $KL_y = 7.5$ ft. Per AISC Table 4-1, the member has adequate strength when braced.

Use $^3/_4$ in diameter A325 bolts at the flange of the WT10.5 × 50.5 and design to transfer 101 kips tension. Per AISC Table 7-2,

$$\phi r_n = 29.8 \text{ kips/bolt}$$
$$n \geq \frac{P_u}{\phi r_n} = \frac{101 \text{ kips}}{29.8 \ \dfrac{\text{kips}}{\text{bolt}}} = 3.4 \text{ bolts} \quad [\text{use 4 bolts}]$$

Weld an end plate to the W12 × 40 to transfer P_u. Use ASTM A36 steel for the plate. For the end plate, 12 in

long, the force transfered by each bolt is 101 kips/ 4 bolts = 25.3 kips/bolt. Using a standard 5.5 in gage between bolts, the bending moment is

$$m_u = \frac{0.25 P_u(0.5g)}{0.5d}$$
$$= \frac{(0.25)(101 \text{ kips})(0.5)(5.5 \text{ in})}{(0.5)(12 \text{ in})}$$
$$= 11.6 \text{ in-kips/in}$$
$$\phi m_p = \frac{\phi F_y t^2}{4} = m_u$$
$$t = \sqrt{\frac{(4)(11.6 \text{ in-kips})}{(0.9)\left(36 \ \dfrac{\text{kips}}{\text{in}^2}\right)}}$$
$$= 1.20 \text{ in}$$

> Use a PL1.25 × 8 × 1 ft 0 in.

19.3. The required details are

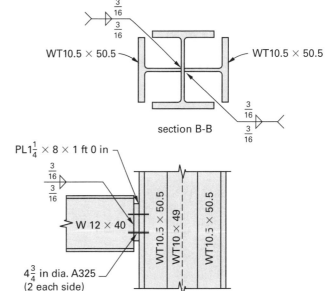

section B-B

detail–joint A
(not to scale)

SOLUTION 20

20.1. For a typical floor beam,

$$A_T = Ls = (20 \text{ ft})(7 \text{ ft}) = 140 \text{ ft}^2$$
$$K_{LL}A_T = (2)(140 \text{ ft}^2)$$
$$= 280 \text{ ft}^2 \quad \begin{bmatrix} < 400 \text{ ft}^2, \text{ so no live load} \\ \text{reduction permitted} \end{bmatrix}$$
$$w_L = w_o s = \frac{\left(100 \ \dfrac{\text{lbf}}{\text{ft}^2}\right)(7 \text{ ft})}{1000 \ \dfrac{\text{lbf}}{\text{kip}}} = 0.7 \text{ kip/ft}$$

The girder supports three beams per span, so

$$A_T = Ls = (28 \text{ ft})(20 \text{ ft}) = 560 \text{ ft}^2$$

$$K_{LL}A_T = (2)(560 \text{ ft}^2)$$

$$= 1120 \text{ ft}^2 \quad \left[\begin{array}{c} > 400 \text{ ft}^2, \text{ so ASCE/SEI7 permits} \\ \text{live load reduction} \end{array} \right]$$

$$w_L = L_o\left(0.25 + \frac{15}{\sqrt{K_{LL}A_T}}\right)$$

$$= \left(100 \text{ } \frac{\text{lbf}}{\text{ft}^2}\right)\left(0.25 + \frac{15}{\sqrt{(2)(560 \text{ ft}^2)}}\right)$$

$$= 70 \text{ lbf/ft}^2$$

$$P_L = w_L Ls = \frac{\left(70 \text{ } \frac{\text{lbf}}{\text{ft}^2}\right)(20 \text{ ft})(7 \text{ ft})}{1000 \text{ } \frac{\text{lbf}}{\text{kip}}} = 9.8 \text{ kips}$$

Use LRFD. For the W10 × 26 beam,

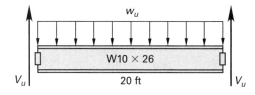

The floor dead load, including partitions, is given as 50 lbf/ft². The factored shear and moment that the strengthened beam must resist are

$$w_D = ws + w_{\text{beam}}$$

$$= \frac{\left(50 \text{ } \frac{\text{lbf}}{\text{ft}^2}\right)(7 \text{ ft}) + 26 \text{ } \frac{\text{lbf}}{\text{ft}}}{1000 \text{ } \frac{\text{lbf}}{\text{kip}}}$$

$$= 0.376 \text{ kip/ft}$$

$$w_u = 1.2w_D + 1.6w_L$$

$$= (1.2)\left(0.376 \text{ } \frac{\text{kip}}{\text{ft}}\right) + (1.6)\left(0.70 \text{ } \frac{\text{kip}}{\text{ft}}\right)$$

$$= 1.57 \text{ kips/ft}$$

$$V_u = 0.5w_u L = (0.5)\left(1.57 \text{ } \frac{\text{kips}}{\text{ft}}\right)(20 \text{ ft})$$

$$= 15.7 \text{ kips}$$

$$M_u = 0.125w_u L^2 = (0.125)\left(1.57 \text{ } \frac{\text{kips}}{\text{ft}}\right)(20 \text{ ft})^2$$

$$= 78.6 \text{ ft-kips}$$

Properly installed metal decking will brace the compression flange, so $L_b = 0$. For the W10 × 26 (compact in A36 steel): $d = 10.33$ in, $t_w = 0.26$ in, $b_f = 5.77$ in, $t_f = 0.44$ in, $Z_x = 31.3$ in³, and $I_x = 144$ in⁴.

Holes drilled in the top flange (compression flange) to attach the clip angles for floor joists should not be

critical; therefore, assume the full plastic moment capacity of the W10 × 26 is available ($Z_x = 31.3$ in³).

$$M_n = M_p = F_y Z_x$$

$$= \frac{\left(36 \text{ } \frac{\text{kips}}{\text{in}^2}\right)(31.3 \text{ in}^3)}{12 \text{ } \frac{\text{in}}{\text{ft}}}$$

$$= 93.9 \text{ ft-kips}$$

$$\phi M_n = (0.9)(93.9 \text{ ft-kips})$$

$$= 85 \text{ ft-kips} \quad [> M_u, \text{ so OK}]$$

$$\phi V_n = \phi(0.6F_y)dt_w$$

$$= (1.0)(0.6)\left(36 \text{ } \frac{\text{kips}}{\text{in}^2}\right)(10.33 \text{ in})(0.26 \text{ in})$$

$$= 58.0 \text{ kips} \quad [> V_u, \text{ OK}]$$

Check the service live load deflection against the limit of $L/360$.

$$\Delta_L = \frac{\left(\frac{5}{384}\right)w_L L^4}{EI}$$

$$= \frac{(0.013)\left(0.7 \text{ } \frac{\text{kip}}{\text{ft}}\right)(20 \text{ ft})^4\left(12 \text{ } \frac{\text{in}}{\text{ft}}\right)^3}{\left(29{,}000 \text{ } \frac{\text{kips}}{\text{in}^2}\right)(144 \text{ in}^4)}$$

$$= 0.6 \text{ in}$$

$$\frac{L}{360} = \frac{(20 \text{ ft})\left(12 \text{ } \frac{\text{in}}{\text{ft}}\right)}{360} = 0.67 \text{ in} > \Delta_L$$

Therefore, the W10 × 26 floor beams $\boxed{\text{are OK}}$ without alteration.

Check the single plate shear tab connecting the W10 × 26 to the W18 × 50 girder. According to the *AISC Manual*, the fasteners for the single plate connection must be designed for direct shear and eccentricity, $e \approx 2^{1}/_4$ in and the shear tab should have fillet welds on both sides. Check the bolts (assume standard 3 in gage).

$$\sum d_i^2 = (2)(1.5 \text{ in})^2 = 4.5 \text{ in}^2$$

$$R_v = \frac{V_u}{n} = \frac{15.7 \text{ kips}}{2} = 7.85 \text{ kips}$$

$$R_m = \frac{V_u e d}{\sum d_i^2} = \frac{(15.7 \text{ kips})(2.25 \text{ in})(1.5 \text{ in})}{4.5 \text{ in}^2}$$

$$= 11.8 \text{ kips}$$

$$R = \sqrt{R_v^2 + R_m^2} = \sqrt{(7.85 \text{ kips})^2 + (11.8 \text{ kips})^2}$$

$$= 14.2 \text{ kips}$$

From AISC Table 7-1, the capacity of a $^5/_8$ in A307 bolt in single shear is only 6.2 kips/bolt, which is inadequate. To remedy the connection,

> add an unstiffened seated connection designed to resist the entire 15.7 kips reaction.

Per AISC Table 10-5, an L4 × 4 × $^3/_8$ × 0 ft 6 in with two $^3/_4$ in diameter bolts through the web of the W18 × 50 girder will provide more than the needed 15.7 kips capacity (AISC Fig. 10-7, type A, gives 35.8 kips capacity). The existing shear tab will prevent rotation of the beam about its longitudinal axis and all vertical force will be assumed to transfer through the seat (conservative).

For the girder (W18 × 50, A36 steel),

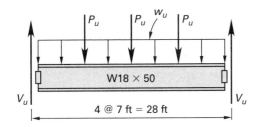

Using the reduced live load on the girder, $P_L = 9.8$ kips, gives

$$P_D = (w_D + w_{\text{beam}})L$$

$$= \left(0.35 \; \frac{\text{kip}}{\text{ft}}\right) + \left(0.026 \; \frac{\text{kip}}{\text{ft}}\right)(20 \text{ ft})$$

$$= 7.52 \text{ kips}$$

$$P_u \geq \begin{cases} 1.4 P_D = (1.4)(7.52 \text{ kips}) \\ \qquad = 10.5 \text{ kips} \\ 1.2 P_D + 1.6 P_L = (1.2)(7.52 \text{ kips}) + (1.6)(9.8 \text{ kips}) \\ \qquad = 24.7 \text{ kips} \quad [\text{controls}] \end{cases}$$

$$V_u = 1.5 P_u + 0.5 w_u L$$

$$= (1.5)(24.7 \text{ kips}) + (0.5)\left(0.06 \; \frac{\text{kip}}{\text{ft}}\right)(28 \text{ ft})$$

$$= 38.0 \text{ kips}$$

$$M_u = 0.125 w_u L^2 + 0.5 P_u L$$

$$= (0.125)\left(0.06 \; \frac{\text{kip}}{\text{ft}}\right)(28 \text{ ft})^2$$

$$\qquad + (0.5)(24.7 \text{ kips})(28 \text{ ft})$$

$$= 352 \text{ ft-kips}$$

For the W18 × 50 (compact in A36 steel): $d = 18.0$ in, $t_w = 0.36$ in, $b_f = 7.50$ in, $t_f = 0.57$ in, $Z_x = 101$ in^3, $S_x = 88.9$ in^3, and $I_x = 800$ in^4. Check live load deflection.

$$P_L = 9.8 \text{ kips}$$

$$\Delta_L = \frac{0.0495 P_L L^3}{EI}$$

$$= \frac{(0.0495)(9.8 \text{ kips})\left((28 \text{ ft})\left(12 \; \frac{\text{in}}{\text{ft}}\right)\right)^3}{\left(29{,}000 \; \frac{\text{kips}}{\text{in}^2}\right)(800 \text{ in}^4)}$$

$$= 0.79 \text{ in}$$

$$\frac{L}{360} = \frac{(28 \text{ ft})\left(12 \; \frac{\text{in}}{\text{ft}}\right)}{360} = 0.93 \text{ in} \quad [> \Delta_L, \text{ so OK}]$$

Check the flexural strength. For the fully braced compression flange with $L_b = 0$,

$$M_n = M_p = F_y Z_x$$

$$= \frac{\left(36 \; \frac{\text{kips}}{\text{in}^2}\right)(101 \text{ in}^3)}{12 \; \frac{\text{in}}{\text{ft}}}$$

$$= 303 \text{ ft-kips}$$

$$\phi M_n = (0.9)(303 \text{ ft-kips})$$

$$= 273 \text{ ft-kips} \quad [< M_u = 353 \text{ ft-kips, so no good}]$$

> Therefore, the flexural capacity of the girder will need to be increased.

20.2. Strengthen the inadequate members. Since the existing floor system is to be removed, both top and bottom cover plates can be added to the W18 × 50 to efficiently strengthen the beam. The increment on flexural capacity required is

$$\Delta M_n = \phi F_y A_{\text{PL}}(d + t_{\text{PL}}) = M_u - \phi M_n$$

$$= 353 \text{ ft-kips} - 273 \text{ ft-kips}$$

$$= 80 \text{ ft-kips}$$

A trial cover plate size can be found by approximating the lever arm between the top and bottom plates as the girder depth, 18.0 in.

$$A_{\text{PL}} \approx \frac{(80 \text{ ft-kips})\left(12 \; \frac{\text{in}}{\text{ft}}\right)}{(0.9)\left(36 \; \frac{\text{kips}}{\text{in}^2}\right)(18 \text{ in})} = 1.65 \text{ in}^2$$

Try a $\text{PL}^3/_8 \times 6$ on top and a $\text{PL}^1/_4 \times 9$ bottom ($A_{\text{PL}} = 2.25$ in² for each plate). The different sizes are chosen to permit downhand welding of the plates to the W18 × 50 ($b_f = 7.5$ in).

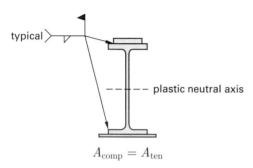

$$A_{\text{comp}} = A_{\text{ten}}$$

The plastic neutral axis is at the middepth of the W18 × 50.

$$\phi \Delta M_n = \phi F_y A_{\text{PL}}\left(d + \frac{t_1 + t_2}{2}\right)$$

$$= \frac{(0.9)\left(36\ \dfrac{\text{kips}}{\text{in}^2}\right)(2.25\ \text{in}^2)}{12\ \dfrac{\text{in}}{\text{ft}}}$$
$$\times \left(18.0\ \text{in} + \dfrac{0.375\ \text{in} + 0.25\ \text{in}}{2}\right)$$

$$= 111\ \text{ft-kips}$$

$$\phi M_n = 273\ \text{ft-kips} + 111\ \text{ft-kips}$$
$$= 384\ \text{ft-kips} \quad [> M_u = 353\ \text{ft-kips, OK}]$$

The cover plates can terminate beyond the points where $M_u = 273$ ft-kips.

$$M_u = V_u x - 0.5 w_u x^2$$
$$273\ \text{ft-kips} = (38.0\ \text{kips})x - (0.5)\left(0.07\ \dfrac{\text{kip}}{\text{ft}}\right)x^2$$
$$x = 7.2\ \text{ft},\ 20.8\ \text{ft}$$

The expression for M_u is valid only where $x < 7$ ft, and the calculated distance to the point where $M_u = 273$ ft-kips slightly exceeds this limit. However, the equation is conservative beyond 7 ft and the location is close enough. To be conservative, extend the plates far enough beyond these points to fully develop the plates.

$$F_{\text{PL}} = F_y A_{\text{PL}} = \left(36\ \dfrac{\text{kips}}{\text{in}^2}\right)(2.25\ \text{in}^2) = 81\ \text{kips}$$

Use $\omega = {}^1/_4$ in, which is greater than minimum per AISC Sec. J2, but recommended for field welds.

$$q \leq \begin{cases} D(\phi f_w) = (4)\left(1.39\ \dfrac{\text{kips}}{\text{in}}\right) \\[2mm] \qquad = 5.6\ \text{kips/in} \\[3mm] \phi(0.6F_y)t = (1.0)(0.6)\left(36\ \dfrac{\text{kips}}{\text{in}^2}\right)(0.25\ \text{in}) \\[2mm] \qquad = 5.4\ \text{kips/in} \quad [\text{controls}] \\[3mm] \phi(0.6F_u)t = (0.75)(0.6)\left(58\ \dfrac{\text{kips}}{\text{in}^2}\right)(0.25\ \text{in}) \\[2mm] \qquad = 6.5\ \text{kips/in} \end{cases}$$

$$L_w = \frac{0.5 F_{\text{PL}}}{q} = \frac{(0.5)(81\ \text{kips})}{5.4\ \dfrac{\text{kips}}{\text{in}}} = 7.5\ \text{in}$$

Therefore, the total length of cover plates is

$$L_{\text{PL}} = L - 2x + 2L_w$$
$$= 28\ \text{ft} - (2)(7.2\ \text{ft}) + (2)\left(\dfrac{7.5\ \text{in}}{12\ \dfrac{\text{in}}{\text{ft}}}\right)$$
$$= 14.9\ \text{ft}$$

Calculate the horizontal shear flow at the cutoff locations. Use elastic theory and treat the top and bottom plates as 0.25 in thick (slightly conservative), which simplifies computation of the section properties.

$$V_{7.2\,\text{ft}} = R - w_u x - P_u$$
$$= 38.0\ \text{kips} - \left(0.07\ \dfrac{\text{kip}}{\text{ft}}\right)(7.2\ \text{ft}) - 24.7\ \text{kips}$$
$$= 12.8\ \text{kips}$$

$$I_x = I_{x,\text{W18}\times 50} + 2A_{\text{PL}}(0.5d + 0.5t_{\text{PL}})^2$$
$$= 800\ \text{in}^4 + (2)(2.25\ \text{in}^2)$$
$$\times \big((0.5)(18.0\ \text{in}) + (0.5)(0.25\ \text{in})\big)^2$$
$$= 1175\ \text{in}^4$$

$$Q = A_{\text{PL}}\bar{y}_{\text{PL}} = (2.25\ \text{in}^2)(9.125\ \text{in}) = 20.5\ \text{in}^3$$

$$q = \frac{V_{7.2\,\text{ft}}Q}{I_x} = \left(\frac{(12.8\ \text{kips})(20.5\ \text{in}^3)}{1175\ \text{in}^4}\right)\left(12\ \dfrac{\text{in}}{\text{ft}}\right)$$
$$= 2.6\ \text{kips/ft}$$

The horizontal shear flow is shared by two parallel welds; therefore, use intermittent welds between ends of minimum size and spacing.

$$L_w \geq \begin{cases} 4\omega = (4)(0.25 \text{ in}) = 1 \text{ in} \\ 1.5 \text{ in} \quad \text{[controls]} \end{cases}$$

The maximum spacing for the compression cover plates is given in AISC Sec. E6.

$$s \leq \begin{cases} t_{\text{PL}}\left(0.75\sqrt{\dfrac{E}{F_y}}\right) = (0.375 \text{ in})(0.75)\sqrt{\dfrac{29{,}000 \dfrac{\text{kips}}{\text{in}^2}}{36 \dfrac{\text{kips}}{\text{in}^2}}} \\ \qquad\qquad = 8 \text{ in} \quad \text{[controls]} \\ 12 \text{ in} \end{cases}$$

For the tension flange plates, spacing is limited by AISC Sec. D4 and Sec. J3.5.

$$s \leq 24t_{\text{PL}} = (24)(0.25 \text{ in}) = 6 \text{ in} \quad \text{[controls]}$$

Use $^1/_4$ in intermittent welds 1.5 in length at 6 in on centers. Thus, the details for strengthening the girders are

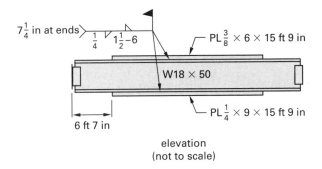

elevation
(not to scale)

Check the framed connections at ends of girder: three $^7/_8$ in diameter A307 bolts in double shear. Using bolt capacities from AISC Table 7-1 and Table 7-4,

$$\phi R = \begin{cases} \phi R_v = 24.4 \text{ kips} \quad \text{[controls]} \\ \phi R_b = \left(91.4 \dfrac{\text{kips}}{\text{in}}\right)t_w = \left(91.4 \dfrac{\text{kips}}{\text{in}}\right)(0.36 \text{ in}) \\ \qquad = 32.9 \text{ kips} \end{cases}$$

$$\begin{aligned} V_u = 38.0 \text{ kips} &< n(\phi R) \\ &= (3)(24.4 \text{ kips}) \\ &= 73.2 \text{ kips} \quad \text{[OK]} \end{aligned}$$

Check the strength of the 2L3 × 3 × $^1/_4$ × 0 ft 9 in.

$$\begin{aligned} \phi R_n &= \phi(0.6F_y)2tL_{\text{angle}} \\ &= (1.0)(0.6)\left(36 \dfrac{\text{kips}}{\text{in}^2}\right)(2)(0.25 \text{ in})(9 \text{ in}) \\ &= 97 \text{ kips} \quad [V_u > \phi R_n, \text{ OK}] \end{aligned}$$

Therefore, the existing girder connection is OK.

SOLUTION 21

21.1. Design the lightest rolled section for the roof girder using plastic design. Use LRFD. To be conservative, consider the mechanical equipment as a live load.

$$\begin{aligned} w_u &= 1.2w_D + 1.6w_L \\ &= 1.2(q_D s + w_{\text{beam}}) + 1.6q_L s \\ &\quad \dfrac{(1.2)\left(\left(32 \dfrac{\text{lbf}}{\text{ft}^2}\right)(30 \text{ ft}) + 50 \dfrac{\text{lbf}}{\text{ft}}\right)}{1000 \dfrac{\text{lbf}}{\text{kip}}} \\[-2pt] &= \dfrac{\qquad + (1.6)\left(12 \dfrac{\text{lbf}}{\text{ft}^2}\right)(30 \text{ ft})}{1000 \dfrac{\text{lbf}}{\text{kip}}} \\ &= 1.788 \text{ kips/ft} \quad (1.8 \text{ kips/ft}) \end{aligned}$$

$$\begin{aligned} P_u &= 1.6P_L \\ &= (1.6)(6 \text{ kips}) \\ &= 9.6 \text{ kips} \end{aligned}$$

Analyze the continuous beam using plastic analysis as permitted by AISC App. 1.3. Superimpose the simple span moment bending diagrams and the moment diagram corresponding to plastic hinge formation over the interior supports and locate the point at which an additional hinge forms to produce a collapse mechanism.

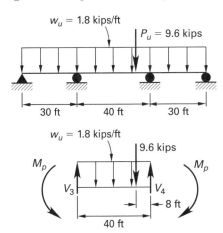

$$\sum M_{\text{line 3}} = V_4 L - M_p + M_p - 0.5 w_u L^2$$
$$- P_u(L - 8 \text{ ft}) = 0$$

$$V_4 = \frac{0.5 w_u L^2 + P_u(L - 8 \text{ ft})}{L}$$

$$= \frac{(0.5)\left(1.8 \dfrac{\text{kips}}{\text{ft}}\right)(40 \text{ ft})^2}{40 \text{ ft}}$$
$$+ (9.6 \text{ kips})(40 \text{ ft} - 8 \text{ ft})$$

$$= 43.7 \text{ kips}$$

$$\sum F_y = V_3 - w_u L - P_u + V_4 = 0$$

$$V_3 = w_u L + P_u - V_4$$

$$= \left(1.8 \frac{\text{kips}}{\text{ft}}\right)(40 \text{ ft}) + 9.6 \text{ kips} - 43.7 \text{ kips}$$

$$= 37.9 \text{ kips}$$

$$V = V_3 - w_u x = 0 \text{ kip}$$

$$x = \frac{V_3}{w_u} = \frac{37.9 \text{ kips}}{1.8 \dfrac{\text{kips}}{\text{ft}}} = 21.07 \text{ ft}$$

$$M_{0,\max} = 0.5 V_3 x$$

$$= (0.5)(37.9 \text{ kips})(21.07 \text{ ft})$$

$$= 399.4 \text{ ft-kips}$$

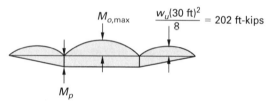

Assign $M_p = 0.5 M_{o,\max} = (0.5)(399.4 \text{ ft-kips}) = 199.7 \text{ ft-kips}$. The moment diagram at collapse is

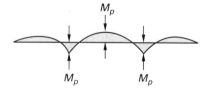

The maximum simple span moment in the end spans is 202 ft-kips, which is only slightly larger than the required M_p. Therefore, superimposing the M_p diagram onto the end spans gives a moment that is well below M_p. Three sufficient hinges form and the moment does not exceed M_p at any point; thus, the failure mechanism has been determined. As a check, use the mechanism method of analysis.

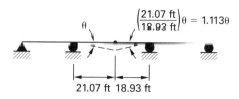

At collapse,

$$W_{\text{ext}} = U_{\text{int}}$$

$$x\theta = (L - x)\theta_{rt}$$

$$\theta_{rt} = \frac{(21.07 \text{ ft})\theta}{40 \text{ ft} - 21.07 \text{ ft}}$$

$$= 1.113\theta$$

$$W_{\text{ext}} = w_u x (0.5 x\theta) + w_u(L - x)\big(0.5(L - x)\big)$$
$$\times 1.113\theta + P_u(8 \text{ ft})1.113\theta$$

$$= \left(1.8 \frac{\text{kips}}{\text{ft}}\right)(21.07 \text{ ft})(0.5)(21.07 \text{ ft})\theta$$

$$+ \left(1.8 \frac{\text{kips}}{\text{ft}}\right)(40 \text{ ft} - 21.07 \text{ ft})(0.5)$$

$$\times (40 \text{ ft} - 21.07 \text{ ft})1.113\theta$$

$$+ (9.6 \text{ kips})(8 \text{ ft})1.113\theta$$

$$= (844 \text{ ft-kips})\theta$$

$$U_{\text{int}} = M_p(\theta + \theta + 1.113\theta + 1.113\theta)$$

$$= 4.226 M_p \theta$$

$$M_p = \frac{(844 \text{ ft-kips})\theta}{4.226\theta}$$

$$= 199.7 \text{ ft-kips} \quad [\text{OK}]$$

The plastic capacity is reached with simultaneous formation of plastic hinges over supports at lines 3 and 4 and at the intermediate point 21.07 ft from line 3. For A992 steel, $\phi = 0.9$.

$$\phi M_p = \phi F_y Z_x = 199.7 \text{ ft-kips}$$

$$Z_x = \frac{(199.7 \text{ ft-kips})\left(12 \dfrac{\text{in}}{\text{ft}}\right)}{(0.9)\left(50 \dfrac{\text{kips}}{\text{in}^2}\right)}$$

$$= 53.3 \text{ in}^3$$

Try a W16 × 31 with the following properties: $Z_x = 54.0 \text{ in}^3$, $b_f/2t_f = 6.28$, $h/t_w = 51.6$, $r_y = 1.17$ in, and $L_p = 4.13$ ft. Check the local buckling provisions from AISC App. 1.2.2.

$$\sqrt{\frac{E}{F_y}} = \sqrt{\frac{29{,}000 \; \frac{\text{kips}}{\text{in}^2}}{50 \; \frac{\text{kips}}{\text{in}^2}}} = 24.1$$

$$\frac{b_f}{2t_f} = 6.28 \quad \begin{bmatrix} < \lambda_{pf} = 0.38\sqrt{E/F_y} \\ = (0.38)(24.1) \\ = 9.15 \end{bmatrix}$$

For $P_u = 0$ kip, the limiting width thickness ratio for the web is

$$\frac{h}{t_w} = 51.6 \quad \begin{bmatrix} < \lambda_{pw} = 3.76\sqrt{E/F_y} \\ = (3.76)(24.1) \\ = 90.6 \end{bmatrix}$$

The section is compact. The plastic hinges form first over the supports where the bottom flange is in compression. The maximum unbraced length of the bottom flange in the vicinity of the supports cannot exceed L_{pd} per AISC Eq. A-1-5.

$$L_{pd} = \left(0.12 - (0.076)\left(\frac{M_1'}{M_2} \right) \right) \left(\frac{E}{F_y} \right) r_y$$

Locate the point of inflection with respect to line 3.

$$M = V_3 x - 0.5 w_u x^2 - M_p \quad [0 \text{ ft} \leq x < 21.07 \text{ ft}]$$

$$= (37.9 \text{ kips})x - (0.5)\left(1.8 \; \frac{\text{kips}}{\text{ft}} \right)x^2 - 199.7 \text{ ft-kips}$$

$$= 0 \text{ ft-kip}$$

$$x = 6.2 \text{ ft}$$

Calculate the distance between braces when M_1'/M_2 equals zero.

$$L_{pd} = \left(0.12 - 0.076\left(\frac{M_1'}{M_2} \right) \right) \left(\frac{E}{F_y} \right) r_y$$

$$= \frac{(0.12 - (0.076)(0))\left(\dfrac{29{,}000 \; \frac{\text{kips}}{\text{in}^2}}{50 \; \frac{\text{kips}}{\text{in}^2}} \right)(1.17 \text{ in})}{12 \; \frac{\text{in}}{\text{ft}}}$$

$$= 6.8 \text{ ft}$$

Therefore, brace the bottom flange of the W16 × 31 at no more than 6.2 ft to either side of lines 3 and 4. The top flange in the vicinity of the last hinge to form in span 3–4 must not exceed the value of $L_p = 4.13$ ft. If lateral bracing in that region is furnished by the joists, the joist spacing is limited to 4.13 ft.

Use W16 × 31 with top flange braced at 4 ft on center and bottom flange braced at 6 ft to either side of interior supports.

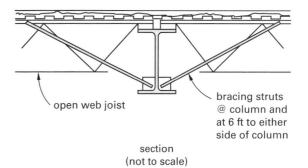

open web joist

bracing struts
@ column and
at 6 ft to either
side of column

section
(not to scale)

SOLUTION 22

22.1. Design the truss connection using ASD. Use $^3/_4$ in diameter A325-SC field bolts and E70XX shop welds.

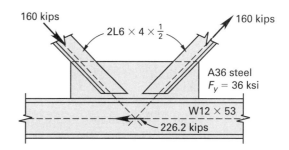

160 kips

2L6 × 4 × $\frac{1}{2}$

160 kips

A36 steel
$F_y = 36$ ksi

W12 × 53

226.2 kips

Assume holes are punched, not drilled.

$$d_{\text{hole}} = d_{\text{bolt}} + 0.125 \text{ in}$$

$$= 0.75 \text{ in} + 0.125 \text{ in}$$

$$= 0.875 \text{ in}$$

Calculate the number of bolts required. Per AISC Table 7-3,

$$\frac{r_n}{\Omega_v} = 12.7 \text{ kips/bolt} \quad \begin{bmatrix} \text{standard holes, double shear,} \\ \text{class A faying surface} \end{bmatrix}$$

Select a gusset plate thickness sufficient to provide bearing strength greater than the shear capacity of the bolts. From AISC Table 7-5 ($F_u = 58$ ksi, $^3/_4$ in bolts, 3 in spacing, 2 in edge distance),

$$\frac{r_n}{\Omega_v} < \left(52.2 \; \frac{\frac{\text{kips}}{\text{bolt}}}{\text{in}} \right) t$$

$$12.7 \; \frac{\text{kips}}{\text{bolt}} < \left(52.2 \; \frac{\frac{\text{kips}}{\text{bolt}}}{\text{in}} \right) t$$

$$t \geq \frac{12.7 \, \dfrac{\text{kips}}{\text{bolt}}}{52.2 \, \dfrac{\text{kips}}{\text{in}}} = 0.24 \text{ in} \quad [\text{say } t \geq \sqrt[3]{8} \text{ in}]$$

$$n \geq \frac{P}{\dfrac{r_n}{\Omega_v}} = \frac{160 \text{ kips}}{12.7 \, \dfrac{\text{kips}}{\text{bolt}}} = 12.6 \text{ bolts} \quad \begin{bmatrix} 12 \text{ bolts in 6 rows} \\ \text{of 2 at 3 in spacing} \\ \text{is close enough.} \end{bmatrix}$$

Check the strength of the 2L6 $\times$ 4 $\times$ $^1/_2$. The gross area is $A_g = (2)(4.75 \text{ in}^2) = 9.50 \text{ in}^2$; with long legs back to back, the connection eccentricity is $\bar{x} = 0.98$ in with a double gage of $^3/_4$ in diameter bolts (without staggering the holes).

$$A_n = 2(l_1 + l_2 - t - 2d_{\text{hole}})t$$
$$= (2)\big(6 \text{ in} + 4 \text{ in} - 0.5 \text{ in} - (2)(0.875 \text{ in})\big)(0.5 \text{ in})$$
$$= 7.75 \text{ in}^2$$

The shear lag factor, U, is computed for a connection length of

$$L = n_{\text{spaces}}s = (5)(3 \text{ in}) = 15 \text{ in}$$
$$U = 1 - \frac{\bar{x}}{L}$$
$$= 1 - \frac{0.98 \text{ in}}{15 \text{ in}}$$
$$= 0.93$$

$$\frac{P_n}{\Omega_t} \leq \begin{cases} \dfrac{F_y A_g}{\Omega_t} = \dfrac{\left(36 \, \dfrac{\text{kips}}{\text{in}^2}\right)(9.50 \text{ in}^2)}{1.67} \\ \qquad = 205 \text{ kips} \quad \begin{bmatrix} \text{controls}; > P = 160 \text{ kips}, \\ \text{so OK} \end{bmatrix} \\ \dfrac{F_u U A_n}{\Omega_t} = \dfrac{\left(58 \, \dfrac{\text{kips}}{\text{in}^2}\right)(0.93)(7.75 \text{ in}^2)}{2.00} \\ \qquad = 209 \text{ kips} \end{cases}$$

Design welds to connect the gusset plate to the flange of the W12 $\times$ 53. Try a weld size of $\omega = {}^1/_4$ in to both sides (exceeding minimum size per AISC Table J2.4).

$$q_{w,\text{each side}} = D\left(0.938 \, \frac{\text{kip}}{\text{in}}\right)$$
$$= (4)\left(0.938 \, \frac{\text{kip}}{\text{in}}\right)$$
$$= 3.75 \text{ kips/in} \quad [\text{each side}]$$
$$q_{w,\text{total}} = (2)\big(q_{w,\text{each side}}\big)$$
$$= (2)\left(3.75 \, \frac{\text{kips}}{\text{in}}\right)$$
$$= 7.5 \text{ kips/in}$$

Make the gusset plate thick enough so that shear of the base material is not the limiting criterion.

$$q_w \leq \begin{cases} \dfrac{(0.6F_y)t}{\Omega_t} = \dfrac{(0.6)\left(36 \, \dfrac{\text{kips}}{\text{in}^2}\right)t}{1.5} \\ \qquad = 14.4t \text{ ksi} \quad [\text{controls}] \\ \dfrac{(0.6F_u)t}{\Omega_t} = \dfrac{(0.6)\left(58 \, \dfrac{\text{kips}}{\text{in}^2}\right)t}{2.0} \\ \qquad = 17.4t \text{ ksi} \end{cases}$$

$$t = \frac{q_{w,\text{total}}}{q_w} = \frac{7.5 \, \dfrac{\text{kips}}{\text{in}}}{14.4 \, \dfrac{\text{kips}}{\text{in}^2}} = 0.51 \text{ in} \quad [\text{say 0.5 in plate}]$$

The gusset plate thickness required for weld strength, 0.5 in, is greater than required for bolt bearing; therefore, use PL$^1/_2$.

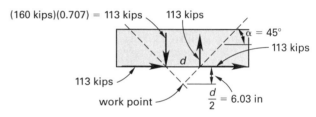

The weld attaching the gusset plate is subject to shear and bending. The shear is

$$V = 2P \cos \alpha = (2)(160 \text{ kips})\cos 45° = 226 \text{ kips}$$

The minimum length required to transfer shear only is

$$L_w = \frac{V}{q_w} = \frac{226 \text{ kips}}{7.4 \, \dfrac{\text{kips}}{\text{in}}} = 31 \text{ in}$$

To transfer both the shear and bending, try $L_w = 40$ in.

$$S_w = \frac{2L_w^2}{6} = \frac{(2)(40 \text{ in})^2}{6} = 534 \text{ in}^2$$
$$M_w = Pd \sin \alpha$$
$$= (160 \text{ kips})(12.1 \text{ in})\sin 45°$$
$$= 1364 \text{ in-kips}$$
$$f_v = \frac{V}{2L_w} = \frac{226 \text{ kips}}{(2)(40 \text{ in})} = 2.8 \text{ kips/in}$$
$$f_b = \frac{M_w}{S_w} = \frac{1364 \text{ in-kips}}{534 \text{ in}^2} = 2.6 \text{ kips/in}$$

$$f = \sqrt{f_v^2 + f_b^2}$$
$$= \sqrt{\left(2.8 \; \frac{\text{kips}}{\text{in}}\right)^2 + \left(2.6 \; \frac{\text{kips}}{\text{in}}\right)^2}$$
$$= 3.8 \text{ kips/in}$$

For the $^1/_4$ in E70XX fillet welds, $q_w = 3.75$ kips/in is close enough. Make the gusset plate 3 ft 4 in wide and $^1/_2$ in thick. Check the strength of the gusset plate at the bolt holes. Bolt pitch $= 3$ in, gage $= 2\,^1/_2$ in.

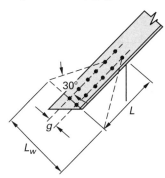

Use the Whitmore section to determine the effective design width. (See AISC Part 9.)

$$L_e = n_{\text{row}}(n_{\text{spaces}}s\tan 30°) + g - 2(0.5d_b)$$
$$= (2)(5)(3 \text{ in})(\tan 30°) + 2.5 \text{ in} - (2)(0.5)(0.875 \text{ in})$$
$$= 18.94 \text{ in}$$

$$A_e = L_e t = (18.9 \text{ in})(0.5 \text{ in}) = 9.45 \text{ in}^2$$

$$\frac{P_n}{\Omega_t} = \frac{F_u A_e}{\Omega_t} = \frac{\left(58 \; \frac{\text{kips}}{\text{in}^2}\right)(9.45 \text{ in}^2)}{2.00} = 274 \text{ kips}$$

Therefore, the $^1/_2$ in gusset plate has adequate tensile fracture strength. Check for block shear rupture (per AISC Sec. J4).

$$A_{nv} = \begin{pmatrix} n_{\text{row}}(n_{\text{spaces}}s) - (n_{\text{row}}) \\ \times (n_{\text{spaces}} + 0.5)d_{\text{hole}} + l_{ev} \end{pmatrix} t$$
$$= \begin{pmatrix} (2)(5)(3 \text{ in}) - (2)(5+0.5) \\ \times (0.875 \text{ in}) + 1.5 \text{ in} \end{pmatrix}(0.5 \text{ in})$$
$$= 10.9 \text{ in}^2$$

$$A_{nt} = (g - d_{\text{hole}})t$$
$$= (2.5 \text{ in} - 0.875 \text{ in})(0.5 \text{ in})$$
$$= 0.81 \text{ in}^2$$

$$A_{gv} = \left(n_{\text{row}}(n_{\text{spaces}}s) + l_{ev}\right)t$$
$$= \left((2)(5)(3 \text{ in}) + 1.5 \text{ in}\right)(0.5 \text{ in})$$
$$= 15.8 \text{ in}^2$$

$$A_{gt} = gt = (2.5 \text{ in})(0.5 \text{ in}) = 1.25 \text{ in}^2$$

$$R_n \leq \begin{cases} 0.6F_u A_{nv} + U_{bs}F_u A_{nt} \\ \quad = (0.6)\left(58 \; \frac{\text{kips}}{\text{in}^2}\right)(10.9 \text{ in}^2) \\ \qquad + (1)\left(58 \; \frac{\text{kips}}{\text{in}^2}\right)(0.81 \text{ in}^2) \\ \quad = 426 \text{ kips} \\ 0.6F_y A_{gv} + U_{bs}F_u A_{nt} \\ \quad = (0.6)\left(36 \; \frac{\text{kips}}{\text{in}^2}\right)(15.0 \text{ in}^2) \\ \qquad + (1)\left(58 \; \frac{\text{kips}}{\text{in}^2}\right)(0.81 \text{ in}^2) \\ \quad = 371 \text{ kips} \quad \text{[controls]} \end{cases}$$

$$\frac{R_n}{\Omega} = \frac{371 \text{ kips}}{2.00} = 186 \text{ kips} \quad [> P = 160 \text{ kips, OK}]$$

Therefore, the plate is adequate. Dimension the plate to accommodate five spaces at 3 in on center plus $1^1/_2$ in edge distance. To allow for the minimum edge distances of 1.25 in (per AISC Table J3.4), the overall width of plate must be

$$w \geq 2s_{\text{edge}} + \left(g + n_{\text{space}}(3 \text{ in})\right)\cos 45°$$
$$\geq (2)(1.25 \text{ in}) + \left(2.5 \text{ in} + (5)(3 \text{ in})\right)\cos 45°$$
$$\geq 14.8 \text{ in}$$

Use PL$^1/_2 \times 15 \times$ 3 ft 4 in.

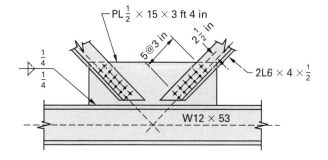

SOLUTION 23

23.1. Design the main girder using LRFD. Consider the effects of unbalanced roof live loads (ASCE/SEI7 Table 4-1, footnote n). Provide lateral braces to the bottom flange at 25 ft intervals and at each end.

$$w_u = 1.2w_D + 1.6w_L$$
$$= (1.2)\left(0.62 \; \frac{\text{kip}}{\text{ft}}\right) + (1.6)\left(0.63 \; \frac{\text{kip}}{\text{ft}}\right)$$
$$= 1.75 \text{ kips/ft}$$
$$P_u = 1.2P_D$$
$$= (1.2)(9.5 \text{ kips})$$
$$= 11.4 \text{ kips}$$

For loading applied over the entire length of the main girder,

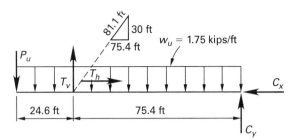

$$\sum M_c = P_u L + 0.5 w_u L^2 - T_v b = 0$$

$$T_v = \frac{P_u L + 0.5 w_u L^2}{b}$$

$$= \frac{(11.4 \text{ kips})(100 \text{ ft}) + (0.5)\left(1.75 \dfrac{\text{kips}}{\text{ft}}\right)(100 \text{ ft})^2}{75.4 \text{ ft}}$$

$$= 131.2 \text{ kips}$$

$$T_h = \left(\frac{T_v}{\sin \alpha}\right)\cos \alpha$$

$$= \left(\frac{131.2 \text{ kips}}{\dfrac{30 \text{ ft}}{81.1 \text{ ft}}}\right)\left(\frac{75.4 \text{ ft}}{81.1 \text{ ft}}\right)$$

$$= 330 \text{ kips}$$

$$\sum F_y = -P_u - w_u L + T_v + C_y = 0$$

$$C_y = P_u + w_u L - T_v$$

$$= 11.4 \text{ kips} + \left(1.75 \dfrac{\text{kips}}{\text{ft}}\right)(100 \text{ ft}) - 131.2 \text{ kips}$$

$$= 55 \text{ kips}$$

$$\sum F_x = T_h + C_x = 0$$

$$C_x = -T_h = -330 \text{ kips} \quad [330 \text{ kips to the left}]$$

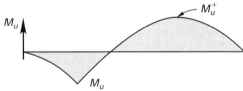

$$M_u^- = P_u a + 0.5 w_u a^2$$

$$= (-11.4 \text{ kips})(24.6 \text{ ft}) - (0.5)\left(1.75 \dfrac{\text{kips}}{\text{ft}}\right)(24.6 \text{ ft})^2$$

$$= -810 \text{ ft-kips}$$

$$x = \frac{C_y}{w_u} = \frac{55.0 \text{ kips}}{1.75 \dfrac{\text{kips}}{\text{ft}}} = 31.4 \text{ ft}$$

$$M_u^+ = 0.5 C_y x$$

$$= (0.5)(55.0 \text{ kips})(31.4 \text{ ft})$$

$$= 864 \text{ ft-kips}$$

For the pattern of live load with live load on the cantilever omitted,

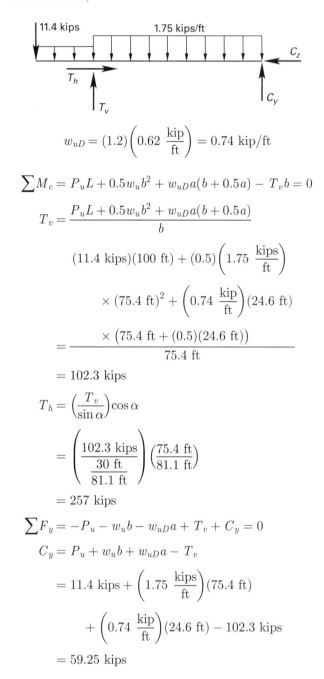

$$w_{uD} = (1.2)\left(0.62 \dfrac{\text{kip}}{\text{ft}}\right) = 0.74 \text{ kip/ft}$$

$$\sum M_c = P_u L + 0.5 w_u b^2 + w_{uD} a (b + 0.5a) - T_v b = 0$$

$$T_v = \frac{P_u L + 0.5 w_u b^2 + w_{uD} a (b + 0.5a)}{b}$$

$$= \frac{\begin{aligned}&(11.4 \text{ kips})(100 \text{ ft}) + (0.5)\left(1.75 \dfrac{\text{kips}}{\text{ft}}\right) \\ &\quad \times (75.4 \text{ ft})^2 + \left(0.74 \dfrac{\text{kip}}{\text{ft}}\right)(24.6 \text{ ft}) \\ &\quad \times \big(75.4 \text{ ft} + (0.5)(24.6 \text{ ft})\big)\end{aligned}}{75.4 \text{ ft}}$$

$$= 102.3 \text{ kips}$$

$$T_h = \left(\frac{T_v}{\sin \alpha}\right)\cos \alpha$$

$$= \left(\frac{102.3 \text{ kips}}{\dfrac{30 \text{ ft}}{81.1 \text{ ft}}}\right)\left(\frac{75.4 \text{ ft}}{81.1 \text{ ft}}\right)$$

$$= 257 \text{ kips}$$

$$\sum F_y = -P_u - w_u b - w_{uD} a + T_v + C_y = 0$$

$$C_y = P_u + w_u b + w_{uD} a - T_v$$

$$= 11.4 \text{ kips} + \left(1.75 \dfrac{\text{kips}}{\text{ft}}\right)(75.4 \text{ ft})$$

$$\quad + \left(0.74 \dfrac{\text{kip}}{\text{ft}}\right)(24.6 \text{ ft}) - 102.3 \text{ kips}$$

$$= 59.25 \text{ kips}$$

$$\sum F_x = T_h + C_x = 0$$

$$C_x = -T_h = -257 \text{ kips} \quad [257 \text{ kips to the left}]$$

$$x = \frac{C_y}{w_u} = \frac{59.25 \text{ kips}}{1.75 \dfrac{\text{kips}}{\text{ft}}} = 33.86 \text{ ft}$$

$$M_u^+ = 0.5 C_y x$$

$$= (0.5)(59.25 \text{ kips})(33.86 \text{ ft})$$

$$= 1003 \text{ ft-kips}$$

Select a trial section to resist $M_u^- = 810$ ft-kips with $L_b = 24.6$ ft and $C_b > 1.0$, and to resist $M_u^+ = 1003$ ft-kips with $P_u = 257$ kips, $C_b = 1.0$, and $L_b = 8.3$ ft (top flange is braced by joists).

A W30 × 90 is adequate for bending only; try something larger for the combined axial compression plus bending; say a W30 × 116. (Section properties of a W30 × 116 include: $A_g = 34.2$ in^2, $d = 30.0$ in, $t_w = 0.565$ in, $b_f = 10.50$ in, $t_f = 0.85$ in, $r_x = 12.0$ in, $r_y = 2.19$ in, $I_x = 4930$ in^4, $I_y = 164$ in^4, $Z_x = 378$ in^3, $S_x = 329$ in^3, $b_f/2t_f = 6.17$, $h/t_w = 47.8$ in, $k = 1.5$ in, $J = 6.43$ in^4, $C_w = 34{,}900$ in^6, $L_p = 7.74$ ft, and $L_r = 22.6$ ft.)

Check the cantilevered portion for $L_b = 24.6$ ft. Compute the bending coefficient, C_b.

$$M_A = 0.25 a P_u + 0.5 w_u (0.25a)^2$$

$$= (0.25)(24.6 \text{ ft})(11.4 \text{ kips})$$

$$\quad + (0.5)\left(1.75 \dfrac{\text{kips}}{\text{ft}}\right)\left((0.25)(24.6 \text{ ft})\right)^2$$

$$= 103 \text{ ft-kips}$$

$$M_B = 0.5 a P_u + 0.5 w_u (0.5a)^2$$

$$= (0.5)(24.6 \text{ ft})(11.4 \text{ kips})$$

$$\quad + (0.5)\left(1.75 \dfrac{\text{kips}}{\text{ft}}\right)\left((0.5)(24.6 \text{ ft})\right)^2$$

$$= 273 \text{ ft-kips}$$

$$M_C = 0.75 a P_u + 0.5 w_u (0.75a)^2$$

$$= (0.75)(24.6 \text{ ft})(11.4 \text{ kips})$$

$$\quad + (0.5)\left(1.75 \dfrac{\text{kips}}{\text{ft}}\right)\left((0.75)(24.6 \text{ ft})\right)^2$$

$$= 508 \text{ ft-kips}$$

$$M_{\max} = a P_u + 0.5 w_u a^2$$

$$= (24.6 \text{ ft})(11.4 \text{ kips}) + (0.5)\left(1.75 \dfrac{\text{kips}}{\text{ft}}\right)(24.6 \text{ ft})^2$$

$$= 810 \text{ ft-kips}$$

$$C_b = \frac{12.5 M_{\max}}{2.5 M_{\max} + 3 M_A + 4 M_B + 3 M_C}$$

$$= \frac{(12.5)(810 \text{ ft-kips})}{(2.5)(810 \text{ ft-kips}) + (3)(103 \text{ ft-kips})}$$
$$\qquad + (4)(273 \text{ ft-kips}) + (3)(508 \text{ ft-kips})$$

$$= 2.05$$

Check local buckling per AISC Table B4.1.

$$\sqrt{\frac{E}{F_y}} = \sqrt{\frac{29{,}000 \dfrac{\text{kips}}{\text{in}^2}}{50 \dfrac{\text{kips}}{\text{in}^2}}} = 24.1$$

$$\frac{b_f}{2t_f} = 6.17 < 0.38\sqrt{\frac{E}{F_y}} = (0.38)(24.1) = 9.15 = \lambda_{pf}$$

$$\frac{h}{t_w} = 47.8 < 3.76\sqrt{\frac{E}{F_y}} = (3.76)(24.1) = 90.6 = \lambda_{pw}$$

The section is compact for bending. Because $L_b = 24.6$ ft is greater than $L_r = 22.6$ ft, check AISC Eq. F2-4.

$$h_o = d - 2k = 30.0 \text{ in} - (2)(1.5 \text{ in}) = 27.0 \text{ in}$$

$$r_{ts} = \sqrt{\frac{\sqrt{I_y C_w}}{S_x}} = \sqrt{\frac{\sqrt{(164 \text{ in}^4)(34{,}900 \text{ in}^6)}}{329 \text{ in}^3}} = 2.70 \text{ in}$$

$$\frac{L_b}{r_{ts}} = \frac{(24.6 \text{ ft})\left(12 \dfrac{\text{in}}{\text{ft}}\right)}{2.70 \text{ in}} = 109$$

$$F_{\text{cr}} = \frac{C_b \pi^2 E}{\left(\dfrac{L_b}{r_{ts}}\right)^2}\sqrt{1 + 0.078\left(\frac{Jc}{S_x h_o}\right)\left(\frac{L_b}{r_{ts}}\right)^2}$$

$$= \left(\frac{2.05\pi^2\left(29{,}000 \dfrac{\text{kips}}{\text{in}^2}\right)}{(109)^2}\right)$$

$$\quad \times \sqrt{1 + (0.078)\left(\frac{(6.43 \text{ in}^4)(1)}{(329 \text{ in}^3)(27.0 \text{ in})}\right)(109)^2}$$

$$= 62.9 \text{ ksi}$$

$$M_n \le \begin{cases} F_{\text{cr}} S_x = \left(62.9 \dfrac{\text{kips}}{\text{in}^2}\right)(329 \text{ in}^3) \\ \qquad = 20{,}700 \text{ in-kips} \\ F_y Z_x = \left(50 \dfrac{\text{kips}}{\text{in}^2}\right)(378 \text{ in}^3) \\ \qquad = 18{,}900 \text{ in-kips} \quad [\text{controls}] \end{cases}$$

$$\phi M_n = \frac{(0.9)(18{,}900 \text{ in-kips})}{12 \frac{\text{in}}{\text{ft}}}$$

$$= 1418 \text{ ft-kips} \quad [> M_u, \text{ so OK}]$$

Section is in the cantilevered region. Check combined axial plus bending in the main span. For axial load only, section is unbraced about the strong axis over the length KL_x, 75.4 ft; assume that joists brace the weak axis at KL_y, 8.3 ft.

$$\frac{KL_x}{r_x} = \frac{(75.4 \text{ ft})\left(12 \frac{\text{in}}{\text{ft}}\right)}{12.0 \text{ in}}$$

$$= 75 \quad [\text{controls axial compression capacity}]$$

$$\frac{KL_y}{r_y} = \frac{(8.3 \text{ ft})\left(12 \frac{\text{in}}{\text{ft}}\right)}{2.19 \text{ in}} = 45$$

$$\frac{KL}{r} = \frac{KL_x}{r_x} = 75 \quad [< 4.71\sqrt{E/F_y} = (4.71)(24.1) = 113]$$

Check local buckling for axial compression only. (See AISC Table B4.1.)

$$\frac{h}{t_w} = 47.8 > 1.49\sqrt{\frac{E}{F_y}} = (1.49)(24.1) = 35.9 = \lambda_{pw}$$

Therefore, the capacity will have to be adjusted per AISC Sec. E7. For a stiffened element, $Q_a = 1$, $f = F_{cr}$, the effective web width is

$$b_e = 1.92 t_w \sqrt{\frac{E}{f}}\left(1 - \frac{0.34}{\frac{h}{t_w}}\sqrt{\frac{E}{f}}\right)$$

$$= (1.92)(0.565 \text{ in})\sqrt{\frac{29{,}000 \frac{\text{kips}}{\text{in}^2}}{33.1 \frac{\text{kips}}{\text{in}^2}}}$$

$$\times \left(1 - \frac{0.34}{47.8}\sqrt{\frac{29{,}000 \frac{\text{kips}}{\text{in}^2}}{33.1 \frac{\text{kips}}{\text{in}^2}}}\right)$$

$$= 25.3 \text{ in}$$

$$A_{\text{eff}} = A_g - t_w(h_w - b_e)$$

$$= 34.2 \text{ in}^2 - (0.565 \text{ in})(27.0 \text{ in} - 25.3 \text{ in})$$

$$= 33.2 \text{ in}^2$$

$$Q = \frac{A_{\text{eff}}}{A_g} = \frac{33.2 \text{ in}^2}{34.2 \text{ in}^2} = 0.97$$

$$F_e = \frac{\pi^2 E}{\left(\frac{KL}{r}\right)^2} = \frac{\pi^2 \left(29{,}000 \frac{\text{kips}}{\text{in}^2}\right)}{(75)^2}$$

$$= 50.8 \text{ ksi}$$

$$F_{cr} = Q(0.658^{QF_y/F_e} F_y)$$

$$= (0.97)\left(0.658^{(0.07)\left(50 \frac{\text{kips}}{\text{in}^2}\right)/50.0 \frac{\text{kips}}{\text{in}^2}}\right)\left(50 \frac{\text{kips}}{\text{in}^2}\right)$$

$$= 32.5 \text{ ksi}$$

$$P_c = \phi F_{cr} A_g$$

$$= (0.9)\left(32.5 \frac{\text{kips}}{\text{in}^2}\right)(34.2 \text{ in}^2)$$

$$= 1000 \text{ kips}$$

For bending only, $C_b = 1.0$ and $L_b = 8.3$ ft, AISC Table 3-10 gives $\phi M_n = 1410$ ft-kips. Amplify the bending moment per AISC Eq. A-8-3.

$$B_1 = \frac{C_m}{1 - \frac{\alpha P_r}{P_{e1}}}$$

$C_m = 1.0$ (single curvature bending), and $\alpha = 1$ when using the LRFD method.

$$P_{e1} = \frac{\pi^2 E I^*}{(K_1 L)^2}$$

$$= \frac{\pi^2 \left(29{,}000 \frac{\text{kips}}{\text{in}^2}\right)(4930 \text{ in}^4)}{\left((75.4 \text{ ft})\left(12 \frac{\text{in}}{\text{ft}}\right)\right)^2}$$

$$= 1720 \text{ kips}$$

$$B_1 = \frac{C_m}{1 - \frac{\alpha P_r}{P_{e1}}}$$

$$= \frac{1.0}{1 - \frac{(1)(257 \text{ kips})}{1720 \text{ kips}}}$$

$$= 1.18$$

$$M_{rx} = B_1 M_u = (1.18)(1003 \text{ ft-kips}) = 1184 \text{ ft-kips}$$

Check using AISC Sec. H1.

$$\frac{P_r}{P_c} = \frac{257 \text{ kips}}{1000 \text{ kips}}$$

$$= 0.26 \quad [> 0.2, \text{ therefore use AISC Eq. H1-1a}]$$

$$\frac{P_r}{P_c} + \frac{8}{9}\left(\frac{M_{rx}}{M_{cx}}\right) = 0.26 + \left(\frac{8}{9}\right)\left(\frac{1184 \text{ ft-kips}}{1410 \text{ ft-kips}}\right) = 1.0$$

$\boxed{\text{Use W30} \times 116.}$

23.2. Per AISC App. 6, take M_r/h_o as $F_y b_f t_f$ to be conservative and design a brace to the lower flange to resist this force. The lower chord of the joists could be used provided that the joists are designed to resist the compression that would develop in the lower chord. An alternative is to provide a brace from the upper chord where the decking will furnish the necessary strength and stiffness. Place the braces on both sides of the W30 × 116 inclined at, say, 30° to the horizontal and design to resist only tension.

$$P_{rb} = 0.02 C_d F_y b_f t_f$$

$$= (0.02)(1.0)\left(50 \ \frac{\text{kips}}{\text{in}^2}\right)(10.5 \text{ in})(0.85 \text{ in})$$

$$= 8.9 \text{ kips}$$

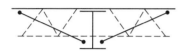

$$P_{\text{brace}} = \frac{8.9 \text{ kips}}{\cos 30°} = 10.2 \text{ kips}$$

Use a single angle.

$$L \approx (6 \text{ ft})\left(12 \ \frac{\text{in}}{\text{ft}}\right) = 72 \text{ in}$$

$$\frac{L}{r} \leq 300 \quad \left[\text{therefore, } r \geq \frac{72 \text{ in}}{300} = 0.24 \text{ in}\right]$$

Try an L2¹⁄₂ × 2¹⁄₂ × ³⁄₁₆ ($A_g = 0.90 \text{ in}^2$). Use A36 steel for the angle. Connect with a single row of two ³⁄₄ in diameter A307 bolts.

$$A_n = A_g - (d_b + 0.125 \text{ in})t$$

$$= 0.90 \text{ in}^2 - (0.75 \text{ in} + 0.125 \text{ in})(0.1875 \text{ in})$$

$$= 0.74 \text{ in}^2$$

$$\phi P_n \leq \begin{cases} \phi F_y A_g = (0.9)\left(36 \ \frac{\text{kips}}{\text{in}^2}\right)(0.90 \text{ in}^2) \\ \qquad = 29.1 \text{ kips} \\ \phi F_u U A_n = (0.75)\left(58 \ \frac{\text{kips}}{\text{in}^2}\right)(0.75)(0.74 \text{ in}^2) \\ \qquad = 24.0 \text{ kips} \quad [\text{controls}] \end{cases}$$

Use L2¹⁄₂ × 2¹⁄₂ × ³⁄₁₆. Attach the brace at upper point using ³⁄₁₆ in plate welded to resist $T_u = 10.2$ kips.

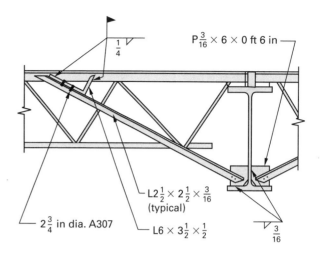

P³⁄₁₆ × 6 × 0 ft 6 in

L2¹⁄₂ × 2¹⁄₂ × ³⁄₁₆ (typical)

2³⁄₄ in dia. A307

L6 × 3¹⁄₂ × ¹⁄₂

brace detail
(braces to be located at the end of cantilever
at the point of suspension and at quarter points
of the 75.4 ft span)
(not to scale)

SOLUTION 24

24.1. The new steel beam must support a concentrated load equal to the reaction on the existing column at line 2. The spacings of the 10 in × 24 in and 6 in × 19 in beams are not specified. From ACI 318 Table 9-5a, the 4 in roof slab will require support spacing no more than

$$l_n \leq 24h = (24)\left(\frac{4 \text{ in}}{12 \ \frac{\text{in}}{\text{ft}}}\right) = 8 \text{ ft}$$

Space the beams at 8 ft on center.

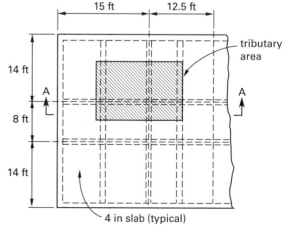

partial roof framing plan

The service load is the weight of the 4 in slab, the roof, the ceiling, and the beams. For the 10 in × 24 in beams at 8 ft on center,

$$w_b = \frac{w_c b (h_b - h_s)}{s}$$

$$= \frac{\left(150 \ \frac{\text{lbf}}{\text{ft}^3}\right)(10 \text{ in})(24 \text{ in} - 4 \text{ in})}{(8 \text{ ft})\left(12 \ \frac{\text{in}}{\text{ft}}\right)^2}$$

$$= 26 \text{ lbf/ft}^2$$

For the 8 in × 19 in beam at $(14 \text{ ft} + 8 \text{ ft})/2 = 11 \text{ ft}$ on center,

$$w_b = \frac{w_c b (h_b - h_s)}{s}$$

$$= \frac{\left(150 \ \frac{\text{lbf}}{\text{ft}^3}\right)(8 \text{ in})(19 \text{ in} - 4 \text{ in})}{(11 \text{ ft})\left(12 \ \frac{\text{in}}{\text{ft}}\right)^2}$$

$$= 11 \text{ lbf/ft}^2$$

Therefore,

$$w_D = w_c h_s + \text{roofing} + \text{ceiling} + \text{beams}$$

$$= \frac{\left(144 \ \frac{\text{lbf}}{\text{ft}^3}\right)(4 \text{ in})}{12 \ \frac{\text{in}}{\text{ft}}} + 10 \ \frac{\text{lbf}}{\text{ft}^2} + 12 \ \frac{\text{lbf}}{\text{ft}^2}$$

$$+ 26 \ \frac{\text{lbf}}{\text{ft}^2} + 11 \ \frac{\text{lbf}}{\text{ft}^2}$$

$$= 107 \text{ lbf/ft}^2$$

$$w_L = 20 \text{ lbf/ft}^2$$

Compute the column reaction based on its tributary area, A_T. The first interior support supports approximately $5/8$ of the load on the exterior span plus $1/2$ of the load on the adjacent interior span.

$$A_T = (0.625 L_1 + 0.5 L_2)(0.625 L_3 + 0.5 L_4)$$

$$= \big((0.625)(15 \text{ ft}) + (0.5)(12.5 \text{ ft})\big)$$

$$\times \big((0.625)(14 \text{ ft}) + (0.5)(8 \text{ ft})\big)$$

$$= 200 \text{ ft}^2$$

$$P_{2D} = w_d A_T = \frac{\left(107 \ \frac{\text{lbf}}{\text{ft}^2}\right)(200 \text{ ft}^2)}{1000 \ \frac{\text{lbf}}{\text{kip}}} = 21.4 \text{ kips}$$

$$P_{2L} = w_L A_T = \frac{\left(20 \ \frac{\text{lbf}}{\text{ft}^2}\right)(200 \text{ ft}^2)}{1000 \ \frac{\text{lbf}}{\text{kip}}} = 4 \text{ kips}$$

Design a new steel beam to span from line 1 to line 3. The span length is

$$L = s_1 + s_2 = 14.7 \text{ ft} + 12.5 \text{ ft} = 27.2 \text{ ft}$$

Use LRFD.

$$P_u \geq \begin{cases} 1.4 P_{2D} = (1.4)(21.4 \text{ kips}) \\ \qquad = 30 \text{ kips} \\ 1.2 P_{2D} + 1.6 P_{2L} = (1.2)(21.4 \text{ kips}) + (1.6)(4 \text{ kips}) \\ \qquad = 32 \text{ kips} \quad \text{[controls]} \end{cases}$$

Estimate the factored weight of a new steel beam as 100 lbf/ft.

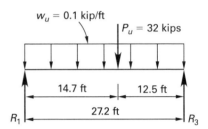

$$\sum M_1 = P_u a + 0.5 w_u L^2 - R_3 L = 0$$

$$R_3 = \frac{P_u a + 0.5 w_u L^2}{L}$$

$$= \frac{(32 \text{ kips})(14.7 \text{ ft}) + (0.5)\left(0.1 \ \frac{\text{kip}}{\text{ft}}\right)(27.2 \text{ ft})^2}{27.2 \text{ ft}}$$

$$= 18.7 \text{ kips}$$

$$\sum F_y = R_1 - w_u L - P_u + R_3 = 0$$

$$R_1 = w_u L + P_u - R_3$$

$$= \left(0.1 \ \frac{\text{kip}}{\text{ft}}\right)(27.2 \text{ ft}) + 32 \text{ kips} - 18.7 \text{ kips}$$

$$= 16.0 \text{ kips}$$

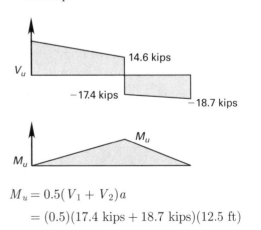

$$M_u = 0.5(V_1 + V_2)a$$

$$= (0.5)(17.4 \text{ kips} + 18.7 \text{ kips})(12.5 \text{ ft})$$

$$= 226 \text{ ft-kips}$$

The beam is laterally unbraced over the full span. $L_b = 27.2$ ft, $C_b = 1.0$, $F_y = 50$ ksi. Using the moment charts from AISC Table 3-10, select W14 × 61 to try. Verify the section properties for W14 × 61: $d = 13.9$ in, $t_w = 0.375$ in, $b_f/2t_f = 7.75$, $h/t_w = 30.4$, $r_y = 2.45$ in, $L_p = 8.65$ ft, $L_r = 27.5$ ft, $Z_x = 102$ in^3, and $S_x = 92.1$ in^3. Check local buckling.

$$\frac{b_f}{2t_f} = 7.75 < 0.38\sqrt{\frac{E}{F_y}} = 0.38\sqrt{\frac{29{,}000\ \frac{\text{kips}}{\text{in}^2}}{50\ \frac{\text{kips}}{\text{in}^2}}} = 9.15$$

$$\frac{h}{t_w} = 30.4 < 3.76\sqrt{\frac{E}{F_y}} = 3.76\sqrt{\frac{29{,}000\ \frac{\text{kips}}{\text{in}^2}}{50\ \frac{\text{kips}}{\text{in}^2}}} = 90.6$$

Therefore, the section is compact. Calculate the flexural strength per AISC Sec. F2.

$$L_p = 8.65\ \text{ft} < L_b = 27.2\ \text{ft} < L_r = 27.5\ \text{ft}$$

Check AISC Eq. F2-2.

$$M_p = F_y Z_x = \left(50\ \frac{\text{kips}}{\text{in}^2}\right)(102\ \text{in}^3) = 5100\ \text{in-kips}$$

$$\phi M_n = \phi C_b\left(M_p - (M_p - 0.7F_y S_x)\left(\frac{L_b - L_p}{L_r - L_p}\right)\right)$$

$$= \frac{(0.9)(1.0)\left(\begin{array}{c} 5100\ \text{in-kips} \\ - \left(\begin{array}{c} 5100\ \text{in-kips} \\ - (0.7)\left(50\ \frac{\text{kips}}{\text{in}^2}\right) \\ \times (92.1\ \text{in}^3) \end{array}\right) \\ \times \left(\dfrac{27.2\ \text{ft} - 8.65\ \text{ft}}{27.5\ \text{ft} - 8.65\ \text{ft}}\right) \end{array}\right)}{12\ \frac{\text{in}}{\text{ft}}}$$

$$= 244\ \text{ft-kips}\quad [> M_u,\ \text{OK}]$$

Check shear per AISC Sec. G2.

$$\phi V_n = \phi(0.6F_y)C_v d t_w$$

$$= (1.0)(0.6)\left(50\ \frac{\text{kips}}{\text{in}^2}\right)(1.0)(13.9\ \text{in})(0.375\ \text{in})$$

$$= 156\ \text{kips}\quad [> V_u,\ \text{OK}]$$

Section is adequate in shear. Since the construction method will allow transferring the concentrated load from the column into the new beam by tightening of rods, the deflection will not be a critical consideration. Therefore, use a W14 × 61.

24.2. Use four hanger rods to support the beam from below.

$$T_u = 0.25P_u = (0.25)\left(32\ \frac{\text{kips}}{\text{rod}}\right) = 8\ \text{kips/rod}$$

Per AISC Table J3.2, for an A36 threaded rod, F_u equals 58 ksi.

$$\phi T_n = \phi(0.75F_u)A_b$$

$$T_u = \phi(0.75F_u)(0.25\pi D^2)$$

$$D = \sqrt{\frac{T_u}{\phi(0.75F_u)(0.25\pi)}}$$

$$= \sqrt{\frac{8\ \text{kips}}{(0.75)(0.75)\left(58\ \frac{\text{kips}}{\text{in}^2}\right)(0.25\pi)}}$$

$$= 0.55\ \text{in}$$

Therefore, use four ⁵⁄₈ in diameter rods, A36 steel.

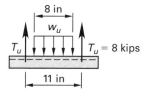

Provide two hanger rods on each side of the beam. Place the hanger rods outside the flange of the W14 × 61 and position one pair of hanger rods to each side of the column that is to be removed as close to face of column as possible. Use C sections to span from side to side and support the 32 kips reaction uniformly on an approximate 11 in span. Each C section will support $0.5P_u = (0.5)(32\ \text{kips}) = 16\ \text{kips}$ and will be bending about its weak axis (y-axis).

$$w_u = \frac{0.5P_u}{b} = \frac{(0.5)(32\ \text{kips})}{8\ \text{in}} = 2\ \text{kips/in}$$

$$M_u = 0.25P_u(0.5L) - w_u(0.5b)(0.25b)$$

$$= (0.25)(32\ \text{kips})(0.5)(11\ \text{in})$$

$$\quad - \left(2\ \frac{\text{kips}}{\text{in}}\right)(0.5)(8\ \text{in})(0.25)(8\ \text{in})$$

$$= 28\ \text{in-kips}$$

$$\phi M_p = \phi Z_y F_y = M_u$$

$$Z_y = \frac{M_u}{\phi F_y} = \frac{28\ \text{in-kips}}{(0.9)\left(36\ \frac{\text{kips}}{\text{in}^2}\right)} = 0.9\ \text{in}^3$$

Use C6 × 8.2.

Connect the rods to the $W14 \times 61$ web with vertical plates using $1/4$ in E70 welds at 5 in eccentricity.

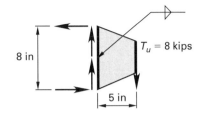

$$f_v = \frac{T_u}{l_w} = \frac{8 \text{ kips}}{(2)(8 \text{ in})} = 0.5 \text{ kip/in}$$

$$f_b = \frac{M}{S_w} = \frac{T_u e}{2\left(\frac{l_w^2}{6}\right)} = \frac{(8 \text{ kips})(5 \text{ in})}{(2)\left(\frac{(8 \text{ in})^2}{6}\right)} = 1.9 \text{ kips/in}$$

$$f = \sqrt{f_v^2 + f_b^2} = \sqrt{\left(0.5 \ \frac{\text{kip}}{\text{in}^2}\right)^2 + \left(1.9 \ \frac{\text{kips}}{\text{in}^2}\right)^2}$$

$$= 1.96 \text{ kips/in}$$

For an E70 fillet weld, the strength is 1.39 kips/in per $1/16$ in of weld. (See AISC Part 8). Therefore, the capacity of a minimum size field weld of $1/4$ in is

$$q_w = D\left(1.39 \ \frac{\text{kips}}{\text{in}}\right) = (4)\left(1.39 \ \frac{\text{kips}}{\text{in}}\right)$$

$$= 5.56 \text{ kips/in} \quad [\text{OK}]$$

Therefore, $\boxed{\text{use } 1/4 \text{ in fillet welds}}$ on each side as shown.

Details of the connection are

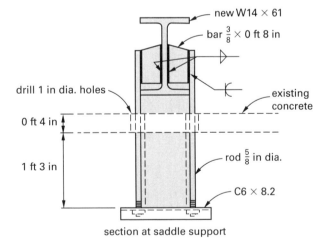

section at saddle support

24.3. The installation procedure is as follows.

First, drill four 1 in diameter holes in the 4 in thick slab.

Second, erect the new $W14 \times 61$ bearing on lines 1 and 3. Let the $5/8$ in diameter rods project through the holes in the roof slab.

Third, place loose $C6 \times 8.2$ members beneath the existing concrete beam. Tighten the nuts on the threaded $5/8$ in rods to transfer the load from the concrete beam to the new steel beam. Monitor the deflection in the steel beam.

Fourth, remove the column.

Seismic Design

PROBLEM 1

The two lines on the graph shown represent the elastic response spectra for a structure with 2% critical viscous damping subjected to the 1940 El Centro earthquake, which had a Richter magnitude of 6.7, and the response spectra used for design per ASCE/SEI7. The structure belonged to seismic design category D, and met the seismic design requirements for a moment-resisting frame building.

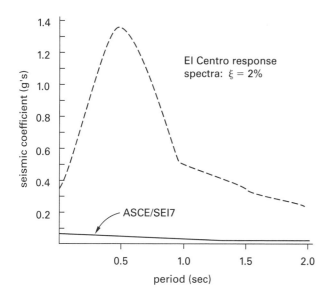

1.1. What are *elastic response spectra*?

1.2. In terms of material behavior, what is the single most important factor that reconciles the apparent discrepancy between earthquake force level and structural design level as illustrated by the preceding graph?

1.3. In addition to the factor referred to in Prob. 1.2, name two other possible factors that might significantly modify the response of a structure to an earthquake.

1.4. In terms of material behavior and the preceding graph, what are two reasons that a building with a bearing wall system is designed with a lateral force coefficient, R, equal to 4, while a building with a special moment-resisting space frame is designed with a coefficient of 8?

1.5. What is the fundamental requirement for the survival of moment-resisting frame structures during major earthquakes if they are designed using ASCE/SEI7 criteria for forces?

1.6. What is the basic difference between one building with a lateral force coefficient, R, of 4.5 and another building where the coefficient is 5.5? Why is the building with the coefficient of 5.5 given the higher coefficient?

1.7. What is the primary purpose of ASCE/SEI7 earthquake regulations? In terms of life safety and property damage, what performance should be expected of structures designed in conformance with these regulations in a

- minor earthquake?
- moderate earthquake?
- major earthquake?

1.8. Some generalizations can be made about how foundation conditions will influence structural response to an earthquake.

(a) What effect does increasing soil thickness have on the predominant period of ground motion and on the amplitude of this motion?

(b) What types of buildings are most likely to be subjected to the greatest seismic force when underlain by firm soils or rock? What period range defines this type of building?

(c) What types of buildings are most likely to be subjected to the greatest seismic force when underlain by soft soil? What period range defines this type of building?

1.9. Earthquakes are measured by various scales. Name two of these scales, and briefly describe the basis for each.

1.10. What minimum requirement must a structural engineer consider in connection with horizontal torsional moments and story shear of a building? How are negative torsional shears considered?

1.11. What is *critical damping*? How is the amount of damping related to the critical damping? In the normal range of accepted values, what are the maxima and minima of damping factors for building? What would be a normal range for steel buildings?

1.12. What is meant by the term *confined concrete*?

1.13. In a typical framed structure, the period for the first mode of vibration is 1.5 sec. How would the period be expected to change for the second mode? How would the period be expected to change for the third mode?

1.14. What is *liquefaction*? How might liquefaction affect buildings?

1.15. What is a *tsunami*?

1.16. A single degree of freedom system is shown. Write the equation of motion for this system when it is subjected to the earthquake ground acceleration $u_g(t)$.

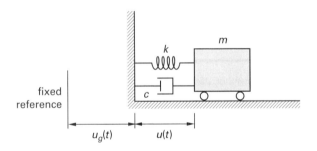

1.17. How would a shear wall structure behave differently from a frame structure with respect to nonstructural seismic damage?

1.18. A building was built in accordance with ASCE/SEI7 earthquake regulations and acceptably withstood a magnitude 7.0 earthquake per these regulations. Briefly discuss three reasons why its performance in a later earthquake might not be acceptable.

1.19. An exterior wall elevation for a symmetrical reinforced concrete building is shown. The elevation shows all vertical and lateral force-resisting elements. The building was designed in accordance with ASCE/SEI7 requirements using a response modification factor, R, of 8.5. Briefly discuss reasons for its failure when subjected to the horizontal ground motion of a magnitude 6.6 earthquake whose epicenter was 6 mi away from the site.

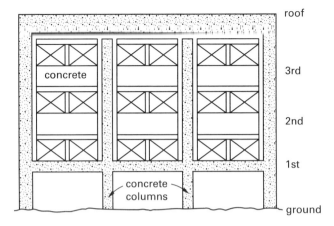

1.20. Dynamic analysis is used to evaluate the behavior of certain types of structures under seismic loadings. Name three types of structures that should be subjected to such analysis.

1.21. What is meant by the *damping characteristics* of a structure?

1.22. Briefly discuss the expected response of a

(a) tall building founded on soft soils

(b) low shear-wall building founded on stiff soils

1.23. Which of the two buildings in Prob. 1.22 will be more significantly affected by a distant earthquake? Explain what the effect might be and why.

1.24. (a) In a 10-story office building, what change in the stiffness (if any) might be expected during extended high-intensity ground motion? What effect accrues to the period, T?

(b) Draw a typical code-specified seismic load distribution for a 10-story building. Draw a typical shear diagram for the same building.

1.25. A choice is to be made between a 20-story building and a 3-story building on the same site. Assuming that the basic frame for each is similar, why is it that the 20-story building may be designed for a lower base shear factor?

1.26. For the joints in a lateral load-resisting concrete frame, why would special care be exercised to confine the concrete?

1.27. What is the approximate proportion of inelastic strain to elastic strain if A36 steel fails in tension at 20% elongation?

1.28. A member in a building (vertical load only) was erected with A325 bearing type bolts loaded in shear. At the time of erection, the impact wrench used was miscalibrated and the applied torque was 50% of the torque specified. What is the resulting effect on the design strength of the connection? Explain.

1.29. Identify three of the most important factors that influence the magnitude of the available ductility in a reinforced concrete frame.

1.30. During the San Fernando earthquake in 1971, many engineered buildings were subjected to a base shear in excess of the code design requirement. Many of these buildings showed only limited distress. Give three reasons why this may have been.

1.31. The second story of a four-story rigid steel frame is infilled with an unreinforced tile wall design as shown. The infill part was not considered in the lateral load design. Describe two important effects on the behavior of the building when subjected to earthquake motion.

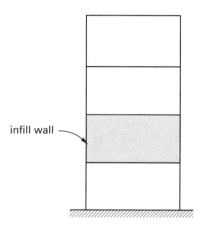

1.32. What are two of the major factors that determine the dynamic characteristics of a building?

1.33. A building site may have a deep, soft soil foundation condition. What might be the effect of this condition on the input ground motion to a tall flexible building, and therefore on the seismic response?

1.34. A proposed shear wall arrangement is shown. Is there anything wrong with this design?

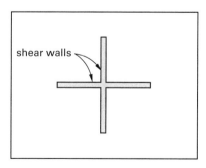

PROBLEM 2

Around the square perimeter of a 12-story building is a series of special steel moment frames. The floor loads and the frame dimensions are shown.

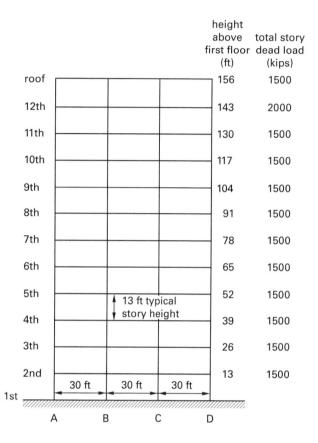

Design Criteria

- risk category II
- site class D
- $I_e = 1.0$
- $S_S = 1.0$, $S_1 = 0.4$
- $T_L = 6$ sec

2.1. Calculate the story shear at each floor.

2.2. Calculate the overturning moment at the first floor due to seismic loads only.

2.3. Calculate the overturning moment at the fourth floor due to seismic loads only.

PROBLEM 3

The two walls shown are joined along the horizontal plane A by an element that is inextensible, and which distributes the force F_Q between the two wall elements.

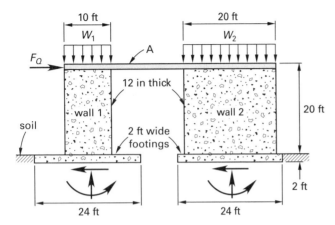

Design Criteria

- walls are 12 in thick reinforced concrete resting on footings

- soil under footings will compress $1/100$ in for each increase of 1000 lbf/ft^2 in soil pressure

- vertical loads W_1 and W_2 are sufficient to maintain compression throughout length of both footings

- $E_c = 3000$ ksi

- $G = 1200$ ksi

3.1. What portion of F_Q is resisted by each wall? Consider flexural and shearing deformations of the wall above the footings, as well as rotation of the footings due to soil compression. Neglect any deformation of the footings themselves.

3.2. What proportion of the force F_Q would each wall resist if flexure and shear in concrete were neglected and only soil compression were considered?

3.3. What proportion of the force F_Q would each wall resist if rotation due to soil yield were neglected and only flexure and shear in concrete were considered?

PROBLEM 4

The design for a symmetrical one story structure is shown. The roof girders have an infinite stiffness relative to the columns. Dimensions are as shown. The structure will be designed for earthquake ground motions that might be expected to occur at the site. These motions are represented by the elastic response spectrum shown. Assume that the structure will remain elastic.

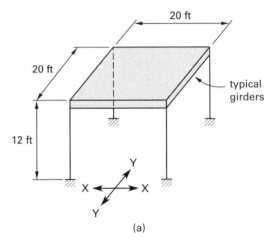

(a)

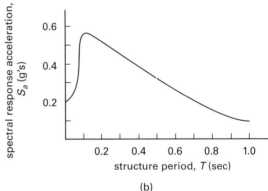

(b)

Design Criteria

- total roof weight $= 40$ lbf/ft^2

- column stiffness is $k_c = 5.0$ kips/in, each way for each column

- weight of walls $= 15$ lbf/ft^2 on all sides of building

4.1. Develop a lumped mass mathematical model of the structure that will be suitable for use in the analysis for the spectrum shown. Assume no damping. Explain what is to be included in mass and stiffness.

4.2. Determine the maximum earthquake force in the X-X direction applied to the roof level.

4.3. Determine the maximum column shear for the force in Prob. 4.2.

4.4. Determine the maximum deflection of the roof in the X-X direction.

PROBLEM 5

A one-story building contains a series of concrete shear walls with their relative stiffnesses, R, as shown. The roof slab is assumed to be infinitely stiff. One-half of the shear wall weight contributes mass at the roof level.

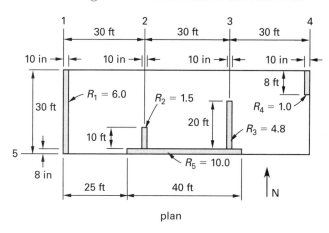

plan

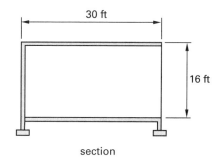

section

Design Criteria

- $w_D = 0.1 \text{ kip/ft}^3$
- $w_c = 0.15 \text{ kip/ft}^3$
- risk category II
- $I_e = 1.0$
- site class D
- $S_S = 0.65$, $S_1 = 0.25$
- $T_L = 6 \text{ sec}$

5.1. For the forces in the north-south direction, calculate the seismic shear carried by each wall.

5.2. For the forces in the east-west direction, calculate the seismic shear carried by each wall.

PROBLEM 6

A two-story concrete building with shear walls at three sides and a concrete rigid frame at the fourth side (line J) is shown.

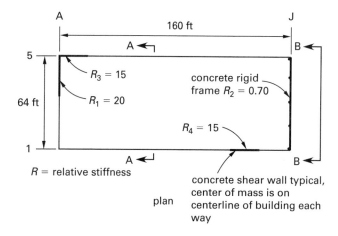

plan

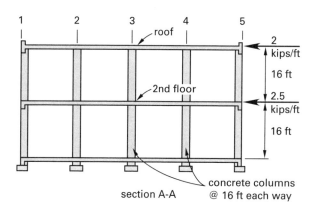

section A-A

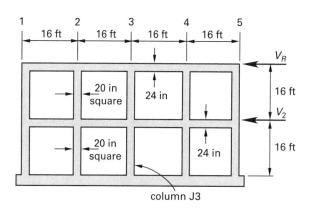

frame elevation B-B

Design Criteria

- rigid diaphragms at second floor and roof levels
- any frame-action resistance involving slabs and columns at all other lines is to be neglected

6.1. Calculate the maximum lateral forces V_R and V_2 that result on the concrete frame due to the lateral loads shown in section A-A. The center of mass coincides with the centerline of the building for both directions. Neglect accidental torsion.

6.2. Using the portal method, find the moment and shear stresses in the first story column J3. Assume that the foundation at line J is sufficiently rigid to fix the column bases.

PROBLEM 7

A steel frame is shown with its fundamental mode shape and a design response spectrum. The frame has a fundamental period of vibration of 0.35 sec.

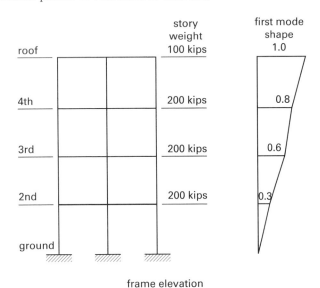

frame elevation

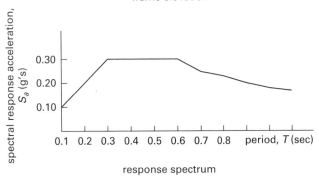

response spectrum

7.1. Determine story forces and base shear.

PROBLEM 8

A two-story steel frame structure is shown. The structure has 7% damping, ζ, and each story is known to deflect 0.1 in under a 10 kip story shear. Answer the following for an x-direction response.

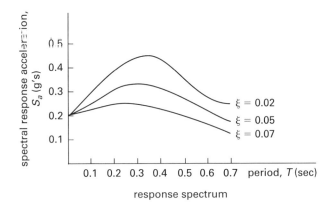

response spectrum

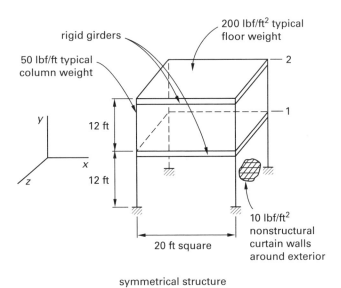

symmetrical structure

8.1. Determine the mathematical model for dynamic analysis. Summarize this in a drawing that indicates story masses and stiffness.

8.2. Draw the approximate first mode shape, and calculate the fundamental period of vibration of the mathematical model. Do not use IBC equations to calculate the period; instead, use recognized approximate methods, such as the Rayleigh method.

Hint: $T_1 \approx 2\pi \sqrt{\dfrac{\sum w\delta^2}{g\sum f\delta}}$

8.3. Assume that for the first mode, $T_1 = 0.5$ sec, and that the mode shape is $\phi_2 = 1.0$ and $\phi_1 = 0.67$. Calculate the first mode story forces.

Hint: $F_i = m_i\phi_i S_a \left(\dfrac{\sum m_i\phi_i}{\sum m_i\phi_i{}^2}\right)$

8.4. What is the peak ground acceleration?

PROBLEM 9

A three-story building with plan dimensions of 100 ft × 100 ft is shown, along with its design response spectra. The effective dead load on each floor is also as shown.

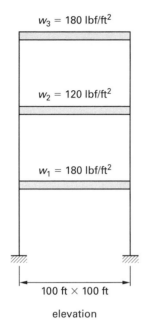

$w_3 = 180 \text{ lbf/ft}^2$

$w_2 = 120 \text{ lbf/ft}^2$

$w_1 = 180 \text{ lbf/ft}^2$

100 ft × 100 ft

elevation

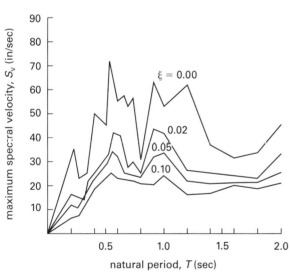

natural period, T (sec)

maximum spectral velocity, S_v (in/sec)

$\xi = 0.00$

0.02

0.05

0.10

Design Criteria

- assume the following matrices

$$\text{mode shape matrix } \phi = \begin{bmatrix} 2.860 & -0.657 & 0.378 \\ 1.959 & 0.725 & -1.610 \\ 1.000 & 1.000 & 1.000 \end{bmatrix}$$

$$\text{modal frequency matrix} = \begin{bmatrix} 15.1 \\ 38.5 \\ 61.7 \end{bmatrix} \text{ rad/sec}$$

9.1. Determine the base shear for each mode by using the design response spectrum with a damping ratio of $\xi = 0.05$.

9.2. Determine the lateral load at each level for each mode.

9.3. What is the most probable design base shear?

PROBLEM 10

The existing steel frame shown is assumed to be restrained perpendicularly to the plane at all joints. The base connections are assumed to be perfectly pinned.

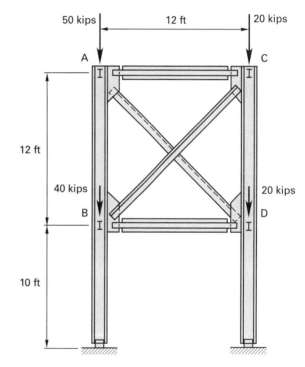

50 kips 12 ft 20 kips

A C

12 ft

40 kips 20 kips

B D

10 ft

Design Criteria

- tributary dead loads at joints A, B, C, and D as shown, weight of frame included
- site class D
- risk category II
- $F_y = 50$ ksi steel
- $S_S = 0.9g$, $S_1 = 0.4g$
- $I_e = 1.0$
- $T_L = 12$ sec

10.1. Calculate forces in the frame due to horizontal seismic loads. Neglect axial deformations.

10.2. Would W10 × 45 columns be adequate? Show calculations.

10.3. How much live load can the frame carry at joint D?

PROBLEM 11

A 9 in thick two-way flat slab system is shown.

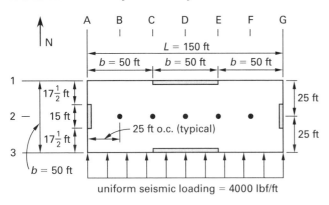

Design Criteria

- shear wall braced building

- shear walls 8 in thick and 10 ft high

- columns 14 in in diameter and 10 ft high

- all walls fixed at bottom, pinned at top

- columns not part of lateral bracing system

- columns fixed at top and bottom

- columns, walls, and diaphragms have the same modulus of elasticity, E_c

- gravity moments in column from frame are 38 ft-kips top and 19 ft-kips bottom

- deflection in shear walls is

$$\Delta = \left(\frac{V}{E_c t}\right)\left(4\left(\frac{h}{d}\right)^3 + 3\left(\frac{h}{d}\right)\right)$$

- deflection in diaphragm is

$$\Delta = \left(\frac{wL}{6.4E_c t}\right)\left(\left(\frac{L}{b}\right)^3 + 2.13\left(\frac{L}{b}\right)\right)$$

- seismic design category D

- $C_d = 5.0$

11.1. Determine the moments at top and bottom of column from seismic loading for column at line 2-D only.

PROBLEM 12

The interior shear wall shown (see *Illustration for Prob. 12*) is of 8 in thick (nominal) concrete block and all cells are fully grouted. A seismic load from a flexible roof diaphragm applied at the top of the wall is determined to be 30 kips.

Design Criteria

- A615 grade 40 reinforcing steel

- $f'_m = 1.5$ ksi

12.1. Determine the lateral load in each of the piers 1 through 6 due to the given load. Neglect weight of the wall for seismic effects.

12.2. Determine the maximum anchorage load from the drag strut to the wall. Neglect pin ends and axial deformation.

12.3. Determine the axial load in pier 5. The seismic load at the top of wall B is 25 kips. Neglect the weight of the wall for seismic effects.

Illustration for Prob. 12

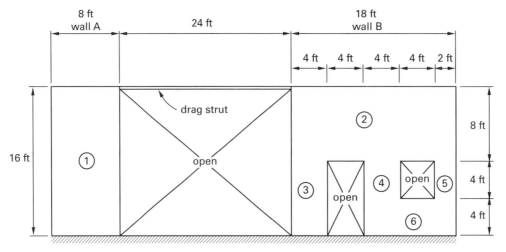

shear wall elevation

SOLUTION 1

1.1. *Elastic response spectra* are plots of maximum response (e.g., displacement, velocity, or acceleration) of a single degree of freedom (SDF) linear oscillator that is subjected to a specific load function (e.g., base motion) over a practical range of periods of the SDF oscillator.

1.2. The most important factor that reconciles the discrepancy between the earthquake and the design level of the response spectra is the inelastic behavior, which results in hysteretic energy dissipation by the structure.

1.3. Other factors that might modify a structure's response to an earthquake are (1) changes in stiffness and (2) changes in damping characteristics as the structure degrades.

1.4. The primary reason that a bearing wall and a special moment resisting space frame will have different lateral force coefficients is the superior energy dissipation capability of the special frame. A secondary reason is that a special moment-resisting frame is typically more redundant and can respond inelastically without jeopardizing vertical capacity and stability. Bearing wall systems have shown variable performance depending on quality of construction.

1.5. In addition to the requirements of strength, stability, and continuous load path, a fundamental requirement for survival of a moment-resisting frame during a strong earthquake is sufficient ductility—that is, inelastic deformation capacity—to dissipate the energy imparted by the ground motion.

1.6. In addition to superior energy dissipation capability, the building with a lateral force coefficient, R, of 5.5 has an essentially complete system to support gravity loads independent of the lateral force resisting system. Therefore, the difference is redundancy. If the lateral force resisting elements fail (that is, where R is 4.5), the structure—or a major part of it—might collapse.

1.7. The primary purpose of the ASCE/SEI7 earthquake regulations is to prevent loss of life by setting minimum standards to prevent structural collapse. According to the *SEAOC Blue Book*, the performance to be expected in structures built to these standards is as follows.

- minor earthquake: no damage

- moderate earthquake: no structural damage, but damage to nonstructural elements is possible

- major earthquake: no collapse, but damage to both structural and nonstructual elements is possible

1.8. (a) Increasing soil thickness generally increases the predominate period of ground motion and amplifies the structural response.

(b) Structures most likely to be subjected to greatest seismic force when underlain by rock are the relatively stiff structures with periods in the range of 0.1 sec to 0.5 sec.

(c) Conversely, relatively flexible structures, those with periods greater than 0.5 sec, are more likely to experience greatest seismic force when underlain by soft soils. The worst situation occurs when the building's period and the predominate period of the ground motion nearly coincide.

1.9. Two scales used to measure earthquake intensity are the *modified Mercalli intensity scale* and the *Richter magnitude scale*. The modified Mercalli scale assigns an earthquake a level of intensity ranging from 1 (not felt except by a few under exceptionally favorable circumstances) to 12 (damage is total). The Richter scale quantifies the earthquake's magnitude using a scale representing the common logarithm (base 10) of the maximum seismic wave amplitude recorded at a distance of 100 km from the earthquake epicenter.

1.10. With respect to ASCE/SEI7 Sec. 12.8.4.2, the minimum accidental eccentricity for torsional moment is 5% of the building dimension perpendicular to the force under consideration. For a building in seismic design categories C through F, with torsional irregularity as defined in ASCE/SEI7 Table 12.3-1, the minimum eccentricity is further amplified by a factor ranging from 1 to 3.

1.11. *Critical damping* is the minimum damping that prevents oscillation of a system when it is disturbed from its equilibrium position. System damping is usually expressed as a percentage of critical damping, called the *damping ratio*. The damping ratio for real structures varies from virtually 0% up to a maximum of about 20%. For most practical building structures, damping in the range of 2% to 7% is more typical of steel structures (2% for rigid, welded structures; 7% for flexible structures riveted or bolted with A307 connectors).

1.12. *Confined concrete* is enclosed by hoops or spirals in such a way that lateral strains are resisted when an element is subjected to uniaxial compression. The compressive strength and ductility are increased over those of an unconfined concrete element.

1.13. A fundamental period of 1.5 sec corresponds to a relatively flexible structure (for example, a 15-story frame). For such structures, a reasonable approximation to the periods of the higher modes is

$$T_1 = 1.5 \text{ sec}$$
$$T_2 \approx \frac{T_1}{3} = \frac{1.5 \text{ sec}}{3} = 0.5 \text{ sec}$$
$$T_3 \approx \frac{T_1}{5} = \frac{1.5 \text{ sec}}{5} = 0.3 \text{ sec}$$

1.14. *Soil liquefaction* is the loss of strength and stiffness that occurs in a loose, saturated soil (usually sand) when the soil particles are jarred and this causes them to compact more densely. The soil loses its shear strength. That is, it behaves as a liquid—and so, loses its bearing capacity. Buildings constructed on this type of soil can settle, tilt, and sustain damage.

1.15. A *tsunami* is a long water wave caused by sudden faulting of the ocean floor.

1.16. The system's equation of motion is

$$m\ddot{u}(t) + c\dot{u}(t) + ku(t) = -m\ddot{u}_g(t)$$

1.17. In comparison to a frame, a shear wall structure will experience significantly less drift and interstory drift when subjected to lateral forces. As a consequence, there is generally less damage to nonstructural items in a shear wall structure than would occur to similar items in a frame structure.

1.18. An earthquake of magnitude 7.0 would likely cause several cycles of inelastic response in the structure. This would likely cause changes to the stiffness and damping characteristics of the building (especially if it is constructed of reinforced concrete). The building's response might then not be acceptable in a subsequent earthquake for several reasons.

- The duration of strong motion in the second earthquake may be significantly longer, and site-specific response more severe, than in the first earthquake.

- The cumulative damage of inelastic response cycles might be too much, causing the building to fail.

- The degradation of strength and stiffness caused by the first earthquake might increase drift, causing instability due to P-Δ effects.

1.19. The structure contains a vertical irregularity called a soft story, which places too large an inelastic rotation demand on the first story columns and beam-column joints. Furthermore, the design based on an R-factor of 8.5 yields a base shear that is too small for a structure with this irregularity.

1.20. Structures that should be analyzed dynamically include

- irregular structures (e.g., unusual plan configurations, abrupt changes in strength and stiffness from floor to floor, large variations in mass, or other irregularities—see ASCE/SEI7 Tables 12.3-1 and 12.3-2)

- tall, regular structures (i.e., those for which response is significantly affected by the higher modes of vibration)

- structures founded on deep deposits of soft soil (e.g., where dynamic response may be amplified by soil-structure interaction)

1.21. *Damping characteristics* relate to the mechanisms that dissipate energy in an oscillating system (examples include Coulomb friction and hysteretic material behavior). To simplify computations, structural damping is usually approximated as viscous (that is, directly proportional to velocity).

1.22. (a) Soft soil amplifies the dynamic response significantly for structures with periods greater than 0.5 sec. Tall buildings generally have periods greater than 0.5 sec; therefore, a tall building on soft soil would experience greater lateral force and displacements than if it were founded on rock.

(b) A low shear-wall building on stiff soil is likely to be so stiff that it would move with the ground, and so would experience nearly the same acceleration as the ground.

1.23. A distant earthquake would more likely affect a tall building founded on soft soil. The effective peak ground accelerations diminish with increased distance from the epicenter of an earthquake. However, the components having longer periods transmit farther. Also, the duration of strong ground motion is greater on soft soil. The periods of motion for soft soil and a tall building are relatively close. All these factors can contribute to create a measurable response in the tall building caused by an earthquake several hundreds of miles away.

1.24. (a) Under extended high-intensity ground motion, the stiffness of the 10 story building would likely degrade. This would result in an increase in its period.

(b) The typical code-specified seismic load distribution and shear diagram for a 10-story building is approximately as shown.

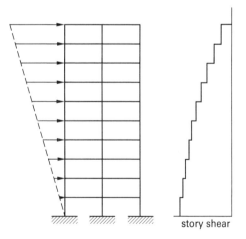

story shear

1.25. The 20-story building may have a smaller base shear coefficient because its period of vibration is generally such that it is on the descending branch of the design response spectrum (that is, a period greater than about 0.5 sec for a building founded on stiff soil).

1.26. Concrete confinement increases the toughness of the joints, and hence their ability to dissipate energy.

1.27. The strain corresponding to 20% elongation is 0.2. The yield strain for A36 steel is

$$\epsilon_y = \frac{F_y}{E_s} = \frac{36 \; \frac{\text{kips}}{\text{in}^2}}{29{,}000 \; \frac{\text{kips}}{\text{in}^2}} = 0.00124$$

Thus, the proportion of inelastic to elastic strain is

$$\frac{\epsilon_{\text{inelastic}}}{\epsilon_y} = \frac{0.2000 - 0.00124}{0.00124} = 160$$

1.28. Bearing type bolts depend on their shear strength rather than on the frictional resistance created when the bolts are tightened. Thus, the bolts in this situation would develop their design strength. The only effect would be slippage in the joint (on the order of $^1/_{16}$ in), which might affect serviceability. In general, bearing type connections are required to be tightened only to a "snug tight" condition, which is roughly the torque that causes an impact wrench to impact.

1.29. Three factors that influence ductility in a reinforced concrete frame are

- the longitudinal steel ratio, $\rho = A_s/bd$ (the higher the steel ratio, the lower the ductility)

- concrete confinement (that is, hoops or spirals)

- shear strength and development of reinforcement

1.30. Many buildings showed only limited distress resulting from the 1971 San Fernando earthquake, despite being subjected to a higher base shear than prescribed. Some reasons for this include the following.

- Resistance was furnished by nonstructural elements, such as partitions and curtain walls.

- Some buildings were designed to earlier, more conservative codes with respect to member strength (for example, using working stress design).

- Some buildings were proportioned for nonseismic loading conditions (for example, low, light-frame structures that were governed by wind design rather than seismic design).

1.31. The infill wall will

- stiffen the frame significantly, causing a change in its dynamic response (for example, shortening its fundamental period)

- create an abrupt change in stiffness, producing high inelastic rotation demand on the columns above and below the in-filled story

1.32. The most influential factors affecting dynamic response are the magnitude and distribution of mass and stiffness.

1.33. The deep, soft soil can amplify the input ground motion. Response is especially severe if the building's period is close to the period of the soil layer, in which case resonance may occur.

1.34. The wall arrangement has poor torsional rigidity. A slight eccentricity in an applied lateral force can cause a large twisting deformation.

SOLUTION 2

2.1. In this solution, all equation and table citations refer to ASCE/SEI7. Calculate story shears due to seismic loads. For site class D, Tables 11.4-1 and 11.4-2 give $F_a = 1.1$ and $F_v = 1.6$. The design spectral response parameters are found using Eqs. 11.4-1 through 11.4-4.

$$S_{MS} = F_a S_S = (1.1)(1.0) = 1.1$$

$$S_{DS} = 0.667 S_{MS} = (0.667)(1.1) = 0.73$$

$$S_{M1} = F_v S_1 = (1.6)(0.4) = 0.64$$

$$S_{D1} = 0.667 S_{M1} = (0.667)(0.64) = 0.43$$

From Tables 11.6-1 and 11.6-2, risk category II, with S_{DS} less than 0.5 and S_{D1} greater than 0.2, the seismic design category is D. From Table 12.8-2 and Eq. 12.8-7, find the approximate fundamental period, T_a, of the steel moment frame. h_n is the building height in feet. (This formula is not dimensionally consistent.)

$$T_a = C_t h_n^x = (0.028)(156 \text{ ft})^{0.8} = 1.6 \text{ sec}$$

Check the applicability of the equivalent lateral force procedure per Table 12.6-1. The limiting period is

$$3.5 T_S = 3.5 \left(\frac{S_{D1}}{S_{DS}} \right) (1 \text{ sec})$$

$$= (3.5) \left(\frac{0.43}{0.73} \right) (1 \text{ sec})$$

$$= 2.07 \text{ sec} \quad [> T_a]$$

The structure is regular and the period is less than the limiting value of $3.5 T_s$, so the equivalent lateral force method applies. For special steel moment-resisting frames, Table 12.2-1 specifies a response modification coefficient, R, of 8. The seismic response coefficient, C_s, must be at least 0.01 and is limited by Eq. 12.8-2. Since

$T = T_a$ is less than T_L, Eq. 12.8-3 applies. (The units of T are ignored.)

$$C_s \leq \begin{cases} \dfrac{S_{DS}I_e}{R} = \dfrac{(0.73)(1.0)}{8} = 0.091 \\[2mm] \dfrac{S_{D1}I_e}{TR} = \dfrac{(0.43)(1.0)}{(1.6)(8)} = 0.034 \quad \text{[controls]} \end{cases}$$

The seismic dead load is the summation of the given dead loads from level 2 to level 13.

$$W = \sum_{i=2}^{13} w_i = (11)(1500 \text{ kips}) + 2000 \text{ kips} = 18{,}500 \text{ kips}$$

From Eq. 12.8-1, the total seismic base shear is

$$V = C_s W = (0.034)(18{,}500 \text{ kips}) = 629 \text{ kips}$$

The vertical distribution of the lateral forces is given by Eqs. 12.8-11 and 12.8-12.

$$F_x = C_{vx} V$$

$$C_{vx} = \frac{w_x h_x^k}{\displaystyle\sum_{i=1}^{n} w_i h_i^k}$$

n is the total number of levels, and i and x represent a particular level. The exponent k is found by linear interpolation between 1.0 at period 0.5 sec and 2.0 at 2.5 sec.

$$k = 0.75 + 0.5 T_a = 0.75 + (0.5)(1.6 \text{ sec}) = 1.55$$

Values of the equivalent lateral forces, F_x, computed by the above provisions are as follows.

level	w_x (kips)	h_x (ft)	$w_x h_x^{1.55}$	F_x (kips)
1	0	0	0	0.0
2	1500	13	79,929	2.4
3	1500	26	234,045	7.1
4	1500	39	438,775	13.3
5	1500	52	685,325	20.8
6	1500	65	968,516	29.4
7	1500	78	1,284,807	39.0
8	1500	91	1,631,569	49.5
9	1500	104	2,006,748	60.9
10	1500	117	2,408,681	73.1
11	1500	130	2,835,982	86.1
12	2000	143	4,383,297	133.1
13	1500	156	3,762,135	114.2
Σ	18,500		20,719,809	629

The lateral force, F, and story shear at each floor are shown.

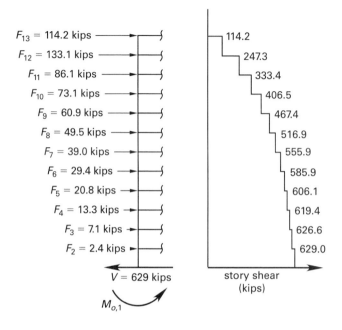

$F_{13} = 114.2$ kips — 114.2
$F_{12} = 133.1$ kips — 247.3
$F_{11} = 86.1$ kips — 333.4
$F_{10} = 73.1$ kips — 406.5
$F_9 = 60.9$ kips — 467.4
$F_8 = 49.5$ kips — 516.9
$F_7 = 39.0$ kips — 555.9
$F_6 = 29.4$ kips — 585.9
$F_5 = 20.8$ kips — 606.1
$F_4 = 13.3$ kips — 619.4
$F_3 = 7.1$ kips — 626.6
$F_2 = 2.4$ kips — 629.0

$V = 629$ kips   story shear (kips)

$M_{o,1}$

2.2. The overturning moment at the first floor due only to seismic loads is

$$\begin{aligned} M_{o,1} &= \sum_{i=2}^{13} F_i h_i \\ &= (114.2 \text{ kips})(156 \text{ ft}) + (133.1 \text{ kips})(143 \text{ ft}) \\ &\quad + (86.1 \text{ kips})(130 \text{ ft}) + (73.1 \text{ kips})(117 \text{ ft}) \\ &\quad + (60.9 \text{ kips})(104 \text{ ft}) + (49.5 \text{ kips})(91 \text{ ft}) \\ &\quad + (39.0 \text{ kips})(78 \text{ ft}) + (29.4 \text{ kips})(65 \text{ ft}) \\ &\quad + (20.8 \text{ kips})(52 \text{ ft}) + (13.3 \text{ kips})(39 \text{ ft}) \\ &\quad + (7.1 \text{ kips})(26 \text{ ft}) + (2.4 \text{ kips})(13 \text{ ft}) \\ &= \boxed{74{,}200 \text{ ft-kips}} \end{aligned}$$

2.3. Calculate the overturning moment at the fourth floor. Use the value for $M_{o,1}$ from Sol. 2.2. Take the area under the shear diagram to get the change in moment from the first floor to the fourth floor.

$$\begin{aligned} M_{o,4} &= M_{o,1} - \sum_{i=2}^{4} V_i \Delta h_i \\ &= 74{,}200 \text{ ft-kips} \\ &\quad - \left(\begin{array}{l} (629 \text{ kips})(13 \text{ ft}) + (626.6 \text{ kips})(13 \text{ ft}) \\ + (619.4 \text{ kips})(13 \text{ ft}) \end{array} \right) \\ &= \boxed{49{,}800 \text{ ft-kips}} \end{aligned}$$

SOLUTION 3

3.1. Calculate the shear in each wall as a function of F_Q. Both shear and flexural deformations in the wall are to be considered.

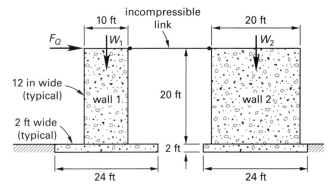

Let αF_Q be the shear in wall 1, where α is a constant. Then $(1 - \alpha)F_Q$ is the shear in wall 2. For wall 1, calculate the peak change in soil bearing pressure, f_1, caused by the lateral force.

$$M_o = \alpha F_Q h = \alpha F_Q (22 \text{ ft})$$

$$S = \frac{bL^2}{6} = \frac{(2 \text{ ft})(24 \text{ ft})^2}{6} = 192 \text{ ft}^3$$

$$f_1 = \frac{M_o}{S} = \frac{\alpha F_Q (22 \text{ ft})}{192 \text{ ft}^3} = 0.1146 \alpha F_Q \text{ ft}^{-2}$$

Calculate the deformation in the soil at the tip of the footing caused by the change in bearing pressure. It is given that the soil can compress 0.01 in for each 1000 lbf/ft^2 in soil pressure.

$$\Delta_{f1} = k f_1 = \left(\frac{0.01 \text{ in}}{1000 \frac{\text{lbf}}{\text{ft}^2}} \right) \left(1000 \frac{\text{lbf}}{\text{kip}} \right)$$

$$\times \left(0.1146 \alpha F_Q \text{ ft}^{-2} \right)$$

$$= 0.001146 \alpha F_Q \text{ in/kip}$$

The deflection at the top of the wall due to support rotation is

$$\Delta_1 = \theta h = \left(\frac{\Delta_{f1}}{0.5L} \right) h$$

$$= \left(\frac{0.001146 \alpha F_Q \frac{\text{in}}{\text{kip}}}{(0.5)(24 \text{ ft})} \right) (22 \text{ ft})$$

$$= 0.0021 \alpha F_Q \text{ in/kip}$$

Similarly, for wall 2 with the same size footing,

$$M_o = \alpha F_Q h = (1 - \alpha) F_Q (22 \text{ ft})$$

$$f_2 = \frac{M_o}{S} = \frac{(1 - \alpha) F_Q (22 \text{ ft})}{192 \text{ ft}^3}$$

$$= 0.1146 \alpha F_Q \text{ ft}^{-2}$$

$$\Delta_{f2} = k f_2 = \left(\frac{0.01 \text{ in}}{1000 \frac{\text{lbf}}{\text{ft}^2}} \right) \left(1000 \frac{\text{lbf}}{\text{kip}} \right)$$

$$\times \left(0.1146 (1 - \alpha) F_Q \text{ ft}^{-2} \right)$$

$$= 0.001146 (1 - \alpha) F_Q \text{ in/kip}$$

$$\Delta_2 = \theta h = \left(\frac{\Delta_{f2}}{0.5L} \right) h$$

$$= \left(\frac{0.001146 (1 - \alpha) F_Q \frac{\text{in}}{\text{kip}}}{(0.5)(24 \text{ ft})} \right) (22 \text{ ft})$$

$$= 0.0021 (1 - \alpha) F_Q \text{ in/kip}$$

The shear and flexural deflection in the walls occur only in the regions above the top of the footing (that is, for $h = 20$ ft). For wall 1,

$$\Delta'_1 = \frac{\alpha F_Q h^3}{3EI} + \frac{1.2 \alpha F_Q h}{GA}$$

$$A = BL = (1 \text{ ft})(10 \text{ ft}) \left(12 \frac{\text{in}}{\text{ft}} \right)^2 = 1440 \text{ in}^2$$

$$I = \frac{BL^3}{12} = \left(\frac{(1 \text{ ft})(10 \text{ ft})^3}{12} \right) \left(12 \frac{\text{in}}{\text{ft}} \right)^4$$

$$= 1,728,000 \text{ in}^4$$

$$\Delta'_1 = \frac{\alpha F_Q \left((20 \text{ ft}) \left(12 \frac{\text{in}}{\text{ft}} \right) \right)^3}{(3) \left(3000 \frac{\text{kips}}{\text{in}^2} \right) (1,728,000 \text{ in}^4)}$$

$$+ \frac{1.2 \alpha F_Q (20 \text{ ft}) \left(12 \frac{\text{in}}{\text{ft}} \right)}{\left(1200 \frac{\text{kips}}{\text{in}^2} \right) (1440 \text{ in}^2)}$$

$$= 0.001055 \alpha F_Q \text{ in/kip}$$

For wall 2,

$$\Delta_2' = \frac{(1-\alpha)F_Q h^3}{3EI} + \frac{1.2(1-\alpha)F_Q h}{GA}$$

$$A = BL = (1\ \text{ft})(20\ \text{ft})\left(12\ \frac{\text{in}}{\text{ft}}\right)^2 = 2880\ \text{in}^2$$

$$I = \frac{BL^3}{12} = \left(\frac{(1\ \text{ft})(20\ \text{ft})^3}{12}\right)\left(12\ \frac{\text{in}}{\text{ft}}\right)^4$$
$$= 13{,}824{,}000\ \text{in}^4$$

$$\Delta_2' = \frac{(1-\alpha)F_Q\left((20\ \text{ft})\left(12\ \frac{\text{in}}{\text{ft}}\right)\right)^3}{(3)\left(3000\ \frac{\text{kips}}{\text{in}^2}\right)(13{,}824{,}000\ \text{in}^4)}$$

$$+ \frac{1.2\alpha F_Q(20\ \text{ft})\left(12\ \frac{\text{in}}{\text{ft}}\right)}{\left(1200\ \frac{\text{kips}}{\text{in}^2}\right)(2880\ \text{in}^2)}$$

$$= 0.0001943(1-\alpha)F_Q\ \text{in/kip}$$

The total deflections are

$$\Delta_{1T} = \Delta_1 + \Delta_1'$$
$$= (0.0021 + 0.001055)\alpha F_Q\ \frac{\text{in}}{\text{kip}}$$
$$= 0.003155\alpha F_Q\ \text{in/kip}$$

$$\Delta_{2T} = \Delta_2 + \Delta_2'$$
$$= (0.0021 + 0.0001943)(1-\alpha)F_Q\ \frac{\text{in}}{\text{kip}}$$
$$= 0.0022943(1-\alpha)F_Q\ \text{in/kip}$$

For consistent displacements, make the deflections equal, cancel the F_Q factor, and solve for α.

$$\Delta_{1T} = \Delta_{2T}$$
$$0.003155\alpha F_Q\ \frac{\text{in}}{\text{kip}} = 0.0022943(1-\alpha)F_Q\ \frac{\text{in}}{\text{kip}}$$
$$\alpha = \frac{0.0022943}{0.003155 + 0.0022943}$$
$$= 0.42$$
$$1 - \alpha = 0.58$$

$\boxed{42\%}$ of the force F_Q is resisted by wall 1, and the remaining $\boxed{58\%}$ is resisted by wall 2.

3.2. If shear and flexural wall deflections are neglected, then, as the footings are of identical size and are assumed to be rigid, $\boxed{50\%}$ of the force F_Q is resisted by wall 1, and $\boxed{50\%}$ is resisted by wall 2.

3.3. If soil compression is neglected, the compatibility equation is

$$\Delta_1' = \Delta_2'$$
$$0.001055\alpha F_Q\ \frac{\text{in}}{\text{kip}} = 0.0001943(1-\alpha)F_Q\ \frac{\text{in}}{\text{kip}}$$
$$\alpha = \frac{0.0001943}{0.001055 + 0.0001943}$$
$$= 0.16$$

Under this assumption, wall 1 resists only $\boxed{16\%}$ of the force F_Q, and wall 2 resists $\boxed{84\%.}$

SOLUTION 4

4.1. Develop a lumped mass mathematical model of the structure. The structure is regular and symmetrical and is assumed to remain elastic, so a two-dimensional single degree of freedom model is sufficient. Lumped masses should include the contribution from the roof plus one-half the mass of the exterior walls (use the total 20 ft × 20 ft roof area).

$$W = w_d BL + w_{\text{wall}}(2L + 2B)(0.5h)$$
$$= \left(40\ \frac{\text{lbf}}{\text{ft}^2}\right)(20\ \text{ft})(20\ \text{ft}) + \left(15\ \frac{\text{lbf}}{\text{ft}^2}\right)$$
$$\times \big((2)(20\ \text{ft}) + (2)(20\ \text{ft})\big)(0.5)(12\ \text{ft})$$
$$= 23{,}200\ \text{lbf}$$

$$m = \frac{W}{g} = \frac{23{,}200\ \text{lbf}}{32.2\ \dfrac{\text{ft}}{\text{sec}^2}} = 720\ \text{lbf-sec}^2/\text{ft}$$

The stiffness of each column is given as 5.0 kips/in per column, so the total stiffness for four columns is

$$k_{\text{total}} = 4k_c = (4)\left(5000\ \frac{\text{lbf}}{\text{in}}\right)\left(12\ \frac{\text{in}}{\text{ft}}\right)$$
$$= 240{,}000\ \text{lbf/ft}$$

The equivalent lumped mass model is

4.2. Determine the maximum earthquake force in the X-X direction. For a single degree of freedom, undamped harmonic oscillator, the natural frequency is

$$\omega = \sqrt{\frac{k_{\text{total}}}{m}} = \sqrt{\frac{240{,}000\ \dfrac{\text{lbf}}{\text{ft}}}{720\ \dfrac{\text{lbf-sec}^2}{\text{ft}}}} = 18.3\ \text{rad/sec}$$

The period is

$$T = \frac{2\pi}{\omega} = \frac{2\pi}{18.3 \; \frac{\text{rad}}{\text{sec}}} = 0.34 \text{ sec}$$

Reading from the given elastic response spectrum, the spectral acceleration is

$$S_a = 0.45g = (0.45)\left(32.2 \; \frac{\text{ft}}{\text{sec}^2}\right) = 14.5 \text{ ft/sec}^2$$

Therefore, the maximum force is

$$V = mS_a = \frac{\left(720 \; \frac{\text{lbf-sec}^2}{\text{ft}}\right)\left(14.5 \; \frac{\text{ft}}{\text{sec}^2}\right)}{1000 \; \frac{\text{lbf}}{\text{kip}}}$$

$$= \boxed{10.4 \text{ kips}}$$

4.3. Determine the maximum column shear corresponding to the force of 10.4 kips determined in Prob. 4.2. Assume a 5% minimum eccentricity to be consistent with building codes. The torsional moment is

$$M_t = Ve_{\min} = V(0.05L)$$

$$= (10.4 \text{ kips})(0.05)(20 \text{ ft})$$

$$= 10.4 \text{ ft-kips}$$

$$\sum(x_i^2 + y_i^2) = (4)\left((10 \text{ ft})^2 + (10 \text{ ft})^2\right)$$

$$= 800 \text{ ft}^2$$

Therefore, the maximum column shear is

$$V_{\text{col}} = 0.5\left(\frac{V}{2} + \frac{M_t x}{\sum(x_i^2 + y_i^2)}\right)$$

$$= (0.5)\left(\frac{10.4 \text{ kips}}{2} + \frac{(10.4 \text{ ft-kips})(10 \text{ ft})}{800 \text{ ft}^2}\right)$$

$$= \boxed{2.7 \text{ kips}}$$

4.4. The spectral displacement is related to the spectral acceleration by the equation $S_d = S_a/\omega^2$. The maximum displacement (deflection) of the roof in the X-X direction is

$$S_d = \frac{S_a}{\omega^2} = \frac{\left(14.5 \; \frac{\text{ft}}{\text{sec}^2}\right)\left(12 \; \frac{\text{in}}{\text{ft}}\right)}{\left(18.3 \; \frac{\text{rad}}{\text{sec}}\right)^2}$$

$$= \boxed{0.52 \text{ in}}$$

SOLUTION 5

5.1. All equation and table citations in this solution refer to ASCE/SEI7. Calculate the seismic shear in the walls for north-south ground motion. Per the design criteria, include one-half the weight of the shear walls in the dead load for computation of seismic base shear.

$$W_{\text{wall}} = w_c(0.5h)\sum_{i=1}^{5} t_i L_i$$

$$= \left(0.15 \; \frac{\text{kip}}{\text{ft}^3}\right)(0.5)(16 \text{ ft})$$

$$\times \left(\begin{array}{l} (0.83 \text{ ft})(30 \text{ ft}) + (0.83 \text{ ft})(20 \text{ ft}) \\ + (0.83 \text{ ft})(10 \text{ ft}) + (0.83 \text{ ft})(8 \text{ ft}) \\ + (0.67 \text{ ft})(40 \text{ ft}) \end{array} \right)$$

$$= 100 \text{ kips}$$

$$W_{\text{roof}} = w_D(BL) = \left(0.1 \; \frac{\text{kip}}{\text{ft}^3}\right)(30 \text{ ft})(90 \text{ ft})$$

$$= 270 \text{ kips}$$

$$W = W_{\text{roof}} + W_{\text{wall}}$$

$$= 270 \text{ kips} + 100 \text{ kips}$$

$$= 370 \text{ kips}$$

For site class D, $S_S = 0.65$ and $S_1 = 0.25$. Tables 11.4-1 and 11.4-2 give $F_a = 1.28$ (by linear interpolation) and $F_v = 1.9$. The design spectral response parameters are found using Eqs. 11.4-1 through 11.4-4.

$$S_{MS} = F_a S_S = (1.28)(0.65) = 0.83$$

$$S_{DS} = 0.667 S_{MS} = (0.667)(0.83) = 0.55$$

$$S_{M1} = F_v S_1 = (1.9)(0.25) = 0.48$$

$$S_{D1} = 0.667 S_{M1} = (0.667)(0.48) = 0.32$$

From Tables 11.6-1 and 11.6-2 for risk category II, with S_{DS} greater than 0.5 and S_{D1} greater than 0.2, the seismic design category is D. From Table 12.8-2 and Eq. 12.8-7, find the approximate period of the concrete bearing wall structure. h_n is the building height in feet. (This formula is not dimensionally consistent.)

$$T = T_a = C_t h_n^x = (0.02)(16 \text{ ft})^{0.75}$$

$$= 0.16 \text{ sec}$$

For a single story building in risk category II, Table 12.6-1 permits the base shear to be found using the equivalent lateral force procedure in Sec. 12.8. For seismic design category D, shear walls must be designed and detailed as special reinforced concrete walls. From Table 12.2-1, $R = 5$. The seismic response coefficient,

C_s, must be at least 0.01 and is limited by Eqs. 12.8-2 and 12.8-3. ($T < T_L$; the units of T are ignored.)

$$C_s \le \begin{cases} \dfrac{S_{DS}I_e}{R} = \dfrac{(0.55)(1.0)}{5} = 0.11 \quad \text{[controls]} \\[2ex] \dfrac{S_{D1}I_e}{TR} = \dfrac{(0.32)(1.0)}{(0.16 \text{ sec})(5)} = 0.40 \end{cases}$$

The base shear is

$$V = C_s W = (0.11)(370 \text{ kips}) = 40.7 \text{ kips}$$

The distribution of shear to elements of the rigid diaphragm-shear wall system will be computed using the centerline-to-centerline dimensions between walls.

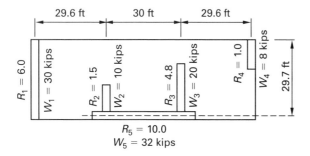

The center of rigidity is at the centerline of wall 5, because it is the only shear-resisting element in the east-west direction. For the other direction,

$$\sum_{i=1}^{5} R_{y,i} = 6R + 1.5R + 4.8R + 1.0R + 0R = 13.3R$$

$$\overline{x} = \dfrac{\displaystyle\sum_{i=1}^{5} x_i R_{y,i}}{\displaystyle\sum_{i=1}^{5} R_{y,i}}$$

$$= \dfrac{\begin{array}{l}(0 \text{ ft})(6R) + (29.6 \text{ ft})(1.5R) + (59.6 \text{ ft}) \\ \times (4.8R) + (89.2 \text{ ft})(1.0R) + (45 \text{ ft})(0R)\end{array}}{13.3R}$$

$$= 31.6 \text{ ft} \quad \text{[from centerline of wall 1]}$$

The center of mass (including one-half the mass of the walls) is located at

$$x_m = \dfrac{w_c(0.5h)\displaystyle\sum_{i=1}^{5} t_i L_i x_i + W_{\text{roof}}\big(0.5(L - t_1)\big)}{W}$$

$$= \dfrac{\begin{array}{l}\left(0.15 \dfrac{\text{kip}}{\text{ft}^3}\right)(0.5)(16 \text{ ft}) \\[1ex] \times \left(\begin{array}{l}(0.83 \text{ ft})(30 \text{ ft})(0 \text{ ft}) + (0.83 \text{ ft})(10 \text{ ft}) \\ \times (29.6 \text{ ft}) + (0.83 \text{ ft})(20 \text{ ft})(59.6 \text{ ft}) \\ + (0.83 \text{ ft})(8 \text{ ft})(89.2 \text{ ft}) + (0.67 \text{ ft}) \\ \times (40 \text{ ft})(45 \text{ ft})\end{array}\right) \\[1ex] + (270 \text{ kips})\big((0.5)(90 \text{ ft} - 0.83 \text{ ft})\big)\end{array}}{370 \text{ kips}}$$

$$= 42.4 \text{ ft}$$

$$y_m = \dfrac{w_c(0.5h)\displaystyle\sum_{i=1}^{5} t_i L_i y_i + W_{\text{roof}}(0.5B)}{W}$$

$$= \dfrac{\begin{array}{l}\left(0.15 \dfrac{\text{kip}}{\text{ft}^3}\right)(0.5)(16 \text{ ft}) \\[1ex] \times \left(\begin{array}{l}(0.83 \text{ ft})(30 \text{ ft})(15 \text{ ft}) + (0.83 \text{ ft})(10 \text{ ft}) \\ \times (5.33 \text{ ft}) + (0.83 \text{ ft})(20 \text{ ft})(10.33 \text{ ft}) \\ + (0.83 \text{ ft})(8 \text{ ft})(25.7 \text{ ft}) + (0.67 \text{ ft}) \\ \times (40 \text{ ft})(0.33 \text{ ft})\end{array}\right) \\[1ex] + (270 \text{ kips})(0.5)(29.4 \text{ ft})\end{array}}{370 \text{ kips}}$$

$$= 13.2 \text{ ft}$$

Tabulate the properties of the shear walls resisting the seismic base shear.

wall	R_x	R_y	d_x (ft)	d_y (ft)	$R_x d_y^2 + R_y d_x^2$
1	–	$6.0R$	-31.6	–	$(5990 \text{ ft}^2)R$
2	–	$1.5R$	-2.0	–	$(6 \text{ ft}^2)R$
3	–	$4.8R$	28.0	–	$(3760 \text{ ft}^2)R$
4	–	$1.0R$	57.6	–	$(3320 \text{ ft}^2)R$
5	$10R$	–	–	0	$(0 \text{ ft}^2)R$
Σ	$10R$	$13.3R$			$(13{,}080 \text{ ft}^2)R$

Check for torsional irregularity by calculating the displacement of the rigid diaphragm due to lateral force plus torsional moment. Include the accidental eccentricity $(0.05L)$, per Sec. 12.8.4.2.

$$e_{\text{acc}} = 0.05L = (0.05)(90 \text{ ft}) = 4.5 \text{ ft}$$

$$e_{\text{act}} = (x_m - \bar{x}) = (42.4 \text{ ft} - 31.6 \text{ ft}) = 10.8 \text{ ft}$$

$$M_t = V(e_{\text{act}} + e_{\text{acc}}) = (40.7 \text{ kips})(10.8 \text{ ft} + 4.5 \text{ ft})$$

$$= 623 \text{ ft-kips}$$

Calculate the shear in walls 1 and 4.

$$V_1 = \frac{VR_{y,1}}{\sum R_y} + \frac{M_t R_{y,1} d_{x,1}}{\sum R_y d_x^2 + R_x d_y^2}$$

$$= \frac{(40.7 \text{ kips})(6R)}{13.3R} + \frac{(623 \text{ ft-kips})(6R)(-31.6 \text{ ft})}{(13,080 \text{ ft}^2)R}$$

$$= 9.3 \text{ kips}$$

$$V_4 = \frac{VR_{y,4}}{\sum R_y} + \frac{M_t R_{y,4} d_{x,4}}{\sum R_y d_x^2 + R_x d_y^2}$$

$$= \frac{(40.7 \text{ kips})(1R)}{13.3R} + \frac{(623 \text{ ft-kips})(1R)(57.6 \text{ ft})}{(13,080 \text{ ft}^2)R}$$

$$= 5.8 \text{ kips}$$

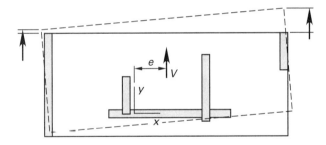

The relative wall rigidities are given in terms of R, which has units of kips per unit displacement; therefore, only relative displacements can be calculated, but these are all that is needed to check for torsional irregularity.

$$\delta_1 = \frac{V_1}{R_1} = \frac{9.3 \text{ kips}}{6 \dfrac{\text{kips}}{\text{unit}}} = 1.55 \text{ units}$$

$$\delta_4 = \frac{V_4}{R_4} = \frac{5.8 \text{ kips}}{1 \dfrac{\text{kip}}{\text{unit}}} = 5.8 \text{ units} \quad [= \delta_{\text{max}}]$$

$$\delta_{\text{ave}} = \frac{\delta_1 + \delta_4}{2} = \frac{1.55 \text{ units} + 5.8 \text{ units}}{2} = 3.7 \text{ units}$$

$$\frac{\delta_{\text{max}}}{\delta_{\text{ave}}} = \frac{5.8 \text{ units}}{3.7 \text{ units}} = 1.57$$

The ratio of maximum to average diaphragm displacements exceeds 1.4, which defines extreme torsional irregularity in Table 12.3-1. Section 12.8.4.3 requires amplification of the accidental torsion by the factor A_x.

$$A_x \leq \begin{cases} \left(\dfrac{\delta_{\text{max}}}{1.2\delta_{\text{ave}}}\right)^2 = \left(\dfrac{1.57}{1.2}\right)^2 = 1.71 \quad [\text{controls}] \\ 3.0 \end{cases}$$

Therefore,

$$e_{\text{acc}} = A_x(0.05L) = (1.71)(0.05)(90 \text{ ft}) = 7.7 \text{ ft}$$

$$M_t = V(e_{\text{act}} \pm e_{\text{acc}})$$

$$= (40.7 \text{ kips})(10.8 \text{ ft} \pm 7.7 \text{ ft})$$

$$= \begin{cases} 753 \text{ ft-kips} \\ 126 \text{ ft-kips} \end{cases}$$

The larger value of M_t applies when the torsional effect increases the direct shear, and the smaller value applies when the torsional effect reduces the direct shear. Recalculating with the amplified torsional moment,

$$V_1 = \frac{VR_{y,1}}{\sum R_y} + \frac{M_t R_{y,1} d_{x,1}}{\sum R_y d_x^2 + R_x d_y^2}$$

$$= \frac{(40.7 \text{ kips})(6R)}{13.3R} + \frac{(126 \text{ ft-kips})(6R)(-31.6 \text{ ft})}{(13,080 \text{ ft}^2)R}$$

$$= \boxed{16.5 \text{ kips}}$$

$$V_2 = \frac{VR_{y,2}}{\sum R_y} + \frac{M_t R_{y,2} d_{x,2}}{\sum R_y d_x^2 + R_x d_y^2}$$

$$= \frac{(40.7 \text{ kips})(1.5R)}{13.3R} + \frac{(126 \text{ ft-kips})(1.5R)(-2.0 \text{ ft})}{(13,080 \text{ ft}^2)R}$$

$$= \boxed{4.6 \text{ kips}}$$

$$V_3 = \frac{VR_{y,3}}{\sum R_y} + \frac{M_t R_{y,3} d_{x,3}}{\sum R_y d_x^2 + R_x d_y^2}$$

$$= \frac{(40.7 \text{ kips})(4.8R)}{13.3R} + \frac{(753 \text{ ft-kips})(4.8R)(28.0 \text{ ft})}{(13,080 \text{ ft}^2)R}$$

$$= \boxed{22.4 \text{ kips}}$$

$$V_4 = \frac{VR_{y,4}}{\sum R_y} + \frac{M_t R_{y,4} d_{x,4}}{\sum R_y d_x^2 + R_x d_y^2}$$

$$= \frac{(40.7 \text{ kips})(1R)}{13.3R} + \frac{(753 \text{ ft-kips})(1R)(57.6 \text{ ft})}{(13,080 \text{ ft}^2)R}$$

$$= \boxed{6.4 \text{ kips}}$$

$$V_5 = \frac{VR_{y,5}}{\sum R_y} + \frac{M_t R_{x,5} d_{y,5}}{\sum R_y d_x^2 + R_x d_y^2}$$

$$= \frac{(40.7 \text{ kips})(0R)}{13.3R} + \frac{(126 \text{ ft-kips})(0R)(0 \text{ ft})}{(13{,}080 \text{ ft}^2)R}$$

$$= \boxed{0 \text{ kips}}$$

5.2. Calculate the distribution of wall shear for east-west ground motion. Check for torsional irregularity.

$$e_{acc} = 0.05B = (0.05)(30 \text{ ft}) = 1.5 \text{ ft}$$

$$e_{act} = y_m - \overline{y} = 13.2 \text{ ft} - 0 \text{ ft} = 13.2 \text{ ft}$$

$$M_t = V(e_{act} + e_{acc})$$

$$= (40.7 \text{ kips})(13.2 \text{ ft} + 1.5 \text{ ft})$$

$$= 598 \text{ ft-kips}$$

$$V_1 = \frac{VR_{x,1}}{\sum R_x} + \frac{M_t R_{y,1} d_{x,1}}{\sum R_y d_x^2 + R_x d_y^2}$$

$$= \frac{(40.7 \text{ kips})(0R)}{10R} + \frac{(598 \text{ ft-kips})(6R)(-31.6 \text{ ft})}{(13{,}080 \text{ ft}^2)R}$$

$$= -8.7 \text{ kips}$$

$$V_4 = \frac{VR_{x,4}}{\sum R_x} + \frac{M_t R_{y,4} d_{x,4}}{\sum R_y d_x^2 + R_x d_y^2}$$

$$= \frac{(40.7 \text{ kips})(0R)}{10R} + \frac{(598 \text{ ft-kips})(1R)(57.6 \text{ ft})}{(13{,}080 \text{ ft}^2)R}$$

$$= 2.6 \text{ kips}$$

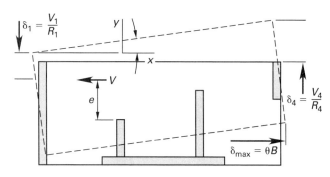

The relative displacements of the diaphragm for east-west seismic force can be calculated from the rotation caused by the applied torsional moment.

$$\delta_1 = \frac{V_1}{R_1} = \frac{-8.7 \text{ kips}}{6 \dfrac{\text{kips}}{\text{unit}}} = -1.45 \text{ units} \quad \text{[downward]}$$

$$\delta_4 = \frac{V_4}{R_4} = \frac{2.6 \text{ kips}}{1 \dfrac{\text{kip}}{\text{unit}}} = 2.6 \text{ units} \quad \text{[upward]}$$

$$\delta_{max} = \theta B = \left(\frac{\delta_4 - \delta_1}{L}\right)B$$

$$= \left(\frac{2.6 \text{ units} - (-1.45 \text{ units})}{90 \text{ ft}}\right)(30 \text{ ft})$$

$$= 1.35 \text{ units}$$

$$\delta_{ave} = \frac{\delta_{max}}{2} = \frac{1.35 \text{ units}}{2} = 0.68 \text{ units}$$

$$\frac{\delta_{max}}{\delta_{ave}} = \frac{1.35 \text{ units}}{0.68 \text{ units}} = 2.0$$

The ratio of maximum to average diaphragm displacements exceeds 1.4, which defines extreme torsional irregularity in Table 12.3-1. Section 12.8.4.3 requires amplification of the accidental torsion by the factor A_x.

$$A_x \leq \begin{cases} \left(\dfrac{\delta_{max}}{1.2\delta_{ave}}\right)^2 = \dfrac{\left(\dfrac{\delta_{max}}{\delta_{ave}}\right)^2}{1.2} = \left(\dfrac{2.0}{1.2}\right)^2 = 2.8 \quad \text{[controls]} \\ \\ 3.0 \end{cases}$$

Therefore,

$$e_{acc} = A_x(0.05B) = (2.8)(0.05)(30 \text{ ft}) = 4.2 \text{ ft}$$

$$M_t = V(e_{act} \pm e_{acc})$$

$$= (40.7 \text{ kips})(13.2 \text{ ft} \pm 4.2 \text{ ft})$$

$$= \begin{cases} 708 \text{ ft-kips} \\ 366 \text{ ft-kips} \end{cases}$$

For east-west ground motion, maximum shear occurs in all walls when M_t equals 708 ft-kips.

$$V_1 = \frac{VR_{x,1}}{\sum R_x} + \frac{M_t R_{y,1} d_{x,1}}{\sum R_y d_x^2 + R_x d_y^2}$$

$$= \frac{(40.7 \text{ kips})(0R)}{10R} + \frac{(708 \text{ ft-kips})(6R)(-31.6 \text{ ft})}{(13{,}080 \text{ ft}^2)R}$$

$$= \boxed{-10.3 \text{ kips}}$$

$$V_2 = \frac{VR_{x,2}}{\sum R_x} + \frac{M_t R_{y,2} d_{x,2}}{\sum R_y d_x^2 + R_x d_y^2}$$

$$= \frac{(40.7 \text{ kips})(0R)}{10R} + \frac{(708 \text{ ft-kips})(1.5R)(-2.0 \text{ ft})}{(13{,}080 \text{ ft}^2)R}$$

$$= \boxed{-0.2 \text{ kips}}$$

$$V_3 = \frac{VR_{x,3}}{\sum R_x} + \frac{M_t R_{y,3} d_{x,3}}{\sum R_y d_x^2 + R_x d_y^2}$$

$$= \frac{(40.7 \text{ kips})(0R)}{10R} + \frac{(708 \text{ ft-kips})(4.8R)(28.0 \text{ ft})}{(13{,}080 \text{ ft}^2)R}$$

$$= \boxed{7.3 \text{ kips}}$$

$$V_4 = \frac{VR_{x,4}}{\sum R_x} + \frac{M_t R_{y,4} d_{x,4}}{\sum R_y d_x^2 + R_x d_y^2}$$

$$= \frac{(40.7 \text{ kips})(0R)}{10R} + \frac{(708 \text{ ft-kips})(1R)(57.7 \text{ ft})}{(13,080 \text{ ft}^2)R}$$

$$= \boxed{3.1 \text{ kips}}$$

$$V_5 = \frac{VR_{x,5}}{\sum R_x} + \frac{M_t R_{x,5} d_{y,5}}{\sum R_y d_x^2 + R_x d_y^2}$$

$$= \frac{(40.7 \text{ kips})(10R)}{10R} + \frac{(708 \text{ ft-kips})(10R)(0 \text{ ft})}{(13,080 \text{ ft}^2)R}$$

$$= \boxed{40.7 \text{ kips}}$$

SOLUTION 6

6.1. Calculate the lateral force acting on the rigid frame at line J. The resultant lateral forces at the center of mass are

$$F_2 = w_2 L$$

$$= \left(2.5 \ \frac{\text{kips}}{\text{ft}}\right)(160 \text{ ft})$$

$$= 400 \text{ kips}$$

$$F_R = w_R L$$

$$= \left(2.0 \ \frac{\text{kips}}{\text{ft}}\right)(160 \text{ ft})$$

$$= 320 \text{ kips}$$

Locate the center of rigidity, using the given rigidities and overall dimensions. From the illustration, the sum of the rigidities in the x-direction is

$$\sum R_x = 15 + 15 = 30$$

The sum in the y-direction is

$$\sum R_y = 20 + 0.70 = 20.7$$

The center of rigidity is

$$\overline{x} = \frac{\sum R_{y,i} x_i}{\sum R_y} = \frac{(20)(0 \text{ ft}) + (0.70)(160 \text{ ft})}{20.7} = 5.4 \text{ ft}$$

e_y is 32 ft (from symmetry).

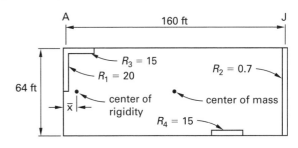

element	R_x	R_y	d_x (ft)	d_y (ft)	Rd^2 (ft^2)
1	–	20	–5.4	–	583
2	–	0.7	154.6	–	16,731
3	15	–	–	32	15,360
4	15	–	–	32	15,360
sum	30	20.7			48,000

The actual eccentricity is

$$e_x = x_m - \overline{x} = 80 \text{ ft} - 5.4 \text{ ft} = 74.6 \text{ ft}$$

Per the problem statement, accidental eccentricity is to be ignored. In reality, an amplified accidental eccentricity would likely apply to this building, considering the apparent torsional irregularity that exists. This solution, however, is based on the actual eccentricity.

$$M_{t,R} = F_R e_x = (320 \text{ kips})(74.6 \text{ ft}) = 23,870 \text{ ft-kips}$$

$$V_R = \frac{F_R R_{y,2}}{\sum R_y} + \frac{M_{t,R} R_{y,2} d_{x,2}}{\sum Rd^2}$$

$$= \frac{(320 \text{ kips})(0.7)}{20.7} + \frac{(23,870 \text{ ft-kips})(0.7)(154.6 \text{ ft})}{48,000}$$

$$= \boxed{64.6 \text{ kips}}$$

Similarly,

$$M_{t,2} = F_2 e_x = (400 \text{ kips})(74.6 \text{ ft})$$

$$= 29,840 \text{ ft-kips}$$

$$V_2 = \frac{F_2 R_{y,2}}{\sum R_y} + \frac{M_{t,2} R_{y,2} d_{x,2}}{\sum Rd^2}$$

$$= \frac{(400 \text{ kips})(0.7)}{20.7} + \frac{(29,840 \text{ ft-kips})(0.7)(154.6 \text{ ft})}{48,000}$$

$$= \boxed{80.8 \text{ kips}}$$

6.2. Analyze column J3 for the lateral forces using the portal method (first story only).

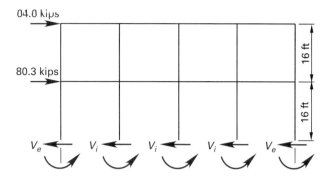

04.0 kips

80.3 kips

V_e V_i V_i V_i V_e

16 ft

16 ft

For a portal analysis, column shears on interior columns are assumed to be twice the shear on the exterior columns, and points of inflection are assumed to occur at the midheight of columns and the midspan of girders. Therefore,

$$V_{\text{ext}} = \frac{V_R + V_2}{n}$$

$$= \frac{(64.6 \text{ kips}) + (80.3 \text{ kips})}{8}$$

$$= 18.2 \text{ kips}$$

$$V_{\text{int}} = 2 V_{\text{ext}} = (2)(18.2 \text{ kips}) = 36.4 \text{ kips}$$

Points of inflection, PI, are assumed to occur at the midheight of columns. The portal analysis yields zero axial force in an interior column.

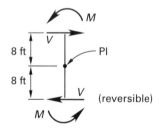

M

V

8 ft

PI

8 ft

V

(reversible)

M

Therefore, the moment and shear stresses are

$$V = V_{\text{int}} = \boxed{36.4 \text{ kips}}$$

$$M = \frac{Vh}{2} = \frac{(36.4 \text{ kips})(16 \text{ ft})}{2}$$

$$= \boxed{291 \text{ ft-kips}}$$

SOLUTION 7

7.1. Data are given only for the first mode. Therefore, the modal analysis will be based on generalized coordinates for a single degree of freedom system (see Naeim, Chap. 4). The generalized mass is

$$\phi = \{1.0,\ 0.8,\ 0.6,\ 0.3\}$$

$$m^* = \sum_{i=1}^{4} m_i \phi_i^2 = \sum_{i=1}^{4} \left(\frac{w_i}{g}\right) \phi_i^2$$

$$= \frac{\begin{array}{c}(100 \text{ kips})(1.0)^2 + (200 \text{ kips})(0.8)^2 \\ + (200 \text{ kips})(0.6)^2 + (200 \text{ kips})(0.3)^2\end{array}}{\left(32.2 \dfrac{\text{ft}}{\text{sec}^2}\right)\left(12 \dfrac{\text{in}}{\text{ft}}\right)}$$

$$= 0.823 \text{ kip-sec}^2/\text{in}$$

The participation factor for the first mode is

$$L = \sum_{i=1}^{4} m_i \phi_i = \sum_{i=1}^{4} \left(\frac{w_i}{g}\right) \phi_i$$

$$= \frac{\begin{array}{c}(100 \text{ kips})(1.0) + (200 \text{ kips})(0.8) \\ + (200 \text{ kips})(0.6) + (200 \text{ kips})(0.3)\end{array}}{\left(32.2 \dfrac{\text{ft}}{\text{sec}^2}\right)\left(12 \dfrac{\text{in}}{\text{ft}}\right)}$$

$$= 1.139 \text{ kip-sec}^2/\text{in}$$

level	ϕ_i	$m_i \phi_i/L$
4	1.0	0.227
3	0.8	0.364
2	0.6	0.272
1	0.3	0.136
Σ		1.000

For the given fundamental period of 0.35 sec, the response spectrum gives

$$S_a = 0.3g = (0.3)\left(32.2 \dfrac{\text{ft}}{\text{sec}^2}\right)\left(12 \dfrac{\text{in}}{\text{ft}}\right) = 115.92 \text{ in/sec}^2$$

The peak lateral force at each level x is given by

$$q(x)_{\text{max}} = \frac{\phi(x) m(x) L S_a}{m^*}$$

The story forces are

$$q(4)_{\text{max}} = \frac{\begin{array}{c}(1.0)\left(\dfrac{100 \text{ kips}}{386.4 \frac{\text{in}}{\text{sec}^2}}\right)\left(1.139 \dfrac{\text{kip-sec}^2}{\text{in}}\right) \\ \times \left(115.92 \dfrac{\text{in}}{\text{sec}^2}\right)\end{array}}{0.823 \dfrac{\text{kip-sec}^2}{\text{in}}}$$

$$= \boxed{41.5 \text{ kips}}$$

$$q(3)_{\max} = \frac{(0.8)\left(\dfrac{200 \text{ kips}}{386.4 \dfrac{\text{in}}{\text{sec}^2}}\right)\left(1.139 \dfrac{\text{kip-sec}^2}{\text{in}}\right)}{0.823 \dfrac{\text{kip-sec}^2}{\text{in}}} \times \left(115.92 \dfrac{\text{in}}{\text{sec}^2}\right)$$

$$= \boxed{66.4 \text{ kips}}$$

$$q(2)_{\max} = \frac{(0.6)\left(\dfrac{200 \text{ kips}}{386.4 \dfrac{\text{in}}{\text{sec}^2}}\right)\left(1.139 \dfrac{\text{kip-sec}^2}{\text{in}}\right)}{0.823 \dfrac{\text{kip-sec}^2}{\text{in}}} \times \left(115.92 \dfrac{\text{in}}{\text{sec}^2}\right)$$

$$= \boxed{49.8 \text{ kips}}$$

$$q(1)_{\max} = \frac{(0.3)\left(\dfrac{200 \text{ kips}}{386.4 \dfrac{\text{in}}{\text{sec}^2}}\right)\left(1.139 \dfrac{\text{kip-sec}^2}{\text{in}}\right)}{0.823 \dfrac{\text{kip-sec}^2}{\text{in}}} \times \left(115.92 \dfrac{\text{in}}{\text{sec}^2}\right)$$

$$= \boxed{24.9 \text{ kips}}$$

The base shear can be found either by summing the story shears calculated above or from the equation

$$V_{\max} = \frac{W^* S_a}{g} = W^* \left(\frac{S_a}{g}\right)$$

$$= \left(\frac{\left(\displaystyle\sum_{i=1}^{4} w_i \phi_i\right)^2}{\displaystyle\sum_{i=1}^{4} w_i \phi_i^2}\right)\left(\frac{0.3g}{g}\right)$$

$$= \left(\frac{\begin{pmatrix} (100 \text{ kips})(1.0) + (200 \text{ kips})(0.8) \\ + (200 \text{ kips})(0.6) \\ + (200 \text{ kips})(0.3) \end{pmatrix}^2}{\begin{matrix} (100 \text{ kips})(1.0)^2 + (200 \text{ kips})(0.8)^2 \\ + (200 \text{ kips})(0.6)^2 \\ + (200 \text{ kips})(0.3)^2 \end{matrix}}\right)(0.3)$$

$$= \boxed{183 \text{ kips}}$$

SOLUTION 8

8.1. Determine an appropriate mathematical model for the structure. For the two-story, symmetrical, regular structure, an appropriate mathematical model is a planar structure undergoing horizontal displacements only (that is, no torsion). Because girders are relatively rigid, a two-story shear building with mass lumped at the story levels is appropriate. Consider one-half of the symmetrical building and take the weight of one-half the exterior walls and columns as contributing mass at each level. The mass and weight of a given level i are

$$m_i = \frac{W_i}{g}$$

$$W_i = W_{\text{floor},i} + W_{\text{wall},i} + W_{\text{col},i}$$

For the second level,

$$\begin{aligned} W_2 &= w_{\text{floor}} BL + w_{\text{wall}}(0.5h)(L + 2B) + w_{\text{col}}(2)(0.5h) \\ &= \left(0.2 \frac{\text{kip}}{\text{ft}^2}\right)(20 \text{ ft})(10 \text{ ft}) + \left(0.01 \frac{\text{kip}}{\text{ft}^2}\right) \\ &\quad \times (0.5)(12 \text{ ft})\big(20 \text{ ft} + (2)(10 \text{ ft})\big) \\ &\quad + \left(0.05 \frac{\text{kip}}{\text{ft}^2}\right)(2)(0.5)(12 \text{ ft}) \\ &= 43.0 \text{ kips} \end{aligned}$$

$$m_2 = \frac{W_2}{g} = \frac{43.0 \text{ kips}}{32.2 \dfrac{\text{ft}}{\text{sec}^2}} = 1.34 \text{ kip-sec}^2/\text{ft}$$

For the first level,

$$\begin{aligned} W_1 &= w_{\text{floor}} BL + w_{\text{wall}} h(L + 2B) + w_{\text{col}}(2h) \\ &= \left(0.2 \frac{\text{kip}}{\text{ft}^2}\right)(20 \text{ ft})(10 \text{ ft}) + \left(0.01 \frac{\text{kip}}{\text{ft}^2}\right) \\ &\quad \times (12 \text{ ft})\big(20 \text{ ft} + (2)(10 \text{ ft})\big) \\ &\quad + \left(0.05 \frac{\text{kip}}{\text{ft}^2}\right)(2)(12 \text{ ft}) \\ &= 46.0 \text{ kips} \end{aligned}$$

$$m_1 = \frac{W_1}{g} = \frac{46.0 \text{ kips}}{32.2 \dfrac{\text{ft}}{\text{sec}^2}} = 1.43 \text{ kip-sec}^2/\text{ft}$$

The column stiffness is such that 10 kips causes a translation of 1 in. The equivalent stiffness of the columns in the model is one-half the stiffness of each story, so

$$\begin{aligned} k_1 = k_2 &= 0.5\left(\frac{\text{story shear}}{\text{drift}}\right) \\ &= (0.5)\left(\frac{10 \text{ kips}}{0.1 \text{ in}}\right) \\ &= 50 \text{ kips/in} \end{aligned}$$

Therefore, the equivalent model may be drawn as

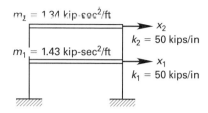

8.2. Approximate the first mode shape and fundamental period of the structure. Use Rayleigh's method. Apply an arbitrary system of lateral forces (for example, 1 kip at the top and 0.5 kip at the first story to give a triangular distribution).

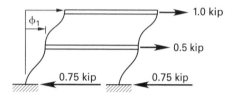

Each story deflects 0.1 in per 10 kips, so each column deflects 0.1 in per 2.5 kips.

$$k_{\text{col}} = \left(\frac{2.5 \text{ kips}}{0.1 \text{ in}}\right)\left(12 \frac{\text{in}}{\text{ft}}\right) = 300 \text{ kips/ft}$$

$$\delta_1 = \frac{V_1}{k_{\text{col}}} = \frac{0.75 \text{ kip}}{300 \frac{\text{kips}}{\text{ft}}} = 0.0025 \text{ ft}$$

$$\delta_2 = \delta_1 + \frac{V_2}{k_{\text{col}}}$$

$$= 0.0025 \text{ ft} + \frac{0.5 \text{ kip}}{300 \frac{\text{kips}}{\text{ft}}}$$

$$= 0.00417 \text{ ft}$$

Using Rayleigh's equation, the fundamental period of vibration is

$$T_1 = 2\pi\sqrt{\frac{\sum w_i \delta_i^2}{g \sum f_i \delta_i}}$$

$$= 2\pi\sqrt{\frac{\begin{array}{c}(46.0 \text{ kip})(0.0025 \text{ ft})^2 \\ + (43.1 \text{ kips})(0.00417 \text{ ft})^2\end{array}}{\left(32.2 \frac{\text{ft}}{\text{sec}^2}\right)\left(\begin{array}{c}(0.5 \text{ kip})(0.0025 \text{ ft}) \\ + (1.0 \text{ kip})(0.00417 \text{ ft})\end{array}\right)}}$$

$$= \boxed{0.48 \text{ sec}}$$

8.3. The period is assumed to be $T_1 = 0.5$ sec, and the normalized displacements for the first mode are assumed to be $\psi = \{0.07, 1.00\}$. The story forces are to be calculated using

$$F_i = m_i \phi_i S_a \frac{\sum m_i \phi_i}{\sum m_i \phi_i^2}$$

S_a is the spectral pseudo-acceleration corresponding to a damping ratio $\zeta = 0.07$. The lowest curve on the given response spectrum corresponds to 7% damping. Thus, for a fundamental period of 0.5 sec,

$$S_a = 0.2g = (0.2)\left(32.2 \frac{\text{ft}}{\text{sec}^2}\right) = 6.44 \text{ ft/sec}^2$$

Calculate the story shears.

$$F_i = m_i \phi_i S_a \left(\frac{\sum m_i \phi_i}{\sum m_i \phi_i^2}\right) = m_i \phi_i S_a \Gamma = \left(\frac{w_i}{g}\right)\phi_i S_a \Gamma$$

$$\Gamma = \frac{\sum m_i \phi_i}{\sum m_i \phi_i^2} = \frac{\sum \left(\frac{w_i}{g}\right)\phi_i}{\sum \left(\frac{w_i}{g}\right)\phi_i^2}$$

$$= \frac{(46 \text{ kips})(0.67) + (43.1 \text{ kips})(1.00)}{(46 \text{ kips})(0.67)^2 + (43.1 \text{ kips})(1.00)^2}$$

$$= 1.16$$

The first mode story forces are

$$F_1 = \left(\frac{46 \text{ kips}}{g}\right)(0.67)(0.2g)(1.16)$$

$$= \boxed{7.2 \text{ kips}}$$

$$F_2 = \left(\frac{43.1 \text{ kips}}{g}\right)(1.00)(0.2g)(1.16)$$

$$= \boxed{10.0 \text{ kips}}$$

8.4. Determine the effective peak ground acceleration. For a rigid structure, T_1 approaches zero, so the acceleration of the structure is the same as that of the ground. For the given response spectra, the curves approach $0.2g$ as T approaches zero. Therefore,

$$S_{g,\text{max}} = 0.2g$$

$$= \boxed{6.44 \text{ ft/sec}^2}$$

SOLUTION 9

9.1. The mode shapes are

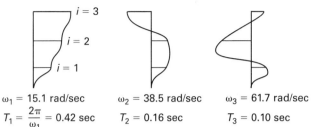

$\omega_1 = 15.1$ rad/sec $\omega_2 = 38.5$ rad/sec $\omega_3 = 61.7$ rad/sec

$T_1 = \dfrac{2\pi}{\omega_1} = 0.42$ sec $T_2 = 0.16$ sec $T_3 = 0.10$ sec

Velocity response spectra are given for ξ-values of zero, 0.02, 0.05, and so on. Use the spectrum for $\xi = 0.05$. For $\omega_1 = 15.1$ rad/sec,

$$T_1 = \frac{2\pi}{\omega_1} = \frac{2\pi}{15.1 \, \dfrac{\text{rad}}{\text{sec}}}$$
$$= 0.42 \text{ sec} \quad [\text{read } S_v = 25 \text{ in/sec}]$$

So, for story 1,

$$S_{a,1} = \omega_1 S_v = \left(15.1 \, \frac{\text{rad}}{\text{sec}}\right)\left(25 \, \frac{\text{in}}{\text{sec}}\right)$$
$$= 378 \text{ in/sec}^2$$

Similarly, for $\omega_2 = 38.5$ rad/sec,

$$T_2 = \frac{2\pi}{\omega_2} = \frac{2\pi}{38.5 \, \dfrac{\text{rad}}{\text{sec}}}$$
$$= 0.16 \text{ sec} \quad [\text{read } S_V = 10 \text{ in/sec}]$$

So, for story 2,

$$S_{a,2} = \omega_2 S_v = \left(38.5 \, \frac{\text{rad}}{\text{sec}}\right)\left(10 \, \frac{\text{in}}{\text{sec}}\right)$$
$$= 385 \text{ in/sec}^2$$

For $\omega_3 = 61.7$ rad/sec,

$$T_3 = \frac{2\pi}{\omega_3} = \frac{2\pi}{61.7 \, \dfrac{\text{rad}}{\text{sec}}}$$
$$= 0.10 \text{ sec} \quad [\text{read } S_V = 8 \text{ in/sec}]$$

So, for story 3,

$$S_{a,3} = \omega_3 S_v = \left(61.7 \, \frac{\text{rad}}{\text{sec}}\right)\left(8 \, \frac{\text{in}}{\text{sec}}\right)$$
$$= 493 \text{ in/sec}^2$$

For the 100 ft × 100 ft floor plan and the given seismic dead loads, the mass at each story is

$$m_1 = \frac{w_1}{g} = \frac{w_{d,1} BL}{g}$$
$$= \frac{\left(0.18 \, \dfrac{\text{kip}}{\text{ft}^2}\right)(100 \text{ ft})(100 \text{ ft})}{g}$$
$$= \frac{1800 \text{ kips}}{g}$$

$$m_2 = \frac{w_2}{g} = \frac{w_{d,2} BL}{g}$$
$$= \frac{\left(0.12 \, \dfrac{\text{kip}}{\text{ft}^2}\right)(100 \text{ ft})(100 \text{ ft})}{g}$$
$$= \frac{1200 \text{ kips}}{g}$$

$$m_3 = m_1$$
$$= \frac{1800 \text{ kips}}{g}$$

Use the response spectrum analysis outlined in App. D in the *SEAOC Blue Book* for a two-dimensional building model, with NP = N = 3 (stories).

$$S_{a,i} = \left(\frac{S_{A,i}}{g_{\text{in/sec}^2}}\right) g$$

$$S_{a,1} = \left(\frac{378 \, \dfrac{\text{in}}{\text{sec}^2}}{386.4 \, \dfrac{\text{in}}{\text{sec}^2}}\right) g = 0.98g$$

$$S_{a,2} = \left(\frac{385 \, \dfrac{\text{in}}{\text{sec}^2}}{386.4 \, \dfrac{\text{in}}{\text{sec}^2}}\right) g = 1.00g$$

$$S_{a,3} = \left(\frac{493 \, \dfrac{\text{in}}{\text{sec}^2}}{386.4 \, \dfrac{\text{in}}{\text{sec}^2}}\right) g = 1.28g$$

$$m = \sum m_i = m_1 + m_2 + m_3$$
$$= \frac{1800 \text{ kips}}{g} + \frac{1200 \text{ kips}}{g} + \frac{1800 \text{ kips}}{g}$$
$$= \frac{4800 \text{ kips}}{g}$$

$\phi_1 = (1.000, 1.959, 2.860)^T$

$$M_1 = \sum_{i=1}^{3} \phi_{1,i}^2 m_i$$

$$= \frac{(1.00)^2(1800 \text{ kips}) + (1.959)^2(1200 \text{ kips}) + (2.86)^2(1800 \text{ kips})}{g}$$

$$= \frac{21{,}128 \text{ kips}}{g}$$

The participation factor for the first mode is

$$p_1 = \frac{\displaystyle\sum_{i=1}^{3} \phi_{1,i} m_i}{\displaystyle\sum_{i=1}^{3} \phi_{1,i}^2 m_i}$$

$$= \frac{\dfrac{(1.00)(1800 \text{ kips}) + (1.952)(1200 \text{ kips}) + (2.86)(1800 \text{ kips})}{g}}{\dfrac{21{,}128 \text{ kips}}{g}}$$

$$= 0.44$$

Similarly, for the second and third modes,

$$\phi_2 = (1.000, 0.725, -0.657)^T$$

$$M_2 = \sum_{i=1}^{3} \phi_{2,i}^2 m_i$$

$$= \frac{(1.00)^2(1800 \text{ kips}) + (0.725)^2(1200 \text{ kips}) + (-0.657)^2(1800 \text{ kips})}{g}$$

$$= \frac{3208 \text{ kips}}{g}$$

$$p_2 = \frac{\displaystyle\sum_{i=1}^{3} \phi_{2,i} m_i}{\displaystyle\sum_{i=1}^{3} \phi_{2,i}^2 m_i}$$

$$= \frac{\dfrac{(1.00)(1800 \text{ kips}) + (0.725)(1200 \text{ kips}) + (-0.657)(1800 \text{ kips})}{g}}{\dfrac{3208 \text{ kips}}{g}}$$

$$= 0.46$$

$\phi_3 = (1.000, -1.610, 0.378)^T$

$$M_3 = \sum_{i=1}^{3} \phi_{3,i}^2 m_i$$

$$= \frac{(1.00)^2(1800 \text{ kips}) + (-1.610)^2(1200 \text{ kips}) + (0.378)^2(1800 \text{ kips})}{g}$$

$$= \frac{5168 \text{ kips}}{g}$$

$$p_3 = \frac{\displaystyle\sum_{i=1}^{3} \phi_{3,i} m_i}{\displaystyle\sum_{i=1}^{3} \phi_{3,i}^2 m_i}$$

$$= \frac{\dfrac{(1.00)(1800 \text{ kips}) + (-1.610)(1200 \text{ kips}) + (0.378)(1800 \text{ kips})}{g}}{\dfrac{5168 \text{ kips}}{g}}$$

$$= 0.11$$

Calculate the base shear for each mode.

$$V_n = p_n^2 M_n S_{a,n}$$

$$V_1 = (0.44)^2 \left(\frac{21{,}128 \text{ kips}}{g}\right)(0.98g) = \boxed{4009 \text{ kips}}$$

$$V_2 = (0.46)^2 \left(\frac{3208 \text{ kips}}{g}\right)(1.00g) = \boxed{679 \text{ kips}}$$

$$V_3 = (0.11)^2 \left(\frac{5168 \text{ kips}}{g}\right)(1.30g) = \boxed{81.3 \text{ kips}}$$

9.2. Calculate the lateral force at each level.

For mode 1,

$$F_{i,n} = m_i p_i \phi_{i,n} S_{a,n}$$

$$F_{1,1} = \left(\frac{1800 \text{ kips}}{g}\right)(0.44)(1.00)(0.98g) = 776 \text{ kips}$$

$$F_{2,1} = \left(\frac{1200 \text{ kips}}{g}\right)(0.44)(1.959)(0.98g) = 1014 \text{ kips}$$

$$F_{3,1} = \left(\frac{1800 \text{ kips}}{g}\right)(0.44)(2.86)(0.98g) = 2220 \text{ kips}$$

Therefore, the lateral load is

$$V_1 = \sum F_{i,1} = 776 \text{ kips} + 1014 \text{ kips} + 2220 \text{ kips}$$

$$= \boxed{4010 \text{ kips}}$$

For mode 2,

$$F_{1,2} = \left(\frac{1800 \text{ kips}}{g}\right)(0.46)(1.00)(1.00g) = 828 \text{ kips}$$

$$F_{2,2} = \left(\frac{1200 \text{ kips}}{g}\right)(0.46)(0.725)(1.00g) = 400 \text{ kips}$$

$$F_{3,2} = \left(\frac{1800 \text{ kips}}{g}\right)(0.46)(-0.657)(1.00g) = -544 \text{ kips}$$

Therefore, the lateral load is

$$V_2 = \sum F_{i,2} = 828 \text{ kips} + 400 \text{ kips} - 544 \text{ kips}$$
$$= \boxed{684 \text{ kips}}$$

For mode 3,

$$F_{1,3} = \left(\frac{1800 \text{ kips}}{g}\right)(0.10)(1.00)(1.30g) = 234 \text{ kips}$$

$$F_{2,3} = \left(\frac{1200 \text{ kips}}{g}\right)(0.10)(-1.610)(1.30g) = -251 \text{ kips}$$

$$F_{3,3} = \left(\frac{1800 \text{ kips}}{g}\right)(0.10)(0.378)(1.30g) = 88 \text{ kips}$$

Therefore, the lateral load is

$$V_3 = \sum F_{i,3} = 234 \text{ kips} - 251 \text{ kips} + 88 \text{ kips}$$
$$= \boxed{71 \text{ kips}}$$

9.3. The most probable design base shear can be estimated by combining the modal base shears, using either the CQC method (from App. D in the *SEAOC Blue Book*) or the square root of the sum of the squares (SRSS) method. For the CQC combination and a 5% damping ratio,

$$\rho_{ij} = \frac{8\xi^2(1+r)r^{3/2}}{(1-r^2)^2 + 4\xi^2 r(1+r)^2}$$

$$\rho_{11} = \rho_{22} = \rho_{33} = 1 \quad [\xi = 0.05]$$

$$\rho_{21} = \frac{(8)(0.05)^2(1+2.625)(2.625)^{3/2}}{\left(1-(2.625)^2\right)^2 + (4)(0.05)^2(2.625)(1+2.625)^2}$$
$$= 0.00880 \quad [r = T_1/T_2 = 0.42/0.16 = 2.625]$$

$$\rho_{12} = \rho_{21} = 0.00880$$

$$\rho_{31} = \frac{(8)(0.05)^2(1+4.2)(4.2)^{3/2}}{\left(1-(4.2)^2\right)^2 + (4)(0.05)^2(4.2)(1+4.2)^2}$$
$$= 0.00322 \quad [r = T_1/T_3 = 0.42/0.10 = 4.2]$$

$$\rho_{13} = \rho_{31} = 0.00322$$

$$\rho_{32} = \frac{(8)(0.05)^2(1+1.6)(1.6)^{3/2}}{\left(1-(1.6)^2\right)^2 + (4)(0.05)^2(1.6)(1+1.6)^2}$$
$$= 0.04140 \quad [r = T_2/T_3 = 0.16/0.10 = 1.6]$$

$$\rho_{23} = \rho_{32} = 0.04140$$

$$\rho = \begin{bmatrix} 1.000 & 0.00880 & 0.00322 \\ 0.00880 & 1.000 & 0.04140 \\ 0.00322 & 0.04140 & 1.000 \end{bmatrix}$$

By the CQC method,

$$V = \sqrt{\sum_{i=1}^{3}\sum_{j=1}^{3} V_i \rho_{ij} V_j}$$

$$= \left[\begin{aligned} &(4010 \text{ kips})(1.000)(4010 \text{ kips}) \\ &+ (4010 \text{ kips})(0.00880)(684 \text{ kips}) \\ &+ (4010 \text{ kips})(0.00322)(71 \text{ kips}) \\ &+ (684 \text{ kips})(0.00880)(4010 \text{ kips}) \\ &+ (684 \text{ kips})(1.000)(684 \text{ kips}) \\ &+ (684 \text{ kips})(0.04140)(71 \text{ kips}) \\ &+ (71 \text{ kips})(0.00322)(4010 \text{ kips}) \\ &+ (71 \text{ kips})(0.04140)(684 \text{ kips}) \\ &+ (71 \text{ kips})(1.000)(71 \text{ kips}) \end{aligned} \right]^{1/2}$$

$$= 4075 \text{ kips}$$

In this problem, the higher modes have little effect on base shear. The response can be obtained accurately enough using the SRSS method of combining modes.

$$V = \sqrt{\sum_{i=1}^{3} V_i^2}$$
$$= \sqrt{(4010 \text{ kips})^2 + (684 \text{ kips})^2 + (71 \text{ kips})^2}$$
$$= \boxed{4069 \text{ kips}}$$

SOLUTION 10

10.1. The given seismic data are $S_S = 0.9$, $S_1 = 0.4$, risk category II, seismic importance factor $I_e = 1.0$, and site class D. For these data, Table 11.4-1 from ASCE/SEI7 gives

$$F_a = 1.14 \quad [\text{by linear interpolation}]$$
$$S_{MS} = F_a S_S = (1.14)(0.9) = 1.026$$
$$S_{DS} = 0.667 S_{MS} = (0.667)(1.026) = 0.68$$

From ASCE/SEI7 Table 11.4-2,

$$F_v = 1.6$$

$$S_{M1} = F_v S_1 = (1.6)(0.4) = 0.64$$

$$S_{D1} = 0.667 S_{M1} = (0.667)(0.64) = 0.43$$

From ASCE/SEI7 Tables 11.6-1 and 11.6-2, for S_{DS} values greater than 0.5 and S_{D1} values greater than 0.2, the seismic design category is D. The trussed upper story prevents significant story drift between levels 1 and 2, so the system can be modeled as a single degree of freedom system.

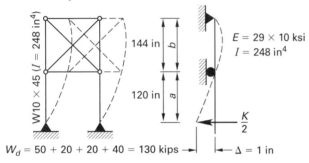

$W_d = 50 + 20 + 20 + 40 = 130$ kips

Axial deformation is to be neglected, so the equivalent stiffness, k, can be computed using the equation for the deflection of the tip of a cantilevered simple beam (AISC Table 3-23, case 26), with the force at the tip taken as $0.5k\Delta$ to represent one of the two unbraced legs.

$$\Delta = \left(\frac{(0.5k\Delta)a^2}{3EI} \right)(l + a)$$

$$k = \frac{6EI}{a^2(l + a)}$$

$$= \frac{(6)\left(29{,}000 \ \dfrac{\text{kips}}{\text{in}} \right)(248 \ \text{in}^4)}{(120 \ \text{in})^2(144 \ \text{in} + 120 \ \text{in})}$$

$$= 11.4 \ \text{kips/in}$$

The seismic dead loads are given as concentrated loads at the joints.

$$m = \frac{\sum P_d}{g}$$

$$= \frac{50 \ \text{kips} + 40 \ \text{kips} + 20 \ \text{kips} + 20 \ \text{kips}}{g}$$

$$= 130 \ \text{kips}/g$$

The fundamental period is

$$T = 2\pi \sqrt{\frac{m}{k}} = 2\pi \sqrt{\frac{130 \ \text{kips}}{386.6 \ \dfrac{\text{in}}{\text{sec}^2}}{11.4 \ \dfrac{\text{kips}}{\text{in}}}} = 1.1 \ \text{sec}$$

From ASCE/SEI7 Table 15.4-2, for an elevated tank, bin, or hopper supported on unbraced legs, $R = 2.0$, $\Omega_o = 2$, and $C_d = 2.5$.

$$C_s = \min \begin{cases} \dfrac{S_{DS}I_e}{R} = \dfrac{(0.63)(1.0)}{2.0} = 0.32 \\[2mm] \dfrac{S_{D1}I_e}{TR} = \dfrac{(0.43)(1.0)}{(1.1)(2.0)} \\[2mm] \quad\quad = 0.20 \quad [\text{controls, } C_s \geq 0.03] \end{cases}$$

$$V = C_s W = (0.20)(130 \ \text{kips}) = 26.0 \ \text{kips}$$

Distribute the base shear in direct proportion to the structure's mass.

$$V_1 = \left(\frac{F_B + F_D}{W} \right) V$$

$$= \left(\frac{40 \ \text{kips} + 20 \ \text{kips}}{130 \ \text{kips}} \right)(26.0 \ \text{kips})$$

$$= 12.0 \ \text{kips}$$

$$V_2 = \left(\frac{F_A + F_C}{W} \right) V$$

$$= \left(\frac{50 \ \text{kips} + 20 \ \text{kips}}{130 \ \text{kips}} \right)(26.0 \ \text{kips})$$

$$= 14.0 \ \text{kips}$$

Analyze for the seismic lateral loads.

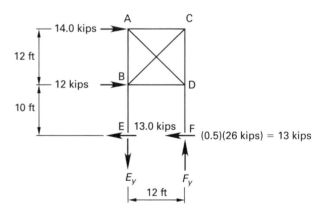

$$\sum M_E = V_1 h_1 + V_2 h_2 - F_y b = 0$$

$$F_y = \frac{V_1 h_1 + V_2 h_2}{b}$$

$$= \frac{(12.0 \ \text{kips})(10 \ \text{ft}) + (14.0 \ \text{kips})(22 \ \text{ft})}{12 \ \text{ft}}$$

$$= 35.7 \ \text{kips}$$

$$\sum F_v = E_y + F_y = 0$$

$$E_y = -F_y = \boxed{-35.7 \ \text{kips}}$$

10.2. For a W10 × 45 column, the following criteria are known. $A_g = 13.3 \ \text{in}^2$, $b_f/2t_f = 6.47$, $h/t_w = 22.5$, $r_x = 4.32 \ \text{in}$, $r_y = 2.01 \ \text{in}$, $Z_x = 54.9 \ \text{in}^3$, $S_x = 49.1 \ \text{in}^3$, $L_p = 7.1 \ \text{ft}$, $L_r = 26.9 \ \text{ft}$, and $I_x = 248 \ \text{in}^4$.

The weak axis is braced at the floor levels, so $K_y = 1.0$. The deformation of the column in its buckled configurations is such that AISC Table C-A-7.1, case f, is a reasonable approximation and gives $K_x = 2.0$.

Find the slenderness ratios.

$$\frac{(KL)_x}{r_x} = \frac{(2.0)(10.0 \text{ ft})\left(12 \frac{\text{in}}{\text{ft}}\right)}{4.32 \text{ in}}$$
$$= 56$$
$$\frac{(KL)_y}{r_y} = \frac{(1.0)(10.0 \text{ ft})\left(12 \frac{\text{in}}{\text{ft}}\right)}{2.01 \text{ in}}$$
$$= 60 \quad \text{[controls axial compression capacity]}$$

For $F_y = 50$ ksi, calculate the strength of the column in the absence of bending (AISC Sec. E3).

$$4.71\sqrt{\frac{E}{F_y}} = 4.71\sqrt{\frac{29{,}000 \frac{\text{kips}}{\text{in}^2}}{50 \frac{\text{kips}}{\text{in}^2}}} = 113$$

$$\frac{KL}{r} = \frac{(KL)_y}{r_y} = 60 \quad \left[< 4.71\sqrt{E/F_y}\right]$$

Use AISC Eq. E3-4 to find the elastic buckling stress.

$$F_e = \frac{\pi^2 E}{\left(\dfrac{KL}{r}\right)^2}$$
$$= \frac{\pi^2\left(29{,}000 \frac{\text{kips}}{\text{in}^2}\right)}{(60)^2}$$
$$= 79.4 \text{ ksi}$$

Use AISC Eq. E3-2 to find the critical stress.

$$F_{\text{cr}} = 0.658^{F_y/F_e} F_y$$
$$= \left(0.658^{50\frac{\text{kips}}{\text{in}^2}/79.4\frac{\text{kips}}{\text{in}^2}}\right)\left(50 \frac{\text{kips}}{\text{in}^2}\right)$$
$$= 38.4 \text{ ksi}$$

Find the available axial strength.

$$P_c = \phi F_{\text{cr}} A_g$$
$$= (0.9)\left(38.4 \frac{\text{kips}}{\text{in}^2}\right)(13.3 \text{ in}^2)$$
$$= 460 \text{ kips}$$

For bending about the x-axis, use the approximate second-order analysis of AISC App. 8.

The story stiffness in Sol. 10.1 was calculated as 11.4 kips/in. For the total lateral force of 26 kips, the story drift is

$$\Delta_H = \frac{H}{k} = \frac{26.0 \text{ kips}}{11.4 \frac{\text{kips}}{\text{in}}} = 2.28 \text{ in}$$

From AISC App. 8,

$$P_{e,\text{story}} = \frac{R_M H L}{\Delta_H}$$
$$= \frac{(0.85)(26.0 \text{ kips})(10 \text{ ft} + 12 \text{ ft})\left(12 \frac{\text{in}}{\text{ft}}\right)}{2.28 \text{ in}}$$
$$= 2560 \text{ kips}$$

$$P_{\text{story}} = (1.2)P_D = (1.2)(130 \text{ kips}) = 156 \text{ kips}$$

$$B_2 = \frac{1}{1 - \dfrac{\alpha P_{\text{story}}}{P_{e,\text{story}}}} = \frac{1}{1 - \dfrac{(1)(156 \text{ kips})}{2560 \text{ kips}}} = 1.06$$

Since connections are pinned, the frame does not translate under the asymmetrical gravity loads and on the more heavily loaded column.

$$P_{nt} = (1.2 + 0.2 S_{DS})(F_A + F_B)$$
$$= (1.2 + (0.2)(0.68))(50 \text{ kips} + 40 \text{ kips})$$
$$= 120 \text{ kips}$$

$$P_r = P_{nt} + B_2 P_{lt}$$
$$= 120 \text{ kips} + (1.06)(35.7 \text{ kips})$$
$$= 158 \text{ kips}$$

$$M_{lt} = Vh = \left(\frac{26 \text{ kips}}{2}\right)(10 \text{ ft}) = 130 \text{ ft-kips}$$

$$M_{rx} = B_1 M_{nt} + B_2 M_{lt}$$
$$= B_1(0 \text{ ft-kips}) + (1.06)(130 \text{ ft-kips})$$
$$= 138 \text{ ft-kips}$$

Check local buckling (AISC Table B4.1, case 10). For the flange,

$$\frac{b_f}{2t_f} = 6.47$$

$$\lambda_p = 0.38\sqrt{\frac{E}{F_y}} = 0.38\sqrt{\frac{29{,}000 \frac{\text{kips}}{\text{in}^2}}{50 \frac{\text{kips}}{\text{in}^2}}}$$
$$= 9.15 \quad \left[\frac{b_f}{2t_f} < \lambda_p\right]$$

For the web (AISC Table B4.1, case 15),

$$\frac{h}{t_w} = 22.5$$

$$\lambda_p = 3.76\sqrt{\frac{E}{F_y}} = 3.76\sqrt{\frac{29{,}000 \frac{\text{kips}}{\text{in}^2}}{50 \frac{\text{kips}}{\text{in}^2}}}$$
$$= 90.6 \quad \left[\frac{h}{t_w} < \lambda_p, \text{ so compact}\right]$$

The section is compact. Check lateral-torsional buckling for $L_b = 12$ ft. Per the problem statement, assume the member is laterally supported at base and joints only. This is greater than $L_p = 7.1$ ft and less than $L_r = 26.9$ ft, so use AISC Eq. F2-2. For the linear moment gradient over the unbraced length, $C_b = 1.67$ (AISC Table 3.1).

$$M_p = F_y Z_x = \left(50 \ \frac{\text{kips}}{\text{in}^2}\right)(54.9 \ \text{in}^3)$$

$$= 2745 \ \text{in-kips} \quad \text{[AISC Eq. F2-1]}$$

$$M_n \leq \begin{cases} C_b\left(M_p - (M_p - 0.7F_y S_x)\left(\dfrac{L_b - L_p}{L_r - L_p}\right)\right) \\[2em] = (1.67)\left(\begin{array}{c} 2745 \ \text{in-kips} \\ -\left(\begin{array}{c} 2745 \ \text{in-kips} \\ -(0.7)\left(50 \ \dfrac{\text{kips}}{\text{in}^2}\right) \\ \times (49.1 \ \text{in}^3) \end{array}\right) \\ \times\left(\dfrac{12.0 \ \text{ft} - 7.1 \ \text{ft}}{26.9 \ \text{ft} - 7.1 \ \text{ft}}\right) \end{array}\right) \\[4em] = 4160 \ \text{in-kips} \\[1em] M_p = 2745 \ \text{in-kips} \quad \text{[controls]} \end{cases}$$

$$M_c = \phi M_p = \frac{(0.9)(2745 \ \text{in-kips})}{12 \ \dfrac{\text{in}}{\text{ft}}} = 206 \ \text{ft-kips}$$

Check combined axial compression and bending (AISC Sec. H1).

$$\frac{P_r}{P_c} = \frac{158 \ \text{kips}}{460 \ \text{kips}}$$

$$= 0.34 \quad [\geq 0.2, \text{ so use AISC Eq. H1-1a}]$$

$$\frac{P_r}{P_c} + \frac{8}{9}\left(\frac{M_{rx}}{M_{cx}}\right) = 0.34 + \left(\frac{8}{9}\right)\left(\frac{138 \ \text{ft-kips}}{206 \ \text{ft-kips}}\right)$$

$$= 0.94 \quad [\leq 1.0, \text{ so OK}]$$

The W10 × 45 is adequate in combined axial compression and bending.

10.3. Determine the live load that could be applied at joint D. Assume that no other live load is imposed and that the nature of the live load is such that a factor 0.5 applies when combined with dead and seismic loads. Use the amplified design moment calculated in Prob. 10.2. Solving for the value of $P_r = P_u$ that corresponds to an interaction value of unity gives

$$\frac{P_r}{P_c} \geq 0.2 \quad [\text{so use AISC Eq. H1-1a}]$$

$$\frac{P_r}{P_c} + \frac{8}{9}\left(\frac{M_{rx}}{M_{cx}}\right) = 1.0$$

$$P_r = P_u = \left(1.0 - \frac{8}{9}\left(\frac{M_{rx}}{M_{cx}}\right)\right)P_c$$

$$= \left(1.0 - \left(\frac{8}{9}\right)\left(\frac{138 \ \text{ft-kips}}{206 \ \text{ft-kips}}\right)\right)(460 \ \text{kips})$$

$$= 186 \ \text{kips}$$

Calculate the live load that the frame can carry.

$$P_r = (1.2 + 0.2S_{DS})P_d + P_e + 0.5P_l$$

$$P_e = E_y = 35.7 \ \text{kips}$$

$$P_l = 2(P_r - ((1.2 + (0.2S_{DS})P_d + P_e)))$$

$$= (2)\left(186 \ \text{kips} - \left(\begin{array}{c}(1.2 + (0.2)(0.68)) \\ \times (40 \ \text{kips}) + 35.7 \ \text{kips}\end{array}\right)\right)$$

$$= \boxed{194 \ \text{kips}}$$

SOLUTION 11

11.1. Column D2 is not part of the lateral force resisting system. The column is at the center of rigidity of the regular structure, so the effects of accidental torsion will not cause any displacement of the member. The displacement at the top of the column is the sum of the displacements of the shear walls and the midspan deflection of the horizontal diaphragm. The lateral force resisted by the shear walls is

$$V = 0.5wL = (0.5)\left(4 \ \frac{\text{kips}}{\text{ft}}\right)(150 \ \text{ft}) = 300 \ \text{kips}$$

Express deflections in terms of the modulus of elasticity, E_c, which from the design criteria is the same for all concrete in the structure. The elastic deflection at the top of the shear wall is

$$\Delta_{we} = \left(\frac{V}{E_c t}\right)\left(4\left(\frac{h}{d}\right)^3 + 3\left(\frac{h}{d}\right)\right)$$

$$= \left(\frac{300 \ \text{kips}}{E_c(8 \ \text{in})}\right)\left(\begin{array}{c}(4)\left(\dfrac{(10 \ \text{ft})\left(12 \ \frac{\text{in}}{\text{ft}}\right)}{(15 \ \text{ft})\left(12 \ \frac{\text{in}}{\text{ft}}\right)}\right)^3 + (3) \\ \times\left(\dfrac{(10 \ \text{ft})\left(12 \ \frac{\text{in}}{\text{ft}}\right)}{(15 \ \text{ft})\left(12 \ \frac{\text{in}}{\text{ft}}\right)}\right)\end{array}\right)$$

$$= \frac{119 \ \dfrac{\text{kips}}{\text{in}}}{E_c}$$

The relative maximum elastic deflection between the diaphragm and shear walls occurs at midspan (line D) and is given by

$$\Delta_{de} = \left(\frac{wL}{6.4E_c t}\right)\left(\left(\frac{L}{b}\right)^3 + 2.13\left(\frac{L}{b}\right)\right)$$

$$= \left(\frac{\left(4 \frac{\text{kips}}{\text{ft}}\right)(150 \text{ ft})}{6.4E_c(9 \text{ in})}\right)\left(\begin{array}{c}\left(\frac{150 \text{ ft}}{50 \text{ ft}}\right)^3 + (2.13) \\ \times \left(\frac{150 \text{ ft}}{50 \text{ ft}}\right)\end{array}\right)$$

$$= \frac{348 \frac{\text{kips}}{\text{in}}}{E_c}$$

The top of the column on line D at line 2 will be displaced.

$$\Delta_e = \Delta_{we} + \Delta_{de} = \frac{348 \frac{\text{kips}}{\text{in}}}{E_c} + \frac{119 \frac{\text{kips}}{\text{in}}}{E_c} = \frac{467 \frac{\text{kips}}{\text{in}}}{E_c}$$

Δ_e is the elastic deflection based on the code-prescribed seismic force. The actual response of the structure during an earthquake will be inelastic and the actual deflection will be significantly greater than Δ_e. ACI 318 Sec. 21.13.2 requires that the column's vertical load-carrying capacity be investigated under the design displacement, δ_u, that is C_d times the displacement caused by the code-prescribed lateral force.

$$\delta_u = C_d\Delta_e = (5.0)\left(\frac{467 \frac{\text{kips}}{\text{in}}}{E_c}\right) = \frac{2335 \frac{\text{kips}}{\text{in}}}{E_c}$$

Assuming columns are fixed top and bottom (per the design criteria), then for the 14 in diameter circular columns,

$$E_c I_g = \frac{E_c \pi D^4}{64} = \frac{E_c \pi (14 \text{ in})^4}{64} = (1885 \text{ in}^4)E_c$$

ACI 318 Sec. 10.10.4.1 permits calculation of the elastic response based on $0.7E_c I_g$, to account for the likely reduction in column stiffness due to cracking. Therefore, for the 10 ft long column at D2, the induced elastic seismic moment is approximated as

$$M_e = \frac{(6)(0.7E_c I_g)\delta_u}{h^2}$$

$$= \frac{(6)(0.7)(1885 \text{ in}^4)E_c\left(\frac{2335 \frac{\text{kips}}{\text{in}}}{E_c}\right)}{\left((10 \text{ ft})\left(12 \frac{\text{in}}{\text{ft}}\right)\right)^2}$$

$$= \boxed{1284 \text{ in-kips}}$$

The moment M_e is combined with the factored gravity moment and axial force and checked against the interaction curve for the column. If the resulting moment is less than the design moment strength under combined axial and bending, and if the shear caused by the induced moments is less than the member's shear strength, then the column is detailed to meet the conditions of ACI 318 Sec. 21.13.3. If the resulting moment exceeds the design moment strength or the induced shear exceeds the design shear capacity, the member must satisfy the conditions of ACI 318 Sec. 21.13.4. In addition, the flat slab must be investigated for transfer of moment and punching shear in accordance with ACI 318 Sec. 21.13.6.

SOLUTION 12

12.1. Let α represent the fraction of the 30 kip lateral force resisted by pier 1. Then $1-\alpha$ is the fraction resisted by piers 2 through 6. Piers are nominal 8 in concrete block that has been grouted solid. Let E_m represent the modulus of elasticity of the piers, and take their shear modulus, G, as $0.4E_m$.

For pier 1,

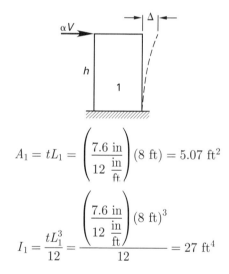

$$A_1 = tL_1 = \left(\frac{7.6 \text{ in}}{12 \frac{\text{in}}{\text{ft}}}\right)(8 \text{ ft}) = 5.07 \text{ ft}^2$$

$$I_1 = \frac{tL_1^3}{12} = \frac{\left(\frac{7.6 \text{ in}}{12 \frac{\text{in}}{\text{ft}}}\right)(8 \text{ ft})^3}{12} = 27 \text{ ft}^4$$

Therefore,

$$\Delta = \alpha V\left(\frac{h^3}{3E_m I_1} + \frac{1.2h}{0.4E_m A_1}\right)$$

$$E_m \Delta = \alpha V\left(\frac{h^3}{3I_1} + \frac{1.2h}{0.4A_1}\right)$$

$$= \alpha V\left(\frac{(16 \text{ ft})^3}{(3)(27 \text{ ft}^4)} + \frac{(1.2)(16 \text{ ft})}{(0.4)(5.07 \text{ ft}^2)}\right)$$

$$= 60\alpha V$$

For piers 2 through 6,

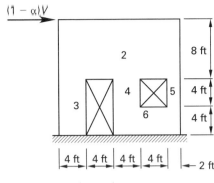

$$A_2 = tL_2 = \left(\frac{7.6 \text{ in}}{12 \frac{\text{in}}{\text{ft}}}\right)(18 \text{ ft}) = 11.4 \text{ ft}^2$$

$$I_2 = \frac{tL_2^3}{12} = \frac{\left(\frac{7.6 \text{ in}}{12 \frac{\text{in}}{\text{ft}}}\right)(18 \text{ ft})^3}{12} = 308 \text{ ft}^4$$

The bending deformation is small relative to the shear deformation in pier 2 and will be neglected.

$$A_3 = tL_3 = \left(\frac{7.6 \text{ in}}{12 \frac{\text{in}}{\text{ft}}}\right)(4 \text{ ft}) = 2.53 \text{ ft}^2$$

$$I_3 = \frac{tL_3^3}{12} = \frac{\left(\frac{7.6 \text{ in}}{12 \frac{\text{in}}{\text{ft}}}\right)(4 \text{ ft})^3}{12} = 3.38 \text{ ft}^4$$

$$A_4 = A_3 = 2.53 \text{ ft}^2$$
$$I_4 = I_3 = 3.38 \text{ ft}^4$$

$$A_5 = tL_5 = \left(\frac{7.6 \text{ in}}{12 \frac{\text{in}}{\text{ft}}}\right)(2 \text{ ft}) = 1.27 \text{ ft}^2$$

$$I_5 = \frac{tL_5^3}{12} = \frac{\left(\frac{7.6 \text{ in}}{12 \frac{\text{in}}{\text{ft}}}\right)(2 \text{ ft})^3}{12}$$
$$= 0.42 \text{ ft}^4$$

$$A_6 = tL_6 = \left(\frac{7.6 \text{ in}}{12 \frac{\text{in}}{\text{ft}}}\right)(10 \text{ ft}) = 6.33 \text{ ft}^2$$

$$I_6 = \frac{tL_6^3}{12} = \frac{\left(\frac{7.6 \text{ in}}{12 \frac{\text{in}}{\text{ft}}}\right)(10 \text{ ft})^3}{12} = 52.8 \text{ ft}^4$$

The dimensions are such that only shear deformation is significant in piers 2 and 6, but both shear and flexural deformations are significant in the other segments. Let β represent the percentage of $(1-\alpha)V$ that is resisted by pier 3; then $1-\beta$ is the percentage resisted by piers 4, 5, and 6. Let r represent the percentage of $(1-\alpha)(1-\beta)$ resisted by pier 4; then $1-r$ is the percentage of $(1-\alpha)(1-\beta)$ resisted by pier 5. Thus, treating piers 4 and 5 as fixed-fixed,

$$r\left(\frac{h_4^3}{12E_mI_4} + \frac{1.2h_4}{0.4E_mA_4}\right) = (1-r)\left(\frac{h_5^3}{12E_mI_5} + \frac{1.2h_5}{0.4E_mA_5}\right)$$

Multiply both sides by E_m and substitute known values.

$$r\left(\frac{(4 \text{ ft})^3}{(12)(3.38 \text{ ft})^4} + \frac{(1.2)(4 \text{ ft})}{(0.4)(2.53 \text{ ft}^2)}\right)$$
$$= (1-r)\left(\frac{(4 \text{ ft})^3}{(12)(0.42 \text{ ft}^4)} + \frac{(1.2)(4 \text{ ft})}{(0.4)(1.27 \text{ ft}^2)}\right)$$
$$r(6.32 \text{ ft}^{-1}) = (1-r)(22.15 \text{ ft}^{-1})$$
$$r = 0.78$$
$$1-r = 0.22$$

Calculate the distribution among piers 3, 4, 5, and 6.

$$\beta\left(\frac{h_3^3}{12E_mI_3} + \frac{1.2h_3}{0.4E_mA_3}\right)$$
$$= (1-\beta)\left(\begin{array}{l} r\left(\dfrac{h_4^3}{12E_mI_4} + \dfrac{1.2h_4}{0.4E_mA_4}\right) \\[8pt] + (1-r)\left(\dfrac{h_5^3}{12E_mI_5} + \dfrac{1.2h_5}{0.4E_mA_5}\right) \\[8pt] + \dfrac{1.2h_6}{0.4E_mA_6} \end{array}\right)$$

Multiplying both sides by E_m and using previously computed results for piers 4 and 5 gives

$$\beta\left(\frac{(8 \text{ ft})^3}{(12)(3.38 \text{ ft}^4)} + \frac{(1.2)(8 \text{ ft})}{(0.4)(2.53 \text{ ft}^2)}\right)$$
$$= (1-\beta)\left(\begin{array}{l} (0.78)(6.32 \text{ ft}^{-1}) \\[6pt] + (0.22)(22.15 \text{ ft}^{-1}) \\[6pt] + \dfrac{(1.2)(4 \text{ ft})}{(0.4)(6.33 \text{ ft}^2)} \end{array}\right)$$
$$\beta(22.1 \text{ ft}^{-1}) = (1-\beta)(11.7 \text{ ft}^{-1})$$
$$\beta = 0.35$$
$$1-\beta = 0.65$$

The final distribution can now be determined.

$$\frac{60\alpha}{E_m} = (1-\alpha) \times \left(\begin{array}{l} \beta\left(\dfrac{h_3^3}{12E_mI_3} + \dfrac{1.2h_3}{0.4E_mA_3}\right) + (1-\beta) \\[2em] \left(\begin{array}{l} r\left(\dfrac{(8\text{ ft})^3}{12E_mI_4} + \dfrac{1.2h_4}{0.4E_mA_4}\right) \\[1.5em] + (1-r) \\[0.5em] \times\left(\dfrac{h_5^3}{12E_mI_5} + \dfrac{1.2h_5}{0.4E_mA_5}\right) \\[1.5em] + \dfrac{1.2h_6}{0.4E_mA_6} \end{array}\right) \\[4em] + \dfrac{1.2h_2}{0.4E_mA_2} \end{array}\right)$$

$$\alpha = \left(\frac{1-\alpha}{60}\right)\left(\begin{array}{l}(0.35)(22.1\text{ ft}^{-1}) + (0.65) \\[1em] \times(11.7\text{ ft}^{-1}) + \dfrac{(1.2)(8\text{ ft})}{(0.4)(11.4\text{ ft}^2)}\end{array}\right)$$

$$= 0.23$$

$$1 - \alpha = 0.77$$

The drag strut is considered inextensible, so the lateral force distributes on the basis of relative translational stiffness. For the given lateral force, $V = 30$ kips,

$$V_1 = \alpha V$$
$$= (0.23)(30\text{ kips})$$
$$= \boxed{6.9\text{ kips}}$$
$$V_2 = (1-\alpha)V$$
$$= (1-0.23)(30\text{ kips})$$
$$= \boxed{23.1\text{ kips}}$$
$$V_3 = \beta(1-\alpha)V$$
$$= (0.35)(1-0.23)(30\text{ kips})$$
$$= \boxed{8.1\text{ kips}}$$
$$V_4 = r(1-\beta)(1-\alpha)V$$
$$= (0.78)(1-0.35)(1-0.23)(30\text{ kips})$$
$$= \boxed{11.7\text{ kips}}$$
$$V_5 = V_2 - V_3 - V_4$$
$$= 23.1\text{ kips} - 8.1\text{ kips} - 11.7\text{ kips}$$
$$= \boxed{3.3\text{ kips}}$$
$$V_6 = V_2 - V_3 = 23.1\text{ kips} - 8.1\text{ kips} = \boxed{15\text{ kips}}$$

12.2. Given that the lateral force transfers to the walls by a flexible diaphragm, the shear is uniformly distributed over the full width of diaphragm.

$$v = \frac{V}{L_A + L_{AB} + L_B} = \frac{30\text{ kips}}{8\text{ ft} + 24\text{ ft} + 18\text{ ft}}$$
$$= 0.6\text{ kip/ft}$$

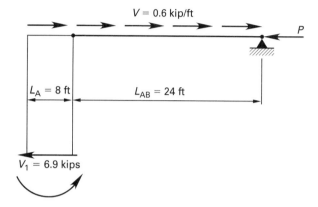

The maximum anchorage load occurs at the connection to wall B.

$$P = v(L_A + L_{AB}) - V_1$$
$$= \left(0.6\ \frac{\text{kip}}{\text{ft}}\right)(8\text{ ft} + 24\text{ ft}) - 6.9\text{ kips}$$
$$= \boxed{12.3\text{ kips}}$$

(For design of the connection of the drag strut, Sec. 12.10.2.1 of ASCE/SEI7 requires application of an overstrength factor per Sec. 12.4.3 for structures located in seismic design categories C through F.)

12.3. Determine the axial load in pier 5, given a 25 kip force at the top of wall B and neglecting wall weight. Cut a free-body diagram through the midpoint of piers 3, 4, and 5 (which are points of inflection and have zero moments).

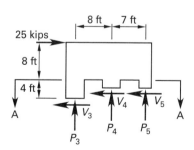

From the distribution factors computed in Prob. 12.1,

$$V_0 = \beta V = (0.35)(25 \text{ kips})$$
$$= 8.8 \text{ kips}$$
$$V_4 = (1 - \beta)r V$$
$$= (1 - 0.35)(0.78)(25 \text{ kips})$$
$$= 12.7 \text{ kips}$$
$$V_5 = (1 - \beta)(1 - r) V$$
$$= (1 - 0.35)(1 - 0.78)(25 \text{ kips})$$
$$= 3.6 \text{ kips}$$

Take moments about line A-A.

$$\sum M = V h_2 - V_3(0.5h_3) - V_4(0.5h_4)$$
$$- V_5(0.5h_5) - \sum p_i d_i = 0$$
$$\sum p_i d_i = V h_2 - V_3(0.5h_3) - V_4(0.5h_4) - V_5(0.5h_5)$$
$$= (25 \text{ kips})(8 \text{ ft}) - (8.8 \text{ kips})(0.5)(8 \text{ ft})$$
$$- (12.7 \text{ kips})(0.5)(4 \text{ ft})$$
$$- (3.6 \text{ kips})(0.5)(4 \text{ ft})$$
$$= 132.2 \text{ ft-kips}$$

Use linear elastic theory to compute the distribution of axial forces in the three piers.

section A-A

The piers have the same thickness, t, so the centroid of the group is located a distance $\bar{x}$ from the left edge of the wall.

$$A = \sum L_i t = (4 \text{ ft} + 4 \text{ ft} + 2 \text{ ft})t = (10 \text{ ft})t$$
$$\bar{x} = \frac{\sum (L_i t)\bar{x}_i}{A}$$
$$= \frac{(4 \text{ ft})t(2 \text{ ft}) + (4 \text{ ft})t(10 \text{ ft}) + (2 \text{ ft})t(17 \text{ ft})}{(10 \text{ ft})t}$$
$$= 8.2 \text{ ft}$$

$$I = \sum \left(\frac{L_i^3 t}{12} + L_i t(\bar{x}_i - \bar{x})^2 \right)$$
$$= \frac{(4 \text{ ft})^3 t}{12} + (4 \text{ ft})t(2 \text{ ft} - 8.2 \text{ ft})^2$$
$$+ \frac{(4 \text{ ft})^3 t}{12} + (4 \text{ ft})t(10 \text{ ft} - 8.2 \text{ ft})^2$$
$$+ \frac{(2 \text{ ft})^3 t}{12} + (2 \text{ ft})t(17 \text{ ft} - 8.2 \text{ ft})^2$$
$$= 333t$$

The axial load for pier 5 is

$$P_5 = \frac{M L_5 t(\bar{x}_5 - \bar{x})}{I}$$
$$= \frac{(132.2 \text{ ft-kips})(2 \text{ ft})t(17 \text{ ft} - 8.2 \text{ ft})}{333t}$$
$$= \boxed{7.0 \text{ kips}}$$

The axial force in pier 5 is 7 kips compression when the 25 kips acts to the right (7 kips tension when the lateral force reverses).

5 Foundations and Retaining Structures

PROBLEM 1

A pile cap and a system of batter piles, as shown, support a horizontal force of 138 kips, which is delivered to the cap by a tension rod with a $3^{1}/_{4}$ in diameter.

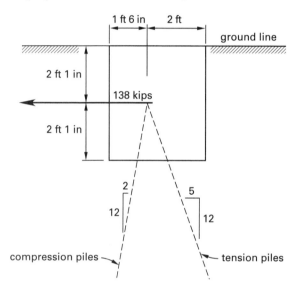

Design Criteria

- two batter tension piles and two batter compression piles tied together by a cap

- each pile is 12 in square

- forces transmitted to soil through skin friction on the pile with values as follows:

    0 ft to 15 ft below bottom of cap $= 500$ lbf/ft^2

    15 ft to 60 ft below bottom of cap $= 900$ lbf/ft^2

- all lateral loads to be resisted by piles

- dead load of pile cap to be neglected

- $f'_c = 3$ ksi

- $f_y = 60$ ksi

1.1. Calculate the load carried by the tension piles, and show a design for a 12 in square precast tension pile.

1.2. Calculate the necessary pile penetration measured from the bottom of the pile cap to resist the forces in tension.

1.3. Draw a section through the pile to show placing of reinforcement. Identify and locate the proper positioning for all bars including ties. Show clearances.

1.4. Draw a detail in elevation to show the method of attaching the cap to the pile in order to transfer the forces in tension adequately.

PROBLEM 2

The concrete retaining wall shown is to be supported by concrete piles.

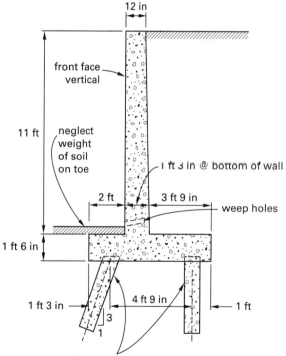

for 12 in diameter concrete piles:
6 ft o.c. for front (battered) piles
12 ft o.c. for rear piles

(a) typical section of
retaining wall

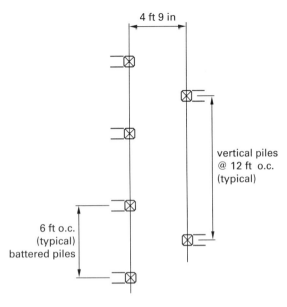

(b) pile layout plan

Design Criteria

- equivalent fluid pressure $= 36$ lbf/ft^2

- concrete weight $= 150$ lbf/ft^3

- soil weight $= 100$ lbf/ft^3

- no surcharge, free-draining soil

- all loads that act on the retaining wall are to be resisted by the piles

- piles take axial loads only

- passive pressure on key to be ignored

- friction on wall face to be ignored

- show all calculations

2.1. Find the axial loads on the piles.

2.2. If $f'_c = 3$ ksi and $f_y = 40$ ksi, what is the maximum spacing for no. 5 vertical bars at the bottom of the stem? Use spacing to nearest $^1/_2$ in.

PROBLEM 3

An exterior 24 in $\times$ 18 in column with a total vertical load of 150 kips, and an interior 24 in $\times$ 24 in column with a total load of 225 kips, are to be supported by a pad footing at each column connected by a 24 in $\times$ 36 in concrete strap as shown.

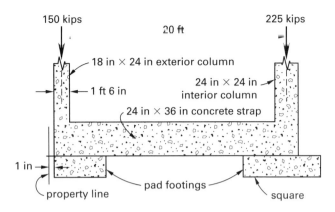

Design Criteria

- strap is placed so as not to bear directly on the soil

- concrete weight is 150 lbf/ft^3

3.1. Determine the dimensions (width and length) for each pad footing that will result in a uniform soil pressure of 3000 lbf/ft^2 under each pad footing. Neglect the weight of the pad footings, but not that of the concrete strap.

3.2. Given the dimensions for the footings as determined in Prob. 3.1, determine and draw the soil pressure profile under the footings for the total load combined with an uplift force of 80 kips at the exterior columns and an uplift force of 25 kips at the interior column.

3.3. Given the loading determined in Prob. 3.2, determine the maximum positive and negative bending moments in the 24 in $\times$ 36 in concrete strap.

PROBLEM 4

The following illustration shows service loads for a combined footing that cannot extend beyond the property line.

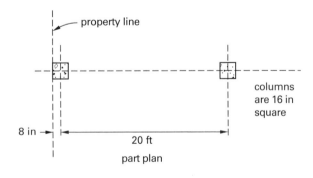

part plan

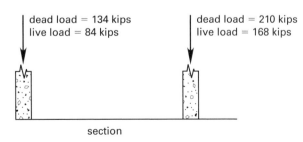

section

Design Criteria

- $f'_c = 3000$ psi

- $f_y = 40,000$ psi

- maximum allowable soil pressure dead load plus live load $= 3000$ lbf/ft^2

- total depth of footing $= 28$ in

- show all calculations

4.1. Design a rectangular footing for the column loads and the allowable soil pressure. Proportion the footing for the total load.

4.2. Determine the size and number of reinforcing bars, and prepare a drawing to show location, size, and number of rebars. Do not calculate or show intermediate cutoffs for the rebar.

SOLUTION 1

1.1. Calculate the load resisted by the two tension piles, and design a 12 in square pile to resist the tension.

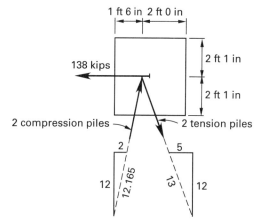

$$\sum F_y = 2C\cos\alpha_C - 2T\cos\alpha_T = 0$$

$$2C\left(\frac{12}{12.165}\right) = 2T\left(\frac{12}{13}\right)$$

$$C = 0.936\,T$$

$$\sum F_x = -H + 2T\sin\alpha_T + 2C\sin\alpha_C = 0$$

$$= -138\text{ kips} + 2T\left(\frac{5}{13}\right) + 2(0.936\,T)\left(\frac{2}{12.165}\right)$$

$$\boxed{T = 128\text{ kips},\ C = 120\text{ kips}}$$

Design the tension pile using grade 60 reinforcement. Assume that handling and driving stresses are not critical. Therefore, design for the tension working load of 128 kips. To control crack width, limit the rebar stress to 24 ksi.

$$A_{st} \geq \frac{T}{f_s} - \frac{128\text{ kips}}{24\,\dfrac{\text{kips}}{\text{in}^2}}$$

$$= 5.33\text{ in}^2$$

$$\rho_g = \frac{A_{st}}{bh} = \frac{5.33\text{ in}^2}{(12\text{ in})(12\text{ in})}$$

$$= 0.037\quad [< 0.04]$$

Therefore, use eight no. 8 bars symmetrically placed. Per IBC Sec. 1810.3.8, use no. 5 gage wire spiral. The pitch is 3 in o.c. for the first 2 ft; use 6 in o.c. elsewhere.

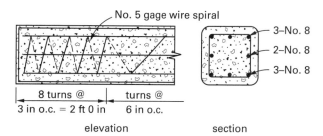

elevation   section

1.2. Calculate the required tension pile penetration. Assume that group action does not affect the given skin friction values. The perimeter of the 12 in × 12 in pile is $b_o = 4h = (4)(1 \text{ ft}) = 4 \text{ ft}$.

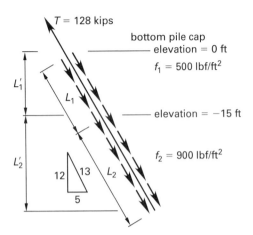

The penetration of the pile in the upper stratum is

$$L_1 = \frac{L}{\cos \alpha_T} = \frac{15 \text{ ft}}{\frac{12}{13}} = 16.25 \text{ ft}$$

For equilibrium,

$$\sum F_{\text{long}} = T - f_1 L_1 b_o - f_2 L_2 b_o = 0$$

$$L_2 = \frac{T - f_1 L_1 b_o}{f_2 b_o}$$

$$= \frac{128 \text{ kips} - \left(0.5 \, \frac{\text{kip}}{\text{ft}^2}\right)(16.25 \text{ ft})(4 \text{ ft})}{\left(0.9 \, \frac{\text{kip}}{\text{ft}^2}\right)(4 \text{ ft})}$$

$$= 26.5 \text{ ft}$$

The required penetration is

$$L_1 + L_2 = 16.25 \text{ ft} + 26.5 \text{ ft} = \boxed{42.8 \text{ ft}}$$

1.3. Draw a section showing bar placement and clear spacing. For a precast concrete element exposed to earth, the minimum concrete cover must be $1^1/4$ in (ACI 318 Sec. 7.7). For three no. 8 bars in a layer enclosed by no. 5 gage wire $(d_b = {}^1/_4 \text{ in})$ and a 12 in wide pile, the clear spacing between longitudinal bars is

$$S = \frac{h - 2(d_{\text{spiral}} + \text{cover}) + 3d_b}{\text{spaces}}$$

$$= \frac{12 \text{ in} - (2)(1.25 \text{ in} + 0.25 \text{ in}) - (3)(1 \text{ in})}{2}$$

$$= 3 \text{ in}$$

The spacing satisfies the limit of ACI 318 Sec. 7.6. The required section through the pile may be drawn as shown.

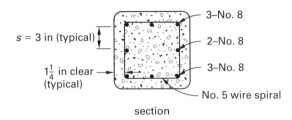

section

1.4. Detail the attachment of the piles to the pile cap. After driving, the longitudinal rebars must be exposed and extended past the point where lines of action intersect for a distance sufficient to develop the tension force in the no. 8 bars. Use the strength design method with $U = 1.6$ (ACI 318 Sec. 9.2.1).

$$T_u = \frac{1.6T}{n} = \frac{(1.6)(128 \text{ kips})}{8}$$

$$= 25.6 \text{ kips}$$

$$\phi T_n = \phi A_b f_y = (0.9)(0.79 \text{ in}^2)\left(60 \, \frac{\text{kips}}{\text{in}^2}\right)$$

$$= 42.7 \text{ kips}$$

Per ACI 318 Sec. 12.2, the embedment length for a no. 8 "top" bar, with $f_y = 60$ ksi and $f'_c = 3$ ksi is

$$l_d = \left(\frac{T_u}{\phi T_n}\right)\left(\frac{f_y \psi_t \psi_e}{20\lambda\sqrt{f'_c}}\right)d_b$$

$$= \left(\frac{25.6 \text{ kips}}{42.7 \text{ kips}}\right)\left(\frac{\left(60{,}000 \, \frac{\text{lbf}}{\text{in}^2}\right)(1.3)(1.0)}{(20)(1.0)\sqrt{3000 \, \frac{\text{lbf}}{\text{in}^2}}}\right)\left(\frac{1.0 \text{ in}}{12 \, \frac{\text{in}}{\text{ft}}}\right)$$

$$= 3.56 \text{ ft}$$

The required length of straight development exceeds available embedment length. Use standard 180° hooks on the tension bars (ACI 318 Sec. 12.5).

$$l_{dh} = \left(\frac{T_u}{\phi T_n}\right)\left(\frac{0.02 f_y \psi_e}{\lambda\sqrt{f'_c}}\right)d_b$$

$$= \left(\frac{25.6 \text{ kips}}{42.7 \text{ kips}}\right)\left(\frac{(0.02)\left(60{,}000 \, \frac{\text{lbf}}{\text{in}^2}\right)(1.0)}{1.0\sqrt{3000 \, \frac{\text{lbf}}{\text{in}^2}}}\right)(1.0 \text{ in})$$

$$= 13 \text{ in}$$

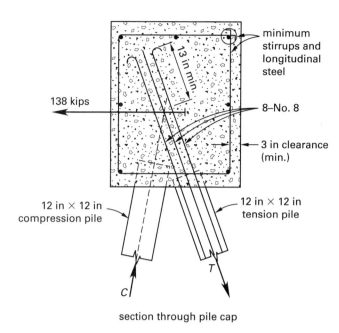

minimum stirrups and longitudinal steel

13 in min

138 kips

8–No. 8

3 in clearance (min.)

12 in × 12 in compression pile

12 in × 12 in tension pile

C T

section through pile cap

In the section shown, the reinforcement for the tension piles must be developed above the line of action of the 138 kip force; otherwise, the concurrent force system is not developed.

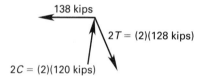

138 kips

$2T = (2)(128 \text{ kips})$

$2C = (2)(120 \text{ kips})$

SOLUTION 2

2.1. For the pile-supported retaining wall, calculate the axial loads on the piles.

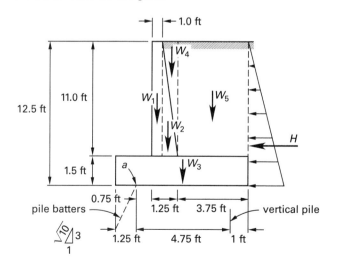

1.0 ft

W_4

11.0 ft

12.5 ft

W_1 W_5

W_2

H

1.5 ft

a W_3

0.75 ft

pile batters

1.25 ft 3.75 ft vertical pile

10 3

1

1.25 ft 4.75 ft 1 ft

Consider a unit length of wall and footing. The lateral force is

$$H = 0.5 p_a h^2 = (0.5)\left(36 \ \frac{\text{lbf}}{\text{ft}^2}\right)(12.5 \ \text{ft})^2$$
$$= 2810 \ \text{lbf}$$

The stem weight is

$$W_1 = w_c(1 \ \text{ft})h_s t_1$$
$$= \left(150 \ \frac{\text{lbf}}{\text{ft}^3}\right)(1 \ \text{ft})(11 \ \text{ft})(1 \ \text{ft})$$
$$= 1650 \ \text{lbf}$$
$$W_2 = w_c(1 \ \text{ft})h_s(0.5 t_2)$$
$$= \left(150 \ \frac{\text{lbf}}{\text{ft}^3}\right)(1 \ \text{ft})(11 \ \text{ft})(0.5)(0.25 \ \text{ft})$$
$$= 200 \ \text{lbf}$$

The footing weight is

$$W_3 = w_c(1 \ \text{ft})B t_3$$
$$= \left(150 \ \frac{\text{lbf}}{\text{ft}^3}\right)(1 \ \text{ft})(7 \ \text{ft})(1.5 \ \text{ft})$$
$$= 1575 \ \text{lbf}$$

The backfill weight is

$$W_4 = w_s(1 \ \text{ft})(0.5 h_s)b_4$$
$$= \left(100 \ \frac{\text{lbf}}{\text{ft}^3}\right)(1 \ \text{ft})(0.5)(11 \ \text{ft})(0.25 \ \text{ft})$$
$$= 138 \ \text{lbf}$$
$$W_5 = w_s(1 \ \text{ft})h_s b_5$$
$$= \left(100 \ \frac{\text{lbf}}{\text{ft}^3}\right)(1 \ \text{ft})(11 \ \text{ft})(3.75 \ \text{ft})$$
$$= 4125 \ \text{lbf}$$

Take moments about the point where the vertical pile centerline intersects the bottom of the footing. Let p_b = the force in the batter piles per foot of wall length.

$$\sum M_{\text{right}} = (p_b \cos \alpha)a - \frac{Hh}{3} - W_1 a_1 - W_2 a_2$$
$$- W_3 a_3 - W_4 a_4 - W_5 a_5 = 0$$

$$p_b = \frac{\dfrac{Hh}{3} + W_1 a_1 + W_2 a_2 + W_3 a_3 + W_4 a_4 + W_5 a_5}{(\cos \alpha)a}$$

$$= \frac{\begin{array}{c}\dfrac{(2810 \ \text{lbf})(12.5 \ \text{ft})}{3} + (1650 \ \text{lbf})(3.5 \ \text{ft}) \\ + (200 \ \text{lbf})(2.92 \ \text{ft}) + (1575 \ \text{lbf})(2.5 \ \text{ft}) \\ + (138 \ \text{lbf})(2.83 \ \text{lbf}) + (4125 \ \text{lbf})(0.875 \ \text{ft})\end{array}}{\left(\dfrac{3}{\sqrt{10}}\right)(4.75 \ \text{ft})}$$

$$= 5770 \ \text{lbf} \quad [\text{per foot of wall}]$$

Equilibrium cannot be satisfied under the constraints of the problem. Resistance to overturning requires a force per foot in the battered piles of 5770 lbf, and resistance to sliding requires a greater force of 8890 lbf. To satisfy equilibrium, one of the following must be done.

- Use additional mechanisms to resist sliding (e.g., passive resistance, adhesion at bottom of footing, lateral resistance of the vertical pile).

- Increase the batter of the piles to match the force polygon of 2810 lbf (horizontal) and $(3/\sqrt{10})$ (5770 lbf) = 5475 lbf (vertical), which is approximately a 1:2 batter.

- Hold the batter at 1:3, and reduce the lever arm between the two rows of piles to satisfy equilibrium.

- Batter both rows of piles, with the heel side piles resisting the complement of the horizontal force that is not resisted by the single row.

This solution uses the first option (use additional mechanisms) to satisfy equilibrium. Therefore, the force in the battered piles is 5770 lbf, and the horizontal force that must be resisted by other mechanisms is

$$\sum F_x = p_b \sin\alpha - H + \Delta P = 0$$
$$\Delta P = H - p_b(\sin\alpha)$$
$$= 2810 \text{ lbf} - (5770 \text{ lbf})\left(\frac{1}{\sqrt{10}}\right)$$
$$= 985 \text{ lbf}$$

The force on the vertical piles is

$$\sum F_y = p_b\cos\alpha + p_v - W_1 - W_2 - W_3 - W_4 - W_5$$
$$= 0$$

$$p_v = W_1 + W_2 + W_3 + W_4 + W_5 - p_b(\cos\alpha)$$
$$= 1650 \text{ lbf} + 200 \text{ lbf} + 1575 \text{ lbf}$$
$$+ 138 \text{ lbf} + 4125 \text{ lbf} - (5770 \text{ lbf})\left(\frac{3}{\sqrt{10}}\right)$$
$$= 2214 \text{ lbf} \quad \text{[per foot of wall length]}$$

The batter piles are spaced at 6 ft on centers and the vertical piles are spaced at 12 ft on centers.

$$P_{\text{batter}} = p_b s_b = \left(5770\ \frac{\text{lbf}}{\text{ft}}\right)(6\text{ ft}) = \boxed{34{,}600 \text{ lbf}}$$

$$P_{\text{vertical}} = p_v s_v = \left(2214\ \frac{\text{lbf}}{\text{ft}}\right)(12\text{ ft}) = \boxed{26{,}600 \text{ lbf}}$$

2.2. Calculate the maximum spacing of no. 5 vertical bars at the bottom of the stem ($f_y = 40{,}000$ psi, $f'_c = 3000$ psi). Use the strength design method with $U = 1.6$ (ACI 318 Sec. 9.2). Calculate the required spacing of no. 5 grade 40 rebars on a per unit length of wall basis.

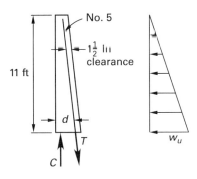

For no. 5 bars exposed to earth, the minimum cover is $1^1/_2$ in (ACI 318 Sec. 7.7).

$$d \approx h - \text{cover} - 0.5d_b$$
$$= 15 \text{ in} - 1.5 \text{ in} - (0.5)(0.625 \text{ in})$$
$$= 13.2 \text{ in}$$

$$w_u = U p_a h_{\text{stem}}$$
$$= (1.6)\left(36\ \frac{\text{lbf}}{\text{ft}^2}\right)(11 \text{ ft})$$
$$= 634 \text{ lbf/ft}$$

$$M_u = w_u\left(\frac{h_{\text{stem}}}{2}\right)\left(\frac{h_{\text{stem}}}{3}\right)$$
$$= \left(634\ \frac{\text{lbf}}{\text{ft}}\right)\left(\frac{11 \text{ ft}}{2}\right)\left(\frac{11 \text{ ft}}{3}\right)$$
$$= 12{,}800 \text{ ft-lbf/ft}$$

Per ACI 318 Sec. 14.1.2, flexural strength must satisfy requirements of ACI 318 Chap. 10.

$$M_u = \phi M_n = \phi\rho bd^2 f_y\left(1 - 0.59\rho\left(\frac{f_y}{f'_c}\right)\right)$$

$$(12{,}800 \text{ ft-lbf})\left(12\ \frac{\text{in}}{\text{ft}}\right)$$
$$= 0.9\rho\left(12\ \frac{\text{in}}{\text{ft}}\right)(13.2 \text{ in})^2\left(40{,}000\ \frac{\text{lbf}}{\text{in}^2}\right)$$
$$\times\left(1 - 0.59\rho\left(\frac{40{,}000\ \frac{\text{lbf}}{\text{in}^2}}{3000\ \frac{\text{lbf}}{\text{in}^2}}\right)\right)$$
$$\rho = 0.00207$$

$$A_s = \rho bd = (0.00207)\left(12\ \frac{\text{in}}{\text{ft}}\right)(13.2 \text{ in})$$
$$= 0.33 \text{ in}^2/\text{ft}$$

The ACI 318 flexural requirements include the minimum steel requirement of Sec. 10.5.1. Therefore,

$$A_{s,\min} \geq \begin{cases} \left(\dfrac{200}{f_y}\right)bd = \left(\dfrac{200 \ \frac{\text{lbf}}{\text{in}^2}}{40{,}000 \ \frac{\text{lbf}}{\text{in}^2}}\right)\left(12 \ \frac{\text{in}}{\text{ft}}\right)(13.2 \ \text{in}) \\ \qquad = 0.79 \ \text{in}^2/\text{ft} \\ \left(\dfrac{3\sqrt{f'_c}}{f_y}\right)bd = \left(\dfrac{3\sqrt{3000 \ \frac{\text{lbf}}{\text{in}^2}}}{40{,}000 \ \frac{\text{lbf}}{\text{in}^2}}\right)\left(12 \ \frac{\text{in}}{\text{ft}}\right)(13.2 \ \text{in}) \\ \qquad = 0.65 \ \text{in}^2/\text{ft} \end{cases}$$

Alternatively, provide one-third more flexural steel than is required by analysis (per ACI 318 Sec. 10.5.3), which is $(^4/_3)(0.33 \ \text{in}^2/\text{ft}) = 0.44 \ \text{in}^2/\text{ft}$.

The best approach would be to reduce the stem thickness at the base (e.g., by using a constant 12 in wall thickness) to lower the effective depth so that ACI 318 Sec. 10.5 does not govern. However, if the given dimensions must be kept, then increase the calculated A_s by $^4/_3$ and use 0.44 in^2/ft.

The minimum horizontal steel per ACI 318 Sec. 14.3.3 is

$$A_s = 0.0025bt = (0.0025)\left(12 \ \frac{\text{in}}{\text{ft}}\right)(15 \ \text{in})$$
$$= 0.45 \ \text{in}^2/\text{ft} \quad [\text{say, no. 5 at 8 in o.c.}]$$

Use no. 5 bars at $\boxed{7.5 \ \text{in o.c.}}$ ($A_s = 0.50 \ \text{in}^2/\text{ft}$) vertical, and no. 5 bars at $\boxed{8 \ \text{in o.c.}}$ horizontal.

SOLUTION 3

3.1. Determine the pad dimensions that will produce uniform bearing pressure under the given loads. Per the problem statement, include the weight of the 3 ft × 2 ft strap in the resultant, but neglect the weight of the footings. From the illustration, the length of the strap is 20 ft plus half the width of each column.

$$L_{\text{strap}} = \left(\tfrac{1}{2}\right)(1.5 \ \text{ft}) + 20 \ \text{ft} + \left(\tfrac{1}{2}\right)(2 \ \text{ft}) = 21.75 \ \text{ft}$$

Using normal weight concrete,

$$W_{\text{strap}} = w_c(bhL)_{\text{strap}}$$
$$= \left(\dfrac{150 \ \frac{\text{lbf}}{\text{ft}^3}}{1000 \ \frac{\text{lbf}}{\text{kip}}}\right)(2 \ \text{ft})(3 \ \text{ft})(21.75 \ \text{ft})$$
$$= 19.6 \ \text{kips}$$

The strap is stiff enough to permit treating the combined footing as rigid. To achieve uniform bearing pressure, the centroid of the two footings must coincide with the center of gravity of the applied forces. Measuring

from the centroid of the exterior column, $\bar{x}_{\text{ext}}$, which is 0.75 ft from the left end of the strap,

$$\bar{x}_{\text{strap}} = 0.5L - 0.75 \ \text{ft}$$
$$= (0.5)(21.75 \ \text{ft}) - 0.75 \ \text{ft}$$
$$= 10.125 \ \text{ft}$$
$$R = P_{\text{ext}} + W_{\text{strap}} + P_{\text{int}}$$
$$= 150 \ \text{kips} + 19.6 \ \text{kips} + 225 \ \text{kips}$$
$$= 394.6 \ \text{kips}$$

$$\bar{x}_{\text{pads}} = \dfrac{P_{\text{ext}}\bar{x}_{\text{ext}} + W_{\text{strap}}\bar{x}_{\text{strap}} + P_{\text{int}}\bar{x}_{\text{int}}}{R}$$
$$= \dfrac{\begin{array}{c}(150 \ \text{kips})(0 \ \text{ft}) + (19.6 \ \text{kips}) \\ \times (10.125 \ \text{ft}) + (225 \ \text{kips})(20 \ \text{ft})\end{array}}{394.6 \ \text{kips}}$$
$$= 11.91 \ \text{ft}$$

Let A_{ext} and A_{int} be the areas of the exterior and interior footings.

$$\sum F_y = -R + A_{\text{ext}}f_p + A_{\text{int}}f_p = 0$$
$$A_{\text{ext}} + A_{\text{int}} = \dfrac{R}{f_p} = \dfrac{394.6 \ \text{kips}}{3 \ \frac{\text{kips}}{\text{ft}^2}} = 131.5 \ \text{ft}^2$$

There are an infinite number of combinations of A_{ext} and A_{int} that can give a total area of 131.5 ft² with the centroid $\bar{x}$. One simple possibility is

$$A_{\text{ext}} = A_{\text{int}} = \dfrac{131.6 \ \text{ft}^2}{2} = 65.8 \ \text{ft}^2$$

In this case, as the interior footing is to be square, it would have a plan dimension of

$$B = \sqrt{A_{\text{int}}} = \sqrt{65.8 \ \text{ft}^2} = 8.11 \ \text{ft}$$

Find the eccentricity of the exterior footing, e_{ext}, again measuring from the centroid of the exterior column.

$$\bar{x}_{\text{pads}} = \dfrac{A_{\text{ext}}(\bar{x}_{\text{ext}} + e_{\text{ext}}) + A_{\text{int}}\bar{x}_{\text{int}}}{A_{\text{int}} + A_{\text{ext}}}$$
$$e_{\text{ext}} = \dfrac{\bar{x}_{\text{pads}}(A_{\text{int}} + A_{\text{ext}}) - A_{\text{int}}\bar{x}_{\text{int}}}{A_{\text{ext}}} - \bar{x}_{\text{ext}}$$
$$= \dfrac{\begin{array}{c}(11.91 \ \text{ft})(65.8 \ \text{ft}^2 + 65.8 \ \text{ft}^2) \\ - (65.8 \ \text{ft}^2)(20 \ \text{ft})\end{array}}{65.8 \ \text{ft}^2} - 0 \ \text{ft}$$
$$= 3.82 \ \text{ft}$$

The exterior footing will extend 0.75 ft to the left of the centroid of the exterior column, so length of the exterior footing is

$$L_{ext} = 2(e_{ext} + 0.75 \text{ ft})$$
$$= (2)(3.82 \text{ ft} + 0.75 \text{ ft})$$
$$= 9.14 \text{ ft}$$

The required width of the exterior footing is

$$B_{ext} = \frac{A_{ext}}{L_{ext}} = \frac{65.8 \text{ ft}^2}{9.14 \text{ ft}} = 7.20 \text{ ft}$$

Use 7.20 ft × 9.14 ft for the exterior footing (as a practical size, say 7 ft 3 in × 9 ft 0 in). Use 8.11 ft × 8.11 ft for the interior footing.

3.2. Determine the soil pressure profile when the forces from Prob. 3.1 are combined with 80 kips of uplift at the exterior column, and 25 kips of uplift at the interior column. For this analysis, use the theoretical dimensions computed in Prob. 3.1 rather than rounding off to practical dimensions.

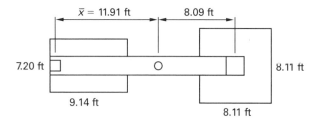

For the combined footing,

$$A = 131.6 \text{ ft}^2$$
$$I = \sum (I_{xo} + A_i d_i^2)$$
$$= \sum \left(\frac{B_i L_i^3}{12} + A_i d_i^2 \right)$$
$$= \left(\frac{(7.20 \text{ ft})(9.14 \text{ ft})^3}{12} + (65.8 \text{ ft}^2) \right.$$
$$\left. \times (11.91 \text{ ft} - 3.82 \text{ ft})^2 \right)$$
$$+ \left(\frac{(8.11 \text{ ft})(8.11 \text{ ft})^3}{12} + (65.8 \text{ ft}^2) \right.$$
$$\left. \times (20 \text{ ft} - 11.91 \text{ ft})^2 \right)$$
$$= 9431 \text{ ft}^4$$

Pressures under the uplift force and moment combine algebraically with the uniform 3 kips/ft^2 pressure from Prob. 3.1.

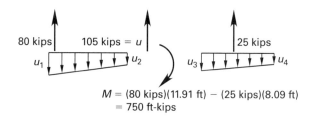

$$M = (80 \text{ kips})(11.91 \text{ ft}) - (25 \text{ kips})(8.09 \text{ ft})$$
$$= 750 \text{ ft-kips}$$

$$u_i = \frac{U}{A} \pm \frac{Mx}{I}$$

$$u_1 = \frac{-105 \text{ kips}}{131.6 \text{ ft}^2} - \frac{(750 \text{ ft-kips})(11.91 \text{ ft} + 0.75 \text{ ft})}{9431 \text{ ft}^4}$$
$$= -1.80 \text{ kips/ft}^2$$

$$u_2 = \frac{-105 \text{ kips}}{131.6 \text{ ft}^2}$$
$$- \frac{(750 \text{ ft-kips})(11.91 \text{ ft} + 0.75 \text{ ft} - 9.14 \text{ ft})}{9431 \text{ ft}^4}$$
$$= -1.08 \text{ kips/ft}^2$$

$$u_3 = \frac{-105 \text{ kips}}{131.6 \text{ ft}^2} + \frac{(750 \text{ ft-kips})(8.09 \text{ ft} - 4.06 \text{ ft})}{9431 \text{ ft}^4}$$
$$= -0.48 \text{ kip/ft}^2$$

$$u_4 \frac{-105 \text{ kips}}{131.6 \text{ ft}^2} + \frac{(750 \text{ ft-kips})(8.09 \text{ ft} + 4.06 \text{ ft})}{9431 \text{ ft}^4}$$
$$= 0.17 \text{ kip/ft}^2$$

Combining forces and pressures, the required pressure diagrams are as shown.

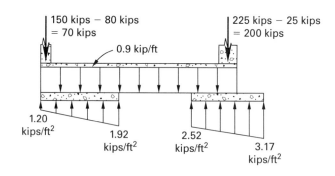

3.3. Calculate the maximum positive and negative bending moments for the loading in Prob. 3.2.

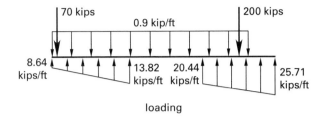

loading

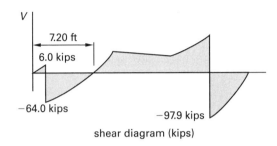

shear diagram (kips)

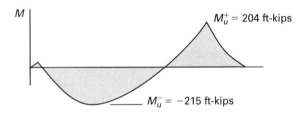

bending moment diagram (ft-kips)

The maximum positive bending moment occurs at the 200 kips concentrated load, where the shear diagram is discontinuous. By representing the linearly varying loading to the right of that location as a rectangle plus a triangle, the resultants of each part, W_1 and W_2, can be calculated.

$$w_1 = \frac{w_{\text{left}} + w_{\text{right}}}{2} = \frac{20.44 \dfrac{\text{kips}}{\text{ft}} + 25.71 \dfrac{\text{kips}}{\text{ft}}}{2}$$

$$= 23.0 \text{ kips/ft}$$

$$W_1 = w_1 a = \left(23.0 \ \frac{\text{kips}}{\text{ft}}\right)(0.5)(8.11 \text{ ft})$$

$$= 93.3 \text{ kips}$$

$$w_2 = w_{\text{right}} - \frac{w_{\text{left}} + w_{\text{right}}}{2}$$

$$= 25.71 \ \frac{\text{kips}}{\text{ft}} - \frac{20.44 \dfrac{\text{kips}}{\text{ft}} + 25.71 \dfrac{\text{kips}}{\text{ft}}}{2}$$

$$= 2.64 \text{ kips/ft}$$

$$W_2 = 0.5 w_2 a = (0.5)\left(2.64 \ \frac{\text{kips}}{\text{ft}}\right)(0.5)(8.11 \text{ ft})$$

$$= 5.4 \text{ kips}$$

$$M_u^+ = W_1(0.5a) + W_2(0.667a)$$

$$= (93.3 \text{ kips})(0.5)(4.055 \text{ ft}) + (5.4 \text{ kips})$$

$$\times (0.667)(4.055 \text{ ft})$$

$$= \boxed{204 \text{ ft-kips}}$$

The free-body diagram is cut at the point of maximum moment (i.e., where V is zero).

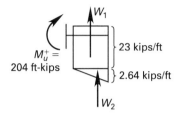

Negative bending is critical at 7.20 ft from the left end where the shear is zero. Replacing the linearly varying load again by equivalent rectangular and triangular loading gives

$$W_1 = w_{\text{strap}} x = \left(0.9 \ \frac{\text{kip}}{\text{ft}}\right)(7.20 \text{ ft})$$

$$= 6.5 \text{ kips} \quad [\text{down}]$$

$$w_2 = w_{\text{left}}$$

$$= 8.64 \text{ kips/ft}$$

$$W_2 = w_2 x = \left(8.64 \ \frac{\text{kips}}{\text{ft}}\right)(7.20 \text{ ft})$$

$$= 62.2 \text{ kips} \quad [\text{up}]$$

$$w_3 = w_{7.2 \text{ ft}} - w_{\text{left}}$$

$$= 12.7 \ \frac{\text{kips}}{\text{ft}} - 8.64 \ \frac{\text{kips}}{\text{ft}}$$

$$= 4.06 \text{ kips/ft}$$

$$W_3 = 0.5 w_3 x = (0.5)\left(4.06 \ \frac{\text{kips}}{\text{ft}}\right)(7.20 \text{ ft})$$

$$= 14.6 \text{ kips}$$

$$M_u^- = -P_1(x - 0.75 \text{ ft}) - W_1(0.5x)$$

$$+ W_2(0.5x) + W_3(0.333x)$$

$$= (-70 \text{ kips})(7.20 \text{ ft} - 0.75 \text{ ft}) - (6.5 \text{ kips})(0.5)$$

$$\times (7.20 \text{ ft}) + (62.2 \text{ kips})(0.5)(7.20 \text{ ft})$$

$$+ (14.6 \text{ kips})(0.333)(7.20 \text{ ft})$$

$$= \boxed{-216 \text{ ft-kips}}$$

The free-body diagram at the point of minimum moment is

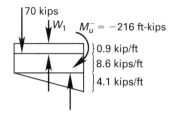

SOLUTION 4

4.1. Design the combined footing for full dead and live load conditions.

$$P_{\text{ext}} = (P_D + P_L)_{\text{ext}} = 134 \text{ kips} + 84 \text{ kips} = 218 \text{ kips}$$

$$P_{\text{int}} = (P_D + P_L)_{\text{int}} = 210 \text{ kips} + 168 \text{ kips} = 378 \text{ kips}$$

$$P = P_{\text{ext}} + P_{\text{int}} = 218 \text{ kips} + 378 \text{ kips} = 596 \text{ kips}$$

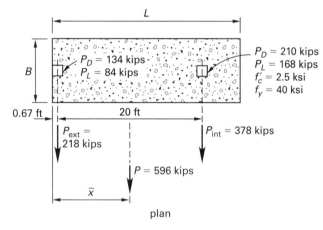

plan

Proportion the footing length to achieve uniform bearing pressure (3 kips/ft^2 net pressure).

$$\bar{x} = \frac{P_{\text{ext}}\bar{x}_{\text{ext}} + P_{\text{int}}\bar{x}_{\text{int}}}{P}$$

$$= \frac{(218 \text{ kips})(0.67 \text{ ft}) + (378 \text{ kips})(20.67 \text{ ft})}{596 \text{ kips}}$$

$$= 13.35 \text{ ft}$$

Therefore,

$$L = 2\bar{x} = (2)(13.35 \text{ ft}) = 26.7 \text{ ft} \quad [\text{say } 26 \text{ ft } 8 \text{ in}]$$

The width that is needed to limit the bearing pressure to 3 kips/ft^2 is

$$B = \frac{P}{f_p L} = \frac{596 \text{ kips}}{\left(3 \dfrac{\text{kips}}{\text{ft}^2}\right)(26.7 \text{ ft})} = 7.4 \text{ ft} \quad [\text{say } 7 \text{ ft } 6 \text{ in}]$$

Use a footing with plan dimensions of 7 ft 6 in × 26 ft 8 in.

4.2. Design the reinforcement using the strength design method in ACI 318, with given values of $f'_c = 3000$, $f_y = 40,000$ psi, and $h = 28$ in.

$$P_{u,\text{ext}} = 1.2P_D + 1.6P_L$$

$$= (1.2)(134 \text{ kips}) + (1.6)(84 \text{ kips})$$

$$= 295 \text{ kips}$$

$$P_{u,\text{int}} = 1.2P_D + 1.6P_L$$

$$= (1.2)(210 \text{ kips}) + (1.6)(168 \text{ kips})$$

$$= 521 \text{ kips}$$

$$P_u = P_{u,\text{ext}} + P_{u,\text{int}}$$

$$= 295 \text{ kips} + 521 \text{ kips}$$

$$= 816 \text{ kips}$$

$$M_u = P_{u,\text{ext}}(\bar{x} - \bar{x}_{\text{ext}}) + P_{u,\text{int}}(\bar{x} - \bar{x}_{\text{int}})$$

$$= (295 \text{ kips})(13.33 \text{ ft} - 0.67 \text{ ft})$$

$$\quad + (521 \text{ kips})(13.33 \text{ ft} - 20.67 \text{ ft})$$

$$= -89.4 \text{ ft-kips}$$

The bearing pressure is nonuniform under the factored loads (slightly higher at the interior end than at the exterior end). It would be reasonable to ignore the slight eccentricity and design for a uniformly distributed loading, but for completeness, the variation is considered in this solution. Calculate the loading in units of kips/ft at the ends of the footing under factored loads.

$$L = 26.7 \text{ ft}$$

$$S = \frac{L^2}{6}$$

$$= \frac{(26.7 \text{ ft})^2}{6}$$

$$= 119 \text{ ft}^2$$

$$w_{u,\text{ext}} = \frac{P_u}{L} + \frac{M_u}{S}$$

$$= \frac{816 \text{ kips}}{26.7 \text{ ft}} + \frac{-89.4 \text{ ft-kips}}{119 \text{ ft}^2}$$

$$= 29.8 \text{ kips/ft}$$

$$w_{u,\text{int}} = \frac{P_u}{L} - \frac{M_u}{S}$$

$$= \frac{816 \text{ kips}}{26.7 \text{ ft}} - \frac{-89.4 \text{ ft-kips}}{119 \text{ ft}^2}$$

$$= 31.3 \text{ kips/ft}$$

This is practically a uniform pressure distribution. However, to satisfy statics, the calculated values are used.

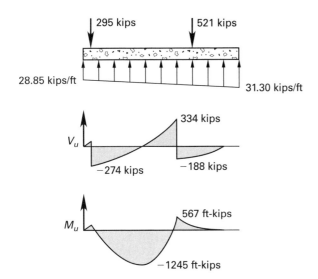

Check the punching shear at the interior column. The intensity of factored bearing pressure at that location is

$$w_u = \frac{w_{u,\text{ext}}}{B} + \frac{(w_{u,\text{int}} - w_{u,\text{ext}})x}{BL}$$

$$= \frac{29.85 \; \frac{\text{kips}}{\text{ft}}}{7.5 \; \text{ft}} + \frac{\left(31.30 \; \frac{\text{kips}}{\text{ft}} - 29.85 \; \frac{\text{kips}}{\text{ft}}\right)(20.67 \; \text{ft})}{(7.5 \; \text{ft})(26.7 \; \text{ft})}$$

$$= 4.13 \; \text{kips/ft}^2$$

$$d = h - \text{cover} - d_b = 28 \; \text{in} - 3 \; \text{in} - 1 \; \text{in} = 24 \; \text{in}$$

$$b_o = 4(h_{\text{column}} + d) = (4)(16 \; \text{in} + 24 \; \text{in}) = 160 \; \text{in}$$

$$V_u = P_{u,\text{int}} - w_u(h_{\text{column}} + d)^2$$

$$= 521 \; \text{kips} - \left(4.77 \; \frac{\text{kips}}{\text{ft}^2}\right)\left(\frac{16 \; \text{in} + 24 \; \text{in}}{12 \; \frac{\text{in}}{\text{ft}}}\right)^2$$

$$= 468 \; \text{kips}$$

$$\phi V_c = 4\phi\lambda\sqrt{f'_c}\,b_o d$$

$$= \frac{(4)(0.75)(1.0)\sqrt{3000 \; \frac{\text{lbf}}{\text{in}^2}}\,(160 \; \text{in})(24 \; \text{in})}{1000 \; \frac{\text{lbf}}{\text{kip}}}$$

$$= 631 \; \text{kips} \quad [>V_u, \; \text{so OK}]$$

Check the punching shear at the exterior column. The intensity of factored bearing pressure at that location is

$$w_u = \frac{w_{u,\text{ext}}}{B} + \frac{(w_{u,\text{int}} - w_{u,\text{ext}})x}{BL}$$

$$= \frac{29.85 \; \frac{\text{kips}}{\text{ft}}}{7.5 \; \text{ft}} + \frac{\left(31.30 \; \frac{\text{kips}}{\text{ft}} - 29.85 \; \frac{\text{kips}}{\text{ft}}\right)(0.67 \; \text{ft})}{(7.5 \; \text{ft})(26.7 \; \text{ft})}$$

$$= 3.98 \; \text{kips/ft}^2$$

$$b_o = (h_{\text{column}} + d) + 2(h_{\text{column}} + 0.5d)$$

$$= (16 \; \text{in} + 24 \; \text{in}) + (2)(16 \; \text{in} + (0.5)(24 \; \text{in}))$$

$$= 96 \; \text{in}$$

$$V_u = P_u - w_u(h_{\text{column}} + d)(h_{\text{column}} + 0.5d)$$

$$= 295 \; \text{kips} - \left(4.60 \; \frac{\text{kips}}{\text{ft}^2}\right)\left(\frac{16 \; \text{in} + 24 \; \text{in}}{12 \; \frac{\text{in}}{\text{ft}}}\right)$$

$$\times \left(\frac{16 \; \text{in} + (0.5)(24 \; \text{in})}{12 \; \frac{\text{in}}{\text{ft}}}\right)$$

$$= 259 \; \text{kips}$$

$$\phi V_c = 4\phi\lambda\sqrt{f'_c}\,b_o d$$

$$= \frac{(4)(0.75)(1.0)\sqrt{3000 \; \frac{\text{lbf}}{\text{in}^2}}\,(96 \; \text{in})(24 \; \text{in})}{1000 \; \frac{\text{lbf}}{\text{kip}}}$$

$$= 379 \; \text{kips} \quad [>V_u, \; \text{so OK}]$$

Check wide beam shear at the critical location, which is at a distance d from the face of the interior column.

$$x = \bar{x}_{\text{int}} - (d + 0.5h_{\text{column}})$$

$$= 20.67 \; \text{ft} - \frac{24 \; \text{in} + (0.5)(16 \; \text{in})}{12 \; \frac{\text{in}}{\text{ft}}}$$

$$= 18.0 \; \text{ft}$$

$$w_{u,18 \; \text{ft}} = w_{u,\text{int}} + \left(\frac{w_{u,\text{ext}} - w_{u,\text{int}}}{L}\right)x$$

$$= 29.85 \; \frac{\text{kips}}{\text{ft}}$$

$$+ \frac{\left(31.30 \; \frac{\text{kips}}{\text{ft}} - 29.85 \; \frac{\text{kips}}{\text{ft}}\right)(18 \; \text{ft})}{26.7 \; \text{ft}}$$

$$= 30.83 \; \text{kips/ft}$$

$$V_u = V_{u,\text{int}} - w_{u,18 \; \text{ft}}(d + 0.5h_{\text{column}})$$

$$= 334 \; \text{kips} - \left(30.83 \; \frac{\text{kips}}{\text{ft}}\right)\left(\frac{24 \; \text{in} + (0.5)(16 \; \text{in})}{12 \; \frac{\text{in}}{\text{ft}}}\right)$$

$$= 252 \; \text{kips}$$

$$V_c = 2\lambda\sqrt{f'_c}\,b_w d = \frac{(2)(1.0)\sqrt{3000 \; \frac{\text{lbf}}{\text{in}^2}} \times (24 \; \text{in})(7.5 \; \text{ft})\left(12 \; \frac{\text{in}}{\text{ft}}\right)}{1000 \; \frac{\text{lbf}}{\text{kip}}}$$

$$= 237 \; \text{kips} \quad [V_u > \phi V_c, \; \text{so stirrups are required}]$$

Try a double set of no. 5 stirrups.

$$A_v = 4A_b = (4)(0.31 \text{ in}^2) = 1.24 \text{ in}^2$$

$$V_s = \frac{V_u}{\phi} - V_c = \frac{252{,}000 \text{ lbf}}{0.75} - 236{,}600 \text{ lbf}$$

$$= 99{,}400 \text{ lbf}$$

$$4\sqrt{f'_c} b_w d = 2V_c = (2)(236{,}600 \text{ lbf}) = 473{,}200 \text{ lbf}$$

$$8\sqrt{f'_c} b_w d = 4V_c = (4)(236{,}600 \text{ lbf})$$

$$= 946{,}400 \text{ lbf}$$

$V_s < 4\sqrt{f'_c} b_w d$, so

$$s \leq \begin{cases} \dfrac{d}{2} = \dfrac{24.0 \text{ in}}{2} = 12.0 \text{ in} \\[2em] \dfrac{A_v f_y}{0.75\sqrt{f'_c} b_w} = \dfrac{(1.24 \text{ in}^2)\left(40{,}000 \text{ } \frac{\text{lbf}}{\text{in}^2}\right)}{0.75\sqrt{3000 \text{ } \frac{\text{lbf}}{\text{in}^2}}(7.5 \text{ ft})\left(12 \text{ } \frac{\text{in}}{\text{ft}}\right)} \\[1em] \qquad = 13.4 \text{ in} \\[2em] \dfrac{A_v f_y}{50 b_w} = \dfrac{(1.24 \text{ in}^2)\left(40{,}000 \text{ } \frac{\text{lbf}}{\text{in}^2}\right)}{\left(50 \text{ } \frac{\text{lbf}}{\text{in}^2}\right)(7.5 \text{ ft})\left(12 \text{ } \frac{\text{in}}{\text{ft}}\right)} \\[1em] \qquad = 11.0 \text{ in} \quad [\text{controls}] \\[2em] \dfrac{A_v f_y d}{V_s} = \dfrac{(1.24 \text{ in}^2)\left(40{,}000 \text{ } \frac{\text{lbf}}{\text{in}^2}\right)(24.0 \text{ in})}{99{,}400 \text{ lbf}} \\[1em] \qquad = 12.0 \text{ in} \end{cases}$$

Use two sets of no. 5 U-stirrups throughout.

For the top steel, the flexural design, $M_u = 1245$ ft-kips, is

$$M_u = \phi M_n = \phi \rho b d^2 f_y \left(1 - 0.59\rho \left(\frac{f_y}{f'_c}\right)\right)$$

$$(1{,}245{,}000 \text{ ft-lbf})\left(12 \text{ } \frac{\text{in}}{\text{ft}}\right)$$

$$= 0.9\rho (90 \text{ in})(24.0 \text{ in})^2 \left(40{,}000 \text{ } \frac{\text{lbf}}{\text{in}^2}\right)$$

$$\times \left(1 - 0.59\rho \left(\frac{40{,}000 \text{ } \frac{\text{lbf}}{\text{in}^2}}{3000 \text{ } \frac{\text{lbf}}{\text{in}^2}}\right)\right)$$

$$\rho = 0.00859$$

$$A_s = \rho b d = (0.00859)(90 \text{ in})(24.0 \text{ in}) = 18.55 \text{ in}^2$$

Use 15 no. 10 top bars. For the bottom steel, the critical moment occurs at the interior column where $M_u = 567$ ft-kips.

$$M_u = \phi M_n = \phi \rho b d^2 f_y \left(1 - 0.59\rho \left(\frac{f_y}{f'_c}\right)\right)$$

$$(567{,}000 \text{ ft-lbf})\left(12 \text{ } \frac{\text{in}}{\text{ft}}\right)$$

$$= 0.9\rho (90 \text{ in})(24.0 \text{ in})^2 \left(40{,}000 \text{ } \frac{\text{lbf}}{\text{in}^2}\right)$$

$$\times \left(1 - 0.59\rho \left(\frac{40{,}000 \text{ } \frac{\text{lbf}}{\text{in}^2}}{3000 \text{ } \frac{\text{lbf}}{\text{in}^2}}\right)\right)$$

$$\rho = 0.00376$$

$$A_s = \rho b d = (0.00376)(90 \text{ in})(24.0 \text{ in}) = 8.12 \text{ in}^2$$

Calculate the minimum footing reinforcement.

$$A_{s,\text{min}} = 0.002 b h = (0.002)(90 \text{ in})(28 \text{ in})$$

$$= 5.04 \text{ in}^2 \quad [\text{not critical}]$$

Use 15 no. 10 top bars, eight no. 9 bottom bars, and double no. 5 stirrups at 11 in o.c.

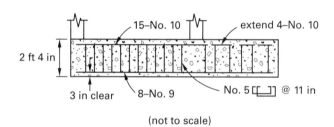

2 ft 4 in · 3 in clear · 15–No. 10 · extend 4–No. 10 · 8–No. 9 · No. 5 [ ] @ 11 in

(not to scale)

6 Timber Design

PROBLEM 1

1.1. On the design specifications for a roof diaphragm, structural I plywood is specified. At the job site, it is found that CDX plywood has been substituted. Is this substitution acceptable? Explain.

1.2. A glulam beam certificate indicates that 24F-E15 hem-fir has been used in lieu of 24F-V8 Douglas-fir members, which are the designed members. The beam is cantilevered past a steel column. What action is indicated and what should be checked? No calculations are required.

1.3. A portion of a wood-framed building wall is shown. The plywood-sheathed wall panel must resist all lateral loads shown. (See *Illustration for Prob. 1.3.*) Sketch where horizontal and vertical ties are needed to transfer the given lateral loads to the resisting elements and into the foundation. Calculate the load for which each tie must be designed. Do not detail the ties.

Illustration for Prob. 1.3

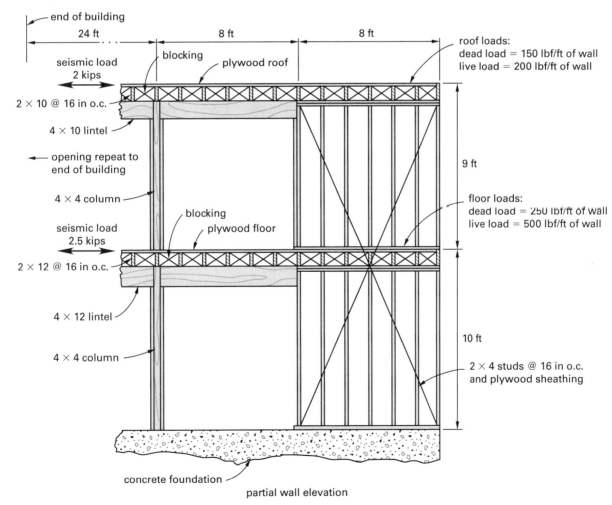

partial wall elevation

1.4. A building with wood roof framing and masonry walls was constructed with roof-to-wall details as shown in details 1, 2, 3, and 4 (See *Illustration for Prob. 1.4.*)

Design Criteria

- wood ledgers have no values in cross-grain bending

- plywood has no value in axial compression

Show how each of the details should be specified.

Illustration for Prob. 1.4

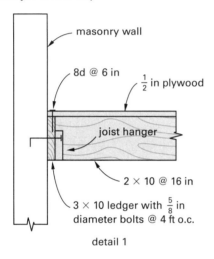

detail 1

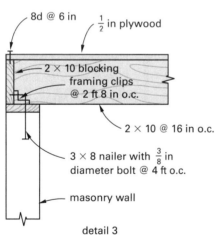

detail 3

1.5. Calculate the allowable load on the designated bolt shown. (See *Illustration for Prob. 1.5.*) Consider wood stresses only.

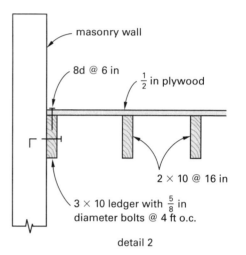

detail 2

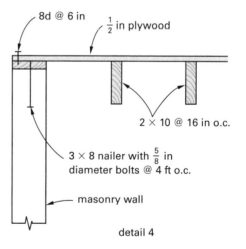

detail 4

Illustration for Prob. 1.5

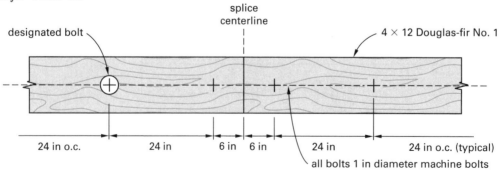

1.6. During the inspection of a building under construction, the details shown are observed. (See *Illustration for Prob. 1.6.*) Comment on the suitability of each member, and identify any potential problems.

1.7. Draw an example of a joint toenail and give the reduction factor for lateral nail load. Do the same for a slant nail.

1.8. How much of a clinched nail must extend beyond the face of the material?

Illustration for Prob. 1.6

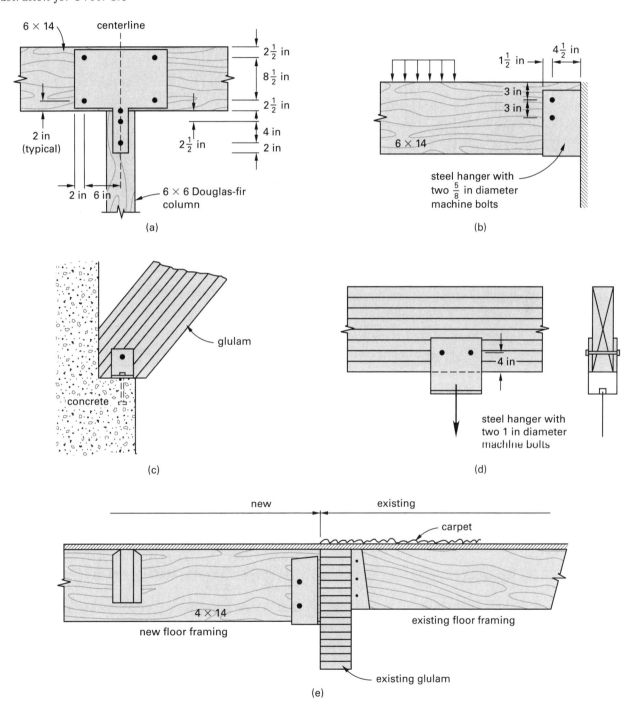

1.9. An anchor bolt designed to be a tie-down was mislocated as shown.

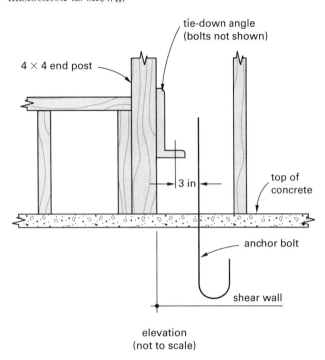

elevation
(not to scale)

Assume the anchor bolt capacity is adequate at the position shown. What action is indicated and what should be checked? No calculations are required.

1.10. Inspection of a plywood shear wall reveals that power-driven nails were used as shown.

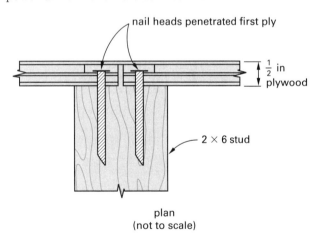

plan
(not to scale)

The plywood has five plies. The existing edge nails are spaced at 4 in o.c. What action is indicated and what should be checked? No calculations are required.

PROBLEM 2

A single-story wood frame building is 40 ft wide × 80 ft long. It has 8 ft long shear walls at each end and a 12 ft long shear wall at the center of the building. The details of the 12 ft shear wall at the building centerline are shown. (See *Illustration for Prob. 2*.)

Design Criteria

- 150 mph basic wind speed

- exposure condition C

- $K_{zt} = 1$

- seismic design category C

- $S_{DS} = 0.45$

- importance factor $= 1$

- diaphragm qualifies as flexible

- weight of exterior walls $= 20 \text{ lbf/ft}^2$

- roof weight $= 15 \text{ lbf/ft}^2$

- roof and shear walls sheathed with C-C grade exterior plywood

- use allowable stress design (ASD)

- Douglas fir-larch framing

2.1. What is the maximum lateral force to be carried by the shear wall at the building centerline?

2.2. What is the nailing required at the building centerline to transfer the loads from the roof into the shear wall?

2.3. A collection member is identified on the drawing that ties into the wall framing at point A. Design and draw a suitable detail for the connection at A.

2.4. During construction, it is discovered that the sheet metal contractor has cut an opening in the shear wall at the location marked X. The top plate has been cut to pass an 18 in × 18 in duct. The top of the duct is set at 4 in below the soffit of the roof plywood. To avoid delay at the job site, it is necessary to redesign a collection member around the duct opening for the transfer of lateral forces from the roof to the shear wall. Design an installation that will transfer the lateral loads around the duct opening.

2.5. Draw a suitable detail that shows the design and intent needed to accomplish the tie.

Illustration for Prob. 2

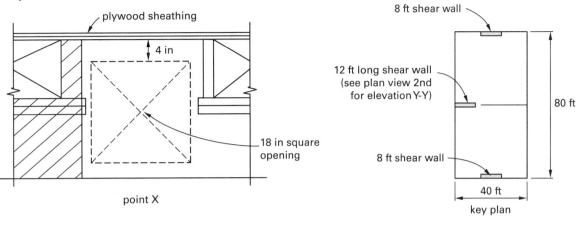

point X

key plan

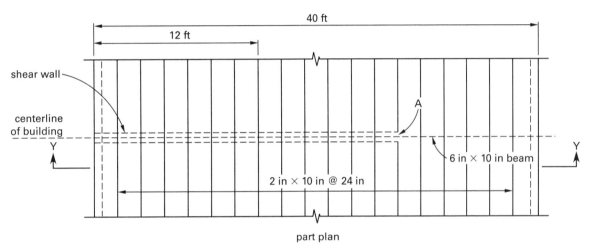

part plan

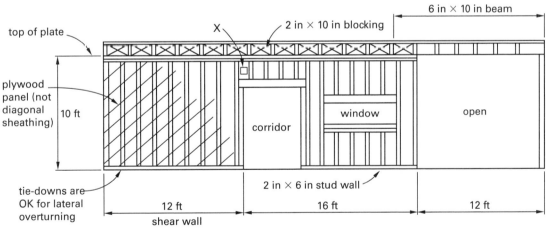

elevation Y-Y

PROBLEM 3

A portion of a wood framing system is shown.

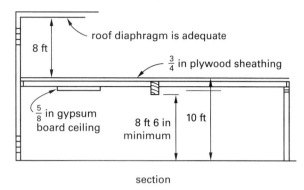

section

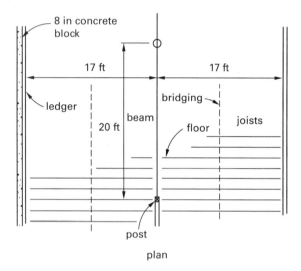

plan

Design Criteria

- no. 1/no. 2 Douglas fir-larch (north) used for floor joists and ledgers

3.1. Design the floor joists for the minimum board footage using standard dimension Douglas-fir and ASD. Give size and spacing, and show calculations.

3.2. Calculate a size for the main wood girder. Assume a simple span. No detail is necessary. Calculate all the principal design considerations that may be critical.

3.3. Provide a suitable wall ledger. Give the size, and show anchorage for vertical and lateral loads. Draw a detail of the ledger showing all the principal elements that will be required to complete the installation.

PROBLEM 4

A plan and cross section of a one-story frame building are shown. (See *Illustration for Prob. 4.*)

Design Criteria

- 114 mph basic wind speed, exposure condition C, $K_{zt} = 1$

- seismic design category C

- $S_{DS} = 0.45$

- $S_{D1} = 0.15$

- importance factor $= 1$

- diaphragm qualifies as flexible

- weight of exterior walls and shear walls $= 20$ lbf/ft^2

- weight of removable partition $= 12$ lbf/ft^2

- weight of main roof $= 16$ lbf/ft^2

- weight of corridor roof $= 12$ lbf/ft^2

- weight of parapet $= 10$ lbf/ft^2

- Douglas fir-larch framing; minimum no. 1

- use allowable stress design (ASD)

4.1. Design the horizontal roof diaphragm for lateral forces in the north-south direction. Show calculations for maximum bending and shear stresses. Show nailing and any other necessary requirements. Show the maximum stress in the chord. Detail an appropriate chord splice.

4.2. Design the shear wall on line B; draw sufficient details to show the principal structural elements. Show calculations for the maximum wall shear. Carry the loads down to the top of the footing. Select a suitable plywood. Show nailing, and any other necessary requirements. Show an analysis of the tie-down requirements, and design and detail a suitable tie-down, where required.

Illustration for Prob. 4

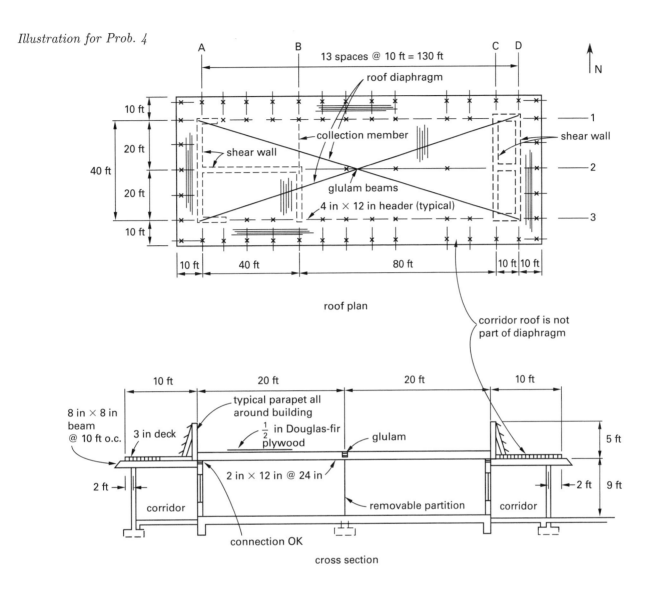

roof plan

corridor roof is not
part of diaphragm

cross section

Table for Prob. 4

allowable nail shear for plywood diaphragms (lbf/ft)

common nail size				6d		8d		10d	
minimum nail penetration in framing (in)				$1^1/_4$		$1^1/_2$		$1^5/_8$	
nominal plywood thickness (in)				$^5/_{16}$		$^3/_8$		$^1/_2$	
lateral force				wind	seismic	wind	seismic	wind	seismic
nominal width of framing lumber (in)				2	3	2	3	2	3
blocked									
nail spacing at	6	nail spacing at	6	188	210	270	300	318	360
diaphragm boundaries	4	other plywood	6	250	280	360	400	425	480
and continuous	$2^1/_2$	panel edges (in)	4	375	420	530	600	640	720
panel edges (in)	2		3	420	475	600	675	730	820
unblocked (nails spaced 6 in max. at supported edge)									
load perpendicular to unblocked edges and									
continuous panel joints				167	187	240	267	283	320
all other configurations				125	140	180	200	212	240

PROBLEM 5

A roof plan of a one-story building and two connection details are shown.

Design Criteria

- seismic forces at working stress level

- shear walls light-frame wood construction

- flexible diaphragms

- redundancy factor $\rho = 1.3$ in both directions

- seismic design category D

5.1. Determine if the connections are adequate for lateral loads. Show calculations.

5.2. Should the details shown be revised or improved? Explain.

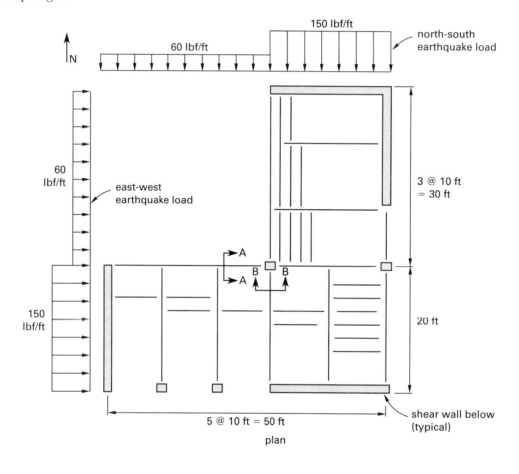

plan

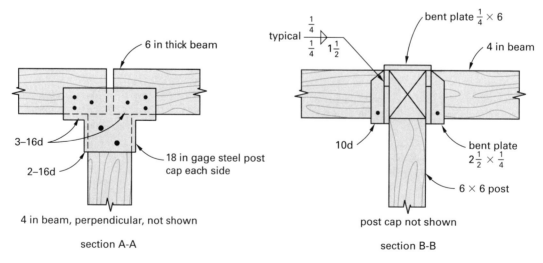

section A-A

section B-B

Illustration for Prob. 6

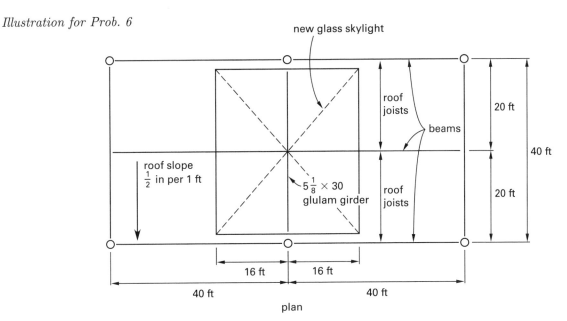

plan

PROBLEM 6

An existing warehouse roof is to be investigated with a prospect to add a new opening for a skylight. (See *Illustration for Prob. 6*.)

Design Criteria

- 24F-1.8E stress class, western species (dry condition), $5^1/_8$ in × 30 in
- roof dead load = 10 lbf/ft^2
- existing connections are adequate
- skylight dead load = 10 lbf/ft^2 (uniformly supported by beams)

6.1. What is the bending stress in the $5^1/_8$ × 30 in glulam beam girder?

6.2. Will the $5^1/_8$ × 30 glulam beam work for the new skylight? Show calculations.

PROBLEM 7

The roof framing plan and the front elevation of a one-story building are shown. The original plans showed the mullion A to be a piece of 4 × 6 stud grade lumber.

The builder has requested to be allowed to substitute two 3 × 4s stud grade in lieu of the 4 × 6 because the 3 × 4s are readily available. Other interested parties have no objection to this substitution, and the engineer is asked to approve the builder's request.

Design Criteria

- Douglas fir-larch stud grade lumber
- gravity loads are concentrically applied to mullions
- column dead load = 3600 lbf
- column live load = 2400 lbf
- consider gravity loads only

7.1. Analyze the structural acceptability of the two 3 × 4s. Show all calculations. Are the two 3 × 4s satisfactory? Should the change be approved?

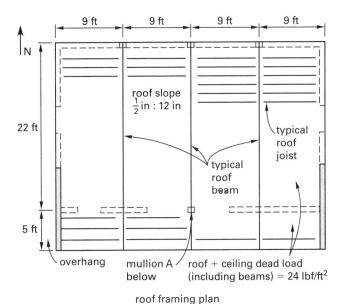

roof framing plan

south elevation

PROBLEM 8

Glulam beams B1 and B2 must be sized according to ASCE/SEI7 requirements to withstand loading. (See *Illustration for Prob. 8.*)

Design Criteria

- roofing = 6.0 lbf/ft^2
- sheathing = 2.0 lbf/ft^2
- purlins = 2.0 lbf/ft^2
- plaster ceiling = 10.0 lbf/ft^2
- dead load (total) = 20.0 lbf/ft^2
- live load = 20.0 lbf/ft^2 (reducible per ASCE/SEI7)
- joists framing perpendicular to glued laminated beams adequately stay beams at points B and C
- use 24F-1.8E western species structural glulam, dry
- use $1\frac{1}{2}$ in laminations

Show all calculations.

8.1. Size beam B2 (span CD) for code requirements.

8.2. Size beam B1 for ASCE/SEI7 requirements. Maximum deflection may be assumed midway between points A and B.

8.3. Determine the camber for beam B2.

PROBLEM 9

A second-floor balcony is shown in plan and elevation. (See *Illustration for Prob. 9.*) The balcony is supported by a building wall and glued laminated beams with wood posts at 10 ft o.c.

Design Criteria

- balcony dead load = 20 lbf/ft^2 (includes rails and plaster ceiling for first floor below)
- balcony live load = 100 lbf/ft^2
- floor joists are spaced at 16 in o.c.
- glued laminated beams are supported by wood posts at 10 ft o.c. except at corner, where beams are cantilevered
- adequate bracing for bottom of beams
- beams are connected at cantilever ends adequate for distribution of vertical loads prior to application of any dead and live load
- Douglas fir-larch (north) joists
- 24F Douglas-fir glulam beams

9.1. Determine the size of floor joists.

9.2. Determine the load to be transmitted through connection at ends of cantilever beams.

9.3. Determine the size of beams with cantilever spans for deflection and strength requirements.

Illustration for Prob. 8

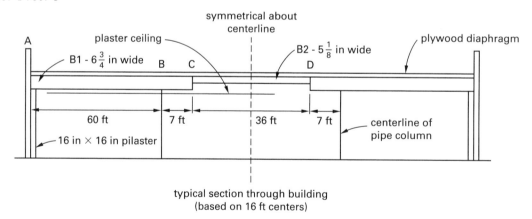

typical section through building
(based on 16 ft centers)

Illustration for Prob. 9

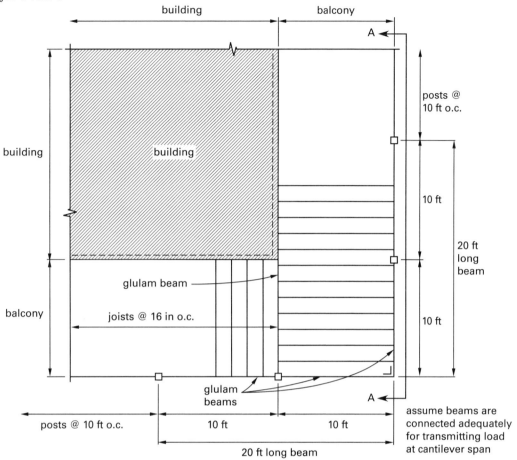

second floor balcony plan

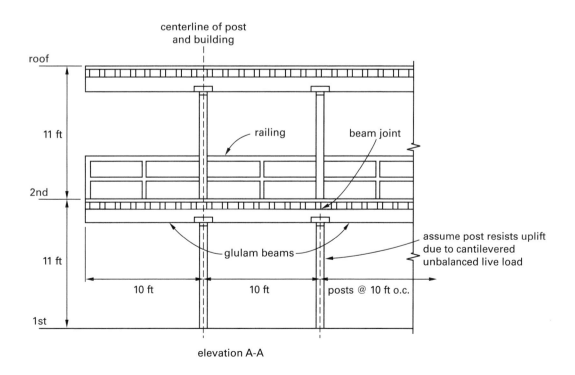

elevation A-A

PROBLEM 10

A typical wood roof truss at 4 ft spacing is shown. The bottom chord is raised above its usual location to satisfy the architectural requirements. Assume supports offer no horizontal restraint.

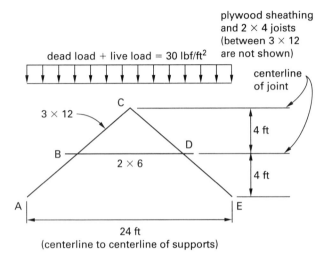

Design Criteria

- dead load = 14 lbf/ft^2 projected upon horizontal plane
- live load = 16 lbf/ft^2 projected upon horizontal plane
- Douglas fir-larch select structural
- 25% increase in allowable stress for roof live loads

10.1. Check the adequacy of member ABC.

10.2. Design and draw a detail of joint B.

PROBLEM 11

Two sections cut through a wood frame warehouse building are shown. The plans and elevation are also shown. (See *Illustration for Prob. 11*.)

Design Criteria

- seismic design category D
- roof dead load = 15 lbf/ft^2
- wall dead load = 12 lbf/ft^2

11.1. Because a lateral load-resisting element (rigid frame or shear wall) is not provided between lines 2 and 3, what structural elements would be needed to maintain diaphragm integrity? Provide at least two.

11.2. Assume the lateral load on the roof in the north-south direction is 190 lbf/ft (working stress level), and no rigid frame or shear wall is to be used below the flat roof at lines 2 and 3. Describe all the structural elements needed to give the structure adequate resistance to lateral loads. Determine the loads that the resisting elements should be designed for.

11.3. Should a stepped diaphragm be used? Observe section A-A. There is no shear wall or rigid frame located where the step in the diaphragm occurs.

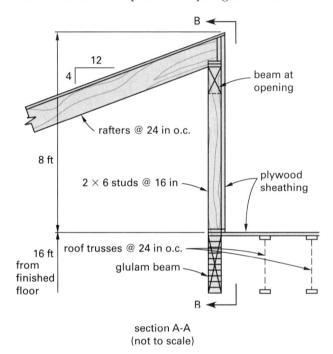

section A-A
(not to scale)

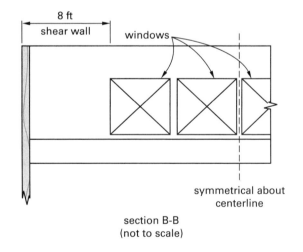

section B-B
(not to scale)

Illustration for Prob. 11

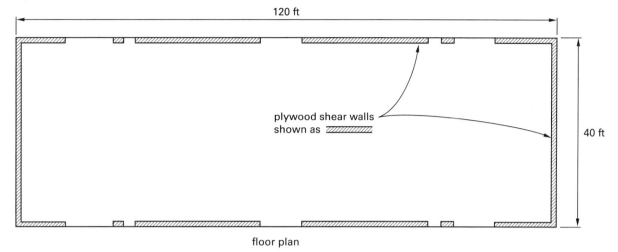

floor plan

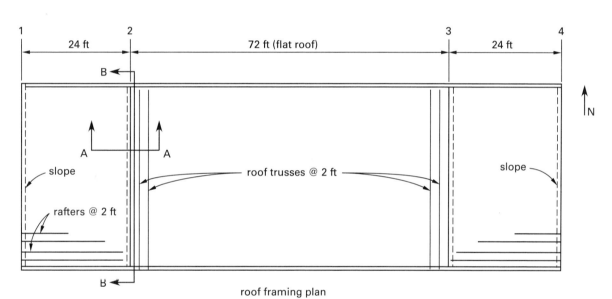

roof framing plan

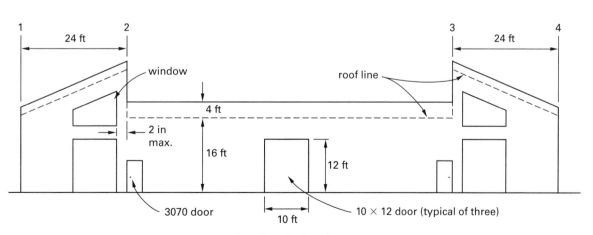

north and south elevation

PROBLEM 12

A partial elevation of a second-story wood frame building with a frame door opening is shown. The plywood sheathed wall must resist all the lateral loads shown.

Design Criteria

- Douglas-fir structural I grade plywood, $^{15}/_{32}$ in

- Douglas-fir construction grade framing

- $F_b = 1000$ psi framing

- seismic forces at the strength level

- seismic design category D, $S_{DS} = 1.0$, $\rho = 1.0$

12.1. Give the nailing of the entire plywood wall sheathing. Show all calculations.

12.2. Design all boundary members below the second floor and describe where they are located. Show all calculations.

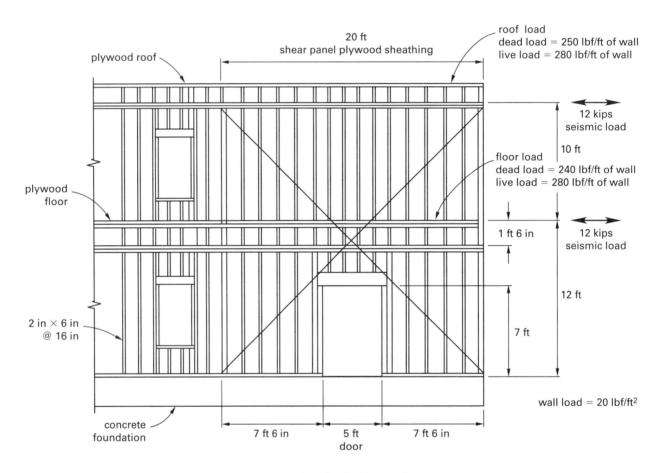

elevation looking north

PROBLEM 13

A pyramid-shaped wood roof is shown. The space enclosed by the pyramid is clear and without any structural members.

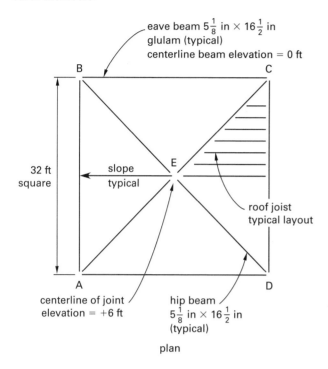

plan

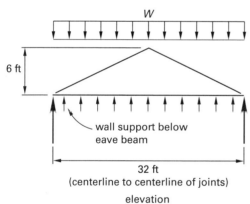

32 ft
(centerline to centerline of joints)

elevation

Design Criteria

- dead load $= 16$ lbf/ft^2 projected upon a horizontal plane
- live load $= 16$ lbf/ft^2 projected upon a horizontal plane
- Douglas fir-larch no. 1 lumber
- 25% increase in allowable stress for roof live load
- unbalanced live load condition to be neglected
- design for vertical loads only

13.1. Find the forces acting at joint A.

13.2. Design and sketch joint A.

PROBLEM 14

An elevation of a glulam roof beam is shown.

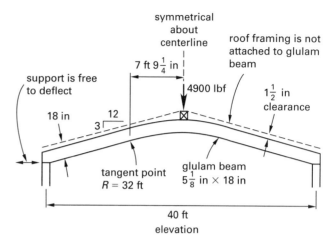

40 ft
elevation

Design Criteria

- $F_b = 2400$ psi glulam
- $5\frac{1}{8}$ in $\times$ 18 in glulam beam
- $1\frac{1}{2}$ in laminations
- $E = 1.8 \times 10^6$ psi
- $S = 276.75$ in^3
- $A = 92.25$ in^2
- radial tension reinforcement available: $^3/_4$ in lag screws
  - 5240 lbf allowable tension
  - 528 lbf/in of penetration allowable withdrawal
- no increases are allowed for load duration
- no need to check deflection
- assume all adjustment factors except volume, stability, and curvature $= 1.0$

14.1. Assuming that lateral bracing is sufficient, what is the allowable bending stress?

14.2. What is the maximum spacing of the top flange bracing to permit loading as shown in the figure?

14.3. Check radial tension (f_t allowable $= 15$ psi).

14.4. How could reinforcement for radial tension be provided? Use a convenient scale.

PROBLEM 15

An elevation of the end wall of a two-story wood frame building is shown. The end wall is sheathed plywood. The doors open out to a second-story balcony. The rough door openings are 3 ft wide × 6 ft high.

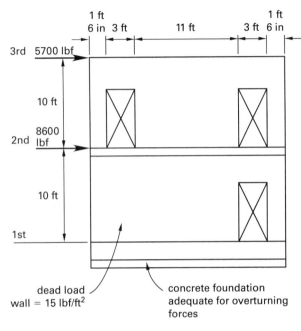

elevation

Design Criteria

- structural I plywood
- seismic loads control
- seismic forces at strength level
- seismic design category D
- $S_{DS} = 1.0$
- $\rho = 1.0$
- roof life load = 200 lbf/ft
- roof dead load = 150 lbf/ft
- second floor live load = 500 lbf/ft
- second floor dead load = 180 lbf/ft

15.1. Compute the shear in the plywood wall sheathing in the first and second stories. Determine the size and spacing of nails.

15.2. Draw an elevation and show the location of tie-down anchors. Show the load each anchor must resist. Use load combinations in ASCE/SEI7.

15.3. Sketch the various types of tie-down anchors that may be used across the second and first floors. It is not necessary to fully design and detail each one.

PROBLEM 16

The one-story wood frame building shown has its center roof portion 4 ft above the building roof. (See *Illustration for Prob. 16.*)

Design Criteria

- all roofs level
- risk category II
- site class D
- $S_S = 1.2$, $S_1 = 0.6$
- $T_L = 6$ sec
- high roof elevation = 16 ft
- low roof elevation = 12 ft
- high roof and low roof dead load = 20 lbf/ft^2 (including beams)
- high roof and low roof walls dead load = 20 lbf/ft^2
- elevations are to bottom of joists
- roof and wall framing is plywood sheathed
- roof joists bear on top of glued laminated beams and to plates of wall studs
- posts do not project above low roof

16.1. Compute and draw the load, shear, and moment diagrams at low roof due to the seismic lateral forces.

16.2. Compute the maximum chord forces at the exterior walls of the low roof.

16.3. Provide a sketch of details required at the corners of the opening for the tension forces due to the lateral forces in both directions. Calculations are required.

16.4. Provide a sketch of details for the chord splices at the low roof exterior walls for the maximum chord forces. Calculations are required.

Illustration for Prob. 16

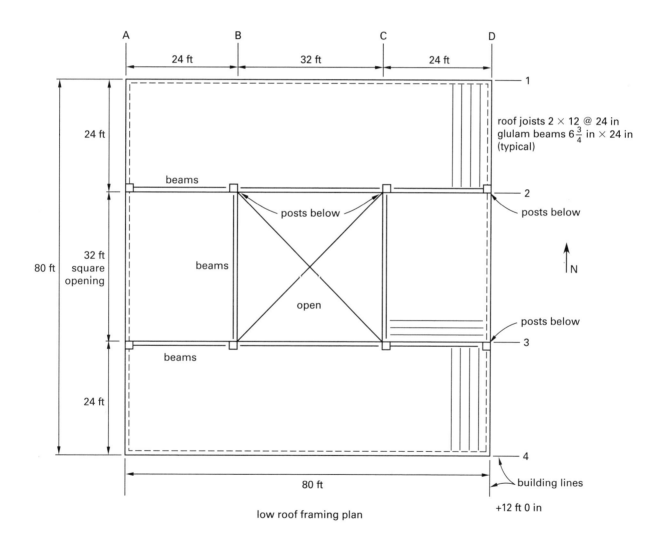

roof joists 2 × 12 @ 24 in
glulam beams $6\frac{3}{4}$ in × 24 in (typical)

B   C

2

post to beam
below

wall below
to low roof

glulam beam
$6\frac{3}{4}$ in × 24 in

32 ft 8 in
square

roof joists
2 × 12 @ 24 in

3

high roof framing plan

A   B   C   D

24 ft   32 ft   24 ft

1

24 ft

roof joists 2 × 12 @ 24 in
glulam beams $6\frac{3}{4}$ in × 24 in
(typical)

beams

2

posts below

posts below

80 ft   32 ft
square
opening

beams

N

open

posts below

beams

3

24 ft

4

80 ft

building lines

+12 ft 0 in

low roof framing plan

PROBLEM 17

An existing building includes one span that has six adjacent roof rafters showing distress. The present condition is shown. (See *Illustration for Prob. 17.*) The six rafters have failed and are sagging.

Design Criteria

- roofing and framing $= 9$ lbf/ft^2
- ceiling dead load $= 10$ lbf/ft^2
- total dead load $= 19$ lbf/ft^2
- live load $= 20$ lbf/ft^2
- total dead $+$ live load $= 39$ lbf/ft^2
- $F_b = 1500$ psi (lumber)
- $F_v = 95$ psi
- $E = 1.8 \times 10^6$ psi (lumber)
- use present lumber dimensions
- total load deflection limited to $L/180$

17.1. Use the design loads given and check the bending stress in the rafters. Assume dry condition of use. Compare the computed stress to the allowable. Is the difference a probable cause of the failure?

17.2. Describe a step-by-step procedure for the repair of the existing framing without removing the existing roof. Assume the broken rafters have not deteriorated and can remain in the repaired structure.

17.3. Give two possible reasons for the rafter failure.

PROBLEM 18

Investigate the glued laminated beams which have failed and which are connected as shown. (See *Illustration for Prob. 18.*)

Design Criteria

- stress grade combination for beams A and B is 24F Douglas fir-larch, glued laminated beam
- beam A is simply supported with a span of 35 ft
- reaction from beam B to connection at midpoint of beam A $= 16{,}000$ lbf (roof total load)
- neglect the effect of the eccentricity of beam B on one side of beam A
- existing connection has been in place for two years
- present moisture content was measured at 9%
- job records show that beams had been exposed to extreme wet conditions before final protection
- all bolts are 1 in diameter machine bolts

18.1. Describe briefly the probable mode of failure in beams A and B. Show all necessary calculations that will confirm your findings.

18.2. Draw a sketch showing a method of reinforcing the connection. Beams must remain in place. No calculations are required.

Illustration for Prob. 17

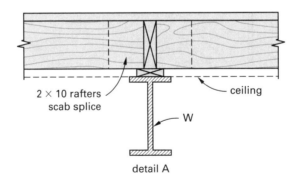

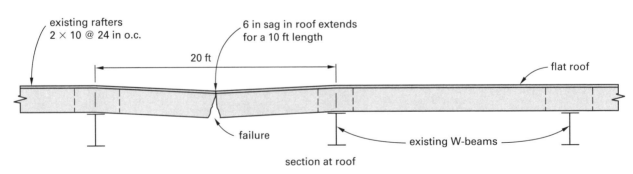

section at roof

Illustration for Prob. 18

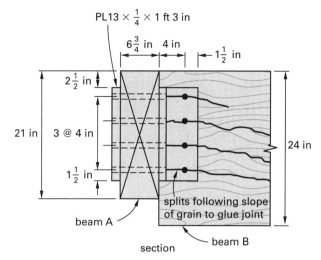

PL13 × $\frac{1}{4}$ × 1 ft 3 in

6$\frac{3}{4}$ in   4 in   1$\frac{1}{2}$ in

2$\frac{1}{2}$ in

21 in   3 @ 4 in   24 in

1$\frac{1}{2}$ in

splits following slope
of grain to glue joint

beam A

section

beam B

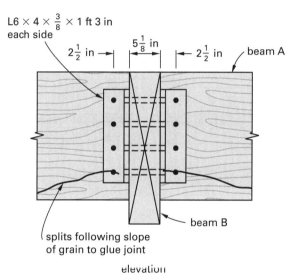

L6 × 4 × $\frac{3}{8}$ × 1 ft 3 in
each side

2$\frac{1}{2}$ in   5$\frac{1}{8}$ in   2$\frac{1}{2}$ in   beam A

beam B

splits following slope
of grain to glue joint

elevation

PROBLEM 19

The design for a two-span glued laminated beam installation is shown. (See *Illustration for Prob. 19*.) The bay
spacing is 18 ft. Loading is from floor purlins connected
across the top, spaced 8 ft from the wall end support 1.
Lateral bracing of the bottom of cantilever beam A is
provided at the interior column 2.

Design Criteria

- 24F-1.7E stress class Douglas fir-larch glulam beam
  (dry use)

- floor occupancy = office with partitions

- total floor dead load = 30 lbf/ft^2 (neglect weight of
  glulam beam)

19.1. Investigate the design of beams A and B for
compliance with the current IBC and NDS.

19.2. What changes or additions, if any, would you
require? Use $P_{\text{dead load}} = 4.3$ kips, $P_{\text{live load}} = 6.8$ kips.

19.3. Specify the camber requirement for beam B. Use
the same loads as in Prob. 19.2.

19.4. Show the camber diagram shape appropriate for
fabrication of beam A. (No magnitudes are required.)

19.5. Draw a detail for connection of beam B to
beam A (joint 3).

Illustration for Prob. 19

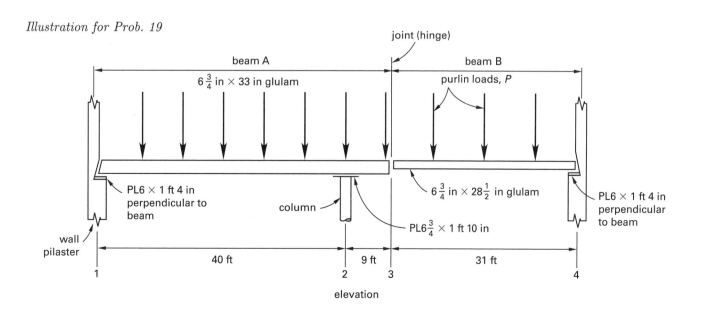

joint (hinge)

beam A   beam B

6$\frac{3}{4}$ in × 33 in glulam   purlin loads, *P*

PL6 × 1 ft 4 in
perpendicular to
beam

column

6$\frac{3}{4}$ in × 28$\frac{1}{2}$ in glulam

PL6$\frac{3}{4}$ × 1 ft 10 in

PL6 × 1 ft 4 in
perpendicular
to beam

wall
pilaster

40 ft   9 ft   31 ft

1   2   3   4

elevation

PROBLEM 20

The framing that supports the ceiling of an existing building is shown. It is necessary to add mechanical equipment in the attic space above the ceiling framing. The equipment loads are shown in elevation B.

Design Criteria

- additional framing limited to space of $5^1/_2$ in over top of existing 4×8s

- existing 4×8s and ceiling joists cannot be removed during construction

- equipment to be installed through existing skylight in roof

- Douglas-fir no. 1 lumber (existing)

- ceiling dead load = 10 lbf/ft^2 (includes framing)

- assume ceiling dead plus equipment loads control, with load duration greater than 10 years ($C_D = 0.9$)

20.1. Design a method to support the new equipment utilizing the existing ceiling framing. Show calculations for new or modified supporting members. Limit total deflection of beam to $L/240$.

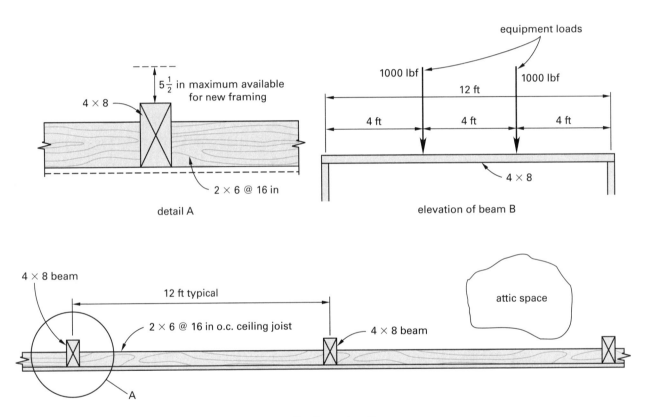

detail A

elevation of beam B

section through ceiling

SOLUTION 1

1.1. Structural I plywood is manufactured using an exterior glue that gives it greater strength, stiffness, and endurance than other grades of sheathing. Quoting from the American Plywood Association's *Engineered Wood Construction Guide,*

> All-veneer APA rated sheathing exposure 1, commonly called CDX in the trade, is frequently mistaken as an exterior panel and erroneously used in applications for which it does not possess the required resistance to weather.

The structural I plywood was probably specified for the roof diaphragm because of its exterior rating, and the CDX plywood is not an acceptable substitution.

1.2. A 24F-E15 combination of hem-fir compared to a 24F-V8 Douglas-fir will have a lower modulus of elasticity about the weak axis (0.78×10^6 psi versus 0.83×10^6 psi), which will increase the lateral buckling tendency. The allowable horizontal shear stress is lower, so horizontal shear must be checked. The 24F-E15 has a much lower allowable bending stress if the compression zone is stressed in tension, as may be the case over the cantilevered support. The species groups differ (per *AITC Manual* Table 6.2), so connections involving mechanical fasteners must be checked. A complete check of the beam is necessary.

1.3. Show the required force transfers for all horizontal and vertical ties in the lateral force resisting system. Neglect gravity loads.

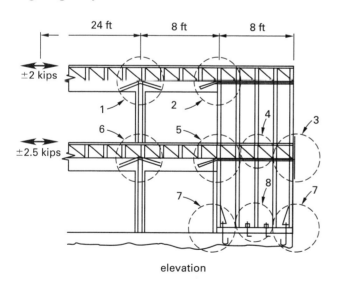

elevation

The 2 kips in the roof diaphragm transfers to the shear walls and collectors at the top.

$$v_2 = \frac{V_2}{B} = \frac{2000 \text{ lbf}}{40 \text{ ft}} = 50 \text{ lbf/ft}$$

The additional force transfers are keyed to the details numbered on the preceding illustration.

First, the collector must be tied across the column to transfer diaphragm shear distributed over the collector length of 24 ft.

$$V = v_2 L_{\text{trib}} = \left(50 \frac{\text{lbf}}{\text{ft}}\right)(24 \text{ ft}) = 1200 \text{ lbf}$$

Second, the collector to top plate of the upper shear wall must transfer the diaphragm shear over a total length of 32 ft.

$$V = v_2 L_{\text{trib}} = \left(50 \frac{\text{lbf}}{\text{ft}}\right)(32 \text{ ft}) = 1600 \text{ lbf}$$

Third, tie across the second floor to connect the chord of shear walls, to transfer the chord force in the upper shear wall.

$$T = C = \frac{M_o}{L} = \frac{V_2 h_2}{L} = \frac{(2000 \text{ lbf})(9 \text{ ft})}{8 \text{ ft}} = 2250 \text{ lbf}$$

Fourth, transfer $V_2 = 2000$ lbf shear from the upper wall into blocking plus an additional $(8 \text{ ft}/40 \text{ ft})(2500 \text{ lbf}) = 500$ lbf from the second-floor diaphragm into blocking between the second-floor diaphragm and the top plate of the lower shear wall.

$$v = \frac{V_2 + \left(\frac{L}{B}\right)V_1}{L} = \frac{2000 \text{ lbf} + \left(\frac{8 \text{ ft}}{40 \text{ ft}}\right)(2500 \text{ lbf})}{8 \text{ ft}}$$
$$= 313 \text{ lbf/ft}$$

Fifth, tie the chords of the upper and lower walls to transfer $T = 2250$ lbf; connect the second-floor collector to the top plate of lower wall to transfer the remainder of the shear from the second-floor diaphragm.

$$v_1 = \frac{V_1}{B} = \frac{2500 \text{ lbf}}{40 \text{ ft}} = 62.5 \text{ lbf/ft}$$
$$V = v_1 L_{\text{trib}} = \left(62.5 \frac{\text{lbf}}{\text{ft}}\right)(32 \text{ ft}) = 2000 \text{ lbf}$$

Sixth, tie the collectors across the column to transfer.

$$V = v_1 L_{\text{trib}} = \left(62.5 \frac{\text{lbf}}{\text{ft}}\right)(24 \text{ ft}) = 1500 \text{ lbf}$$

Seventh, tie the outside chords of the lower shear wall to the foundation to transfer.

$$T = C = \frac{M_o}{L} = \frac{V_2(h_2 + h_1) + V_1 h_1}{L}$$
$$= \frac{(2000 \text{ lbf})(9 \text{ ft} + 10 \text{ ft}) + (2500 \text{ lbf})(10 \text{ ft})}{8 \text{ ft}}$$
$$= 7880 \text{ lbf}$$

Eighth, transfer $V_2 + V_1 = 2000 \text{ lbf} + 2500 \text{ lbf} = 4500 \text{ lbf}$ of shear from the lower shear wall to the foundation using anchor bolts.

1.4. The appropriate details for the given criteria are as shown. Only the additional items required are listed; other items are omitted for clarity.

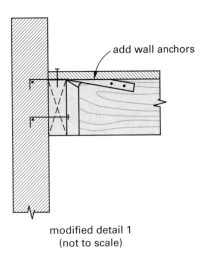

modified detail 1
(not to scale)

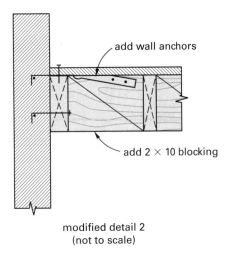

modified detail 2
(not to scale)

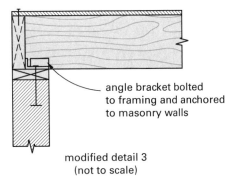

modified detail 3
(not to scale)

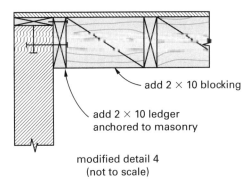

modified detail 4
(not to scale)

1.5. Calculate the capacity of the designated 1 in diameter bolt. From NDS Sec. 11.5.1, when the end distance for parallel-to-grain loading is less than required for a full load (i.e., $7d$), the design value for all bolts in the group must be taken as the lowest bolt design value in the group. Assume dry use, and normal load duration, single shear with $3^{1}/_{2}$ in main and secondary member thickness, and no group action reduction. For the designated bolt, per NDS Table 11.5.1A and Sec. 11.5.1,

$$C_\Delta = \frac{\text{actual end distance}}{\text{minimum end distance for } C_V = 1.0}$$
$$= \frac{6.0 \text{ in}}{7.0 \text{ in}}$$
$$= 0.86$$
$$Z'_\| = C_\Delta Z_\| = (0.86)(2260 \text{ lbf}) = 1940 \text{ lbf/bolt}$$

1.6. Comment on the given construction details (per AITC 104).

(a) The primary problem is that the $8^{1}/_{2}$ in gage fasteners in the 6×14 will restrain cross-grain shrinkage and likely cause splitting. Also, high support of the 6×14 may cause loss of bearing and hence overload of fasteners as shrinkage occurs. A better detail—assuming load transfer is primarily bearing rather than uplift—is the T-plate shown in Fig. 5.3 of the *AITC Manual*.

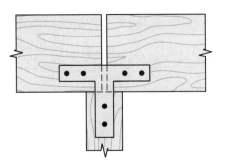

(b) The location of bolts high in the beam will likely result in loss of bearing in the beam seat. The effective depth resisting shear, d_e, is 6 in. The end joint shear capacity is

$$V'_r = \tfrac{2}{3} F'_v b d_e \left(\frac{d_e}{d}\right)^2$$

This would likely be too small, resulting in horizontal shear failure at the level of the lower bolt. Better details are shown in Figs. 4.1, 4.2, and 4.3 of the *AITC Manual*, in which the fastener is located near the bearing seat.

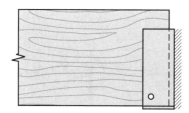

(c) Lack of support of the outer lams will create large shear stresses, which will likely lead to splitting in the glulam. Support should be provided over the full width of the tapered cut, as shown in Fig. 2.6 of the *AITC Manual*.

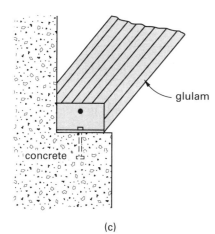

(c)

(d) The connection induces high cross-grain tension, so splitting failure at the level of the bolts is likely. The best connection would be a hanger that crosses the top of the beam and transfers force by bearing perpendicular to the grain. If a bolted connection must be used, the connection should be made well above the neutral axis. Preferred types are shown in Fig. 10.1 of the *AITC Manual*.

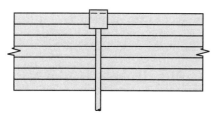

(e) The new connection has the same potential problem as in part b, with bolts located too high. An additional potential problem is differential vertical deflection of the new floor where it abuts the existing floor. This can be caused by shrinkage of the new 4×14 if, as is generally the case, the moisture content of the 4×14 is higher at time of erection than it will be during service. Either the beam should be allowed to reach its final moisture content before the flooring is attached, or provision should be made to compensate for the shrinkage.

1.7. For a toenail,

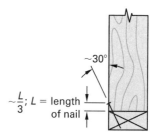

From NDS Sec. 11.5.4, the reduction factor, C_{tn}, equals 0.83 for lateral loading and 0.67 for withdrawal. A slant nail is driven at a 30° slant perpendicular rather than parallel to the grain (see Breyer et al.). The NDS does not distinguish between toenails and slant nails so the same reduction factor, C_{tn}, applies in both cases.

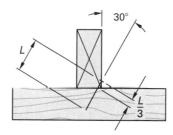

1.8. From the *AITC Manual*, the minimum dimensions for a clinched nail are as follows.

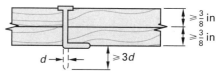

1.9. If the wall is long it may be practical to shift the 4×4 post 3 in to the right to connect the post to the anchor bolt. The connection will shorten the wall, and therefore, it will increase the unit wall shear (pounds per linear foot) and axial force in the boundary member. If the increase is small, and there was some extra margin of safety in the original design (as is frequently the case), the movement would not overstress the components. If this is the case, two 2×4 studs can be used to fill the 3 in gap caused by shifting the 4×4. If it is not possible to move the post, a stiffened connection angle could be designed to handle the force transfer with the additional 3 in eccentricity.

1.10. The penetration of the nail head past the outer ply reduces the lateral resistance of the nail to about half of its intended value. If the same problem exists at the top plate and sills, providing companion studs and renailing will not solve it. Since the nailing shown is 4 in o.c., additional nails driven into the nominal 2 in studs might cause splitting (need 3 in or wider for spacings of 3 in or less). The only obvious solution, if the sheathing is on only one side, is to sheath the other side to resist one half or more of the wall shear. If this is not possible, alternative force paths or replacement of the wall would need to be considered.

SOLUTION 2

2.1. Compute the lateral force resisted by the center wall using the loading criteria from ASCE/SEI7.

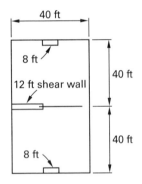

Use the envelope procedure from ASCE/SEI7 Chap. 28, Part 2 to compute wind forces on the diaphragm. Assume that exterior walls are 11 ft tall from ground to roof level. The end zone extends a distance of $2a$, where a is defined as

$$a \leq \begin{cases} 0.1 \times \text{least building dimension} \\ \quad = (0.1)(40 \text{ ft}) \\ \quad = 4 \text{ ft} \quad \text{[controls]} \\ 0.4h = (0.4)(11 \text{ ft}) \\ \quad = 4.4 \text{ ft} \end{cases}$$

From ASCE/SEI7 Fig. 28.6-1, a roof with a slope of less than 5° in winds of 150 mph has simplified design wind pressures of

$$p_{s30,A} = 35.7 \text{ lbf/ft}^2$$
$$p_{s30,C} = 23.7 \text{ lbf/ft}^2$$

For exposure C, $\lambda = 1.21$.

For ASD, the wind forces to the diaphragm at the end region A and at middle region C are

$$w_A = (0.6)(0.5h\lambda I K_{zt} p_{s30,A})$$
$$= (0.6)(0.5)(11 \text{ ft})(1.21)(1.0)(1.0)\left(35.7 \frac{\text{lbf}}{\text{ft}^2}\right)$$
$$= 143 \text{ lbf/ft}$$
$$w_C = (0.6)(0.5h\lambda I K_{zt} p_{s30,C})$$
$$= (0.6)(0.5)(11 \text{ ft})(1.21)(1.0)(1.0)\left(23.7 \frac{\text{lbf}}{\text{ft}^2}\right)$$
$$= 95 \text{ lbf/ft}$$

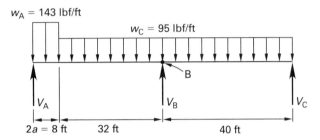

Calculate the diaphragm shear to the left of the wall's midpoint, B.

$$\sum M_A = \left((w_A - w_C)2a\right)a + (w_C L)(0.5L) - V_{B,\text{left}}L = 0$$
$$V_{B,\text{left}} = \frac{(w_A - w_C)2a^2 + 0.5w_C L^2}{L}$$
$$= \frac{\left(143 \frac{\text{lbf}}{\text{ft}} - 95 \frac{\text{lbf}}{\text{ft}}\right)(2)(4 \text{ ft})^2 + (0.5)\left(95 \frac{\text{lbf}}{\text{ft}}\right)(40 \text{ ft})^2}{40 \text{ ft}}$$
$$= 1940 \text{ lbf}$$

The shear to the right of B is

$$\sum M_C = -(w_C L)(0.5L) + V_{B,\text{right}}L = 0$$
$$V_{B,\text{right}}L = \frac{0.5w_C L^2}{L} = 0.5w_C L$$
$$= (0.5)\left(95 \frac{\text{lbf}}{\text{ft}}\right)(40 \text{ ft}) = 1900 \text{ lbf}$$

Thus, the total wind force to the middle wall is

$$V_{B,W} = V_{B,\text{left}} + V_{B,\text{right}} = 1940 \text{ lbf} + 1900 \text{ lbf}$$
$$= 3840 \text{ lbf}$$

For seismic loading in the longitudinal direction, the seismic dead load to the diaphragm, per foot, is

$$w = 2w_{\text{wall}}(0.5h) + w_{\text{roof}}B$$
$$= (2)\left(20 \frac{\text{lbf}}{\text{ft}^2}\right)(0.5)(11 \text{ ft}) + \left(15 \frac{\text{lbf}}{\text{ft}^2}\right)(40 \text{ ft})$$
$$= 820 \text{ lbf/ft}$$

From ASCE/SEI7 Eq. 12.8-7, the approximate period is

$$T_a = C_t h_n^x = (0.02)(11 \text{ ft})^{0.75} = 0.12 \text{ sec}$$

For this short period, ASCE/SEI7 Eq. 12.8-2 controls the seismic base shear.

$$C_s = \frac{S_{DS}}{\dfrac{R}{I_e}} = \frac{0.45}{\dfrac{6.5}{1.0}} = 0.07 \quad \text{[design level]}$$

The seismic force to the diaphragm on a working load basis is

$$w_E = 0.7 C_s w = (0.7)(0.07)\left(820 \; \frac{\text{lbf}}{\text{ft}}\right) = 40 \text{ lbf/ft}$$

The seismic shear transferred to the middle wall is

$$V_{\text{B},E} = w_E L = \left(40 \; \frac{\text{lbf}}{\text{ft}}\right)(40 \text{ ft}) = 1600 \text{ lbf} \quad [< V_{\text{B},W}]$$
$$V_{\text{B}} = V_{\text{B},W} = 3840 \text{ lbf}$$

Thus, wind controls lateral force on the middle wall.

2.2. Determine the nailing required to transfer the shear from the roof diaphragm to the center shear wall. The diaphragm shear to the top plate is transferred in part through the connection of the collector (see Sol. 2.3) and from the plywood-to-blocking-to-plate by toenails, which must transfer 3840 lbf/40 ft = 96 lbf/ft. Use 8d nails with joists spaced at 2 ft on centers.

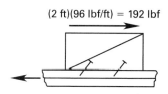

(2 ft)(96 lbf/ft) = 192 lbf

For 8d common nails loaded laterally in Douglas fir-larch framing (NDS Table 11N), with adjustment for toenails ($C_{\text{tn}} = 0.83$) and load duration ($C_D = 1.6$) and a projected penetration of greater than $10d$,

$$Z' = C_{\text{tn}} C_D Z = (0.83)(1.6)\left(97 \; \frac{\text{lbf}}{\text{nail}}\right)$$
$$= 129 \text{ lbf/nail}$$
$$n \geq \frac{V}{Z'} = \frac{202 \text{ lbf}}{129 \; \dfrac{\text{lbf}}{\text{nail}}} = 1.6 \text{ nails}$$

IBC Table 2304.9.1 requires a minimum of three 8d common toenails to fasten the blocking to the top plate.

Use 3 × 8d (toenails) for each 2 × 10 blocking over the center wall.

2.3. Design the connection of the collector at point A. Assume a 2 ft lap length between the 6 × 10 beam and

the top plates (total length = 14 ft). The shear force transferred is

$$P = \left(\frac{V_{\text{B}}}{B}\right) b = \left(\frac{3840 \text{ lbf}}{40 \text{ ft}}\right)(12 \text{ ft} + 2 \text{ ft}) = 1344 \text{ lbf}$$

The force is eccentric to the top plate by a distance equal to the depth of the nominal 10 in beam. Assuming an effective span of 13 ft and neglecting dead weight, the uplift force experienced by the connection can be found from the equation

$$\sum M_{\text{end}} = Pd - WL = 0$$
$$W = \frac{Pd}{L} = \frac{(1344 \text{ lbf})(9.25 \text{ in})}{(13 \text{ ft})\left(12 \; \dfrac{\text{in}}{\text{ft}}\right)} = 80 \text{ lbf}$$

Try $^7/_{16}$ in diameter lag screws laterally loaded parallel to the grain in Douglas fir-larch. Assume a group factor of $C_g = 0.95$ (conservative for five or fewer screws), a load duration of $C_D = 1.6$, and adequate penetration into the main member. From NDS Table 11J,

$$Z' = C_g C_D Z_{\|} = (0.95)(1.6)\left(310 \; \frac{\text{lbf}}{\text{screw}}\right)$$
$$= 471 \text{ lbf/screw}$$
$$n \geq \frac{P}{Z'} = \frac{1344 \text{ lbf}}{471 \; \dfrac{\text{lbf}}{\text{screw}}} = 3.0 \text{ screws}$$

The reference withdrawal value for a $^7/_{16}$ in lag screw in Douglas fir-larch per inch of penetration is 342 lbf/in (NDS Table 11.2A). Thus, compared to the uplift force of 80 lbf shared by the three lag screws, the withdrawal effect is negligible and it is not necessary to check combined lateral and withdrawal forces (NDS Eq. 11.4-2). Provide a minimum end distance of $7d = (7)\left(^7/_{16} \text{ in}\right) = 3.0$ in and a minimum spacing of 3 in o.c.

Use three $\frac{7}{16}$ in diameter × 6 in lag screws to connect the top plates to the collector.

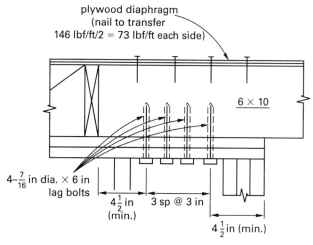

plywood diaphragm
(nail to transfer
146 lbf/ft/2 = 73 lbf/ft each side)

6 × 10

$4-\frac{7}{16}$ in dia. × 6 in
lag bolts

$4\frac{1}{2}$ in
(min.)

3 sp @ 3 in

$4\frac{1}{2}$ in (min.)

detail–connection at A
(not to scale)

2.4. Design to transfer the force at the cut top plate (point X)

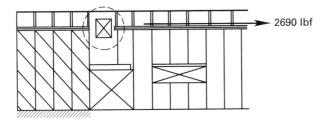

$$P = vb = \left(96 \; \frac{\text{lbf}}{\text{ft}}\right)(28 \text{ ft}) = 2690 \text{ lbf}$$

The more direct path for force transfer is above the duct. However, at this stage of construction framing over the duct is probably difficult; therefore, transfer the 2690 lbf (reversible) force into the shear wall beneath the duct. The proposed scheme is as follows.

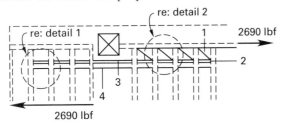

The duct extends 9.75 in below the bottom of the top plates. Install 2×12 blocking beneath the top plate of the 2×6 stud wall 16 ft long (effectively 14 ft long with the cutout for the duct). Install a new 2×6 beneath the blocking face nailed into the blocking. Thus, each 2 ft long blocking piece must resist a shear force of

$$F = \frac{Ps}{L} = \frac{(2690 \text{ lbf})(2 \text{ ft})}{14 \text{ ft}} = 384 \text{ lbf}$$

Nominal shear stress in the blocking is

$$f_v = \frac{F}{bs'} = \frac{384 \text{ lbf}}{(1.5 \text{ in})(24 \text{ in} - 2.5 \text{ in})} = 12 \text{ psi}$$

$$F'_v = C_D F_v = (1.6)\left(180 \; \frac{\text{lbf}}{\text{in}^2}\right) = 288 \text{ psi} \quad [> f_v]$$

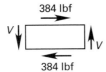

Toenail the top and sides of the blocking with 8d common nails. $Z' = 129$ lbf/nail (from Sol. 3.2).

$$\sum M_{\text{left}} = Fd - Vs' = 0$$

$$V = \frac{Fd}{s'} = \frac{(384 \text{ lbf})(11.25 \text{ in})}{(24 \text{ in} - 2.5 \text{ in})} = 201 \text{ lbf}$$

$$n_v \geq \frac{201 \text{ lbf}}{129 \; \frac{\text{lbf}}{\text{nail}}} = 1.6 \text{ nails}$$

Use two 8d nails on each side face of the blocking into the studs.

$$n_h \geq \frac{384 \text{ lbf}}{129 \; \frac{\text{lbf}}{\text{nail}}} = 3.0 \text{ nails}$$

Use three 8d nails between the blocking and the top plates. Add a 2×6 block beneath each 2×12 face nailed with 8d nails.

$$Z' = C_D Z_{\parallel} = (1.6)\left(97 \; \frac{\text{lbf}}{\text{nail}}\right) = 155 \text{ lbf/nail}$$

$$n \geq \frac{F}{Z'} = \frac{384 \text{ lbf}}{155 \; \frac{\text{lbf}}{\text{nail}}} = 2.5 \text{ nails}$$

Continue the 2×6 blocking beneath the duct and into the shear wall at the same elevation. This blocking must extend far enough across the shear wall to transfer the force $P = 2690$ lbf. If the same nail pattern used is the same as for the shear wall boundary, the required length is

$$L = \left(\frac{b}{B}\right)L_{\text{wall}} = \left(\frac{28 \text{ ft}}{40 \text{ ft}}\right)(12 \text{ ft}) = 8.4 \text{ ft}$$

The 2×6 block transfers force in compression only.

$$f_c = \frac{P}{A} = \frac{2690 \text{ lbf}}{(1.5 \text{ in})(5.5 \text{ in})}$$

$$= 326 \text{ psi} \quad [< F_{c\perp} = 625 \text{ psi, so OK}]$$

Provide a steel rod to resist the 2690 lbf tensile force. Drill through the center of the studs as close to the new 2×6 as practicable. For an ASTM A36 rod with threaded ends, AISC Table J3.2 gives the nominal tensile strength as $0.75F_u$, which is converted to working stress by the factor $\Omega = 2.0$. Thus,

$$A_{\text{req}} = \frac{P\Omega}{0.75F_u} = \frac{(2690 \text{ lbf})(2.0)}{(0.75)\left(58 \; \frac{\text{kips}}{\text{in}^2}\right)\left(1000 \; \frac{\text{lbf}}{\text{kip}}\right)} = 0.12 \text{ in}^2$$

$$= 0.25\pi D^2$$

$$D = \sqrt{\frac{0.12 \text{ in}^2}{0.25\pi}} = 0.39 \text{ in} \quad [\text{say } 0.5 \text{ in}]$$

Use a $1/2$ in diameter rod and try a 2 in diameter plate washer.

$$f_{c\perp} = \frac{P}{A} = \frac{P}{0.25\pi(D_o^2 - D_i^2)}$$

$$= \frac{2690 \text{ lbf}}{0.25\pi\left((2 \text{ in})^2 - (0.625 \text{ in})^2\right)}$$

$$= 950 \text{ psi}$$

$$F'_{c\perp} = C_b F_{c\perp} = (1.19)\left(650 \; \frac{\text{lbf}}{\text{in}^2}\right)$$

$$= 775 \text{ psi} \quad [\text{not OK}]$$

$$A_{\text{req}} \geq \frac{P}{F_{c\perp}} = \frac{2690 \text{ lbf}}{650 \; \frac{\text{lbf}}{\text{in}^2}} = 4.14 \text{ in}^2$$

Use 3 in wide (9 in²) plate washers with the ¹/₂ in diameter rod. In addition to the force transfer, there is a moment created by the 2 ft lever arm between the two forces to transfer into the wall. One possibility is to use the header over the corridor opening to produce a counteracting couple.

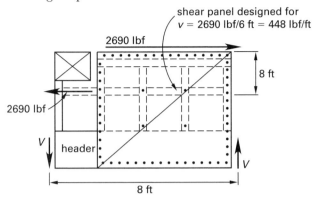

For example, assuming an 8 ft corridor width, the shear in the header needed to counteract the couple is

$$\sum M_{\text{left}} = Pd - VL = 0$$

$$V = \frac{Pd}{L} = \frac{(2690 \text{ lbf})(2 \text{ ft})}{8 \text{ ft}}$$

$$= 673 \text{ lbf} \quad [\text{reversible}]$$

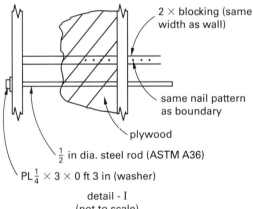

detail - I
(not to scale)

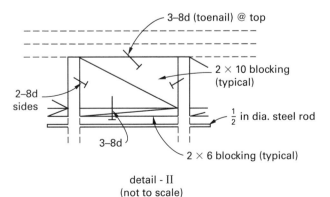

detail - II
(not to scale)

This force can be transferred into the adjacent chord of the shear wall using bearing blocks above and below the header, and in combination with the lateral forces gives the same overturning moment as would exist if the original top plates had not been cut. However, the equal and opposite shear at the opposite end of the header would have to be investigated and the column supporting the header might require supplementary tie-down anchorage if the gravity loads are not sufficient to counter the uplift. Since the span of the 2 × 10 joists is not given, this check will be omitted.

SOLUTION 3

3.1. Design the floor joist using no. 1/no. 2 Douglas fir-larch (north). Take the allowable stresses for visually graded dimension lumber no. 1/no. 2 (NDS Table 4A).

$$F_b = 850 \text{ psi}$$

$$F_v = 180 \text{ psi}$$

$$F_{C\perp} = 625 \text{ psi}$$

$$E = 1,600,000 \text{ psi}$$

Estimate the dead loads.

$$w_D = {}^3/_4 \text{ in plywood} + \text{floor finish} + \text{ceiling}$$

$$+ \text{partitions} + \text{joists}$$

$$= 3 \frac{\text{lbf}}{\text{ft}^2} + 1 \frac{\text{lbf}}{\text{ft}^2} + 1 \frac{\text{lbf}}{\text{ft}^2} + 8 \frac{\text{lbf}}{\text{ft}^2} + 2 \frac{\text{lbf}}{\text{ft}^2}$$

$$= 15 \frac{\text{lbf}}{\text{ft}^2} \quad [\text{to joists}]$$

The live load (residential) from ASCE/SEI7 Table 4-1, w_L, equals 40 lbf/ft².

Try joists spaced at 12 in on center. The influence area for a joist, $K_{LL}A_T$, is less than 400 ft²; therefore, no reduction in live load is permitted for the joist.

$$w = (w_D + w_L)s = \left(15 \frac{\text{lbf}}{\text{ft}^2} + 40 \frac{\text{lbf}}{\text{ft}^2}\right)(1 \text{ ft})$$

$$= 55 \text{ lbf/ft}$$

Assume the size factor, C_F, is 1.0. For joists spaced 12 in on center, the repetitive use factor, C_r, is 1.15. For this application, all other adjustment factors are equal to one and the adjusted allowable bending stress is

$$F'_b = C_r F_b = (1.15)\left(850 \frac{\text{lbf}}{\text{in}^2}\right) = 978 \text{ psi}$$

The section modulus required to resist bending is

$$M_{\text{max}} = 0.125wL^2 = (0.125)\left(55 \ \frac{\text{lbf}}{\text{ft}}\right)(17 \ \text{ft})^2$$
$$= 1987 \ \text{ft-lbf}$$

$$S_{\text{req}} = \frac{M_{\text{max}}}{F_b'} = \frac{(1987 \ \text{ft-lbf})\left(12 \ \frac{\text{in}}{\text{ft}}\right)}{978 \ \frac{\text{lbf}}{\text{in}^2}} = 24.3 \ \text{in}^3$$

Limit the live load deflection to $L/360$. ($E' = E = 1{,}600{,}000$ psi.)

$$\Delta_L = \frac{0.013w_L L^4}{E' I_{\text{req}}} \le \frac{L}{360}$$

$$I_{\text{req}} = \frac{(0.013)360w_L L^3}{E'}$$

$$= \frac{(0.013)(360)\left(40 \ \frac{\text{lbf}}{\text{ft}}\right)(17 \ \text{ft})^3\left(12 \ \frac{\text{in}}{\text{ft}}\right)^2}{1{,}600{,}000 \ \frac{\text{lbf}}{\text{in}^2}}$$

$$= 83 \ \text{in}^4$$

Therefore, try 2×12 (per NDS Table 1B, $S = 31.6 \ \text{in}^3$ and $I = 178 \ \text{in}^4$). Assume the size factor is 1. Check shear.

$$V = w(0.5L - d) = \left(55 \ \frac{\text{lbf}}{\text{ft}}\right)\left((0.5)(17 \ \text{ft}) - \frac{11.25 \ \text{in}}{12 \ \frac{\text{in}}{\text{ft}}}\right)$$

$$= 416 \ \text{lbf}$$

$$f_v = \frac{1.5V}{bd} = \frac{(1.5)(416 \ \text{lbf})}{(1.5 \ \text{in})(11.25 \ \text{in})} = 37 \ \text{psi} \quad [< F_v' = 180 \ \text{psi}]$$

Check bearing length, N.

$$R = 0.5wL = (0.5)\left(55 \ \frac{\text{lbf}}{\text{ft}}\right)(17 \ \text{ft}) = 468 \ \text{lbf}$$

$$f_{c\perp} = \frac{R}{bN} = \frac{468 \ \text{lbf}}{(1.5 \ \text{in})N} \le F_{c\perp} = 625 \ \text{psi}$$

$$N \ge 0.5 \ \text{in}$$

Any reasonably sized joist seat or ledger will provide this length of bearing.

Therefore, use 2×12 at 12 in o.c. for the joists.

3.2. Design the girder assuming a simple span. For the span and loading, a glulam or other type of member might be preferable, but the problem statement implies a solid sawn member. From NDS Table 4D, for beams and stringers, $F_b = 1300$ psi (single member use), $F_v = 170$ psi, $F_{c\perp} = 625$ psi, and $E = 1{,}600{,}000$ psi for no. 1 grade Douglas fir-larch (north).

$$A_T = (17 \ \text{ft})(20 \ \text{ft}) = 340 \ \text{ft}^2$$

$$K_{LL}A_T = (2)(340 \ \text{ft}^2)$$
$$= 680 \ \text{ft}^2 \quad \begin{bmatrix} > 400 \ \text{ft}^2, \text{ so reduce live} \\ \text{load on girder} \end{bmatrix}$$

Per ASCE/SEI7 Sec. 4.7.2,

$$w_L = L_o\left(0.25 + \frac{15}{\sqrt{K_{LL}A_T}}\right)$$
$$= \left(40 \ \frac{\text{lbf}}{\text{ft}^2}\right)\left(0.25 + \frac{15 \ \text{ft}}{\sqrt{680 \ \text{ft}^2}}\right)$$
$$= 33 \ \text{lbf/ft}^2$$

The load to the girder is essentially uniformly distributed. Estimate the girder weight as 25 lbf/ft.

$$w_{D,\text{girder}} = w_D s + w_{\text{girder}} = \left(15 \ \frac{\text{lbf}}{\text{ft}^2}\right)(17 \ \text{ft}) + 25 \ \frac{\text{lbf}}{\text{ft}}$$
$$= 280 \ \text{lbf/ft}$$

$$w_{L,\text{girder}} = w_L s = \left(33 \ \frac{\text{lbf}}{\text{ft}^2}\right)(17 \ \text{ft}) = 561 \ \text{lbf/ft}$$

$$w = w_{D,\text{girder}} + w_{L,\text{girder}} = 280 \ \frac{\text{lbf}}{\text{ft}} + 561 \ \frac{\text{lbf}}{\text{ft}}$$
$$= 841 \ \text{lbf/ft}$$

$$M_{\text{max}} = 0.125wL^2 = (0.125)\left(841 \ \frac{\text{lbf}}{\text{ft}}\right)(20 \ \text{ft})^2$$
$$= 42{,}000 \ \text{ft-lbf}$$

Assume C_F equals 0.95 as a trial value; thus,

$$F_b' = F_b C_F = \left(1300 \ \frac{\text{lbf}}{\text{in}^2}\right)(0.95) = 1235 \ \text{psi}$$

$$S_{\text{req}} = \frac{M_{\text{max}}}{F_b'} = \frac{(42{,}000 \ \text{ft-lbf})\left(12 \ \frac{\text{in}}{\text{ft}}\right)}{1235 \ \frac{\text{lbf}}{\text{in}^2}} = 408 \ \text{in}^3$$

$$\Delta_L = \frac{0.013w_L L^4}{E' I_{\text{req}}} \le \frac{L}{360}$$

$$I_{\text{req}} = \frac{(0.013)360w_L L^3}{E'}$$

$$= \frac{(0.013)(360)\left(561 \ \frac{\text{lbf}}{\text{ft}}\right)(20 \ \text{ft})^3\left(12 \ \frac{\text{in}}{\text{ft}}\right)^2}{1{,}600{,}000 \ \frac{\text{lbf}}{\text{in}^2}}$$

$$= 1890 \ \text{in}^4$$

Per NDS Table 1B, an 8×20 is adequate ($S = 475 \ \text{in}^3$ and $I = 4634 \ \text{in}^4$). Check the size factor.

$$C_F = \left(\frac{12 \ \text{in}}{d}\right)^{1/9} = \left(\frac{12 \ \text{in}}{19.5 \ \text{in}}\right)^{1/9} = 0.95$$

The assumed allowable stress is correct. Check shear.

$$V = w(0.5L - d)$$

$$= \left(841 \ \frac{\text{lbf}}{\text{ft}}\right)\left((0.5)(20 \ \text{ft}) - \frac{15.25 \ \text{in}}{12 \ \frac{\text{in}}{\text{ft}}}\right)$$

$$= 7340 \ \text{lbf}$$

$$f_v = \frac{1.5V}{bd} = \frac{(1.5)(7340 \ \text{lbf})}{(7.5 \ \text{in})(19.5 \ \text{in})} = 75 \ \text{psi} \quad [< F_v' = 170 \ \text{psi}]$$

Check bearing length, N.

$$R = 0.5wL = (0.5)\left(841 \ \frac{\text{lbf}}{\text{ft}}\right)(20 \ \text{ft}) = 8410 \ \text{lbf}$$

$$f_{c\perp} = \frac{R}{bN} = \frac{8410 \ \text{lbf}}{(7.5 \ \text{in})N} \le F_{c\perp} = 625 \ \text{psi}$$

$$N \ge 1.8 \ \text{in} \quad [\text{no problem}]$$

Therefore, use 8×20 no. 1 Douglas fir-larch (north) for the girders.

3.3. Design a wall ledger to support the joists. Try a 3×12 no. 1 Douglas fir-larch (north) and $^7/_8$ in diameter anchor bolts. For gravity loads, the reaction from joists spaced 12 in o.c. gives $R = 467$ lbf/ft of ledger length. Provide joist hangers to support the 467 lbf reaction from joists. For a $^7/_8$ in diameter bolt bearing perpendicular to the grain (per NDS Table 11E and assuming bolts embed a minimum of 6 in into masonry or concrete with a minimum of 2500 psi compressive strength), the single shear capacity is 920 lbf/bolt.

$$Z_\perp' = 1000 \ \text{lbf/bolt}$$

The spacing of bolts is

$$s = \frac{Z_\perp'}{R} = \frac{1000 \ \text{lbf}}{467 \ \frac{\text{lbf}}{\text{ft}}} = 2.14 \ \text{ft} \quad [\text{say 2 ft o.c.}]$$

Check the shear stress near the end of member for a 3×12 ledger with the bolt supporting its maximum capacity of 1000 lbf.

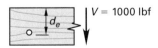

Assume the worst case with the bolt positioned five diameters from the end of the member. NDS Eq. 3.4-6 gives

$$V_r' = \tfrac{2}{3}F_v' b d_e \left(\frac{d_e}{d}\right)^2 = 1000 \ \text{lbf}$$

$$d_e = \sqrt[3]{\frac{V_r' d^2}{\tfrac{2}{3}F_v' b}} = \sqrt[3]{\frac{(1000 \ \text{lbf})(11.25 \ \text{in})^2}{\left(\tfrac{2}{3}\right)\left(170 \ \frac{\text{lbf}}{\text{in}^2}\right)(2.5 \ \text{in})}}$$

$$= 7.6 \ \text{in}$$

Therefore, position the $^7/_8$ in diameter anchor bolts at 7.6 in below the top of the ledger and space them at 2 ft o.c. No information is given regarding the lateral loads in this problem. However, if diaphragm shear is transferred into the top of the ledger by nails through plywood, then the anchor bolts would resist the shear by parallel-to-grain loading. Lateral forces in the other direction (i.e., out of plane) would be resisted by ties anchored into the reinforced masonry wall that are designed to resist the larger of the code specified forces (ASCE/SEI7 Sec. 12.11, for seismic forces). An acceptable detail is

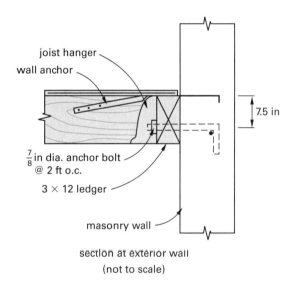

section at exterior wall
(not to scale)

SOLUTION 4

4.1. Design the horizontal roof diaphragm using the loading criteria from ASCE/SEI7. To account for wind loading on the parapets, use provisions from ASCE/SEI7 Sec. 28.4.2. Given V equal to 114 mph and ASCE/SEI7 Table 26.6-1, $K_d = 0.85$. For exposure condition C, ASCE/SEI7 Table 28.3-1 gives K_z as 0.85 when z is between 0 ft and 15 ft. For the design of the main lateral force resisting system, only external pressure coefficients are required. For the building, $L/B = 40 \ \text{ft}/130 \ \text{ft} < 1$; therefore, C_{pe} is $+0.8$ for the windward face and -0.5 for the leeward face. The combined net pressure coefficient for the parapets, GC_{pn}, is given in ASCE/SEI7 Sec. 28.4-2 as $+1.5$ on the windward parapet and -1.0 for the leeward parapet. Since the tops of the parapets and roof are less than 15 ft above grade, the same velocity pressure, q_z, given by ASCE/SEI7

Eq. 28.3-1, applies to all elements. Convert loads to working load basis by multiplying by 0.6.

$$q_{15\,\text{ft}} = (0.6)(0.00256 K_{15\,\text{ft}} K_{zt} K_d V^2)$$

$$= (0.6)\left(0.00256\ \frac{\text{lbf-hr}^2}{\text{ft}^2\text{-mi}^2}\right)(0.85)(1.0)$$

$$\times (0.85)\left(114\ \frac{\text{mi}}{\text{hr}}\right)^2$$

$$= 15\ \text{lbf/ft}^2$$

$$q_p = q_{14} = q_h = q_{10} = q_{15} = 15\ \text{lbf/ft}^2$$

Assuming exterior walls span 10 ft from base to roof and that all force on the parapets transfers to the roof diaphragm, the force per foot on the diaphragm due to wind is

$$w_w = 0.5hG(C_{\text{pe,windward}} - C_{\text{pe,leeward}})q_{15}$$

$$+ h_p(GC_{\text{pn,windward}} - GC_{\text{pn,leeward}})q_{15}$$

$$= (0.5)(10\ \text{ft})(0.85)\big(0.8 - (-0.5)\big)\left(15\ \frac{\text{lbf}}{\text{ft}^2}\right)$$

$$+ (4\ \text{ft})\big(1.5 - (-1.0)\big)\left(15\ \frac{\text{lbf}}{\text{ft}^2}\right)$$

$$= 233\ \text{lbf/ft}$$

For seismic dead load in the north-south direction, include one-half of the weight of exterior walls and the movable partition plus the weight of the roof and parapet. Express per foot of diaphragm length.

$$w_D = 2aw_{\text{corridor}} + Lw_{\text{main roof}} + 2h_p w_{\text{parapet}}$$

$$+ 0.5h(2w_{\text{ext wall}} + w_{\text{partition}})$$

$$= (2)(10\ \text{ft})\left(12\ \frac{\text{lbf}}{\text{ft}^2}\right) + (40\ \text{ft})\left(16\ \frac{\text{lbf}}{\text{ft}^2}\right)$$

$$+ (2)(4\ \text{ft})\left(10\ \frac{\text{lbf}}{\text{ft}^2}\right) + (0.5)(10\ \text{ft})$$

$$\times \left((2)\left(20\ \frac{\text{lbf}}{\text{ft}^2}\right) + 12\ \frac{\text{lbf}}{\text{ft}^2}\right)$$

$$= 1220\ \text{lbf/ft}$$

The approximate period, from ASCE/SEI7 Eq. 12.8-7, is

$$T_a = C_t h_n^x = (0.02)(10\ \text{ft})^{0.75} = 0.11\ \text{sec}$$

For this short period, ASCE/SEI7 Eq. 12.8-2 controls the seismic base shear.

$$C_s = \frac{S_{DS}}{\dfrac{R}{I_e}} = \frac{0.45}{\dfrac{6.5}{1.0}} = 0.07\quad[\text{design level}]$$

The seismic force to the diaphragm on a working load basis is

$$w_E = 0.7C_s w = (0.7)(0.07)\left(1220\ \frac{\text{lbf}}{\text{ft}}\right) = 60\ \text{lbf/ft}$$

Thus, wind controls the design of the diaphragm in the north-south direction. For the flexible diaphragm, the critical regions are adjacent to shear walls on lines B and C, where the diaphragm spans 80 ft. The shear per foot at these locations is

$$v = \frac{0.5w_w s}{L} = \frac{(0.5)\left(233\ \dfrac{\text{lbf}}{\text{ft}}\right)(80\ \text{ft})}{40\ \text{ft}} = 233\ \text{lbf/ft}$$

From the table provided with the problem statement, a $^5/_{16}$ in exterior grade sheathing nailed to 2 in nominal framing in a blocked diaphragm furnishes 250 lbf/ft allowable shear with 6d nails at 4 in on center on the boundary and continuous panel edges. Blocking can be eliminated where the shear is less than 167 lbf/ft. This occurs in the 80 ft span at a distance of

$$v = 233\ \frac{\text{lbf}}{\text{ft}} - \frac{\left(233\ \dfrac{\text{lbf}}{\text{ft}}\right)x}{40\ \text{ft}} = 167\ \text{lbf/ft}$$

$$x = 11.3\ \text{ft}\quad[\text{from lines B and C}]$$

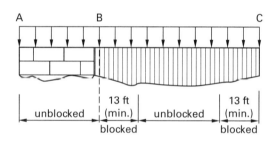

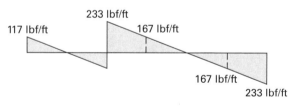

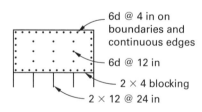

The maximum chord force is

$$M_{\max} = 0.125 w_w s^2 = (0.125)\left(233\ \frac{\text{lbf}}{\text{ft}}\right)(80\ \text{ft})^2$$

$$= 186{,}400\ \text{ft-lbf}$$

$$T = C = \frac{M_{\max}}{L} = \frac{186{,}400\ \text{ft-lbf}}{40\ \text{ft}}$$

$$= 4660\ \text{lbf}$$

For the chords, try 4×12 dimension lumber Douglas fir-larch ($A = 39.4$ in^2). Assume $5/8$ in diameter bolts with steel side plates are used to splice the chords. For the 4×12 no. 1 Douglas fir-larch in tension, C_F equals 1.0, C_D equals 1.6, and all other adjustment factors equal 1.0.

$$F'_t = C_D F_t = (1.6)\left(800 \frac{\text{lbf}}{\text{in}^2}\right) = 1280 \text{ psi}$$

$$A_n = A - bd_{\text{hole}} = 39.4 \text{ in}^2 - (2.5 \text{ in})(0.69 \text{ in}) = 37.7 \text{ in}^2$$

$$T = F'_t A_n = \left(1280 \frac{\text{lbf}}{\text{in}^2}\right)(37.7 \text{ in}^2) = 48,300 \text{ lbf}$$

The chord is laterally braced about the weak axis and is subjected to combined bending and axial compression that should be checked at its midspan. This check is omitted in this solution.

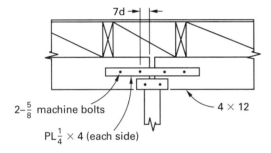

2-$\frac{5}{8}$ machine bolts

PL$\frac{1}{4} \times 4$ (each side)

4×12

Try PL$^1/_4 \times 3$ each side (ASTM A36 steel). Check allowable tension.

$$A_n = A_g - (d_b + 0.125 \text{ in})t$$
$$= 0.75 \text{ in}^2 - (0.625 \text{ in} + 0.125 \text{ in})(0.25 \text{ in})$$
$$= 0.56 \text{ in}^2$$

$$\frac{P_n}{\Omega_t} \leq \begin{cases} \dfrac{F_y A_g}{\Omega_t} = \dfrac{\left(36 \frac{\text{kips}}{\text{in}^2}\right)(0.75 \text{ in}^2)}{1.67} \\ \qquad = 16.1 \text{ kips} \quad [\text{controls}] \\ \dfrac{F_u U A_n}{\Omega_t} = \dfrac{\left(58 \frac{\text{kips}}{\text{in}^2}\right)(1.0)(0.56 \text{ in}^2)}{2.0} \\ \qquad = 16.2 \text{ kips} \end{cases}$$

Check compression strength of the side plates assuming an end spacing of 7 bolt diameters, $1/2$ in clearance between members, and an effective length factor for fixed-fixed condition ($K = 0.65$).

$$\frac{KL}{r} = \frac{K(14d_b + \text{clearance})}{0.25t}$$
$$= \frac{(0.65)\big((14)(0.625 \text{ in}) + 0.5 \text{ in}\big)}{(0.25)(0.25 \text{ in})}$$
$$= 96$$

From AISC Table 4-22, $F_{\text{cr}}/\Omega_c = 13.3$ ksi.

$$\frac{P_n}{\Omega_c} = \left(\frac{F_{\text{cr}}}{\Omega_c}\right)A_g = \left(13.3 \frac{\text{kips}}{\text{in}^2}\right)(0.75 \text{ in}^2) = 10.0 \text{ kips}$$

Compression in the splice plates controls strength. Each plate can resist 10.0 kips; thus, two plates will be more than adequate to resist the total 10 kips. From NDS Table 11G, with 2.5 in main member and $G = 0.5$, each bolt can resist

$$Z' = C_D Z_{\parallel} = (1.6)(2190 \text{ lbf}) = 3500 \text{ lbf}$$

$$n > \frac{C}{Z'} = \frac{4660 \text{ lbf}}{3500 \frac{\text{lbf}}{\text{bolt}}} = 1.3 \text{ bolts} \quad [\text{use 2 bolts each side}]$$

Use two $\frac{5}{8}$ in diameter bolts with $3 \times \frac{1}{4}$ in steel side plates.

4.2. Design the shear wall at line B.

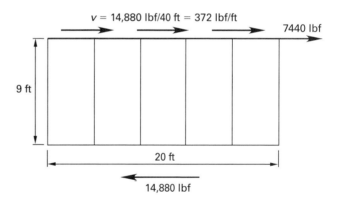

The total shear contributed by the diaphragms on either side of line B is

$$V_B = (v_{B,\text{left}} + v_{B,\text{right}})L$$
$$= \left(124 \frac{\text{lbf}}{\text{ft}} + 248 \frac{\text{lbf}}{\text{ft}}\right)(40 \text{ ft})$$
$$= 14,880 \text{ lbf}$$

The unit shear at the base of the 20 ft long shear wall is

$$v = \frac{V_B}{L_{\text{wall}}} = \frac{14,880 \text{ lbf}}{20 \text{ ft}} = 744 \text{ lbf/ft}$$

Per SDPWS Table 4.3A for $^{15}/_{32}$ in wood structural panels with edges nailed using 10d nails at 6 in on center, sheathed on two sides, the allowable (ASD) shear capacity is

$$\frac{(2)\left(870 \frac{\text{lbf}}{\text{ft}}\right)}{2} = 870 \text{ lbf/ft}$$

Nailing into 2 in nominal studs, sills, and blocking is OK. Nail spacing on intermediate supports at 12 in o.c.

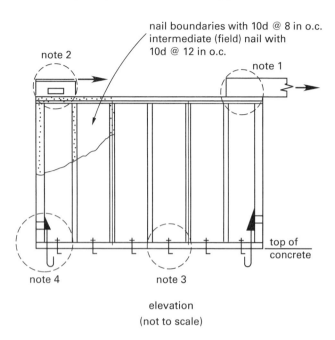

elevation
(not to scale)

Note 1: Connect the collector to the top plate to transfer 744 lbf.

Note 2: Provide blocking from the horizontal diaphragm to the top plates to transfer 372 lbf/ft.

Note 3: Provide anchor bolts to transfer the total base shear of 14,880 lbf. (For $^5/_8$ in diameter bolts in a 2 in sill plate, from NDS Table 11E, the single shear parallel to the grain for $t_s = 1.5$ in is $Z' = (1.6)(930 \text{ lbf}) = 1490$ lbf/bolt. This is not critical in concrete per IBC Table 1908.2.) Therefore,

$$n \geq \frac{14,880 \text{ lbf}}{1490 \dfrac{\text{lbf}}{\text{bolt}}} = 10 \text{ bolts}$$

Note 4: Design the end posts and tie-down anchors to resist the overturning moment, $M_0 = (14,880 \text{ lbf})(9 \text{ ft}) = 133,900$ ft-lbf. Neglect gravity loads.

$$\begin{aligned} T = C &= \frac{M_0}{L_{\text{wall}}} \\ &= \frac{V_B h}{L_{\text{wall}}} \\ &= \frac{(14,880 \text{ ft-lbf})(9 \text{ ft})}{20 \text{ ft}} \\ &= 6696 \text{ lbf} \quad (6700 \text{ lbf}) \end{aligned}$$

Therefore, the anchorage and end post must be designed for 6700 lbf.

SOLUTION 5

5.1. Check the adequacy of the given connection details. For the existing connection A under north-south lateral forces,

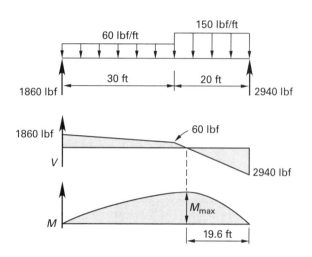

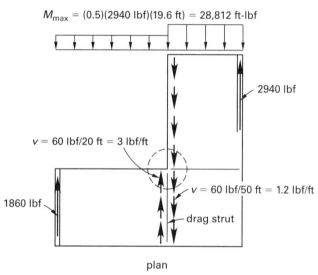

plan

The force transfer across the connection shown in section A-A must tie the collector for resistance to north-south lateral forces and tie the chord for resistance to east-west lateral forces. Since the structure is braced entirely by light frame shear walls, the collector elements and splices are exempt from the load combinations with overstrength factor (ASCE/SEI7 Sec. 12.10.2.1). However, the splices occur at a reentrant corner irregularity as defined in ASCE/SEI7 Table 12.3-1, horizontal irregularity type 2, which requires that the computed forces in collectors and their splices must be increased by 25%. Furthermore, since the splices are not designed based on the overstrength factor, the redundancy factor of 1.3 must be applied (ASCE/SEI7 Sec. 12.3.4.1). For the forces acting in the east-west direction, the force transfer across the connection shown in section A-A functions to tie the diaphragm chord, where the moment essentially is the maximum.

$$F_{A,chord} = \pm 1.25\rho \left(\frac{M_{max}}{2s} \right)$$

$$= (\pm 1.25)(1.3)\left(\frac{28{,}812 \text{ ft-lbf}}{(2)(10 \text{ ft})} \right)$$

$$= \pm 2340 \text{ lbf}$$

When lateral forces act in the north-south direction, the tie force that must transfer across the connection shown in section A-A can be estimated from a free-body diagram of the wing above the diaphragm. Lateral seismic forces are distributed according to the mass distribution of the structure rather than concentrated along one edge, as is typically shown for simple diaphragm analysis. For the purpose of computing the tie force, split the given lateral force into distributed forces that act on opposite sides of the building.

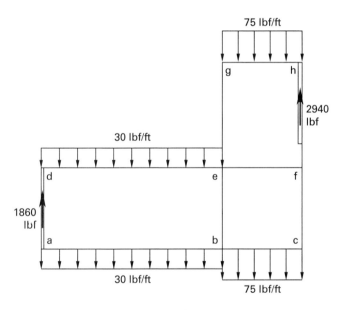

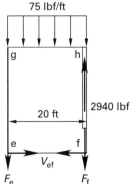

free-body diagram of portion efgh

Taking moments about point f,

$$\sum M_f = F_e L_{hg} + 0.5 w_{75} L_{hg}^2 = 0$$

$$F_e = \frac{-0.5 w_{75} L_{hg}^2}{L_{hg}} = -0.5 w_{75} L_{hg}$$

$$= (-0.5)\left(75 \ \frac{\text{lbf}}{\text{ft}} \right)(20 \text{ ft})$$

$$= -750 \text{ lbf} \quad [750 \text{ lbf upward}]$$

$$\sum F_{vert} = V + F_e - F_f - w_{75} L_{hg} = 0$$

$$F_f = V + F_e - w_{75} L_{hg}$$

$$= 2940 \text{ lbf} + 750 \text{ lbf} - \left(75 \ \frac{\text{lbf}}{\text{ft}} \right)(20 \text{ ft})$$

$$= 2190 \text{ lbf} \quad [\text{downward}]$$

As a check, applying equal and opposite forces F_e and F_f to a free body of the lower portion, abcdef, and taking moments about corner d gives

$$\sum M_d = F_f L_{df} - F_e L_{ef} - 0.5 w_{60} L_{de}^2$$

$$\qquad - w_{75} L_{ef}(L_{df} - 0.5 L_{ef})$$

$$= (2190 \text{ lbf})(50 \text{ ft}) - (750 \text{ lbf})(30 \text{ ft})$$

$$\qquad - (0.5)\left(60 \ \frac{\text{lbf}}{\text{ft}} \right)(30 \text{ ft})^2 - \left(75 \ \frac{\text{lbf}}{\text{ft}} \right)$$

$$\qquad \times (20 \text{ ft})\big(50 \text{ ft} - (0.5)(20 \text{ ft}) \big)$$

$$= 0$$

Thus, the computed tie force satisfies equilibrium. Applying the redundancy and 25% increase required by ASCE/SEI7 Sec. 12.3.3.4 gives

$$F_{A,drag} = \pm 1.25\rho F_e = \pm (1.25)(1.3)(750 \text{ lbf})$$

$$= \pm 1219 \text{ lbf} < F_{A,chord}$$

Therefore, the controlling force occurs when the member acts as a diaphragm chord. The dimensions and applied forces are such that the forces for the connection shown in B-B are the same; that is, the forces are obtained by simply interchanging the above results. Thus, the controlling force is

$$F_{A,chord} = F_{B,chord} = \pm 2340 \text{ lbf}$$

Both connections shown must provide a tie for ± 2340 lbf. Check the splice shown in section A-A. Tie strength is furnished by six 16d nails (three on each side) through 18-gage steel side plates. For this condition, NDS Table 11P gives

$$Z' = C_D Z = (1.6)\left(137 \ \frac{\text{lbf}}{\text{nail}} \right) = 219 \text{ lbf/nail}$$

$$P_{max} = nZ' = (6 \text{ nails})\left(219 \ \frac{\text{lbf}}{\text{nail}} \right) = 1314 \text{ lbf}$$

Therefore, the connection in A is inadequate and needs an additional tie capacity of 2340 lbf − 1315 lbf − 1025 lbf.

For the connection shown in B-B, there is no mechanism to transfer the axial tension out of the 4 in beam into the PL6 × $^1/_4$ that serves as a bearing cap.

Therefore, connection B needs a tension tie to develop the full 2340 lbf.

5.2. The following revisions are proposed.

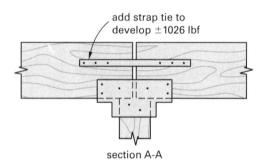

section A-A

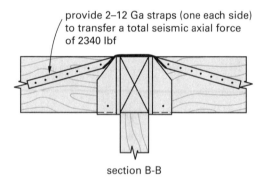

section B-B

SOLUTION 6

6.1. Calculate the maximum bending stress in the $5^1/_8 \times 30$ glulam beam. Per ASCE/SEI7 Table 4-1, take L_o as equal to 20 lbf/ft^2. Using ASCE/SEI7 Sec. 4.8.1, the load is reducible. The tributary area to the glulam girder is

$$A_T = wb = (40 \text{ ft})(40 \text{ ft}) = 1600 \text{ ft}^2$$

The roof slopes $^1/_2$ in per 12 in; therefore, F equals $^1/_2$.

$$R_1 = 0.6 \quad [\text{for } A_T > 600 \text{ ft}^2]$$
$$R_2 = 1.0 \quad [\text{for } F = 1/2]$$
$$L_r = R_1 R_2 L_o = (0.6)(1.0)\left(20 \ \frac{\text{lbf}}{\text{ft}^2}\right) = 12 \ \text{lbf/ft}^2$$

Given that all dead loads are uniformly distributed, and assuming that the given 10 lbf/ft^2 roof load includes the beam weight,

$$w = (w_D + L_r)L = \left(10 \ \frac{\text{lbf}}{\text{ft}^2} + 12 \ \frac{\text{lbf}}{\text{ft}^2}\right)(20 \text{ ft})$$
$$= 440 \text{ lbf/ft}$$

From NDS Supplement Table 1C, for the $5^1/_8 \times 30$ girder,

$$A = 154 \text{ in}^2$$
$$S = 769 \text{ in}^3$$
$$I = 11{,}530 \text{ in}^4$$

$$w_{\text{girder}} = A w_{\text{wood}} = \left(\frac{(5.125 \text{ in})(30 \text{ in})}{\left(12 \ \frac{\text{in}}{\text{ft}}\right)^2}\right)\left(36 \ \frac{\text{lbf}}{\text{ft}^3}\right)$$
$$= 38 \text{ lbf/ft}$$
$$P = wL = \left(440 \ \frac{\text{lbf}}{\text{ft}}\right)(40 \text{ ft}) = 17{,}600 \text{ lbf}$$

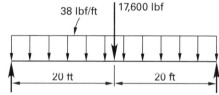

$$M_{\text{max}} = 0.125 w_{\text{girder}} L^2 + 0.25 PL$$
$$= (0.125)\left(38 \ \frac{\text{lbf}}{\text{ft}}\right)(40 \text{ ft})^2$$
$$\quad + (0.25)(17{,}600 \text{ lbf})(40 \text{ ft})$$
$$= 183{,}600 \text{ ft-lbf}$$

The maximum bending stress in the girder is

$$f_b = \frac{M_{\text{max}}}{S} = \frac{(183{,}600 \text{ ft-lbf})\left(12 \ \frac{\text{in}}{\text{ft}}\right)}{769 \text{ in}^3} = \boxed{2865 \text{ psi}}$$

6.2. Check the adequacy of the glulam girder as modified by the skylight. The skylight eliminates the lateral support to the top (compression) flange. Lateral support is now present only at its ends and at midspan. Check the slenderness effect (NDS Sec. 3.3.3) with $l_u = 20$ ft, $C_D = 1.25$, normal temperature, and dry use. For a beam laterally supported at ends and midspan and subjected primarily to a concentrated force at midspan, the effective length of the compression flange is found from NDS Table 3.3.3.

$$l_e = 1.11 l_u = (1.11)(20 \text{ ft})\left(12 \ \frac{\text{in}}{\text{ft}}\right) = 266 \text{ in}$$

Lateral buckling occurs about the weak axis (i.e., y-axis); therefore, use $E_{y,\min}$.

$$R_B = \sqrt{\frac{l_e d}{b^2}} = \sqrt{\frac{(266 \text{ in})(30 \text{ in})}{(5.125 \text{ in})^2}} = 17.4$$

$$E'_{y,\min} = E_{y,\min} = 850{,}000 \text{ psi}$$

$$F^*_{bx} = C_D F_b = (1.25)\left(2400 \ \frac{\text{lbf}}{\text{in}^2}\right) = 3000 \text{ psi}$$

$$F_{bE} = \frac{1.20 E'_{y,\min}}{R_B^2} = \frac{(1.20)\left(850{,}000 \ \dfrac{\text{lbf}}{\text{in}^2}\right)}{(17.4)^2} = 3369 \text{ psi}$$

$$\frac{F_{bE}}{F^*_{bx}} = \frac{3369 \ \dfrac{\text{lbf}}{\text{in}^2}}{3000 \ \dfrac{\text{lbf}}{\text{in}^2}} = 1.12$$

$$C_L = \frac{1 + \dfrac{F_{bE}}{F^*_{bx}}}{1.9} - \sqrt{\left(\frac{1 + \dfrac{F_{bE}}{F^*_{bx}}}{1.9}\right)^2 - \frac{\dfrac{F_{bE}}{F^*_{bx}}}{0.95}}$$

$$= \frac{1 + 1.12}{1.9} - \sqrt{\left(\frac{1 + 1.12}{1.9}\right)^2 - \frac{1.12}{0.95}}$$

$$= 0.86$$

Compute the volume factor, C_V, and compare to the stability factor, C_L. From NDS Sec. 5.3.6,

$$C_V = \left(\left(\frac{21}{L}\right)\left(\frac{12}{d}\right)\left(\frac{5.125}{b}\right)\right)^{1/x} \leq 1.0$$

$$= \left(\left(\frac{21 \text{ ft}}{40 \text{ ft}}\right)\left(\frac{12 \text{ in}}{30 \text{ in}}\right)\left(\frac{5.125 \text{ in}}{5.125 \text{ in}}\right)\right)^{1/10}$$

$$= 0.86 \quad [> C_L]$$

The stability factor controls. Therefore,

$$F'_b = C_D C_L F_b = (1.25)(0.85)\left(2400 \ \frac{\text{lbf}}{\text{in}^2}\right) = 2550 \text{ psi}$$

$$f_b = 2865 \text{ psi} > F'_b \quad [14\% \text{ overstress}]$$

> Therefore, the proposed skylight modification is no good.

SOLUTION 7

7.1. Determine whether the proposed two 3×4s have axial compression strength equivalent to that of the original 4×6. Use the allowable stress design method. Assume the given roof live load qualifies for a duration of load factor, $C_D = 1.25$. Per NDS Table 4A, the 4×6 and 3×4 do not have the same size factor for compression parallel to grain: $C_F = 1.05$ for the 3×4 and $C_F = 1.0$ for the 4×6. All other adjustment factors except the stability factor are equal to 1.0.

Per NDS Sec. 15.3, for the built-up column, assume the most favorable case, which is the two 3×4s bolted in accordance with NDS Sec. 15.3.4. For this case, the stability factor will be computed for an equivalent solid section and reduced by the factor $K_f = 0.75$ to account for diminished shear transfer in the built-up member. For stud-grade Douglas fir-larch,

$$F_c = 850 \text{ psi}$$

$$E_{\min} = 510{,}000 \text{ psi}$$

For the 4×6 in the original plans,

$$A = 19.25 \text{ in}^2$$

The column is unsupported over 9 ft about its strong axis and over the 5 ft window opening about its weak axis. The controlling slenderness ratio is

$$\frac{l_e}{d} \geq \begin{cases} \dfrac{(9 \text{ ft})\left(12 \ \dfrac{\text{in}}{\text{ft}}\right)}{5.5 \text{ in}} = 19.6 \quad [\text{controls}] \\[4mm] \dfrac{(5 \text{ ft})\left(12 \ \dfrac{\text{in}}{\text{ft}}\right)}{3.5 \text{ in}} = 17.1 \end{cases}$$

$$F^*_c = C_D C_F F_c = (1.25)(1.0)\left(850 \ \frac{\text{lbf}}{\text{in}^2}\right) = 1063 \text{ psi}$$

$$E'_{\min} = E_{\min} = 510{,}000 \text{ psi}$$

From NDS Sec. 3.7.1, with $c = 0.8$ for sawn lumber,

$$F_{cE} = \frac{0.822 E'_{\min}}{\left(\dfrac{l_e}{d}\right)^2} = \frac{(0.822)\left(510{,}000 \ \dfrac{\text{lbf}}{\text{in}^2}\right)}{(19.6)^2} = 1090 \text{ psi}$$

$$\frac{F_{cE}}{F^*_c} = \frac{1090 \ \dfrac{\text{lbf}}{\text{in}^2}}{1063 \ \dfrac{\text{lbf}}{\text{in}^2}} = 1.03$$

$$C_P = \frac{1 + \dfrac{F_{cE}}{F^*_c}}{2c} - \sqrt{\left(\frac{1 + \dfrac{F_{cE}}{F^*_c}}{2c}\right)^2 - \frac{\dfrac{F_{cE}}{F^*_c}}{c}}$$

$$= \frac{1 + 1.03}{(2)(0.8)} - \sqrt{\left(\frac{1 + 1.03}{(2)(0.8)}\right)^2 - \frac{1.03}{0.8}}$$

$$= 0.70$$

$$F'_c = C_P C_D C_F F_c = (0.70)(1.25)(1.0)\left(850 \ \frac{\text{lbf}}{\text{in}^2}\right)$$

$$= 744 \text{ psi}$$

$$P = F'_c A = \left(744 \ \frac{\text{lbf}}{\text{in}^2}\right)(19.25 \text{ in}^2) = 14{,}300 \text{ lbf}$$

$$P_{D+L_r} = 3600 \text{ lbf} + 2400 \text{ lbf}$$

$$= 6000 \text{ lbf} \quad [P > P_{D+L_r}, \text{ so OK}]$$

> Therefore, the original 4×6 is adequate.

Check the proposed two 3×4 substitution.

$$A = 2bd = (2)(2.5 \text{ in})(3.5 \text{ in}) = 17.5 \text{ in}^2$$

The area of the built-up column is about 90% that of the 4×6. Check the bearing perpendicular to grain on the sill plate and beam above.

$$f_{c\perp} = \frac{P_{D+L_r}}{A} = \frac{6000 \text{ lbf}}{17.5 \text{ in}^2}$$
$$= 343 \text{ psi} \quad [< F_{c\perp} = 625 \text{ psi, so OK}]$$

The equivalent depth about the strong axis is $(2)(2.5 \text{ in}) = 5.0 \text{ in}$, which is smaller than the corresponding depth of the 4×6 and which results in a larger slenderness ratio. NDS Sec. 15.3 requires a further reduction in the stability factor by an adjustment of $K_f = 0.75$ for a properly bolted built-up member. Thus,

$$\frac{l_e}{d} = \frac{(9 \text{ ft})\left(12 \frac{\text{in}}{\text{ft}}\right)}{5.0 \text{ in}} = 21.6$$

$$F_c^* = C_D C_F F_c = (1.25)(1.05)\left(850 \frac{\text{lbf}}{\text{in}^2}\right) = 1116 \text{ psi}$$

$$E'_{\min} = E_{\min} = 510{,}000 \text{ psi}$$

Per NDS Sec. 3.7.1, with $c = 0.8$ for sawn lumber,

$$F_{cE} = \frac{0.822 E'_{\min}}{\left(\frac{l_e}{d}\right)^2} = \frac{(0.822)\left(510{,}000 \frac{\text{lbf}}{\text{in}^2}\right)}{(21.6)^2} = 899 \text{ psi}$$

$$\frac{F_{cE}}{F_c^*} = \frac{899 \frac{\text{lbf}}{\text{in}^2}}{1116 \frac{\text{lbf}}{\text{in}^2}} = 0.81$$

$$C_P = K_f \left(\frac{1 + \dfrac{F_{cE}}{F_c^*}}{2c} - \sqrt{\left(\frac{1 + \dfrac{F_{cE}}{F_c^*}}{2c}\right)^2 - \dfrac{\dfrac{F_{cE}}{F_c^*}}{c}} \right)$$

$$= (0.75)\left(\frac{1 + 0.81}{(2)(0.8)} - \sqrt{\left(\frac{1 + 0.81}{(2)(0.8)}\right)^2 - \frac{0.81}{0.8}} \right)$$

$$= 0.46$$

$$F_c' = C_P C_D C_F F_c = (0.46)(1.25)(1.05)\left(850 \frac{\text{lbf}}{\text{in}^2}\right)$$

$$= 513 \text{ psi}$$

$$P = F_c' A = \left(513 \frac{\text{lbf}}{\text{in}^2}\right)(17.5 \text{ in}^2)$$

$$= 8980 \text{ lbf} \quad [P > P_{D+L_r} = 6000 \text{ lbf, so OK}]$$

The proposed substitution significantly reduces the column's axial compression strength (8980 lbf versus 14,300 lbf); however, the compression strength is adequate for the given axial loads.

| The change should be approved. |

SOLUTION 8

8.1. Design beam B2. Per ASCE/SEI7 Sec. 4.8.2 for a flat roof, $R_2 = 1.0$. For member CD,

$$A_T = Ls = (36 \text{ ft})(16 \text{ ft}) = 576 \text{ ft}^2$$

For $200 \text{ ft}^2 \le A_T \le 600 \text{ ft}^2$,

$$R_1 = 1.2 - 0.001 A_T = 1.2 - \left(\frac{0.001}{1 \text{ ft}^2}\right)(576 \text{ ft}^2)$$
$$= 0.624$$

$$L_r = L_o R_1 R_2 = \left(20 \frac{\text{lbf}}{\text{ft}^2}\right)(0.624)(1.0)$$
$$= 12.5 \text{ lbf/ft}^2$$

Estimate the member weight as 50 lbf/ft for beam B2.

$$w_{CD} = w_D s + w_{\text{beam}} + w_L$$
$$= \left(20 \frac{\text{lbf}}{\text{ft}^2}\right)(16 \text{ ft}) + 50 \frac{\text{lbf}}{\text{ft}} + \left(12.5 \frac{\text{lbf}}{\text{ft}^2}\right)(16 \text{ ft})$$
$$= 570 \text{ lbf/ft}$$

Beam B2 has full lateral support. Take $C_D = 1.25$ (roof live) and estimate the volume factor, $C_V \approx 0.85$.

$$F_b' = C_D C_V F_b = (1.25)(0.85)\left(2400 \frac{\text{lbf}}{\text{in}^2}\right) = 2550 \text{ psi}$$

$$M_{\max} = 0.125 w_{CD} L_{CD}^2 = (0.125)\left(570 \frac{\text{lbf}}{\text{ft}}\right)(36 \text{ ft})^2$$
$$= 92{,}340 \text{ ft-lbf}$$

$$S \ge \frac{M_{\max}}{F_b'} = \frac{(92{,}340 \text{ ft-lbf})\left(12 \frac{\text{in}}{\text{ft}}\right)}{2550 \frac{\text{lbf}}{\text{in}^2}} = 435 \text{ in}^3$$

Try $5\frac{1}{8} \times 22.5 \text{ in}$ ($S = 432 \text{ in}^3$, $I = 4865 \text{ in}^4$). Check the assumed F_b'.

$$C_V = \left(\left(\frac{21}{L}\right)\left(\frac{12}{d}\right)\left(\frac{5.125}{b}\right)\right)^{1/x} \le 1.0$$

$$= \left(\left(\frac{21 \text{ ft}}{36 \text{ ft}}\right)\left(\frac{12 \text{ in}}{22.5 \text{ in}}\right)\left(\frac{5.125 \text{ in}}{5.125 \text{ in}}\right)\right)^{1/10}$$

$$= 0.89$$

$$F_b' = C_D C_V F_b = (1.25)(0.89)\left(2400 \frac{\text{lbf}}{\text{in}^2}\right) = 2670 \text{ psi}$$

$$f_b = \frac{M_{\max}}{S} = \frac{(92{,}340 \text{ ft-lbf})\left(12 \frac{\text{in}}{\text{ft}}\right)}{432 \text{ in}^3} = 2565 \text{ psi} \quad [< F_b']$$

Check the deflection between the hinge points. Per IBC Table 1604.3, the limit is $\Delta_L = L/360$. For plastered ceilings, the limit is $\Delta_{TL} \le L/240$; however, footnote d

permits using the deflection from $0.5D + L$ in lieu of the deflection from the total load.

$$w_{BC} = w_L s = \left(12.5 \; \frac{\text{lbf}}{\text{ft}^2}\right)(16.0 \text{ ft}) = 200 \text{ lbf/ft}$$

$$E' = E = 1{,}800{,}000 \text{ psi}$$

$$\frac{L}{360} = \frac{(36 \text{ ft})\left(12 \; \frac{\text{in}}{\text{ft}}\right)}{360} = 1.2 \text{ in}$$

$$\begin{aligned}
\Delta_L &= \frac{0.013 w_{BC} L_{BC}^4}{E' I} \\
&= \frac{(0.013)\left(200 \; \frac{\text{lbf}}{\text{ft}}\right)(36 \text{ ft})^4 \left(12 \; \frac{\text{in}}{\text{ft}}\right)^3}{\left(1{,}800{,}000 \; \frac{\text{lbf}}{\text{in}^2}\right)(4865 \text{ in}^4)} \\
&= 0.86 \text{ in} \quad [< L/360, \text{ so OK}]
\end{aligned}$$

$$\begin{aligned}
w_{0.5D+L} &= (0.5)\left(\left(20 \; \frac{\text{lbf}}{\text{ft}^2}\right)(16 \text{ ft}) + 50 \; \frac{\text{lbf}}{\text{ft}}\right) + 200 \; \frac{\text{lbf}}{\text{ft}} \\
&= 385 \text{ lbf/ft}
\end{aligned}$$

$$\frac{L}{240} = \frac{(36 \text{ ft})\left(12 \; \frac{\text{in}}{\text{ft}}\right)}{240} = 1.8 \text{ in}$$

$$\begin{aligned}
\Delta_{TL} &= \left(\frac{w_{0.5D+L}}{w_L}\right)\Delta_L = \left(\frac{385 \; \frac{\text{lbf}}{\text{ft}}}{200 \; \frac{\text{lbf}}{\text{ft}}}\right)(0.86 \text{ in}) \\
&= 1.66 \text{ in} \quad [< L/240, \text{ so OK}]
\end{aligned}$$

Check the assumed weight (36 lbf/ft^3).

$$\begin{aligned}
w_{\text{beam}} &= bd w_{\text{wood}} = \left(\frac{(5.125 \text{ in})(22.5 \text{ in})}{\left(12 \; \frac{\text{in}}{\text{ft}}\right)^2}\right)\left(36 \; \frac{\text{lbf}}{\text{ft}^3}\right) \\
&= 29 \text{ lbf/ft} \quad [< 50 \text{ lbf/ft}, \text{ so OK}]
\end{aligned}$$

Check shear.

$$\begin{aligned}
V &= V_{CD} - w_{BC} d = 10{,}260 \text{ lbf} - \left(570 \; \frac{\text{lbf}}{\text{ft}}\right)(2 \text{ ft}) \\
&= 9120 \text{ lbf}
\end{aligned}$$

$$f_v = \frac{1.5 V}{bd} = \frac{(1.5)(9120 \text{ lbf})}{(5.125 \text{ in})(22.5 \text{ in})} = 119 \text{ psi}$$

$$F_v' = C_D F_v = (1.25)\left(265 \; \frac{\text{lbf}}{\text{in}^2}\right) = 331 \text{ psi} \quad [> f_v]$$

Therefore, $\boxed{\text{use } 5\frac{1}{8} \times 22\frac{1}{2} \text{ for beam B2.}}$

8.2. Design the cantilevered member B1. For member ABC, the tributary area is

$$A_T = Ls = (60 \text{ ft})(16 \text{ ft}) = 960 \text{ ft}^2$$

For $A_T > 600 \text{ ft}^2$,

$$R_1 = 0.6$$

$$L_r = L_o R_1 R_2 = \left(20 \; \frac{\text{lbf}}{\text{ft}^2}\right)(0.6)(1.0) = 12 \text{ lbf/ft}^2$$

Per ASCE/SEI7 Table 4.1, footnote n, the reduced live loads must be applied to adjacent and to alternate spans to produce the most unfavorable effects.

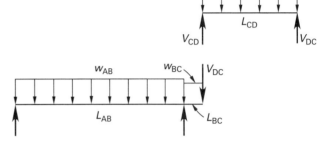

Estimate the member weight as 100 lbf/ft for beam B1. The maximum positive bending occurs with the dead load only on span B to D.

$$\begin{aligned}
w_{AB} &= w_D s + w_{\text{beam}} + w_L s \\
&= \left(20 \; \frac{\text{lbf}}{\text{ft}^2}\right)(16 \text{ ft}) + 100 \; \frac{\text{lbf}}{\text{ft}} + \left(12 \; \frac{\text{lbf}}{\text{ft}^2}\right)(16 \text{ ft}) \\
&= 612 \text{ lbf/ft}
\end{aligned}$$

$$\begin{aligned}
w_{BC} &= w_D s + w_{\text{beam}} = \left(20 \; \frac{\text{lbf}}{\text{ft}^2}\right)(16 \text{ ft}) + 100 \; \frac{\text{lbf}}{\text{ft}} \\
&= 420 \text{ lbf/ft}
\end{aligned}$$

$$\begin{aligned}
w_{CD} &= w_D s + w_{\text{beam}} = \left(20 \; \frac{\text{lbf}}{\text{ft}^2}\right)(16 \text{ ft}) + 50 \; \frac{\text{lbf}}{\text{ft}} \\
&= 370 \text{ lbf/ft}
\end{aligned}$$

$$V_{CD} = 0.5 w_{CD} L_{CD} = (0.5)\left(370 \; \frac{\text{lbf}}{\text{ft}}\right)(36 \text{ ft}) = 6660 \text{ lbf}$$

$$\begin{aligned}
\sum M_A &= R_B L_{AB} - 0.5 w_{AB} L_{AB}^2 - w_{BC} L_{BC}(L_{AC} - 0.5 L_{BC}) \\
&\quad - V_{CD} L_{AC} = 0
\end{aligned}$$

$$R_B = \frac{\begin{array}{c} 0.5 w_{AB} L_{AB}^2 + w_{BC} L_{BC}(L_{AC} - 0.5 L_{BC}) \\ + V_{CD} L_{AC} \end{array}}{L_{AB}}$$

$$= \frac{\begin{array}{c}(0.5)\left(612 \; \frac{\text{lbf}}{\text{ft}}\right)(60 \text{ ft})^2 + \left(420 \; \frac{\text{lbf}}{\text{ft}}\right)(7 \text{ ft}) \\ \times (67 \text{ ft} - (0.5)(7 \text{ ft})) + (6660 \text{ lbf})(67 \text{ ft})\end{array}}{60 \text{ ft}}$$

$$= 28{,}900 \text{ lbf}$$

$$\sum F_y = R_A + R_B - w_{AB}L_{AB} - w_{BC}L_{BC} - V_{CD} = 0$$

$$R_A = w_{AB}L_{AB} + w_{BC}L_{BC} + V_{CD} - R_B$$

$$= \left(612 \frac{\text{lbf}}{\text{ft}}\right)(60 \text{ ft}) + \left(420 \frac{\text{lbf}}{\text{ft}}\right)(7 \text{ ft})$$

$$+ 6660 \text{ lbf} - 28,900 \text{ lbf}$$

$$= 17,420 \text{ lbf}$$

Maximum positive bending moment occurs where shear is zero.

$$V = R_A - w_{AB}x = 0 \text{ lbf}$$

$$x = \frac{R_A}{w_{AB}} = \frac{17,420 \text{ lbf}}{612 \frac{\text{lbf}}{\text{ft}}} = 28.46 \text{ ft}$$

$$M_{\text{max}}^+ = 0.5R_A x = (0.5)(17,420 \text{ lbf})(28.46 \text{ ft})$$

$$= 247,900 \text{ ft-lbf}$$

The distance from the left support to the point of inflection for this loading condition is $2x = (2)(28.46 \text{ ft}) = 57 \text{ ft}$. The critical negative moment and maximum shear occur with the live load on both spans.

$$w_{BC} = w_{AB} = 612 \text{ lbf/ft}$$

$$w_{CD} = w_D s + w_{\text{beam}} + w_L$$

$$= \left(20 \frac{\text{lbf}}{\text{ft}^2}\right)(16 \text{ ft}) + 50 \frac{\text{lbf}}{\text{ft}} + \left(12.5 \frac{\text{lbf}}{\text{ft}^2}\right)(16 \text{ ft})$$

$$= 570 \text{ lbf/ft}$$

$$V_{CD} = 0.5w_{CD}L_{CD} = (0.5)\left(570 \frac{\text{lbf}}{\text{ft}}\right)(36 \text{ ft}) = 10,260 \text{ lbf}$$

$$\sum M_A = R_B L_{AB} - 0.5w_{AB}L_{AC}^2 - V_{CD}L_{AC} = 0$$

$$R_B = \frac{0.5w_{AB}L_{AC}^2 + V_{CD}L_{AC}}{L_{AB}}$$

$$= \frac{(0.5)\left(612 \frac{\text{lbf}}{\text{ft}}\right)(67 \text{ ft})^2 + (10,260 \text{ lbf})(67 \text{ ft})}{60 \text{ ft}}$$

$$= 34,350 \text{ lbf}$$

$$\sum F_y = R_A + R_B - w_{AB}L_{AC} - V_{CD} = 0$$

$$R_A = w_{AB}L_{AC} + V_{CD} - R_B$$

$$= \left(612 \frac{\text{lbf}}{\text{ft}}\right)(67 \text{ ft}) + 10,260 \text{ lbf} - 34,350 \text{ lbf}$$

$$= 16,910 \text{ lbf}$$

Maximum negative bending moment occurs over the support.

$$M_{\text{max}}^- = V_{CD}L_{BC} + 0.5w_{BC}L_{BC}^2$$

$$= (10,260 \text{ lbf})(7 \text{ ft}) + (0.5)\left(612 \frac{\text{lbf}}{\text{ft}}\right)(7 \text{ ft})^2$$

$$= 86,800 \text{ ft-lbf}$$

There are two flexural checks: First, with $l_u = 0$ ft, $M_{\text{max}}^+ = 247,900$ ft-lbf; second, with $l_u = 7$ ft (and $l_e = (1.87)(7 \text{ ft}) = 13.09 \text{ ft}$), $M_{\text{max}} = 86,800$ ft-lbf. Select a trial section based on the positive moment and F_b' approximated as 2200 psi.

$$S \geq \frac{M_{\text{max}}^+}{F_b'} = \frac{(247,900 \text{ ft-lbf})\left(12 \frac{\text{in}}{\text{ft}}\right)}{2200 \frac{\text{lbf}}{\text{in}^2}} = 1352 \text{ in}^3$$

Try $6^3/_4 \times 36$ in ($S = 1458 \text{ in}^3$, $I = 26,240 \text{ in}^4$). The tension region extends from the end support to the point of inflection, but it is conservative and practically correct to use the full span length, 60 ft, to compute the volume factor.

$$C_V = \left(\left(\frac{21}{L}\right)\left(\frac{12}{d}\right)\left(\frac{5.125}{b}\right)\right)^{1/x} \leq 1.0$$

$$= \left(\left(\frac{21 \text{ ft}}{60 \text{ ft}}\right)\left(\frac{12 \text{ in}}{36 \text{ in}}\right)\left(\frac{5.125 \text{ in}}{6.75 \text{ in}}\right)\right)^{1/10}$$

$$= 0.78$$

$$F_b' = C_D C_V F_b = (1.25)(0.78)\left(2400 \frac{\text{lbf}}{\text{in}^2}\right)$$

$$= 2340 \text{ psi}$$

$$f_b = \frac{M_{\text{max}}^+}{S} = \frac{(247,900 \text{ ft-lbf})\left(12 \frac{\text{in}}{\text{ft}}\right)}{1458 \text{ in}^3}$$

$$= 2040 \text{ psi} \quad [< F_b']$$

Check the overhang. For a cantilever with $l_u/d = 84 \text{ in}/36 \text{ in} = 2.3$, and primarily loaded by a concentrated force at its unsupported end, NDS Table 3.3.3 gives $l_e = 187l_u = (1.87)(84 \text{ in}) = 157 \text{ in}$.

$$R_B = \sqrt{\frac{l_e d}{b^2}} = \sqrt{\frac{(157 \text{ in})(36 \text{ in})}{(6.75 \text{ in})^2}} = 11.1$$

$$E_{y,\text{min}}' = E_{y,\text{min}} = 830,000 \text{ psi}$$

$$F_{bx}^* = C_D F_b = (1.25)\left(2400 \frac{\text{lbf}}{\text{in}^2}\right) = 3000 \text{ psi}$$

$$F_{bE} = \frac{1.20E_{y,\text{min}}'}{R_B^2} = \frac{(1.20)\left(830,000 \frac{\text{lbf}}{\text{in}^2}\right)}{(11.1)^2} = 8080 \text{ psi}$$

$$\frac{F_{bE}}{F_{bx}^*} = \frac{8080 \frac{\text{lbf}}{\text{in}^2}}{3000 \frac{\text{lbf}}{\text{in}^2}} = 2.69$$

$$C_L = \frac{1 + \dfrac{F_{bE}}{F_{bx}^*}}{1.9} - \sqrt{\left(\frac{1 + \dfrac{F_{bE}}{F_{bx}^*}}{1.9}\right)^2 - \frac{\dfrac{F_{bE}}{F_{bx}^*}}{0.95}}$$

$$= \frac{1 + 2.69}{1.9} - \sqrt{\left(\frac{1 + 2.69}{1.9}\right)^2 - \frac{2.69}{0.95}} = 0.97$$

The tension region for negative bending is the length of overhang plus distance from support to point of inflection.

$$L = L_{BC} + L_{PI} = 7 \text{ ft} + 3 \text{ ft} = 10 \text{ ft}$$

$$C_V = \left(\left(\frac{21}{L}\right)\left(\frac{12}{d}\right)\left(\frac{5.125}{b}\right)\right)^{1/x} \leq 1.0$$

$$= \left(\left(\frac{21 \text{ ft}}{10 \text{ ft}}\right)\left(\frac{12 \text{ in}}{36 \text{ in}}\right)\left(\frac{5.125 \text{ in}}{6.75 \text{ in}}\right)\right)^{1/10}$$

$$= 0.94 \quad [< C_L, \text{ so } C_V \text{ controls}]$$

A 24F-1.8E stress class glulam for a simple beam layout has an allowable bending stress of 1450 psi when the normal compression region is stressed in tension. On this basis,

$$F_b' = C_D C_V F_b = (1.25)(0.94)\left(1450 \ \frac{\text{lbf}}{\text{in}^2}\right)$$

$$= 1700 \text{ psi}$$

$$f_b = \frac{M_{\max}^-}{S} = \frac{(86,800 \text{ ft-lbf})\left(12 \ \dfrac{\text{in}}{\text{ft}}\right)}{1458 \text{ in}^3}$$

$$= 714 \text{ psi} \quad [< F_b']$$

Therefore, bending is OK at the overhang. On span AB (simple span condition), check deflection with live load only.

$$w_{AB} = w_L s = \left(12.0 \ \frac{\text{lbf}}{\text{ft}^2}\right)(16.0 \text{ ft}) = 192 \text{ lbf/ft}$$

$$E' = E = 1,800,000 \text{ psi}$$

$$\Delta_L = \frac{0.013 w_{BC} L_{BC}^4}{E' I}$$

$$= \frac{(0.013)\left(192 \ \dfrac{\text{lbf}}{\text{ft}}\right)(60 \text{ ft})^4 \left(12 \ \dfrac{\text{in}}{\text{ft}}\right)^3}{\left(1,800,000 \ \dfrac{\text{lbf}}{\text{in}^2}\right)(26,244 \text{ in}^4)}$$

$$= 1.18 \text{ in}$$

$$\frac{L}{360} = \frac{(60 \text{ ft})\left(12 \ \dfrac{\text{in}}{\text{ft}}\right)}{360} = 2.0 \text{ in} \quad [> \Delta_L, \text{ so OK}]$$

The total load deflection is affected by the dead load bending moment at the support.

$$M_B^- = 0.5 w_{BC} L_{BC}^2 + V_{CD} L_{BC}$$

$$= (0.5)\left(429 \ \frac{\text{lbf}}{\text{ft}}\right)(7 \text{ ft})^2 + (6660 \text{ lbf})(7 \text{ ft})$$

$$= 57,100 \text{ ft-lbf}$$

The total load deflection is found by superposition of the deflections caused by uniform total load minus the upward deflection caused by negative bending over the support, which can be approximated at midspan.

$$w_{AB} = w_D s + w_{\text{beam}} + w_L s$$

$$= \left(20 \ \frac{\text{lbf}}{\text{ft}^2}\right)(16 \text{ ft}) + 100 \ \frac{\text{lbf}}{\text{ft}} + \left(12 \ \frac{\text{lbf}}{\text{ft}^2}\right)(16 \text{ ft})$$

$$= 612 \text{ lbf/ft}$$

$$\Delta_{TL} = \frac{0.013 w_{AB} L_{AB}^4}{E' I} - \frac{M_{BA}^- L_{AB}^2}{16 E I}$$

$$= \frac{(0.013)\left(612 \ \dfrac{\text{lbf}}{\text{ft}}\right)(60 \text{ ft})^4 \left(12 \ \dfrac{\text{in}}{\text{ft}}\right)^3}{\left(1,800,000 \ \dfrac{\text{lbf}}{\text{in}^2}\right)(26,244 \text{ in}^4)}$$

$$\quad - \frac{(57,100 \text{ lbf})(60 \text{ ft})^2 \left(12 \ \dfrac{\text{in}}{\text{ft}}\right)^2}{(16)\left(1,800,000 \ \dfrac{\text{lbf}}{\text{in}^2}\right)(26,244 \text{ in}^4)}$$

$$= 3.73 \text{ in}$$

$$\frac{L}{240} = \frac{(60 \text{ ft})\left(12 \ \dfrac{\text{in}}{\text{ft}}\right)}{240} = 3 \text{ in} \quad [< \Delta_{TL}, \text{ not OK}]$$

Therefore, the beam needs to be stiffened.

$$I_{\text{req}} \geq \frac{\Delta_{TL} I_{\text{trial}}}{\Delta_{\text{allowable}}} = \frac{(3.73 \text{ in})(26,244 \text{ in}^4)}{3.0 \text{ in}} = 32,630 \text{ in}^4$$

Select $6\frac{3}{4} \times 39$ in ($I = 33,370$ in^4). Check the assumed weight (36 lbf/ft^3).

$$w_{\text{beam}} = bd w_{\text{wood}} = \left(\frac{6.75 \text{ in}}{12 \ \dfrac{\text{in}}{\text{ft}}}\right)\left(\frac{39 \text{ in}}{12 \ \dfrac{\text{in}}{\text{ft}}}\right)\left(36 \ \frac{\text{lbf}}{\text{ft}^3}\right)$$

$$= 66 \text{ lbf/ft} \quad [< 100 \text{ lbf/ft}, \text{ therefore OK}]$$

Check shear. Maximum shear occurs adjacent to support B for load case B above.

$$V_{max} = R_A - w_{AC}L_{AB} = 16{,}910 \text{ lbf} - \left(612 \frac{\text{lbf}}{\text{ft}}\right)(60 \text{ ft})$$

$$= 19{,}810 \text{ lbf}$$

$$V = V_{max} - w_{AB}d = 19{,}810 \text{ lbf} - \left(612 \frac{\text{lbf}}{\text{ft}}\right)(3.25 \text{ ft})$$

$$= 17{,}820 \text{ lbf}$$

$$f_v = \frac{1.5V}{bd} = \frac{(1.5)(17{,}820 \text{ lbf})}{(6.75 \text{ in})(39.0 \text{ in})} = 102 \text{ psi}$$

$$F'_v = C_D F_v = (1.25)\left(265 \frac{\text{lbf}}{\text{in}^2}\right) = 331 \text{ psi} \quad [>f_v, \text{ so OK}]$$

Therefore, use $6\frac{3}{4} \times 39$ for beam B1.

At this stage, it would be reasonable to revise the design of beam B2 to match the 6.75 in width of beam B1 in order to simplify the connection at the splice. This requires a $6\frac{3}{4} \times 22\frac{1}{2}$, which furnishes $I_x = 6407$ in^4 and $S_x = 569$ in^3.

8.3. Compute the required camber for beam B1. Per *AITC Manual* Table 5.9, the recommended camber is $1.5\Delta_d$.

$$w_{AB} = w_D s + w_{beam} + w_L s$$

$$= \left(20 \frac{\text{lbf}}{\text{ft}^2}\right)(16 \text{ ft}) + 66 \frac{\text{lbf}}{\text{ft}}$$

$$= 386 \text{ lbf/ft}$$

$$\Delta_{TL} = \frac{0.013 w_{AB}L_{AB}^4}{E'I} - \frac{M_{BA}^- L_{AB}^2}{16E'I}$$

$$= \frac{(0.013)\left(386 \frac{\text{lbf}}{\text{ft}}\right)(60 \text{ ft})^4\left(12 \frac{\text{in}}{\text{ft}}\right)^3}{\left(1{,}800{,}000 \frac{\text{lbf}}{\text{in}^2}\right)(33{,}370 \text{ in}^4)}$$

$$- \frac{(57{,}100 \text{ lbf})(60 \text{ ft})^2\left(12 \frac{\text{in}}{\text{ft}}\right)^2}{(16)\left(1{,}800{,}000 \frac{\text{lbf}}{\text{in}^2}\right)(33{,}370 \text{ in}^4)}$$

$$= 1.8 \text{ in}$$

Therefore, the camber for beam B2 is $(1.5)(1.8 \text{ in}) = 2.7 \text{ in } (2.75 \text{ in})$.

SOLUTION 9

9.1. Design the joists spaced at 16 in on center. Try a 2×10 no. 1 Douglas fir-larch (north). From NDS Table 4A, $F_b = 1150$ psi, $F_v = 180$ psi, and $E = 1{,}800{,}000$ psi. For normal load duration, $C_D = 1.0$. From Table 4A, the repetitive use factor is $C_r = 1.15$ and the size factor for flexure is $C_F = 1.1$. Members are exposed to weather and $C_F F_b = (1.1)(1150 \text{ psi}) > 1150 \text{ psi}$;

therefore, from Table 4A, the wet service factors are $C_M = 0.85$ (for flexure), 0.97 (shear), 0.67 (compression perpendicular to grain), and 0.9 (modulus of elasticity). Thus,

$$F'_b = C_D C_M C_F C_r F_b$$

$$= (1.0)(0.85)(1.1)(1.15)\left(1150 \frac{\text{lbf}}{\text{in}^2}\right)$$

$$= 1237 \text{ psi}$$

The tributary area to the joists is less than 200 ft^2; therefore, no reduction in live loads is permitted.

$$w = (w_D + w_L)s = \left(100 \frac{\text{lbf}}{\text{ft}^2} + 20 \frac{\text{lbf}}{\text{ft}^2}\right)(1.33 \text{ ft})$$

$$= 160 \text{ lbf/ft}$$

$$M_{max} = 0.125 wL^2 = (0.125)\left(160 \frac{\text{lbf}}{\text{ft}}\right)(10 \text{ ft})^2$$

$$= 2000 \text{ ft-lbf}$$

$$f_b = \frac{M_{max}}{S} = \frac{(2000 \text{ ft-lbf})\left(12 \frac{\text{in}}{\text{ft}}\right)}{21.4 \text{ in}^3}$$

$$= 1120 \text{ psi} \quad [<F'_b]$$

Check shear (reduce to $d = 0.83$ ft from support).

$$L_e = L - 2d \approx 10 \text{ ft} - (2)(0.83 \text{ ft}) = 8.34 \text{ ft}$$

$$V = 0.5 wL_e = (0.5)\left(160 \frac{\text{lbf}}{\text{ft}}\right)(8.34 \text{ ft}) = 667 \text{ lbf}$$

$$f_v = \frac{1.5V}{bd} = \frac{(1.5)(667 \text{ lbf})}{(1.5 \text{ in})(9.25 \text{ in})} = 72 \text{ psi}$$

$$F'_v = C_M F_v = (0.97)\left(180 \frac{\text{lbf}}{\text{in}^2}\right)$$

$$= 174.6 \text{ psi} \quad [f_v < F'_v, \text{ so OK}]$$

Therefore, OK in shear. Limit the live load deflection to $L/360$.

$$w = w_L s = \left(100 \frac{\text{lbf}}{\text{ft}^2}\right)(1.33 \text{ ft}) = 133 \text{ lbf/ft}$$

$$E' = C_M E = (0.9)\left(1{,}800{,}000 \frac{\text{lbf}}{\text{in}^2}\right) = 1{,}620{,}000 \text{ psi}$$

$$\Delta_L = \frac{0.013 wL^4}{E'I}$$

$$= \frac{(0.013)\left(133 \frac{\text{lbf}}{\text{ft}}\right)(10 \text{ ft})^4\left(12 \frac{\text{in}}{\text{ft}}\right)^3}{\left(1{,}620{,}000 \frac{\text{lbf}}{\text{in}^2}\right)(98.9 \text{ in}^4)}$$

$$= 0.19 \text{ in}$$

$$\frac{L}{360} = \frac{120 \text{ in}}{360} = 0.33 \text{ in} \quad [>\Delta_L, \text{ so OK}]$$

Serviceability is OK. Check bearing length required at supports.

$$R = 0.5wL = (0.5)\left(160\ \frac{\text{lbf}}{\text{ft}}\right)(10\ \text{ft}) = 800\ \text{lbf}$$

$$F'_{c\perp} = C_M F_{c\perp} = (0.67)\left(625\ \frac{\text{lbf}}{\text{in}^2}\right) = 419\ \text{psi}$$

$$N \geq \frac{R}{F'_{c\perp} b} = \frac{800\ \text{lbf}}{\left(419\ \frac{\text{lbf}}{\text{in}^2}\right)(1.5\ \text{in})} = 1.3\ \text{in}\quad[\text{say 1.5 in}]$$

Therefore, use 2×10 ($S = 21.4\ \text{in}^3$, $I = 98.9\ \text{in}^4$) with a minimum 1.5 in bearing.

9.2. Compute the shear that must transfer between the two cantilevered beams at their intersection.

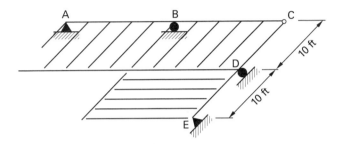

To simplify the representation, develop the two cantilevered beams into a planar idealization.

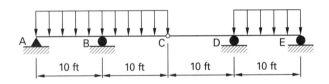

The influence line (qualitative sketch) for maximum shear at C is

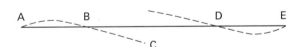

Thus, to produce maximum shear transfer, apply live and dead load on spans $\overline{BC}$ and $\overline{DE}$, and apply the dead load only on span $\overline{AB}$. The tributary area, A_T, is $(10\ \text{ft})(10\ \text{ft}) = 100\ \text{ft}^2$; $K_{LL}A_T = (1)(100\ \text{ft}^2) < 400\ \text{ft}^2$; therefore, no live load reduction is allowed (ASCE/SEI7 Sec. 4.7.1). The dead load reaction from the joists on span $\overline{AB}$ is

$$w = 0.5w_D L = (0.5)\left(20\ \frac{\text{lbf}}{\text{ft}^2}\right)(10\ \text{ft}) = 100\ \text{lbf/ft}$$

Joists span parallel to $\overline{CD}$; therefore, the load on member $\overline{CD}$ is zero. For spans, the combined dead and live load is

$$w = 0.5(w_D + w_L)L$$
$$= (0.5)\left(20\ \frac{\text{lbf}}{\text{ft}^2} + 100\ \frac{\text{lbf}}{\text{ft}^2}\right)(10\ \text{ft})$$
$$= 600\ \text{lbf/ft}$$

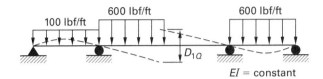

The structure is statically indeterminate to one degree. Treat the shear between the two cantilever tips as unknown, and use the consistent displacement method to solve for the shear. Release the connection at point C, letting D_{1Q} equal the relative deflection between tips of the cantilevers caused by the applied forces. The deflection at the tip is

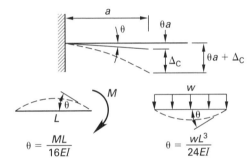

Using the deflection equations for cases 24, 25, and 27 from the AISC Table 3-23, and expressing in terms of $a = L = 10$ ft, the relative deflection between the released tips is

$$D_{1Q} = \frac{w_{AC}(6a^4) + w_{BC}(7a^4) + w_{DE}a^4}{24EI}$$

$$= \frac{\left(100\ \frac{\text{lbf}}{\text{ft}}\right)(6)(10\ \text{ft})^4 + \left(500\ \frac{\text{lbf}}{\text{ft}}\right)(7)}{24EI}$$
$$\times (10\ \text{ft})^4 + \left(600\ \frac{\text{lbf}}{\text{ft}}\right)(10\ \text{ft})^4$$

$$= \frac{1.958333 \times 10^6}{EI}\ \text{ft}^3\text{-lbf}$$

Apply a unit shear at C in the released structure.

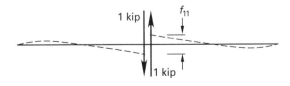

The relative displacement between the tips is found using case 26 from the *AISC Manual*, with $P = 1$ lbf and $a = L = 10$ ft.

$$f_{11} = 2\left(\frac{P(4a^3)}{6EI}\right) = (2)\left(\frac{(1 \text{ lbf})(4)(10 \text{ ft})^3}{6EI}\right)$$

$$= \frac{1333.33}{EI} \text{ ft}^3\text{-lbf (per lbf)}$$

For consistent displacement,

$$D_{1Q} + V_C f_{11} = 0 \text{ ft}^3\text{-lbf}$$

$$V_C = \frac{-D_{1Q}}{f_{11}} = \frac{\dfrac{-1.958333 \times 10^6}{EI} \text{ ft}^3\text{-lbf}}{\dfrac{1333.33}{EI} \text{ ft}^3\text{-lbf (per lbf)}}$$

$$= -1470 \text{ lbf}$$

Therefore, the required shear transfer is

$$V_C = \boxed{1.47 \text{ kips}}$$

9.3. Design the glulam beams for appropriate strength and deflection limits. Critical bending moment occurs at support B with dead load plus live load on spans $\overline{AB}$ and $\overline{BC}$, and with dead load only on span $\overline{DE}$.

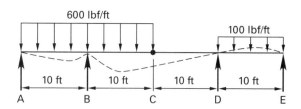

600 lbf/ft

100 lbf/ft

10 ft   10 ft   10 ft   10 ft

A   B   C   D   E

As in Sol. 9.2, release shear at joint C, and use the equations from AISC Table 3-23.

$$D_{1Q} = \frac{w_{AC}(6a^4) + w_{DE}a^4}{24EI}$$

$$= \frac{\left(600 \dfrac{\text{lbf}}{\text{ft}}\right)(6)(10 \text{ ft})^4 + \left(100 \dfrac{\text{lbf}}{\text{ft}}\right)(10 \text{ ft})^4}{24EI}$$

$$= \frac{1{,}541{,}666.7}{EI} \text{ ft}^3\text{-lbf}$$

Thus, for consistent displacement,

$$D_{1Q} + V_C f_{11} = 0 \text{ ft}^3\text{-lbf}$$

$$V_C = \frac{-D_{1Q}}{f_{11}} = \frac{\dfrac{-1{,}541{,}666.7}{EI} \text{ ft}^3\text{-lbf}}{\dfrac{1333.33}{EI} \text{ ft}^3\text{-lbf (per lbf)}}$$

$$= -1156 \text{ lbf}$$

Thus,

$$M_B = 0.5w_{AC}a^2 - V_C a$$

$$= (0.5)\left(600 \dfrac{\text{lbf}}{\text{ft}}\right)(10 \text{ ft})^2 - (1156 \text{ lbf})(10 \text{ ft})$$

$$= 18{,}440 \text{ ft-lbf}$$

Per NDS Table 5.3.1, for bending under normal load duration, the smaller of C_L and C_V applies in combination with a wet service factor that is 0.8 for bending and 0.833 for $E_{y,\text{min}}$ (NDS Table 5A). Compute a trial size assuming an adjusted bending stress of

$$F_b' = (0.8)(0.9)\left(2400 \dfrac{\text{lbf}}{\text{in}^2}\right) = 1730 \text{ psi}$$

$$S_{\text{trial}} = \frac{M_B}{F_b'} = \frac{(18{,}440 \text{ ft-lbf})\left(12 \dfrac{\text{in}}{\text{ft}}\right)}{1730 \dfrac{\text{lbf}}{\text{in}^2}} = 128 \text{ in}^3$$

Try a $3^1/2 \times 15$ glulam ($S_x = 131 \text{ in}^3$, $I_x = 984 \text{ in}^4$). For a uniformly loaded, cantilevered 10 ft span, NDS Table 3.3.3 gives

$$\frac{l_u}{d} = \frac{120 \text{ in}}{15 \text{ in}} = 8 \quad [> 7]$$

$$l_e = 0.9l_u + 3d = (0.9)(120 \text{ in}) + (3)(15 \text{ in}) = 153 \text{ in}$$

$$R_B = \sqrt{\frac{l_e d}{b^2}} = \sqrt{\frac{(153 \text{ in})(15 \text{ in})}{(3.5 \text{ in})^2}} = 13.7$$

$$E_{y,\text{min}}' = C_M E_{y,\text{min}} = (0.833)\left(850{,}000 \dfrac{\text{lbf}}{\text{in}^2}\right)$$

$$= 708{,}000 \text{ psi}$$

$$F_{bx}^* = C_D C_M F_b = (1.0)(0.8)\left(2400 \dfrac{\text{lbf}}{\text{in}^2}\right)$$

$$= 1920 \text{ psi}$$

$$F_{bE} = \frac{1.20 E_{y,\text{min}}'}{R_B^2} = \frac{(1.20)\left(708{,}000 \dfrac{\text{lbf}}{\text{in}^2}\right)}{(13.7)^2}$$

$$= 4530 \text{ psi}$$

$$\frac{F_{bE}}{F_{bx}^*} = \frac{4530 \dfrac{\text{lbf}}{\text{in}^2}}{1920 \dfrac{\text{lbf}}{\text{in}^2}} = 2.36$$

$$C_L = \frac{1 + \dfrac{F_{bE}}{F_{bx}^*}}{1.9} - \sqrt{\left(\frac{1 + \dfrac{F_{bE}}{F_{bx}^*}}{1.9}\right)^2 - \frac{\dfrac{F_{bE}}{F_{bx}^*}}{0.95}}$$

$$= \frac{1 + 2.36}{1.9} - \sqrt{\left(\frac{1 + 2.36}{1.9}\right)^2 - \frac{2.36}{0.95}}$$

$$= 0.97$$

Compute the volume factor, C_V, assuming that the tension flange extends essentially the full length (20 ft) of the member (conservative).

$$C_V = \left(\left(\frac{21}{L}\right)\left(\frac{12}{d}\right)\left(\frac{5.125}{b}\right)\right)^{1/x}$$

$$= \left(\left(\frac{21\text{ ft}}{20\text{ ft}}\right)\left(\frac{12\text{ in}}{15\text{ in}}\right)\left(\frac{5.125\text{ in}}{3.5\text{ in}}\right)\right)^{1/10}$$

$$= 1.02 \quad [\geq 1.0]$$

The adjusted allowable bending stress is

$$F_b' = C_D C_M C_L F_b = (1.0)(0.8)(0.97)\left(2400\ \frac{\text{lbf}}{\text{in}^2}\right)$$

$$= 1862\text{ psi}$$

Thus, bending is OK. Check shear (per NDS Table 5A, $C_M = 0.875$).

$$V = w_{AC}(a - d) - V_C$$

$$= \left(600\ \frac{\text{lbf}}{\text{ft}}\right)(10\text{ ft} - 1.25\text{ ft}) - 1156\text{ lbf}$$

$$= 4094\text{ lbf}$$

$$f_v = \frac{1.5V}{bd} = \frac{(1.5)(4094\text{ lbf})}{(3.5\text{ in})(15.0\text{ in})} = 117\text{ psi}$$

$$F_v' = C_D C_M F_v = (1.0)(0.875)\left(265\ \frac{\text{lbf}}{\text{in}^2}\right)$$

$$= 232\text{ psi} \quad [> f_v]$$

Limit live load deflection to $L'/360$. For live load only on span BC of the released structure, AISC Table 3-23, case 26, gives

$$D_{1Q} = \frac{w_{BC}(7a^4)}{24EI} = \frac{\left(500\ \frac{\text{lbf}}{\text{ft}}\right)(7)(10\text{ ft})^4}{24EI}$$

$$= \frac{1{,}458{,}333.33}{EI}\text{ ft}^3\text{-lbf}$$

$$D_{1Q} + V_C f_{11} = 0\text{ ft}^3\text{-lbf}$$

$$V_C = \frac{-D_{1Q}}{f_{11}} = \frac{-\dfrac{1{,}458{,}333.33}{EI}\text{ ft}^3\text{-lbf}}{\dfrac{1333.33}{EI}\text{ ft}^3\text{-lbf (per lbf)}}$$

$$= -1093\text{ lbf}$$

The live load deflection at the tip of the cantilever is found by superimposing the deflections caused by the uniform live load with the shear at the tip.

$$\Delta_L = \frac{w_{BC}(7a^4)}{24EI} - \frac{V_C(4a^3)}{6EI}$$

$$= \frac{\left(500\ \frac{\text{lbf}}{\text{ft}}\right)(7)(10\text{ ft})^4}{24EI} - \frac{(1093\text{ lbf})(4)(10\text{ ft})^3}{6EI}$$

$$= \frac{729{,}666.67}{EI}\text{ ft}^3\text{-lbf}$$

For deflection, use $E' = C_M E = (0.833)(1{,}800{,}000\text{ psi}) = 1{,}500{,}000\text{ psi}$. For the cantilever, limit the deflection to $L'/360$, where $L' = 2L = (2)(10\text{ ft}) = 20$ ft. Find the required moment of inertia.

$$\Delta_L = \frac{729{,}666.67}{E'I}\text{ ft}^3\text{-lbf} \leq \frac{L'}{360}$$

$$I \geq \frac{(729{,}666.67)(360)}{E'L'}\text{ ft}^3\text{-lbf}$$

$$\geq \frac{(729{,}666.67)(360)\left(12\ \dfrac{\text{in}}{\text{ft}}\right)^3}{\left(1{,}500{,}000\ \dfrac{\text{lbf}}{\text{in}^2}\right)(20\text{ ft})\left(12\ \dfrac{\text{in}}{\text{ft}}\right)}\text{ ft}^3\text{-lbf}$$

$$\geq 1261\text{ in}^4$$

Thus, serviceability governs design. Select a stiffer $3\frac{1}{2} \times 16\frac{1}{2}$ glulam ($I_x = 1310\text{ in}^4$), which is adequate for bending and shear.

$$\boxed{\text{Use } 3\tfrac{1}{2} \times 16\tfrac{1}{2} \text{ stress class 24F-1.8E.}}$$

SOLUTION 10

10.1. Check member ABC of the roof truss.

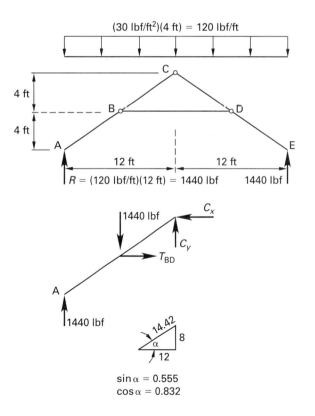

The dead and roof live loads are given as service loads per lineal foot of horizontal projection. The equivalent load on the truss is

$$w = (w_{D+L})s = \left(30 \; \frac{\text{lbf}}{\text{ft}^2}\right)(4 \text{ ft}) = 120 \text{ lbf/ft}$$

From a free-body diagram of the left rafter,

$$\sum M_C = 0.5w(0.5L)^2 - R_A(0.5L) - T_{BD}(0.5h) = 0$$

$$T_{BD} = \frac{0.5w(0.5L)^2 - R_A(0.5L)}{0.5h}$$

$$= \frac{(0.5)\left(120 \; \frac{\text{lbf}}{\text{ft}}\right)\left((0.5)(24 \text{ ft})\right)^2 - (1440 \text{ lbf})(0.5)(24 \text{ ft})}{(0.5)(8 \text{ ft})}$$

$$= 2160 \text{ lbf} \quad [\text{tension}]$$

$$\sum F_x = C_x + T_{BD} = 0$$

$$C_x = -T_{BD} = -2160 \text{ lbf} \quad [\text{to the left}]$$

$$\sum F_y = R_A - w(0.5L) + C_y = 0$$

$$C_y = w(0.5L) - R_A$$

$$= \left(120 \; \frac{\text{lbf}}{\text{ft}}\right)(0.5)(24 \text{ ft}) - 1440 \text{ lbf}$$

$$= 0$$

Resolve the forces into components parallel to and perpendicular to the rafter.

$$\sin \alpha = \frac{\text{opp}}{\text{hyp}} = \frac{8 \text{ ft}}{\sqrt{8 \text{ ft}^2 + 12 \text{ ft}^2}} = 0.555$$

$$\cos \alpha = \frac{\text{adj}}{\text{hyp}} = \frac{12 \text{ ft}}{\sqrt{8 \text{ ft}^2 + 12 \text{ ft}^2}} = 0.832$$

$$w_\perp = w \cos^2 \alpha = \left(120 \; \frac{\text{lbf}}{\text{ft}}\right)(0.832)^2 = 83 \text{ lbf/ft}$$

$$w_\parallel = w \cos \alpha \sin \alpha = \left(120 \; \frac{\text{lbf}}{\text{ft}}\right)(0.832)(0.555)$$

$$= 55.4 \text{ lbf/ft}$$

Critical axial compression and bending moment occurs at midlength where the crosstie connects to the rafter. Shear, moment, and axial forces for the left rafter are as follows.

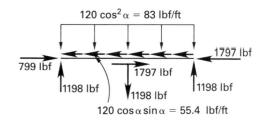

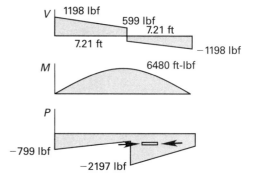

Check the rafter using ASD. For select structural Douglas fir-larch, NDS Table 4A gives $F_b = 1500$ psi, $F_v = 180$ psi, $F_t = 1000$ psi, $F_{c\perp} = 625$ psi, $F_c = 1700$ psi, $E = 1,900,000$ psi, and $E_{min} = 690,000$ psi. The trussed rafters are spaced 4 ft o.c.; therefore, they do not qualify for the repetitive use factor, C_r, for establishing bending stress. For the roof live load, $C_D = 1.25$ and $C_F = 1.0$ for all quantities for 3×12. For the 3×12 rafter, $b = 2.5$ in, $d - 11.25$ in, $A = 28.1$ in^3, $S_x = 52.7$ in^3, and $I_x = 297$ in^4. Check shear.

$$V = V_A - w_\perp d = 1198 \text{ lbf} - \left(83 \; \frac{\text{lbf}}{\text{ft}}\right)\left(\frac{11.25 \text{ in}}{12 \; \frac{\text{in}}{\text{ft}}}\right)$$

$$= 1120 \text{ lbf}$$

$$f_v = \frac{1.5 V}{bd} = \frac{(1.5)(1120 \text{ lbf})}{(2.5 \text{ in})(11.25 \text{ in})}$$

$$= 60 \text{ psi} \quad [< F'_v, \text{ so OK}]$$

Check combined axial and bending stress at the critical point (point B). The rafter is braced about its weak axis and at its ends and midspan for bending about the strong axis.

$$l_e = 0.5l = (0.5)(14.42 \text{ ft}) = 7.21 \text{ ft}$$

$$\frac{l_e}{d} = \frac{(7.21 \text{ ft})\left(12 \; \frac{\text{in}}{\text{ft}}\right)}{11.25 \text{ in}} = 7.7$$

$$F_c^* = C_D C_F F_c = (1.25)(1.0)\left(1700 \; \frac{\text{lbf}}{\text{in}^2}\right) = 2125 \text{ psi}$$

$$E'_{min} = E_{min} = 690,000 \text{ psi}$$

For NDS Sec. 3.7.1, with $c = 0.8$ for sawn lumber,

$$F_{cE} = \frac{0.822 E'_{\min}}{\left(\frac{l_e}{d}\right)^2} = \frac{(0.822)\left(690{,}000 \ \frac{\text{lbf}}{\text{in}^2}\right)}{(7.7)^2} = 9566 \text{ psi}$$

$$\frac{F_{cE}}{F^*_c} = \frac{9566 \ \frac{\text{lbf}}{\text{in}^2}}{2125 \ \frac{\text{lbf}}{\text{in}^2}} = 4.50$$

$$C_P = \frac{1 + \frac{F_{cE}}{F^*_c}}{2c} - \sqrt{\left(\frac{1 + \frac{F_{cE}}{F^*_c}}{2c}\right)^2 - \frac{\frac{F_{cE}}{F^*_c}}{c}}$$

$$= \frac{1 + 4.50}{(2)(0.8)} - \sqrt{\left(\frac{1 + 4.50}{(2)(0.8)}\right)^2 - \frac{4.50}{0.8}}$$

$$= 0.95$$

$$F'_c = C_P C_D C_F F_c = (0.95)(1.25)(1.0)\left(1700 \ \frac{\text{lbf}}{\text{in}^2}\right)$$

$$= 2019 \text{ psi}$$

For bending about the strong axis, $F_{cE1} = F_{cE} = 9566$ psi. NDS Sec. 3.9.2 requires

$$\left(\frac{f_c}{F'_c}\right)^2 + \frac{f_{b1}}{F'_{b1}\left(1 - \frac{f_c}{F_{cE1}}\right)} \le 1.0$$

$$f_c = \frac{P}{A} = \frac{2197 \text{ lbf}}{28.1 \text{ in}^2} = 78.2 \text{ psi}$$

$$f_{b1} = \frac{M_{\max}}{S_x} = \frac{(6480 \text{ ft-lbf})\left(12 \ \frac{\text{in}}{\text{ft}}\right)}{52.7 \text{ in}^3} = 1476 \text{ psi}$$

$$F'_{b1} = (1.25)\left(1500 \ \frac{\text{lbf}}{\text{in}^2}\right) = 1875 \text{ psi}$$

$$\left(\frac{f_c}{F'_c}\right)^2 + \frac{f_{b1}}{F'_{b1}\left(1 - \frac{f_c}{F_{cE1}}\right)}$$

$$= \left(\frac{78.2 \ \frac{\text{lbf}}{\text{in}^2}}{2019 \ \frac{\text{lbf}}{\text{in}^2}}\right)^2 + \frac{1476 \ \frac{\text{lbf}}{\text{in}^2}}{\left(1875 \ \frac{\text{lbf}}{\text{in}^2}\right)\left(1 - \frac{78.2 \ \frac{\text{lbf}}{\text{in}^2}}{9566 \ \frac{\text{lbf}}{\text{in}^2}}\right)}$$

$$= 0.80 \quad [< 1.0, \text{ so OK}]$$

Check transverse deflection under the total load.

$$\text{limit} = \frac{L}{120} = \frac{(14.42 \text{ ft})\left(12 \ \frac{\text{in}}{\text{ft}}\right)}{120} = 1.44 \text{ in}$$

$$\Delta_{\text{TL}} = \frac{0.013 wL^4 + 0.02083 PL^3}{EI}$$

$$= \frac{\left(\begin{array}{c}(0.013)\left(83 \ \frac{\text{lbf}}{\text{ft}}\right)(14.42 \text{ ft})^4 \\ + (0.02083)(1198 \text{ lbf})(14.42 \text{ ft})^3\end{array}\right)\left(12 \ \frac{\text{in}}{\text{ft}}\right)^3}{\left(1{,}900{,}000 \ \frac{\text{lbf}}{\text{in}^2}\right)(297 \text{ in}^4)}$$

$$= 0.37 \text{ in} \quad [< L/120 = 1.44 \text{ in}]$$

Therefore, the 3×12 rafter is OK.

10.2. Design the connection of the 2×6 to the 3×12 rafter. Try $2^1/_2$ in diameter split rings in single shear.

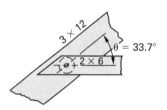

Per NDS Table 12.2A, for group B species, bearing 33.7° to grain,

$$Z_{\parallel} = 2730 \text{ lbf}$$

$$Z_{\perp} = 1940 \text{ lbf}$$

$$Z = \frac{Z_{\parallel} Z_{\perp}}{Z_{\parallel} \sin^2 \theta + Z_{\perp} \cos^2 \theta}$$

$$= \frac{(2730 \text{ lbf})(1940 \text{ lbf})}{(2730 \text{ lbf})(\sin^2 33.3°) + (1940 \text{ lbf})(\cos^2 33.3°)}$$

$$= 2432 \text{ lbf}$$

$$Z' = C_D Z = (1.25)(2432 \text{ lbf})$$

$$= 3040 \text{ lbf} \quad [> T_{\text{BD}} = 2160 \text{ lbf}]$$

$$\% \text{ loaded} = \frac{2160 \text{ lbf}}{3040 \text{ lbf}} \times 100\%$$

$$= 71\% \quad [4 \text{ in end distance is adequate}]$$

Use one $2\frac{1}{2}$ in diameter split ring.

Check the end joint shear in the rafter (per NDS Sec. 3.4.3).

$$d_e = 0.5(d + d_{\text{ring}}) = (0.5)(11.25 \text{ in} + 2.5 \text{ in}) = 6.9 \text{ in}$$

$$f_v = \frac{1.5V}{bd_e} = \frac{(1.5)(599 \text{ lbf})}{(2.5 \text{ in})(6.9 \text{ in})}$$

$$= 52 \text{ psi} \quad [< F'_v, \text{ so OK}]$$

Check tension in the collar tie (per NDS App. K, $^9/_{16}$ in bolt hole diameter and 1.10 in^2 projected area of split ring).

$$A_n = A - A_{ring} - (b - t_{ring})d_h$$
$$= (1.5 \text{ in})(5.5 \text{ in}) - 1.1 \text{ in}^2$$
$$- (1.5 \text{ in} - 0.375 \text{ in})(0.56 \text{ in})$$
$$= 6.5 \text{ in}^2$$
$$f_t = \frac{T}{A_n} = \frac{2160 \text{ lbf}}{6.5 \text{ in}^2} = 333 \text{ psi} \quad [< F'_t, \text{ so OK}]$$

The connection is adequate.

SOLUTION 11

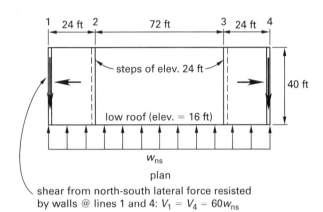

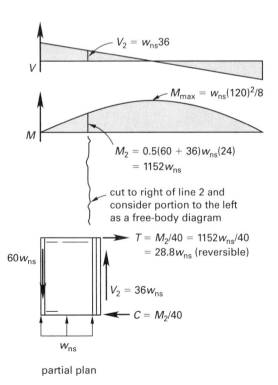

The stepped diaphragm has a vertical irregularity, an in-plane discontinuity in vertical lateral force resistance, as defined in ASCE/SEI7 Table 12.3-2. The elements that support the discontinuous walls at lines 2 and 3, as well as the connections, must be designed for load combinations with overstrength factors per ASCE/SEI7 Sec. 12.3.3.4. Forces calculated in the following solutions are based on the given lateral forces with the understanding that the overstrength and redundancy factors will be applied in assessing the load combinations for element design.

11.1. Name two structural elements needed to maintain diaphragm integrity at the step (lines 2 and 3).

First, a vertical shear wall to transfer the shear from the low roof diaphragm, $V_2 = 36w_{ns}$, up to the sloped roof diaphragm at elevation 24 ft.

Second, the chord force, $T = C = 28.8w_{ns}$, must transfer from the low roof at elevation 16 ft up to the sloped boundary member 8 ft above.

11.2. For a given lateral load $w_{ns} = 190$ lbf/ft (working stress level), determine the forces to transfer at lines 2 and 3.

$$V_{\text{line 1}} = V_{\text{line 4}} = 0.5wL = (0.5)\left(190 \frac{\text{lbf}}{\text{ft}}\right)(120 \text{ ft})$$
$$= 11,400 \text{ lbf}$$
$$V_{\text{line 2}} = V_{\text{line 3}} = 0.5wL' = (0.5)\left(190 \frac{\text{lbf}}{\text{ft}}\right)(72 \text{ ft})$$
$$= 6840 \text{ lbf}$$
$$M_{max} = 0.125w_{ns}L^2 = (0.125)\left(190 \frac{\text{lbf}}{\text{ft}}\right)(120 \text{ ft})^2$$
$$= 342,000 \text{ ft-lbf}$$
$$M_2 = M_3 = 0.5wLx - 0.5wx^2$$
$$= (0.5)\left(190 \frac{\text{lbf}}{\text{ft}}\right)(120 \text{ ft})(24 \text{ ft})$$
$$- (0.5)\left(190 \frac{\text{lbf}}{\text{ft}}\right)(24 \text{ ft})^2$$
$$= 219,000 \text{ ft-lbf}$$

The required force transfers for elements of the lateral force resisting system (without increases for overstrength and redundancy factors) are as follows.

First, for low roof diaphragm at lines 2 and 3,

$$v = \frac{V_{\text{line 2}}}{B} = \frac{6840 \text{ lbf}}{40 \text{ ft}} = 171 \text{ lbf/ft}$$

Second, beams at lines 2 and 3 serve as collectors for the 171 lbf/ft low roof diaphragm shear and they support the shear walls (gravity load plus overturning moments) that transfer the shear up to elevation 28 ft.

Third, two shear walls (each 8 ft long) between elevations 16 ft and 24 ft on lines 2 and 3 must transfer a total horizontal shear in each wall equal to

$$V_{\text{wall}} = 0.5vB = (0.5)\left(171\ \frac{\text{lbf}}{\text{ft}}\right)(40\ \text{ft}) = 3420\ \text{lbf}$$

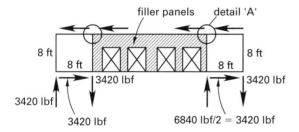

The nominal shear in each shear wall is

$$v = \frac{V_{\text{wall}}}{L_{\text{wall}}} = \frac{3420\ \text{lbf}}{8\ \text{ft}} = 428\ \text{lbf/ft}$$

The top plate over the full 40 ft width will transfer the diaphragm shear, 171 lbf/ft, into the sloped diaphragm. Connection of the top plate to the shear wall (detail A) must transfer

$$\begin{aligned}
F_{\text{drag}} &= v(0.5B - L_{\text{wall}}) \\
&= \left(171\ \frac{\text{lbf}}{\text{ft}}\right)\big((0.5)(40\ \text{ft}) - 8\ \text{ft}\big) \\
&= 2050\ \text{lbf}
\end{aligned}$$

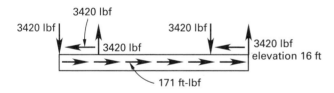

Fourth, end walls (north and south between lines 1 and 2) must transfer the chord force from the low roof diaphragm to the higher roof at elevation 24 ft. The chord forces at the step locations are

$$T = C = \frac{M_2}{B} = \frac{219{,}000\ \text{ft-lbf}}{40\ \text{ft}} = 5480\ \text{lbf}$$

The transfer results in an overturning moment of (8 ft)(5480 lbf) = 43,840 ft-lbf that must be resisted (e.g., by a couple F_1 and F_2 as indicated in the figure below). This will be difficult given the location of the window and door openings and that the load combinations with overstrength factor apply.

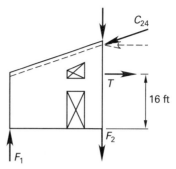

Fifth, the sloped diaphragms between lines 1 and 2 and between lines 3 and 4 must resist

$$v = \frac{V_{\text{line 1}}}{B} = \frac{11{,}400\ \text{lbf}}{40\ \text{ft}} = 285\ \text{lbf/ft}$$

Sixth, the end shear wall at lines 1 and 4 must transfer the 11,400 lbf shear from elevation 16 ft to the base.

11.3. The stepped diaphragm amplifies and complicates the force transfers needed to bring the lateral force from the roof to the ground. Costs and failure risks are greater for the stepped diagram than they would be for a horizontal diaphragm at one level. The ideal case would be to maintain the diaphragm at the level of the low roof. If this cannot be done (as is the case for this problem), a better structural solution is to provide shear transfer to the ground at lines 2 and 3 in addition to lines 1 and 4, perhaps by rigid frames spanning 40 ft at those locations.

SOLUTION 12

12.1. Determine the nailing required for the plywood wall sheathing. Use the ASD design method.

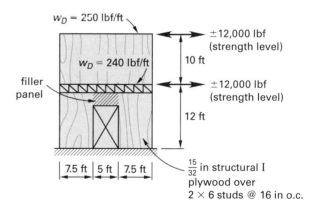

For the upper level, with the full length of the wall resisting shear,

$$v = \frac{0.7V_2}{L_2} = \frac{(0.7)(12{,}000\ \text{lbf})}{20\ \text{ft}} = 420\ \text{lbf/ft}$$

From SDPWS Table 4.3A, for $^{15}/_{32}$ in structural I sheathing, use 10d nails at 4 in o.c. along panel edges and 10d nails at 12 in o.c. at intermediate supports to give an allowable shear, v, of 510 lbf/ft for seismic. For a 10d common nail through $\frac{1}{2}$ in sheathing, penetration is $2\frac{1}{2}$ in, so this is OK. When the allowable shear value exceeds 350 lbf/ft for seismic design category D, E, or F, the SDPWS requires the framing members receiving nailing from abutting panels to have at least a nominal 3 in width. Therefore, change the 2×6 studs to 3×6 studs within the length of the shear wall, and use 3×6 blocking at the horizontal panel edges.

For the bottom story,

$$v = \frac{0.7(V_2 + V_1)}{2L_1} = \frac{(0.7)(12{,}000 \text{ lbf} + 12{,}000 \text{ lbf})}{(2)(7.5 \text{ ft})}$$

$$= 1120 \text{ lbf/ft}$$

The shear is too large to resist with sheathing on only one side, so sheath both sides ($v = (0.5)(1120 \text{ lbf/ft}) = 560 \text{ lbf/ft}$ each side). For 3×6 studs, with $^{15}/_{32}$ in structural I sheathing, use 10d nails at 3 in o.c. along panel edges and 10d nails at 12 in o.c. at intermediate supports to give $v = 665$ lbf/ft. There are several alternatives for designing the lower shear wall containing the door opening (see Breyer et al., Chap. 10). Generally, it is possible to couple the two walls by designing the panel over the door opening to transfer forces between the two side panels. Considering the magnitude of the overturning moment on the lower wall and the resulting chord forces, it is unlikely that the 5 ft × 5 ft panel could transfer the vertical shear between the two side panels. Therefore, design each side panel to act independently to resist one-half of the total lateral force and use the panel over the door as a filler panel with sheathing on one side and minimum nailing (6d nails at 6 in o.c. at edges and at 12 in o.c. at intermediate supports).

12.2. Design the boundary members for the lower walls.

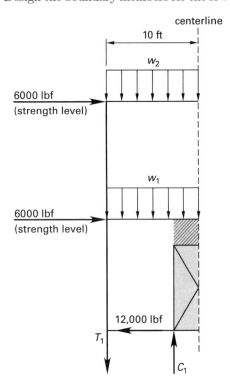

For combined seismic and gravity forces, from ASCE/SEI7 Sec. 12.4.2.3, the maximum axial compression in the shear wall chord is controlled by either case 5 ($D + 0.7E$) or case 6 ($D + 0.75(0.7E) + 0.75(L_r$ or L)). For the lower level, case 5 (with $\rho = 1.0$) gives

$$w_2 = w_{D2}(1 + 0.14S_{DS}) = \left(250 \ \frac{\text{lbf}}{\text{ft}}\right)\big(1.0 + (0.14)(1.0)\big)$$

$$= 285 \text{ lbf/ft}$$

$$w_1 = w_{D1}(1 + 0.14S_{DS}) = \left(240 \ \frac{\text{lbf}}{\text{ft}}\right)\big(1.0 + (0.14)(1.0)\big)$$

$$= 274 \text{ lbf/ft}$$

$$C_1 = \frac{0.5(w_2 + w_1)L^2 + 0.7V_2(h_2 + h_1) + 0.7V_1h_1}{L_1}$$

$$= \frac{\begin{array}{c}(0.5)\left(285 \ \dfrac{\text{lbf}}{\text{ft}} + 274 \ \dfrac{\text{lbf}}{\text{ft}}\right)(10 \text{ ft})^2 \\ + (0.7)(6000 \text{ lbf})(10 \text{ ft} + 12 \text{ ft}) \\ + (0.7)(6000 \text{ lbf})(12 \text{ ft})\end{array}}{7.5 \text{ ft}}$$

$$= 22{,}800 \text{ lbf}$$

Load case 6 with floor live load taken as $0.5L = 140$ lbf/ft gives

$$w_2 = w_{D2}\big(1 + (0.75)(0.14S_{DS})\big) + 0.75w_r$$

$$= \left(250 \ \frac{\text{lbf}}{\text{ft}}\right)\big(1.0 + (0.75)(0.14)(1.0)\big)$$

$$\quad + (0.75)\left(280 \ \frac{\text{lbf}}{\text{ft}}\right)$$

$$= 486 \text{ lbf/ft}$$

$$w_1 = w_{D1}\big(1 + (0.75)(0.14S_{DS})\big) + 0.75w_L$$

$$= \left(240 \ \frac{\text{lbf}}{\text{ft}}\right)\big(1.0 + (0.75)(0.14)(1.0)\big)$$

$$\quad + (0.75)\left(140 \ \frac{\text{lbf}}{\text{ft}}\right)$$

$$= 370 \text{ lbf/ft}$$

$$C_1 = \frac{\begin{array}{c}0.5(w_2 + w_1)L^2 + 0.7(0.75V_2)(h_2 + h_1) \\ + 0.7(0.75V_1h_1)\end{array}}{L_1}$$

$$= \frac{\begin{array}{c}(0.5)\left(486 \ \dfrac{\text{lbf}}{\text{ft}} + 370 \ \dfrac{\text{lbf}}{\text{ft}}\right)(10 \text{ ft})^2 \\ + (0.7)(0.75)(6000 \text{ lbf})(10 \text{ ft} + 12 \text{ ft}) \\ + (0.7)(0.75)(6000 \text{ lbf})(12 \text{ ft})\end{array}}{7.5 \text{ ft}}$$

$$= 20{,}000 \text{ lbf}$$

Case 5 gives the higher value and thus controls axial compressions in chords. Loading combination 8 controls uplift on the wall.

$$w_2 = w_{D2}(0.6 - 0.14 S_{DS})$$
$$= \left(250 \ \frac{\text{lbf}}{\text{ft}}\right)(0.6 - (0.14)(1.0))$$
$$= 115 \ \text{lbf/ft}$$

$$w_1 = w_{D1}(0.6 - 0.14 S_{DS})$$
$$= \left(240 \ \frac{\text{lbf}}{\text{ft}}\right)(0.6 - (0.14)(1.0))$$
$$= 110 \ \text{lbf/ft}$$

$$T_1 = \frac{0.7 V_2(h_2 + h_1) + 0.7 V_1 h_1 - 0.5(w_2 + w_1)L^2}{L_1}$$

$$= \frac{\begin{array}{c}(0.7)(6000 \ \text{lbf})(10 \ \text{ft} + 12 \ \text{ft}) \\ + (0.7)(6000 \ \text{lbf})(12 \ \text{ft}) \\ - (0.5)\left(115 \ \dfrac{\text{lbf}}{\text{ft}} + 110 \ \dfrac{\text{lbf}}{\text{ft}}\right)(10 \ \text{ft})^2\end{array}}{7.5 \ \text{ft}}$$

$$= 17{,}500 \ \text{lbf} \quad [\text{controls uplift}]$$

Design the diaphragm chords for the controlling conditions.

$$P = \begin{cases} 17{,}500 \ \text{lbf} & [\text{tension}] \\ 22{,}800 \ \text{lbf} & [\text{compression}] \end{cases}$$

From NDS Table 4A, for construction grade Douglas fir-larch, 6 in and wider,

$$F_c = 1650 \ \text{psi}$$
$$F_t = 650 \ \text{psi}$$
$$E_{min} = 550{,}000 \ \text{psi}$$
$$F_{c\perp} = 625 \ \text{psi}$$

For seismic loading, $C_D = 1.6$, $C_F = 1.0$ for all sizes of construction graded dimension lumber, and all other adjustment factors $= 1.0$. For compression parallel to the grain controlled by a nominal 6 in wide member, $l_u =$ clear height $= 10.5$ ft.

$$\frac{l_e}{d} = \frac{(10.5 \ \text{ft})\left(12 \ \dfrac{\text{in}}{\text{ft}}\right)}{5.5 \ \text{in}} = 22.9$$

$$F_c^* = C_D C_F F_c = (1.6)(1.0)\left(1650 \ \frac{\text{lbf}}{\text{in}^2}\right) = 2640 \ \text{psi}$$

$$E'_{min} = E_{min} = 550{,}000 \ \text{psi}$$

From NDS Sec. 3.7.1, with $c = 0.8$ for sawn lumber,

$$F_{cE} = \frac{0.822 E'_{min}}{\left(\dfrac{l_e}{d}\right)^2} = \frac{(0.822)\left(550{,}000 \ \dfrac{\text{lbf}}{\text{in}^2}\right)}{(22.9)^2} = 862 \ \text{psi}$$

$$\frac{F_{cE}}{F_c^*} = \frac{862 \ \dfrac{\text{lbf}}{\text{in}^2}}{2640 \ \dfrac{\text{lbf}}{\text{in}^2}} = 0.33$$

$$C_P = \frac{1 + \dfrac{F_{cE}}{F_c^*}}{2c} - \sqrt{\left(\frac{1 + \dfrac{F_{cE}}{F_c^*}}{2c}\right)^2 - \frac{\dfrac{F_{cE}}{F_c^*}}{c}}$$

$$= \frac{1 + 0.33}{(2)(0.8)} - \sqrt{\left(\frac{1 + 0.33}{(2)(0.8)}\right)^2 - \frac{0.33}{0.8}}$$

$$= 0.30$$

$$F_c' = C_P C_D C_F F_c = (0.30)(1.6)(1.0)\left(1650 \ \frac{\text{lbf}}{\text{in}^2}\right)$$
$$= 792 \ \text{lbf/in}^2$$

$$P = F_c' A = \left(792 \ \frac{\text{lbf}}{\text{in}^2}\right) A = 22{,}800 \ \text{lbf}$$
$$= C_{max}$$
$$A = 28.8 \ \text{in}^2 = bd$$
$$b = \frac{A}{d} = \frac{28.8 \ \text{in}^2}{5.5 \ \text{in}} = 5.25 \ \text{in} \quad [\text{say two } 4 \times 6\text{s}]$$

Check the bearing perpendicular to grain against the 3×6 sill.

$$F_{c\perp}' = F_{c\perp} = 625 \ \text{lbf/in}^2$$
$$f_{c\perp} = \frac{P}{A} = \frac{22{,}800 \ \text{lbf}}{38.5 \ \text{in}^2} = 592 \ \text{lbf/in}^2 \quad [< F_{c\perp}', \text{ so OK}]$$

The member is adequate in axial compression. Check tension, assuming $3/4$ in bolts for hold-down anchors through the $(2)(3.5 \ \text{in})$ dimension (hole diameter $= 0.81$ in).

$$F_t' = C_D F_t = (1.6)\left(650 \ \frac{\text{lbf}}{\text{in}^2}\right) = 1040 \ \text{lbf/in}^2$$
$$A_n = A - b d_h = 38.5 \ \text{in}^2 - (7.00 \ \text{in})(0.81 \ \text{in}) = 32.8 \ \text{in}^2$$
$$T = F_t' A_n = \left(1040 \ \frac{\text{lbf}}{\text{in}^2}\right)(32.8 \ \text{in}^2)$$
$$= 34{,}000 \ \text{lbf} \quad [> T_{max} = 17{,}500 \ \text{lbf}]$$

> Use two 4×6 construction grade DF-L for shear wall chords, and use 3×6 sill and top plates.

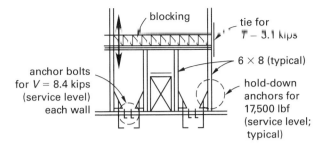

SOLUTION 13

13.1. Find the forces at joint A of the pyramidal roof under a symmetrical gravity load of 32 lbf/ft^2 on the horizontal projection. Neglect unbalanced loading per problem statement. Consider the hip rafter AE and assume that roof joists are spaced closely enough so that the load on the rafter is uniformly distributed. The load on AE varies linearly from its maximum at the top to zero at the bottom.

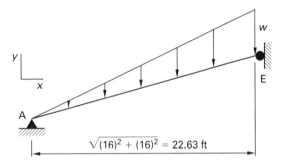

The roof joists are simply supported and span from the exterior walls to the hip rafters. They support an equivalent 32 lbf/ft^2 over the horizontal projection. One-half the reaction from each joist bears on the exterior wall; the other half bears on the rafter. Each quadrant measures 16 ft × 16 ft; therefore, the horizontal projection of the hip rafter is

$$L_p = \sqrt{2(0.5L)^2} = \sqrt{(2)\big((0.5)(32\ \text{ft})\big)^2} = 22.63\ \text{ft}$$

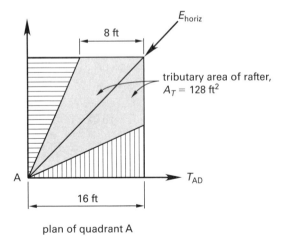

plan of quadrant A

The resultant force acting on each hip rafter is

$$A_T = 0.5(0.5L)^2 = (0.5)\big((0.5)(32\ \text{ft})\big)^2 = 128\ \text{ft}^2$$

$$W = w_{D+L}A_T = \left(32\ \frac{\text{lbf}}{\text{ft}^2}\right)(128\ \text{ft}^2) = 4096\ \text{lbf}$$

For the symmetrical loading, there is no shear transfer at the peak E; only horizontal force acting parallel to the horizontal projection of the hip rafter acts at E. Thus, equilibrium of the hip rafter gives

$$\sum M_A = E_{\text{horiz}}h - W(0.667 L_{\text{proj}}) = 0$$

$$E_{\text{horiz}} = \frac{W(0.667 L_{\text{proj}})}{h}$$

$$= \frac{(4096\ \text{lbf})(0.667)(22.63\ \text{ft})}{6\ \text{ft}}$$

$$= 10{,}304\ \text{lbf}$$

$$\sum F_y = A_y - W = 0$$

$$A_y = W = 4096\ \text{lbf}$$

Summing forces in the direction of the hip rafter projection gives

$$\sum F_{AE} = T_{AD}\cos 45° + T_{AD}\cos 45° - E_{\text{horiz}} = 0$$

$$T_{AD} = \frac{E_{\text{horiz}}}{2\cos 45°} = \frac{10{,}304\ \text{lbf}}{(2)(0.707)} = 7287\ \text{lbf}$$

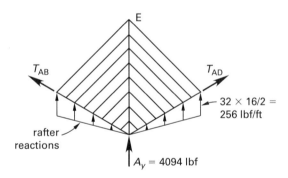

Check the vertical equilibrium in one quadrant.

$$A_{\text{quad}} = (0.5L)^2 = \big((0.5)(32\ \text{ft})\big)^2 = 256\ \text{ft}^2$$

$$\sum F_y = -w_{D+L}A_{\text{quad}} + A_y + 2\big(0.5w_{\text{wall}}(0.5L)\big)$$

$$= \left(-32\ \frac{\text{lbf}}{\text{ft}^2}\right)(256\ \text{ft}^2) + 4096\ \text{lbf}$$

$$\quad + (2)(0.5)\left(256\ \frac{\text{lbf}}{\text{ft}}\right)(0.5)(32\ \text{ft})$$

$$= 0$$

Equilibrium is satisfied.

13.2. Design the connection at A using ASD. Use a welded steel assembly with the rafter cut to transfer 4096 lbf vertically and 10,300 lbf horizontally in the direction of the rafter projection. Weld steel side plates and transfer 7277 lbf into the eave beam. For the rafter bearing,

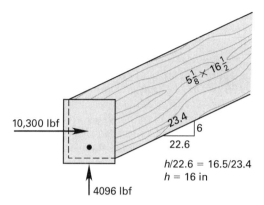

The slope of the rafter is

$$\alpha = \arctan \frac{h}{L_{\text{proj}}} = \arctan \frac{6 \text{ ft}}{22.63 \text{ ft}} = 14.9°$$

Maximum axial compression in the hip rafter occurs at the support

$$P = \frac{E_{\text{horiz}}}{\cos \alpha} = \frac{10{,}300 \text{ lbf}}{\cos 14.9°} = 10{,}660 \text{ lbf}$$

$$f_c = \frac{P}{bd} = \frac{10{,}660 \text{ lbf}}{(5.125 \text{ in})(16 \text{ in})} = 130 \text{ psi} \quad [< F_c']$$

Compression parallel to grain is not a problem. For the vertical reaction, the angle of load to grain is 75.1°, but use $F_{c\perp}$, which is conservative and close enough.

$$F'_{c\perp} = 625 \text{ psi} \quad [\text{no increase for load duration}]$$

$$N \geq \frac{A_y}{bF'_{c\perp}} = \frac{4096 \text{ lbf}}{(5.125 \text{ in})\left(625 \dfrac{\text{lbf}}{\text{in}^2}\right)} = 1.3 \text{ in}$$

Use a minimum bearing length of 6 in as a practical size. Connect across the back with steel plate, one side only, using two $5/8$ in diameter shear plates. Per NDS Table 12.2B, group B wood has

$$Z_{\parallel} = 2670 \text{ lbf}$$

$$Z' = C_D Z = (1.25)(2670 \text{ lbf}) = 3338 \text{ lbf}$$

$$N > \frac{T}{Z'} = \frac{7280 \text{ lbf}}{3338 \dfrac{\text{lbf}}{\text{connector}}}$$

$$= 2.2 \text{ connectors} \quad [\text{therefore, use 3 shear connectors}]$$

Check tension in the $5^{1}/_{8} \times 16^{1}/_{2}$ glulam (per NDS App. K, $^{13}/_{16}$ in bolt hole diameter and 1.18 in^2 projected area of split ring).

$$\begin{aligned}
A_n &= A - A_{\text{ring}} - (b - t_{\text{ring}})d_h \\
&= (5.125 \text{ in})(16.5 \text{ in}) - 1.18 \text{ in}^2 \\
&\quad - (5.125 \text{ in} - 0.42 \text{ in})(0.81 \text{ in}) \\
&= 79.6 \text{ in}^2 \\
f_t &= \frac{T}{A_n} = \frac{7280 \text{ lbf}}{79.6 \text{ in}^2} = 91 \text{ psi} \quad [< F'_t, \text{ so OK}]
\end{aligned}$$

Design the steel plates using A36 steel and the AISC ASD method. Try PL$^{3}/_{8} \times 5$ with the single gage of $^{3}/_{4}$ in bolts.

$$\begin{aligned}
A_n &= A_g - (d_b + 0.125 \text{ in})t \\
&= 1.875 \text{ in}^2 - (0.75 \text{ in} + 0.125 \text{ in})(0.375 \text{ in}) \\
&= 1.55 \text{ in}^2
\end{aligned}$$

$$\frac{P_n}{\Omega_t} \leq \begin{cases} \dfrac{F_y A_g}{\Omega_t} = \dfrac{\left(36 \dfrac{\text{kips}}{\text{in}^2}\right)(1.875 \text{ in}^2)}{1.67} \\ \qquad = 40.4 \text{ kips} \quad [\text{controls}] \\[4pt] \dfrac{F_u U A_n}{\Omega_t} = \dfrac{\left(58 \dfrac{\text{kips}}{\text{in}^2}\right)(1.0)(1.55 \text{ in}^2)}{2.00} \\ \qquad = 45.0 \text{ kips} \end{cases}$$

Strength is adequate. Detail the connection as follows.

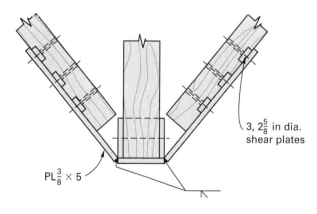

plan view
hip rafter to eave beam connection

SOLUTION 14

14.1. Compute the allowable bending stress given a 2400F-1.8E stress grade glulam with $1^{1}/_{2}$ in laminations, $E = 1.8 \times 10^{6}$ psi, and no increase for load duration.

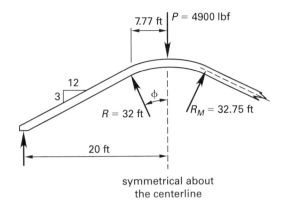

symmetrical about
the centerline

Because lateral bracing is assumed to be sufficient, $C_V < C_L$; therefore, reduce for volume effect and curvature (NDS Sec. 5.3.8).

$$C_c = 1 - 2000\left(\frac{t}{R}\right)^2$$

$$= 1 - (2000)\left(\frac{1.5 \text{ in}}{(32 \text{ ft})\left(12 \frac{\text{in}}{\text{ft}}\right)}\right)^2$$

$$= 0.97$$

$$C_V = \left(\left(\frac{21}{L}\right)\left(\frac{12}{d}\right)\left(\frac{5.125}{b}\right)\right)^{1/x} \leq 1.0$$

$$= \left(\left(\frac{21 \text{ ft}}{40 \text{ ft}}\right)\left(\frac{12 \text{ in}}{18 \text{ in}}\right)\left(\frac{5.125 \text{ in}}{5.125 \text{ in}}\right)\right)^{1/10}$$

$$= 0.90$$

The allowable bending stress is $F'_b = C_c C_V F_b = (0.97)(0.90)(2400 \text{ psi}) = \boxed{2100 \text{ psi.}}$

14.2. Check the bending stress including the member weight based on $\gamma = 36$ lbf/ft^3. Approximate the equivalent dead load per horizontal foot and ignore the slight variance due to the curvature in the central region.

$$w_{\text{beam}} = \left(\frac{\text{hyp}}{\text{adj}}\right) bd\gamma$$

$$= \left(\frac{12.4 \text{ ft}}{12 \text{ ft}}\right)\left(\frac{(5.125 \text{ in})(18 \text{ in})}{\left(12 \frac{\text{in}}{\text{ft}}\right)^2}\right)\left(36 \frac{\text{lbf}}{\text{ft}^3}\right)$$

$$= 24 \text{ lbf/ft}$$

$$M_{\text{max}} = 0.125 w_{\text{beam}} L^2 + 0.25 PL$$

$$= (0.125)\left(24 \frac{\text{lbf}}{\text{ft}}\right)(40 \text{ ft})^2 + (0.25)(4900 \text{ lbf})(40 \text{ ft})$$

$$= 53,800 \text{ ft-lbf}$$

For a pitched and curved beam of constant depth,

$$f_b = \frac{M_{\text{max}}}{S} = \frac{(53,800 \text{ ft-lbf})\left(12 \frac{\text{in}}{\text{ft}}\right)}{277 \text{ in}^3} = 2330 \text{ psi}$$

$f_b > F'_b$ (11% overstress), so in a strict sense there is no value of l_u for which the given loading is acceptable ($F'_b = 2100$ psi is an upper bound to the allowable stress). However, the problem can be regarded as finding the maximum unsupported length for which $C_L \geq C_V$. For a 24F-1.8E stress class glulam, NDS Table 5A gives $E_{y,\text{min}} = 850,000$ psi. Setting C_L equal to 0.90 gives

$$C_L = \frac{1 + \frac{F_{bE}}{F^*_{bx}}}{1.9} - \sqrt{\left(\frac{1 + \frac{F_{bE}}{F^*_{bx}}}{1.9}\right)^2 - \frac{\frac{F_{bE}}{F^*_{bx}}}{0.95}} = 0.90$$

$$\left(\frac{1 + \frac{F_{bE}}{F^*_{bx}}}{1.9}\right)^2 - \frac{\frac{F_{bE}}{F^*_{bx}}}{0.95} = \left(\frac{1 + \frac{F_{bE}}{F^*_{bx}}}{1.9} - 0.90\right)^2$$

$$= \left(\frac{1 + \frac{F_{bE}}{F^*_{bx}}}{1.9}\right)^2$$

$$- 1.8\left(\frac{1 + \frac{F_{bE}}{F^*_{bx}}}{1.9}\right) + 0.81$$

$$1.8\left(1 + \frac{F_{bE}}{F^*_{bx}}\right) - \frac{1.9\left(\frac{F_{bE}}{F^*_{bx}}\right)}{0.95} = (1.9)(0.81)$$

$$\frac{F_{bE}}{F^*_{bx}} = 1.305$$

$$F^*_{bx} = C_c F_b = (0.97)\left(2400 \frac{\text{lbf}}{\text{in}^2}\right) = 2330 \text{ psi}$$

$$F_{bE} = 1.305 F^*_{bx} = (1.305)\left(2330 \frac{\text{lbf}}{\text{in}^2}\right) = 3040 \text{ psi}$$

$$E'_{y,\text{min}} = E_{y,\text{min}} = 850,000 \text{ psi}$$

$$R_B = \sqrt{\frac{l_e d}{b^2}} = \sqrt{\frac{l_e (18 \text{ in})}{(5.125 \text{ in})^2}} = 0.828\sqrt{l_e}$$

$$F_{bE} = \frac{1.20 E'_{y,\text{min}}}{R_B^2} = \frac{1.20 E'_{y,\text{min}}}{(0.828\sqrt{l_e})^2} = 3040 \text{ psi}$$

$$l_e = \frac{1.20E'_{y,\text{min}}}{\left(3040 \frac{\text{lbf}}{\text{in}^2}\right)(0.828)^2}$$

$$= \frac{(1.20)\left(850{,}000 \frac{\text{lbf}}{\text{in}^2}\right)}{\left(3040 \frac{\text{lbf}}{\text{in}^2}\right)(0.828)^2}$$

$$= 489 \text{ in}$$

Assuming lateral bracing at the ends and load point, NDS Table 3.3.3 gives

$$l_e = 1.11l_u$$

$$l_u = \frac{l_e}{1.11} = \frac{489 \text{ in}}{(1.11)\left(12 \frac{\text{in}}{\text{ft}}\right)} = 36.7 \text{ ft}$$

The actual unbraced length of 20 ft is well below the maximum 36.7 ft for which stability would control.

Bracing at the ends and at midspan is adequate.

14.3. Compute the radial tension stress (per the *AITC Manual*). The problem statement stipulates that there is to be no increase in allowable stresses for load duration; therefore, $F'_r = F_r = 15$ psi.

$$f_r = \frac{3M_{\text{max}}}{2R_m b d} = \frac{(3)(53{,}800 \text{ ft-lbf})}{(2)(32.75 \text{ ft})(5.125 \text{ in})(18 \text{ in})}$$

$$= 26.7 \text{ psi} \quad [> F'_r = 15 \text{ psi}]$$

14.4. Radial tension stress is excessive. Use lag screws to reinforce the curved central region. Orient the lag bolts radially, and install from the top flange into pre-drilled holes to within 2 in of the soffit. In this way, the allowable radial tension stress is $0.33F'_v = (0.33)(265 \text{ psi}) = 88.3$ psi, which is acceptable. The effective penetration of the $^3/_4$ in lag bolt into the member is given in NDS Table L2 as $T - E = 5.5$ in, which includes the threaded portion but excludes the tapered tip. Thus, the allowable tensile strength of the radial reinforcement is

$$T \leq \begin{cases} \text{tensile strength of bolt} = 5240 \text{ lbf} \\[6pt] \text{given withdrawal strength} \\[4pt] \quad \times \text{ penetration of threads} \\[4pt] \quad = \left(513 \frac{\text{lbf}}{\text{in}}\right)(T - E) = \left(513 \frac{\text{lbf}}{\text{in}}\right)(5.5 \text{ in}) \\[6pt] \quad = 2820 \text{ lbf} \quad [\text{controls}] \end{cases}$$

Let s be the spacing of lag bolts measured along the arc of radius R_m. Provide bolts to resist the total 26.7 psi computed radial tension stress.

$$\sum F_{\text{radial}} = T - f_r b s = 0$$

$$s = \frac{T}{b f_r} = \frac{2820 \text{ lbf}}{(5.125 \text{ in})\left(26.7 \frac{\text{lbf}}{\text{in}^2}\right)} = 20.6 \text{ in}$$

For the curved central region,

$$\phi = \arcsin \frac{\text{opp}}{\text{hyp}}$$

$$= \arcsin \frac{7.8 \text{ ft}}{32 \text{ ft}}$$

$$= 0.246 \text{ rad}$$

$$\text{arc length} = 2R_m \phi$$

$$= (2)(32.75 \text{ ft})\left(12 \frac{\text{in}}{\text{ft}}\right)(0.246 \text{ rad})$$

$$= 193 \text{ in}$$

$$n \geq \frac{\text{arc length}}{s}$$

$$= \frac{193 \text{ in}}{20.6 \text{ in}}$$

$$= 9.4 \quad [\text{say 10 screws}]$$

$$s_o = \frac{2R_o \phi}{n}$$

$$= \frac{(2)(402 \text{ in})(0.246 \text{ rad})}{10}$$

$$= 20 \text{ in}$$

Use 10 fully threaded $\frac{3}{4}$ in diameter lag screws in the central region spaced symmetrically at 20 in o.c. along the top flange.

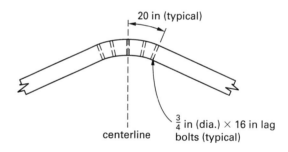

20 in (typical)

$\frac{3}{4}$ in (dia.) × 16 in lag bolts (typical)

centerline

partial elevation

SOLUTION 15

15.1. Compute the wall shears in the upper and lower stories and determine the required nailing. The maximum length available to resist shear is 11 ft in the upper wall and 15.5 ft in the lower wall. However, if the maximum length is used for each level, the ratio of their lengths is 15.5 ft/11 ft = 1.41, which exceeds the 130% limit that triggers a vertical structural irregularity, type 3, in ASCE/SEI7 Table 12.3-2. To simplify design and to provide a direct load path for the chord forces from the upper level, align the upper and lower shear walls with an overall length of 11 ft in each and treat the remaining regions as filler panels sheathed on one side with minimum nailing.

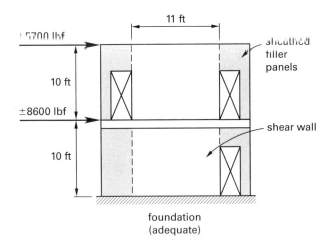

For the upper wall, using ASD, the shear is

$$v_2 = \frac{0.7\,V_2}{L} = \frac{(0.7)(5700\text{ lbf})}{11\text{ ft}} = 363\text{ lbf/ft}$$

For the lower wall,

$$v_1 = \frac{0.7(V_1 + V_2)}{L} = \frac{(0.7)(8600\text{ lbf} + 5700\text{ lbf})}{11\text{ ft}}$$
$$= 910\text{ lbf/ft}$$

From SDPWS, for seismic design category D and higher, walls with $v > 350$ lbf/ft require nominal $3 \times$ framing where panels abut; therefore, use $3 \times$ studs and blocking within the confine of the shear walls.

For the top wall ($v = 363$ lbf/ft), use a single sheathing of $\frac{15}{32}$ in structural I plywood with 10d nails at 6 in o.c. along panel edges (furnishes 340 lbf/ft; close enough); 12 in o.c. along intermediate supports.

For the lower wall ($v = 910$ lbf/ft), use $\frac{15}{32}$ in structural I plywood on both sides with 10d nails at 4 in o.c. on panel edges (furnishes 510 lbf/ft on each side for a total of 1020 lbf/ft); 12 in o.c. along intermediate supports.

15.2. Determine loads on tie-down anchors. For combined seismic and gravity forces (using ASD), per ASCE/SEI7 Sec. 12.4.2.3, the maximum axial compression in the shear wall chord is controlled by either case 5, $D + 0.7E$, or case 6, $D + 0.75(0.7E) + 0.75(L_r$ or $L)$.

For the upper level, combination 5 (with $\rho = 1.0$) gives

$$w_2 = (1 + 0.14S_{DS})(w_{D2} + w_{wall}h_2)$$
$$= \big(1.0 + (0.14)(1.0)\big)\left(150\,\frac{\text{lbf}}{\text{ft}} + \left(15.0\,\frac{\text{lbf}}{\text{ft}^2}\right)(10\text{ ft})\right)$$
$$= 342\text{ lbf/ft}$$

$$C_2 = \frac{0.5w_2L_2^2 + 0.7\,V_2h_2}{L_2}$$
$$= \frac{(0.5)\left(342\,\frac{\text{lbf}}{\text{ft}}\right)(11\text{ ft})^2 + (0.7)(5700\text{ lbf})(10\text{ ft})}{11\text{ ft}}$$
$$= 5510\text{ lbf}$$

Load combination 6 gives

$$w_2 = (1 + 0.1S_{DS})(w_{D2} + w_{wall}h_2) + 0.75w_r$$
$$= \big(1 + (0.1)(1.0)\big)$$
$$\times \left(150\,\frac{\text{lbf}}{\text{ft}} + \left(15\,\frac{\text{lbf}}{\text{ft}^2}\right)(10\text{ ft})\right)$$
$$+ (0.75)\left(200\,\frac{\text{lbf}}{\text{ft}}\right)$$
$$= 480\text{ lbf/ft}$$

$$C_2 = \frac{0.5w_2L_2^2 + 0.7(0.75\,V_2h_2)}{L_2}$$
$$= \frac{(0.5)\left(480\,\frac{\text{lbf}}{\text{ft}}\right)(11\text{ ft})^2 + (0.7)}{11\text{ ft}}$$
$$\times (0.75)(5700\text{ lbf})(10\text{ ft})$$
$$= 5360\text{ lbf}\quad \left[\begin{array}{l}< 5510\text{ lbf, case 5 controls}\\ \text{axial compression in chords}\end{array}\right]$$

Loading combination 8 controls uplift on the wall.

$$w_2 = (0.6 - 0.14S_{DS})(w_{D2} + w_{wall}h_2)$$
$$= \big(0.6 - (0.14)(1.0)\big)\left(150\,\frac{\text{lbf}}{\text{ft}} + \left(15\,\frac{\text{lbf}}{\text{ft}^2}\right)(10\text{ ft})\right)$$
$$= 138\text{ lbf/ft}$$

$$T_2 = \frac{0.7\,V_2h_2 - 0.5w_2L_2^2}{L_2}$$
$$= \frac{(0.7)(5700\text{ lbf})(10\text{ ft}) - (0.5)\left(138\,\frac{\text{lbf}}{\text{ft}}\right)(11\text{ ft})^2}{11\text{ ft}}$$
$$= 2870\text{ lbf}\quad[\text{controls uplift}]$$

For the lower level, combination 5 (with $\rho = 1.0$) gives

$$w_1 = (1 + 0.14S_{DS})(w_{D2} + w_{D1} + w_{wall}h_{1+2})$$
$$= \big(1.0 + (0.14)(1.0)\big)$$
$$\times \left(150\,\frac{\text{lbf}}{\text{ft}} + 180\,\frac{\text{lbf}}{\text{ft}} + \left(15\,\frac{\text{lbf}}{\text{ft}^2}\right)(20\text{ ft})\right)$$
$$= 718\text{ lbf/ft}$$

$$C_1 = \frac{0.5w_1 L^2 + 0.7\left(V_2(h_2 + h_1) + V_1 h_1\right)}{L_2}$$

$$\begin{aligned}&\frac{(0.5)\left(718\,\frac{\text{lbf}}{\text{ft}}\right)(11\text{ ft})^2 + (0.7)}{}\\&= \frac{\times \left(\begin{array}{c}(5700\text{ lbf})(10\text{ ft} + 10\text{ ft})\\ + (8600\text{ lbf})(10\text{ ft})\end{array}\right)}{11\text{ ft}}\\&= 16{,}700\text{ lbf}\end{aligned}$$

Load combination 6 gives

$$\begin{aligned}w_1 &= (1 + 0.1S_{DS})(w_{D2} + w_{D1} + w_{\text{wall}}h_{1+2})\\ &\quad + 0.75w_r\\ &= \left(1 + (0.1)(1.0)\right)\\ &\quad \times \left(150\,\frac{\text{lbf}}{\text{ft}} + 180\,\frac{\text{lbf}}{\text{ft}} + \left(15\,\frac{\text{lbf}}{\text{ft}^2}\right)(20\text{ ft})\right)\\ &\quad + (0.75)\left(200\,\frac{\text{lbf}}{\text{ft}} + 500\,\frac{\text{lbf}}{\text{ft}}\right)\\ &= 1220\text{ lbf/ft}\end{aligned}$$

$$C_1 = \frac{0.5w_1 L_1^2 + (0.7)(0.75)\left(V_2(h_2 + h_1) + V_1 h_1\right)}{L_1}$$

$$\begin{aligned}&\frac{(0.5)\left(1220\,\frac{\text{lbf}}{\text{ft}}\right)(11\text{ ft})^2 + (0.7)(0.75)}{}\\&= \frac{\times \left((5700\text{ lbf})(10\text{ ft} + 10\text{ ft}) + (8600)(10\text{ ft})\right)}{11\text{ ft}}\\&= 16{,}300\text{ lbf} \quad \left[\begin{array}{l}< 16{,}700\text{ lbf, case 5 controls}\\ \text{axial compression in chords}\end{array}\right]\end{aligned}$$

Loading combination 8 controls uplift on the wall.

$$\begin{aligned}w_1 &= (0.6 - 0.14S_{DS})(w_{D2} + w_{D1} + w_{\text{wall}}h_{1+2})\\ &= \left(0.6 - (0.14)(1.0)\right)\left(\begin{array}{c}150\,\frac{\text{lbf}}{\text{ft}} + 180\,\frac{\text{lbf}}{\text{ft}}\\ + \left(15\,\frac{\text{lbf}}{\text{ft}^2}\right)(20\text{ ft})\end{array}\right)\\ &= 290\text{ lbf/ft}\end{aligned}$$

$$T_1 = \frac{0.7\left(V_2(h_2 + h_1) + V_1 h_1\right) - 0.5w_1 L_1^2}{L_1}$$

$$\begin{aligned}&\frac{(0.7)\left((5700\text{ lbf})(10\text{ ft} + 10\text{ ft}) + (8600\text{ lbf})(10\text{ ft})\right)}{}\\&= \frac{- (0.5)\left(290\,\frac{\text{lbf}}{\text{ft}}\right)(11\text{ ft})^2}{11\text{ ft}}\\&= 11{,}130\text{ lbf} \quad [\text{controls uplift}]\end{aligned}$$

The lateral forces are reversible; therefore, design chords to resist both 16,700 lbf compression and 11,000 lbf tension at the lower level.

15.3. Two types of tie-down anchors across the second floor ($T = 2870$ lbf) are shown as follows.

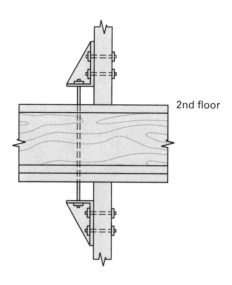

2nd floor

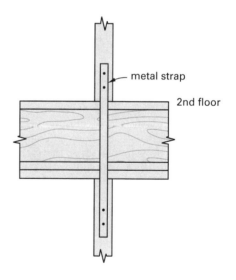

metal strap

2nd floor

Two types of tie-down anchors at the first floor ($T = 11,130$ lbf) are shown as follows.

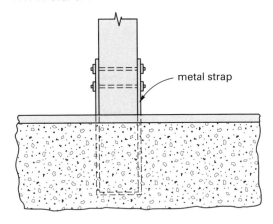

metal strap

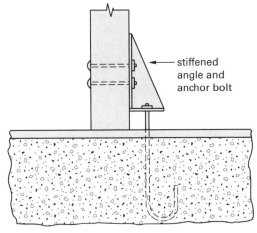

stiffened angle and anchor bolt

SOLUTION 16

16.1. Compute the load, shear, and moment diagrams for the low roof due to seismic forces. The structure is symmetrical about its principal axes; therefore, consider the seismic forces in the north-south direction and use identical loads for east-west ground movement. For the north-south direction, the seismic dead load to the diaphragm consists of the weight of the high roof and the walls between the low and high roofs, plus the weight of the low-roof diaphragm and one-half the exterior walls each side. For the low roof portion,

$$w_{AB} = w_{CD} = 2(0.5 w_{wall} h) + w_{roof} L$$

$$= (2)(0.5)\left(20 \; \frac{lbf}{ft^2}\right)(12 \; ft) + \left(20 \; \frac{lbf}{ft^2}\right)(80 \; ft)$$

$$= 1840 \; lbf/ft$$

$$w_{BC} = 2(0.5 w_{wall} h) + w_{roof}(L - L_{open})$$

$$= (2)(0.5)\left(20 \; \frac{lbf}{ft^2}\right)(12 \; ft)$$

$$\quad + \left(20 \; \frac{lbf}{ft^2}\right)(80 \; ft - 32 \; ft)$$

$$= 1200 \; lbf/ft$$

For the high roof and walls, take one-half the total weight as concentrated on each line B and C.

$$W_B = W_C = w_{wall} h_2 L_{open} + 2 w_{wall} h_2 (0.5 L_{open})$$

$$\quad + w_{roof} L_{open}(0.5 L_{open})$$

$$= \left(20 \; \frac{lbf}{ft^2}\right)(4 \; ft)(32 \; ft) + (2)\left(20 \; \frac{lbf}{ft^2}\right)(4 \; ft)(0.5)$$

$$\quad \times (32 \; ft) + \left(20 \; \frac{lbf}{ft^2}\right)(32 \; ft)(0.5)(32 \; ft)$$

$$= 15,360 \; lbf$$

Calculate the seismic forces per ASCE/SEI7. For site class D, $S_S = 1.20$ and $S_1 = 0.60$, ASCE/SEI7 Table 11.4-1 and Table 11.4-2 give $F_a = 1.02$ (by linear interpolation) and $F_v = 1.5$. Thus, the design spectral response parameters are found using ASCE/SEI7 Eq. 11.4-1 through Eq. 11.4-4.

$$S_{MS} = F_a S_S = (1.02)(1.20) = 1.22$$

$$S_{DS} = 0.667 S_{MS} = (0.667)(1.22) = 0.82$$

$$S_{M1} = F_v S_1 = (1.5)(0.60) = 0.90$$

$$S_{D1} = 0.667 S_{M1} = (0.667)(0.90) = 0.60$$

From ASCE/SEI7 Table 1.5-2, for risk category II, the importance factor is 1.0. From ASCE/SEI7 Table 11.6-1 and Table 11.6-2, with risk category II, $S_{DS} > 0.5$, and $S_{D1} > 0.2$, the seismic design category is D. From ASCE/SEI7 Table 12.8-2 and Eq. 12.8-7, the approximate period of the light-frame-bearing wall structure is

$$T = T_a = C_t h_n^x = (0.02)(16 \; ft)^{0.75} = 0.16 \; sec$$

For a single-story building in risk category II, ASCE/SEI7 Table 12.6-1 permits the equivalent lateral force procedure of Sec. 12.8 and that method will be used to calculate the base shear. For light-framed walls sheathed with wood structural panels rated for shear capacity, ASCE/SEI7 Table 12.2-1 specifies $R = 6.5$, $\Omega_0 = 3$, and $C_d = 4$. The seismic response coefficient, C_s, must be at least 0.01 and is limited by ASCE/SEI7 Eq. 12.8-2 and Eq. 12.8-3 to the following.

$$C_s \leq \begin{cases} \dfrac{S_{DS}}{\dfrac{R}{I_e}} = \dfrac{0.82}{\dfrac{6.5}{1}} = 0.13 \quad \text{[controls]} \\[4mm] \dfrac{S_{D1}}{T\left(\dfrac{R}{I_e}\right)} = \dfrac{0.60}{(0.16 \; sec)\left(\dfrac{6.5}{1}\right)} = 0.58 \end{cases}$$

The seismic loading to the diaphragm is uniformly distributed over the low roof and concentrated shear wall reactions at lines B and C from the upper roof. Calculating first the uniform loads,

$$w_1 = C_s w_{AB} = (0.13)\left(1840 \; \frac{lbf}{ft}\right) = 239 \; lbf/ft$$

$$V_{hr} = C_s W_B = (0.13)(15,360 \; lbf) = 2000 \; lbf$$

$$w_2 = C_s w_{BC} = (0.13)\left(1200 \; \frac{lbf}{ft}\right) = 156 \; lbf/ft$$

The ratio of the area of opening in the low roof to the gross enclosed diaphragm area is

$$\frac{A_{\text{open}}}{A_{\text{gross}}} = \frac{L_{\text{open}}^2}{L^2} = \frac{(32 \text{ ft})^2}{(80 \text{ ft})^2} = 0.16 \quad [< 0.5]$$

Therefore, the diaphragm is not penalized for a horizontal structural irregularity type 3 per ASCE/SEI7 Table 12.3-1. The discontinuous shear walls from the high roof will impose penalties on their supporting posts and connections, but not on the diaphragm and chords. The distribution of shears to elements of the diaphragm shear wall system will be as shown.

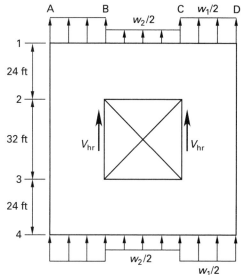

The resulting shears and moments in the flexible diaphragm are

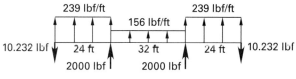

north-south seismic forces on low roof

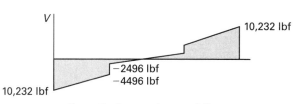

north-south shear on low roof diagram

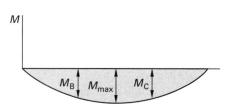

north-south moment on low roof diagram

$$V_{\text{AB}} = w_1 L_{\text{AB}} + V_{\text{hr}} + 0.5 w_2 L_{\text{BC}}$$
$$= \left(239 \, \frac{\text{lbf}}{\text{ft}}\right)(24 \text{ ft}) + 2000 \text{ lbf}$$
$$+ (0.5)\left(156 \, \frac{\text{lbf}}{\text{ft}}\right)(32 \text{ ft})$$
$$= 10{,}232 \text{ lbf}$$

$$V_{\text{BA}} = V_{\text{AB}} - w_1 L_{\text{AB}} = 10{,}232 \text{ lbf} - \left(239 \, \frac{\text{lbf}}{\text{ft}}\right)(24 \text{ ft})$$
$$= 4496 \text{ lbf}$$

$$V_{\text{BC}} = V_{\text{BA}} - V_{\text{hr}} = 4496 \text{ lbf} - 2000 \text{ lbf} = 2496 \text{ lbf}$$

$$M_{\text{B}} = M_{\text{C}} = 0.5(V_{\text{AB}} + V_{\text{BA}})L_{\text{AB}}$$
$$= (0.5)(10{,}232 \text{ lbf} + 4496 \text{ lbf})(24 \text{ ft})$$
$$= 176{,}740 \text{ ft-lbf}$$

$$M_{\text{max}} = M_{\text{B}} + 0.5 V_{\text{BC}}(0.5 L_{\text{BC}})$$
$$= 176{,}740 \text{ ft-lbf} + (0.5)(2496 \text{ ft-lbf})(0.5)(32 \text{ ft})$$
$$= 196{,}200 \text{ ft-lbf}$$

16.2. Compute the chord forces at lines B and C (and 2 and 3 by symmetry).

$$T_{\text{B}} = C_{\text{B}} = \frac{M_{\text{B}}}{L} = \frac{176{,}740 \text{ ft-lbf}}{80 \text{ ft}} = 2209 \text{ lbf}$$

Draw a free-body diagram for the portion of the roof between lines B and C. The shear just to the left of line B is

$$V = \left(\frac{V_{\text{BA}}}{L}\right)L_{12} = \left(\frac{4496 \text{ lbf}}{80 \text{ ft}}\right)(24 \text{ ft}) = 1348 \text{ lbf}$$

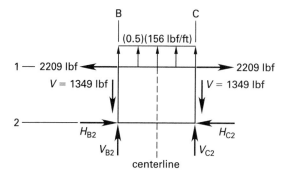

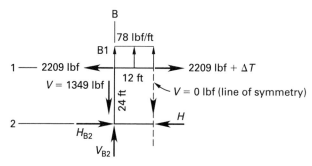

$$\sum M_{B2} = 0.5w_2(0.5L_{BC})(0.25L_{BC})$$
$$+ TL_{21} - (T + \Delta T)L_{21} = 0$$
$$\Delta T = \frac{0.5w_2(0.5L_{BC})(0.25L_{BC})}{L_{21}}$$
$$= \frac{(0.5)\left(156\ \frac{\text{lbf}}{\text{ft}}\right)(0.5)(32\ \text{ft})(0.25)(32\ \text{ft})}{24\ \text{ft}}$$
$$= 416\ \text{lbf}$$

Thus, the maximum chord force is

$$C_{\max} = T_{\max} = T_B + \Delta T = 2209\ \text{lbf} + 416\ \text{lbf}$$
$$= 2625\ \text{lbf}$$

16.3. From a free-body diagram taken through the middle of the building (between lines B and C), sum moments about a point on line A, say at A1.

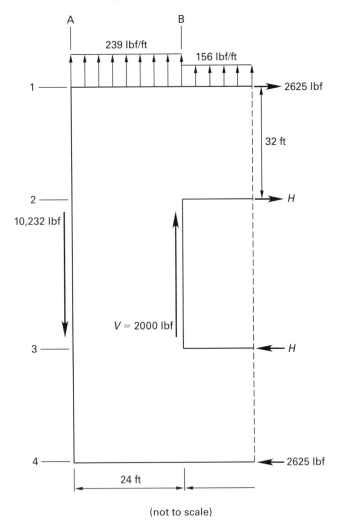

(not to scale)

$$\sum M_{A1} = w_1 L_{AB}(0.5L_{AB}) + w_2(0.5L_{BC})(L_{AB} + 0.25L_{BC})$$
$$+ V_{hr}L_{AB} - C_{\max}L + HL_{12} = 0$$
$$C_{\max}L - w_1 L_{AB}(0.5L_{AB}) - w_2(0.5L_{BC})$$
$$H = \frac{\times (L_{AB} + 0.25L_{BC}) - V_{hr}L_{AB}}{L_{12}}$$
$$= \frac{\begin{array}{c}(2625\ \text{lbf})(80\ \text{ft}) - \left(239\ \frac{\text{lbf}}{\text{ft}}\right)(24\ \text{ft})(0.5)\\ \times (24\ \text{ft}) - \left(156\ \frac{\text{lbf}}{\text{ft}}\right)(0.5)(32\ \text{ft})\\ \times (24\ \text{ft} + (0.25)(32\ \text{ft}))\\ - (2000\ \text{lbf})(24\ \text{ft})\end{array}}{32\ \text{ft}}$$
$$= 416\ \text{lbf}$$

Thus, from the free-body diagram in Prob. 16.2,

$$\sum F_x = T_B - T_{\max} + H - H_{B2} = 0$$
$$H_{B2} = T_B - T_{\max} + H$$
$$= 2209\ \text{lbf} - 2625\ \text{lbf} + 416\ \text{lbf}$$
$$= 0$$

There is no horizontal tie force at B for north-south ground motion. However, summing forces vertically gives

$$\sum F_y = 0.5w_2(0.5L_{BC}) - V - V_{B2} = 0$$
$$V_{B2} = (0.5)\left(156\ \frac{\text{lbf}}{\text{ft}}\right)(0.5)(32\ \text{ft}) - 1348\ \text{lbf}$$
$$= 100\ \text{lbf}$$

By symmetry and considering both north-south and east-west ground motion, a tie force of 100 lbf is required in both directions at all corners of the low roof opening.

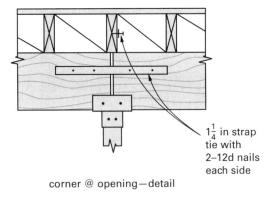

corner @ opening—detail

16.4. Splice the top chords for $T_{\max} = 2625$ lbf. Specify chords as two 2×6 no. 1 Douglas fir-larch and splice using $^3/_4$ in diameter bolts. If the length of the chord pieces is relatively long and precautions are taken to ensure maximum distance between splice points, then it is reasonable to assume that each piece resists an equal share of the axial force between splice points and the connections can be designed to transfer one-half the total force. However, if only minimum lap lengths are used (see Breyer et al.), then the connection must transfer the full force. Assume the latter, more conservative situation and design a single shear lap splice to transfer the total 2625 lbf force.

Per NDS Table 4A, $F_t = 675$ psi, $C_F = 1.3$, and $C_D = 1.6$.

$$F'_t = C_F C_D F_t = (1.3)(1.6)\left(675 \ \frac{\text{lbf}}{\text{in}^2}\right) = 1400 \text{ psi}$$

$$A_n = bd - d(d_b + 0.0625 \text{ in})$$
$$= (1.5 \text{ in})(5.5 \text{ in}) - (1.5 \text{ in})(0.75 \text{ in} + 0.0625 \text{ in})$$
$$= 7.03 \text{ in}^2$$

$$f_t = \frac{T}{A_n} = \frac{2625 \text{ lbf}}{7.03 \text{ in}^2} = 373 \text{ psi} \quad [< F'_t]$$

Per NDS Table 11A, for a 1.5 in main member, estimate $C_g = 0.85$ (group action) and $C_D = 1.6$.

$$Z'_{\parallel} = C_D C_g Z_{\parallel} = (1.6)(0.85)\left(720 \ \frac{\text{lbf}}{\text{bolt}}\right)$$
$$= 979 \text{ lbf/bolt}$$

$$N > \frac{T}{Z'_{\parallel}} = \frac{2625 \text{ lbf}}{979 \ \frac{\text{lbf}}{\text{bolt}}} = 2.7 \text{ bolts} \quad [\text{use 3 bolts}]$$

Provide minimum $7d_b$ end spacing and minimum $4d_b$ between bolts. From NDS Table 10.3.6A, for $A_s/A_m = 1$, $D < 1$ in, $A_s = 8$ in², and $C_g < 0.98$; therefore, the assumed value is conservative.

Use three $\frac{3}{4}$ in diameter bolts each side of splice.

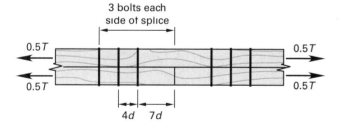

3 bolts each side of splice

0.5T   0.5T
0.5T   0.5T

4d   7d

SOLUTION 17

17.1. Check the bending stress in the rafters using ASD and the given allowable stresses. The scab splice provides negligible restraint to rotation; therefore, assume that rafters are simply supported. For the given allowable stresses,

$$F_b = 1500 \text{ psi}$$
$$F_V = 95 \text{ psi}$$
$$E = 1.8 \times 10^6 \text{ psi}$$

For the 2 × 10 rafters,

$$b = 1.5 \text{ in}$$
$$d = 9.25 \text{ in}$$
$$S = 21.4 \text{ in}^3$$
$$I = 99 \text{ in}^4$$

The distressed rafters have fractured at midspan where shear is zero and bending moment is maximum. This suggests a flexural failure. Check flexural stresses under the design load of 39 lbf/ft². For a 20 lbf/ft² live load, a load duration factor $C_D = 1.25$ should apply; repetitive members spaced at 24 in o.c. qualify for a factor $C_r = 1.15$; for a nominal 2 × 10, NDS Table 4A permits a shape factor $C_F = 1.1$; all other adjustment factors are assumed to be 1.0. Therefore,

$$F'_b = C_D C_F C_r F_b = (1.25)(1.1)(1.15)\left(1500 \ \frac{\text{lbf}}{\text{in}^2}\right)$$
$$= 2370 \text{ psi}$$

$$w = w_{D+L} s = \left(39 \ \frac{\text{lbf}}{\text{ft}^2}\right)(2 \text{ ft}) = 78 \text{ lbf/ft}$$

$$M_{\max} = 0.125 w L^2 = (0.125)\left(78 \ \frac{\text{lbf}}{\text{ft}}\right)(20 \text{ ft})^2$$
$$= 3900 \text{ ft-lbf}$$

$$f_b = \frac{M_{\max}}{S} = \frac{(3900 \text{ ft-lbf})\left(12 \ \frac{\text{in}}{\text{ft}}\right)}{21.4 \text{ in}^3}$$
$$= 2187 \text{ psi} \quad [< F'_b, \text{ so OK in flexure}]$$

Check deflection. The type of supported ceiling is not specified, but whether it is plastered or another type of ceiling would not be relevant to the failure. Total load deflection is limited to $L/180$ (IBC Table 1604.3).

$$E' = E = 1,800,000 \text{ psi}$$

$$\Delta_{\text{TL}} = \frac{0.013 w L^4}{E' I}$$

$$= \frac{(0.013)\left(78 \ \frac{\text{lbf}}{\text{ft}}\right)(20 \text{ ft})^4\left(12 \ \frac{\text{in}}{\text{ft}}\right)^3}{\left(1,800,000 \ \frac{\text{lbf}}{\text{in}^2}\right)(99 \text{ in}^4)}$$

$$= 1.57 \text{ in}$$

$$\frac{L}{180} = \frac{240 \text{ in}}{180} = 1.33 \text{ in} \quad [< \Delta_{\text{TL}}]$$

The deflection is more than is usually allowed ($L/152$ compared to $L/180$). The roof shown is flat and ponding is a potential problem. A detailed check of ponding cannot be performed because the complete framing system is not specified. However, assuming that the stiffness of the steel beams is adequate, a simple check on ponding is that the deflection under a 5 lbf/ft² loading does not exceed 0.25 in (see the *AITC Manual*), where 5 lbf/ft² is the weight of 1 in of water uniformly distributed over the full length of the member. By proportioning the deflection computed above for 39 lbf/ft², the deflection caused by 5 lbf/ft² is

$$\Delta_{5\,psf} = \left(\frac{5 \ \frac{lbf}{ft^2}}{39 \ \frac{lbf}{ft^2}} \right) \Delta_{TL} = (0.128)(1.57 \text{ in})$$

$$= 0.20 \text{ in} \quad [< 0.25 \text{ in}]$$

Therefore, ponding should not be a problem.

> The failure was not likely caused by application of the design loads.

17.2. Describe a repair scheme. Assume that the broken rafters have not deteriorated and can remain in the repaired structure. The repair will consist of providing a companion 2×10 rafter for each broken rafter.

Check the shear and bearing stresses in the 2×10s.

$$V = 0.5wL - wd$$
$$= (0.5)\left(78 \ \frac{lbf}{ft}\right)(20 \text{ ft}) - \left(78 \ \frac{lbf}{ft}\right)(0.77 \text{ ft})$$
$$= 720 \text{ lbf}$$

$$f_v = \frac{1.5V}{bd} = \frac{(1.5)(720 \text{ lbf})}{(1.5 \text{ in})(9.25 \text{ in})} = 78 \text{ psi}$$

$$F_v' = C_D F_v = (1.25)\left(95 \ \frac{lbf}{in^2}\right) = 119 \text{ psi} \quad [> f_v]$$

Check the required bearing length for a nominal 2 in width distributed over the bearing length, N.

$$R = 0.5wL = (0.5)\left(78 \ \frac{lbf}{ft}\right)(20 \text{ ft}) = 780 \text{ lbf}$$

Take $F_{c\perp} = 625$ psi.

$$N \geq \frac{R}{bF_{c\perp}} = \frac{780 \text{ lbf}}{(1.5 \text{ in})\left(650 \ \frac{lbf}{in^2}\right)} = 0.8 \text{ in}$$

End bearing should not be a problem. For properly installed rafters, lateral support for a $d/b = 10/2 = 5$ (per NDS Sec. 4.4.1) requires that only one edge be held in line. Thus, it may not be necessary to remove the intermediate bridging to install the new 2×10s from below.

First, insert new 2×10s, flat, adjacent to damaged rafters. Use the maximum length (approximately $20 \text{ ft} - 1.5 \text{ ft} = 18.5 \text{ ft}$) that can be tilted to clear the flange of the supporting W section.

Second, use two-point jacking to raise the damaged 2×10s to the horizontal. Force tilt the new 2×10s to the vertical (so that they will be loaded to bend about the strong axis) and position them alongside the damaged members. Jack the members to provide a midspan camber of about $1/2$ in.

Third, nail the damaged 2×10s using 20d nails (clinched) in the following pattern.

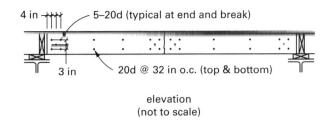

elevation
(not to scale)

Fourth, remove the jacks.

17.3. There are several likely causes of the localized failure.

- Local overload (e.g., construction materials may have been concentrated over a relatively small area during a roofing operation).

- Cuts or holes (e.g., for electrical or mechanical installation) that weakened a series of rafters and were not detected until after failure.

- Water damage caused by roof or drain leakage that significantly weakened several rafters.

- Substitution of an inferior grade of dimension lumber.

SOLUTION 18

18.1. Describe the probable mode of failure in beams A and B. Check the shear stress in beam B. Assume the most favorable case of a 24F-1.8E stress class for which $F_v = 265$ psi. For the roof live load, take $C_D = 1.25$. For a connection less than $5d$ from member end, NDS Eq. 3.4-6 gives

$$d_e = 16 \text{ in}$$
$$d = 24 \text{ in}$$
$$F_v' = C_D F_v = (1.25)\left(265 \ \frac{lbf}{in^2}\right) = 331 \text{ psi}$$

$$V_r' = \frac{2}{3} F_v' b d_e \left(\frac{d_e}{d}\right)^2$$
$$= \left(\frac{2}{3}\right)\left(331 \ \frac{lbf}{in^2}\right)(5.125 \text{ in})(16 \text{ in})\left(\frac{16 \text{ in}}{24 \text{ in}}\right)^2$$
$$= 8042 \text{ lbf}$$

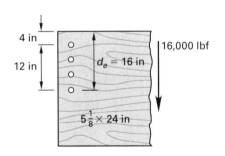

Thus, even for the most favorable assumption for an allowable horizontal shear stress, the applied shear, $V = 16{,}000$ lbf, is nearly double the allowable and this is the likely cause of splitting of beam B (i.e., a poor connection detail). Restraint to cross-grain shrinkage as the wood dried from its as-built condition to its present moisture content would be an additional contributing factor.

For splitting in beam A,

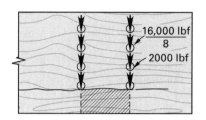 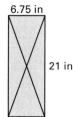

Consider each bolt to resist an equal share of the reaction from beam B.

$$r = \frac{R}{n} = \frac{16{,}000 \text{ lbf}}{8 \text{ bolts}} = 2000 \text{ lbf/bolt}$$

Take a free-body diagram of the crosshatched region.

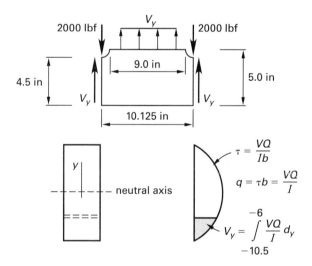

For the 16,000 lbf reaction from beam B at the midspan of beam A, the shear to either side of the connection is 8000 lbf. The portion of the resisting shears in beam A below the lowest bolt is obtained by integrating the shear flow, $q = \tau b = VQ/I$, from the lowest fiber, $y = -10.5$ in, to the level at the bottom of the bolt, $y = -6.0$ in.

$$V_y = \int_{-10.5\,\text{in}}^{-6.0\,\text{in}} \frac{VQ}{I} \, dy = \frac{V}{I} \int_{-10.5\,\text{in}}^{-6.0\,\text{in}} Q \, dy$$

$$I = \frac{bd^3}{12} = \frac{(6.75 \text{ in})(21.0)^3}{12} = 5209 \text{ in}^4$$

$$\overline{y} = y + \frac{10.5 \text{ in} - y}{2} = \frac{10.5 \text{ in} + y}{2}$$

$$Q = A\overline{y} = \left((6.75 \text{ in})(10.5 \text{ in} - y)\right)\left(\frac{10.5 \text{ in} + y}{2}\right)$$

$$= \left(\frac{6.75 \text{ in}}{2}\right)(10.5 \text{ in} - y)(10.5 \text{ in} + y)$$

$$= 372 \text{ in}^3 - (3.375 \text{ in})y^2$$

Thus,

$$V_y = \frac{V}{I} \int_{-10.5\,\text{in}}^{-6.0\,\text{in}} \left(372 \text{ in}^3 - (3.375 \text{ in})y^2\right) dy$$

$$= \left(\frac{8000 \text{ lbf}}{5209 \text{ in}^4}\right)\left((372 \text{ in}^3)y - \frac{(3.375 \text{ in})y^3}{3}\right)\Bigg|_{-10.5\,\text{in}}^{-6.0\,\text{in}}$$

$$= 944 \text{ lbf}$$

Take a summation of the vertical forces.

$$\sum F_y = 2V_y - 2r + f_t bL = 0$$

$$f_t = \frac{2r - 2V_y}{bL}$$

$$= \frac{(2)(2000 \text{ lbf}) - (2)(944 \text{ lbf})}{(6.75 \text{ in})(9 \text{ in})}$$

$$= 35 \text{ psi}$$

> The cross-grain tensile stress is about twice what is allowable for Douglas-fir (from the *AITC Manual*). This is the likely cause of the splitting in beam A.

18.2. Proposed in-place repair: Fabricate a bearing seat assembly for beam B that uses the same bolts as the present web connection. Reinforce the web of beam A using lag bolts.

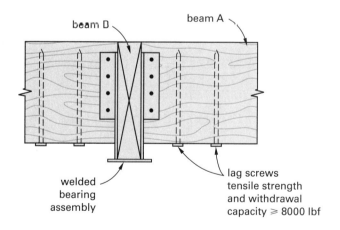

detail of proposed repair
(not to scale)

SOLUTION 19

19.1. Investigate the compliance of beams A and B with the current IBC and NDS. Include a 20 lbf/ft^2 movable partition load as part of the total live load

(IBC Sec. 1607.5): $w_L = 50$ lbf/ft^2 + 20 lbf/ft^2 = 70 lbf/ft^2, $w_n = 30$ lbf/ft^2. Reduce live loads in accordance with the alternate method of IBC Sec. 1607.10.2.

$$A_t = sL = (18 \text{ ft})(40 \text{ ft}) = 720 \text{ ft}^2 \quad [> 150 \text{ ft}^2]$$

$$L_o = 70 \text{ lbf/ft}^2 \quad [< 100 \text{ lbf/ft}^2]$$

Therefore, live loads are reducible.

$$R \leq \begin{cases} r(A_t - 150 \text{ ft}^2) = (0.0008)(720 \text{ ft}^2 - 150 \text{ ft}^2) \\ \qquad\qquad = 0.46 \\ 0.23\left(1 + \dfrac{w_D}{w_L}\right) = (0.23)\left(1 + \dfrac{30 \dfrac{\text{lbf}}{\text{ft}^2}}{70 \dfrac{\text{lbf}}{\text{ft}^2}}\right) \\ \qquad\qquad = 0.33 \quad [\text{controls}] \end{cases}$$

$$P = sL(w_D + (1 - R)w_L)$$
$$= (8 \text{ ft})(18 \text{ ft})\left(30 \frac{\text{lbf}}{\text{ft}^2} + (1 - 0.33)\left(70 \frac{\text{lbf}}{\text{ft}^2}\right)\right)$$
$$= 11{,}070 \text{ lbf} \quad [\text{say } 11.1 \text{ kips}]$$

For the beam between lines 3 and 4,

11.1 kips   11.1 kips   11.1 kips

7 ft | 8 ft | 8 ft | 8 ft

V_3 R_4

$$\sum M_4 = V_3 L - P(s_1 + s_2 + s_3) = 0$$
$$V_3 = \frac{(11.1 \text{ kips})(8 \text{ ft} + 16 \text{ ft} + 24 \text{ ft})}{31 \text{ ft}} = 17.2 \text{ kips}$$
$$\sum F_y = V_4 + V_3 - 3P = 0$$
$$V_4 = (3)(11.1 \text{ kips}) - 17.2 \text{ kips} = 16.1 \text{ kips}$$

Maximum moment occurs where V is zero (that is, at 16 ft).

$$M_{34}^+ = V_4 s_2 - P s_1$$
$$= (16.1 \text{ kips})(16 \text{ ft}) - (11.1 \text{ kips})(8 \text{ ft})$$
$$= 169 \text{ ft-kips}$$

For the cantilevered span from lines 1 to 3, the member is loaded by concentrated forces at joist locations plus the shear from member 3–4. Per the problem statement, the beam weight may be neglected. Maximum positive bending moment occurs when the dead load only acts over the region from 2 to 4; critical negative moment occurs when the live and dead loads act on that region.

For the dead load only, compute the loads on the region from 2 to 4 by proportion.

$$\%D = \frac{w_D}{w_D + (1 - R)w_L}$$
$$= \frac{30 \frac{\text{lbf}}{\text{ft}^2}}{30 \frac{\text{lbf}}{\text{ft}^2} + (1 - 0.33)\left(70 \frac{\text{lbf}}{\text{ft}^2}\right)} = 0.39$$

$$V_{3D} = (\%D)V_3 = (0.39)(17.2 \text{ kips}) = 6.7 \text{ kips}$$
$$P_D = (\%D)P = (0.39)(11.1 \text{ kips}) = 4.3 \text{ kips}$$

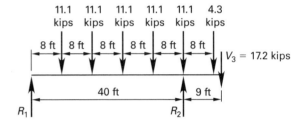

11.1 kips   11.1 kips   11.1 kips   11.1 kips   11.1 kips   4.3 kips

8 ft | 8 ft | 8 ft | 8 ft | 8 ft | 8 ft $V_3 = 17.2$ kips

40 ft   9 ft

R_1 R_2

$$\sum M_1 = R_2 L - P(s_1 + s_2 + s_3 + s_4 + s_5)$$
$$\qquad\quad - P_D s_6 - V_{3D}(L + a) = 0$$
$$R_2 = \frac{P(s_1 + s_2 + s_3 + s_4 + s_5) + P_D s_6 + V_{3D}(L + a)}{L}$$
$$= \frac{\begin{array}{c}(11.1 \text{ kips})(8 \text{ ft} + 16 \text{ ft} + 24 \text{ ft} + 32 \text{ ft} + 40 \text{ ft}) \\ + (4.3 \text{ kips})(48.0 \text{ ft}) \\ + (6.7 \text{ kips})(40 \text{ ft} + 9 \text{ ft})\end{array}}{40 \text{ ft}}$$
$$= 46.7 \text{ kips}$$

$$\sum F_y = R_1 + R_2 - 5P - P_D - V_{3D} = 0$$
$$R_1 = 5P + P_D + V_{3D} - R_2$$
$$= (5)(11.1 \text{ kips}) + 4.3 \text{ kips} + 6.7 \text{ kips} - 46.7 \text{ kips}$$
$$= 19.8 \text{ kips}$$

Maximum positive moment occurs where $V = 0$ (that is, at 16 ft from the left support).

$$M_{12}^+ = R_1 s_2 - P s_1$$
$$= (19.8 \text{ kips})(16 \text{ ft}) - (11.1 \text{ kips})(8 \text{ ft})$$
$$= 228 \text{ ft-kips}$$

Critical negative bending occurs over the support with live and dead loads on the region from 2 to 3.

$$M_2^- = P s_1 + V_3 a$$
$$= (11.1 \text{ kips})(8 \text{ ft}) + (17.2 \text{ kips})(9 \text{ ft})$$
$$= 243.6 \text{ ft-kips}$$

$$V_{\max} \leq \begin{cases} P + V_3 = 11.1 \text{ kips} + 17.2 \text{ kips} \\ \qquad = 28.3 \text{ kips} \quad \text{[controls]} \\ R_2 - P - P_D - V_{3D} = 46.7 \text{ kips} - 11.1 \text{ kips} \\ \qquad\qquad\qquad - 4.3 \text{ kips} - 6.7 \text{ kips} \\ \qquad = 24.6 \text{ kips} \end{cases}$$

For the proposed $6^{3}/_{4} \times 33$, $S = 1225$ in^3. For the top fiber of a 24F-1.7E stress class (top stressed in tension), $F_b = 1450$ psi, $E = 1.7 \times 10^6$ psi, load duration is normal, and all other adjustment factors except stability and volume are taken as 1. The critical condition for flexure is at the overhang. From NDS Table 3.3.3,

$$\frac{l_u}{d} = \frac{(9 \text{ ft})\left(12 \,\frac{\text{in}}{\text{ft}}\right)}{33 \text{ in}} = 3.3 \quad [< 7]$$

Therefore,

$$l_e = 1.87 l_u = (1.87)(9 \text{ ft})\left(12 \,\frac{\text{in}}{\text{ft}}\right) = 202 \text{ in}$$

$$R_B = \sqrt{\frac{l_e d}{b^2}} = \sqrt{\frac{(202 \text{ in})(33 \text{ in})}{(6.75 \text{ in})^2}} = 12.1$$

$$E'_{y,\min} = E_{y,\min} = 690{,}000 \text{ psi}$$

$$F^*_{bx} = C_D F_b = (1.0)\left(1450 \,\frac{\text{lbf}}{\text{in}^2}\right) = 1450 \text{ psi}$$

$$F_{bE} = \frac{1.20 E'_{y,\min}}{R_B^2} = \frac{(1.20)\left(690{,}000 \,\frac{\text{lbf}}{\text{in}^2}\right)}{(12.1)^2}$$

$$\qquad = 5660 \text{ psi}$$

$$\frac{F_{bE}}{F^*_{bx}} = \frac{5660 \,\dfrac{\text{lbf}}{\text{in}^2}}{1450 \,\dfrac{\text{lbf}}{\text{in}^2}} = 3.90$$

$$C_L = \frac{1 + \dfrac{F_{bE}}{F^*_{bx}}}{1.9} - \sqrt{\left(\dfrac{1 + \dfrac{F_{bE}}{F^*_{bx}}}{1.9}\right)^2 - \dfrac{\dfrac{F_{bE}}{F^*_{bx}}}{0.95}}$$

$$\qquad = \frac{1 + 3.90}{1.9} - \sqrt{\left(\frac{1 + 3.90}{1.9}\right)^2 - \frac{3.90}{0.95}}$$

$$\qquad = 0.98$$

From NDS Sec. 5.3.6, check the volume factor. Take the length of tension region as the span AB (conservative).

$$C_V = \left(\left(\frac{21}{L}\right)\left(\frac{12}{d}\right)\left(\frac{5.125}{b}\right)\right)^{1/x} \leq 1.0$$

$$\quad = \left(\left(\frac{21 \text{ ft}}{40 \text{ ft}}\right)\left(\frac{12 \text{ in}}{33 \text{ in}}\right)\left(\frac{5.125 \text{ in}}{6.75 \text{ in}}\right)\right)^{1/10}$$

$$\quad = 0.82 \quad [< C_L, \text{ so } C_V \text{ controls}]$$

$$F'_b = C_V F_b = (0.82)\left(1450 \,\frac{\text{lbf}}{\text{in}^2}\right) = 1190 \text{ psi}$$

$$f_b = \frac{M_2^-}{S} = \frac{(243{,}500 \text{ ft-lbf})\left(12 \,\frac{\text{in}}{\text{ft}}\right)}{1225 \text{ in}^3}$$

$$\quad = 2385 \text{ psi} \quad [> F'_b, 100\% \text{ overstress}]$$

Check shear.

$$f_v = \frac{1.5 V}{bd} = \frac{(1.5)(28{,}290 \text{ lbf})}{(6.75 \text{ in})(33.0 \text{ in})} = 190 \text{ psi}$$

$$F'_v = F_v = 210 \text{ psi} \quad [> f_v]$$

Shear is OK. Check the bearing perpendicular to grain. The maximum bearing occurs with live and dead loads on both spans. From a free-body diagram of the region from 1 to 3,

$$\sum M_1 = R_2 L - P(s_1 + s_2 + s_3 + s_4 + s_5 + s_6)$$

$$\qquad\quad - V_3(L + a) = 0$$

$$R_2 = \frac{P(s_1 + s_2 + s_3 + s_4 + s_5 + s_6) + V_3(L + a)}{L}$$

$$\quad = \frac{(11.1 \text{ kips})\begin{pmatrix} 8 \text{ ft} + 16 \text{ ft} + 24 \text{ ft} \\ + 32 \text{ ft} + 40 \text{ ft} + 48 \text{ ft} \end{pmatrix}}{40 \text{ ft}}$$

$$\qquad\qquad\quad \frac{+ (17.2 \text{ kips})(40 \text{ ft} + 9 \text{ ft})}{40 \text{ ft}}$$

$$\quad = 67.7 \text{ kips}$$

$$F'_{c\perp} = F_{c\perp} = 500 \text{ psi}$$

$$N \geq \frac{R}{F'_{c\perp} b} = \frac{67{,}700 \text{ lbf}}{\left(500 \,\frac{\text{lbf}}{\text{in}^2}\right)(6.75 \text{ in})}$$

$$\quad = 20 \text{ in} \quad [\text{at support 2}]$$

For the exterior supports, support 1 with R_1 is more critical than at 4 with $R_4 = 16.1$ kips.

$$N \geq \frac{R}{F'_{c\perp} b} = \frac{19{,}800 \text{ lbf}}{\left(500 \,\frac{\text{lbf}}{\text{in}^2}\right)(6.75 \text{ in})}$$

$$\quad = 5.9 \text{ in} \quad [\text{at support 1}]$$

Thus, the specified bearing lengths of 6 in at exterior supports and 20 in at the interior support are adequate assuming plate thickness is adequate and length provided at exterior supports is adequate for masonry bearing.

For beam B, $6^3/4$ in $\times$ 28.5 in, $I_x = 13{,}020$ in^4, and $S_x = 914$ in^3.

$$C_V = \left(\left(\frac{21}{L}\right)\left(\frac{12}{d}\right)\left(\frac{5.125}{b}\right)\right)^{1/x} \leq 1.0$$

$$= \left(\left(\frac{21 \text{ ft}}{31 \text{ ft}}\right)\left(\frac{12 \text{ in}}{28.5 \text{ in}}\right)\left(\frac{5.125 \text{ in}}{6.75 \text{ in}}\right)\right)^{1/10}$$

$$= 0.86 \quad [\text{controls}]$$

$$F'_b = C_V F_b = (0.86)\left(2400 \ \frac{\text{lbf}}{\text{in}^2}\right) = 2064 \text{ psi}$$

$$f_b = \frac{M_{34}^+}{S} = \frac{(169 \text{ ft-kips})\left(12 \ \frac{\text{in}}{\text{ft}}\right)\left(1000 \ \frac{\text{lbf}}{\text{kip}}\right)}{914 \text{ in}^3}$$

$$= 2220 \text{ psi} \quad [> F'_b, \ 8\% \text{ overstress}]$$

$$f_v = \frac{1.5 \, V}{bd} = \frac{(1.5)(17{,}200 \text{ lbf})}{(6.75 \text{ in})(28.5 \text{ in})} = 134 \text{ psi}$$

$$F'_v = F_v = 210 \text{ psi} \quad [> f_v]$$

> Both beams A and B are overstressed in flexure.

19.2. Recommended revisions include the following: Shortening the length of overhang will reduce the controlling negative moment and increase the positive moment in beam A, and result in an increase in bending moment in beam B. This may make more efficient use of material but considering the magnitude of the overstress (100% in beam A) it will not resolve the problem. Changing the specified 24F-E4 to 24F-E13 alleviates problems at the cantilever by making the top and bottom allowable fiber stresses equal to 2400 psi (rather than the 1450 psi limit for the compression zone stressed in tension in a 24F-E4).

Relocating the splice to 8 ft to the right of line 2 changes the shears and moments as follows,

$$\sum M_4 = V_3 L - P(s_1 + s_2 + s_3) = 0$$

$$V_3 = \frac{P(s_1 + s_2 + s_3)}{L}$$

$$= \frac{(11.1 \text{ kips})(8 \text{ ft} + 16 \text{ ft} + 24 \text{ ft})}{32 \text{ ft}}$$

$$= 16.7 \text{ kips} \quad [V_{3D} = 0.39 V_3 = 6.5 \text{ kips}]$$

$$\sum F_y = V_4 + V_3 - 3P = 0$$

$$V_4 = 3P - V_3$$

$$= (3)(11.1 \text{ kips}) - 16.7 \text{ kips}$$

$$= 16.6 \text{ kips}$$

$$M_{34}^+ = V_4 s_2 - P s_1$$

$$= (16.6 \text{ kips})(16 \text{ ft}) - (11.1 \text{ kips})(8 \text{ ft})$$

$$= 177 \text{ ft-kips}$$

$$\sum M_1 = R_2 L - P(s_1 + s_2 + s_3 + s_4 + s_5)$$
$$- P_D s_6 - V_{3D}(L + a) = 0$$

$$R_2 = \frac{P(s_1 + s_2 + s_3 + s_4 + s_5) + P_D s_6 + V_{3D}(L + a)}{L}$$

$$= \frac{\begin{array}{c}(11.1 \text{ kips})(8 \text{ ft} + 16 \text{ ft} + 24 \text{ ft} + 32 \text{ ft} + 40 \text{ ft}) \\ + (4.3 \text{ kips})(48.0 \text{ ft}) \\ + (6.5 \text{ kips})(40 \text{ ft} + 8 \text{ ft})\end{array}}{40 \text{ ft}}$$

$$= 46.3 \text{ kips}$$

$$\sum F_y = R_1 + R_2 - 5P - P_D - V_{3D} = 0$$

$$R_1 = 5P + P_D + V_{3D} - R_2$$

$$= (5)(11.1 \text{ kips}) + 4.3 \text{ kips} + 6.5 \text{ kips} - 46.3 \text{ kips}$$

$$= 20.0 \text{ kips}$$

Maximum positive moment is at 16 ft from the left support.

$$M_{12}^+ = R_1 s_2 - P s_1$$

$$= (20.0 \text{ kips})(16 \text{ ft}) - (11.1 \text{ kips})(8 \text{ ft})$$

$$= 231 \text{ ft-kips}$$

$$M_2^- = P s_1 + V_3 a$$

$$= (11.1 \text{ kips})(8 \text{ ft}) + (16.7 \text{ kips})(8 \text{ ft})$$

$$= 222.4 \text{ ft-kips}$$

The positive and negative bending moments are more nearly equal in magnitude in beam A and now controlled by positive bending. The required section modulus is

$$S = \frac{M_{12}^+}{F'_b} = \frac{(231{,}000 \text{ ft-lbf})\left(12 \ \frac{\text{in}}{\text{ft}}\right)}{1970 \ \frac{\text{lbf}}{\text{in}^2}} = 1407 \text{ in}^3$$

> Increase member size for beam A to $6\frac{3}{4}$ in $\times$ 36 in ($S_x = 1458$ in^3, $I_x = 26{,}240$ in^4).

$$C_V = \left(\left(\frac{21}{L} \right) \left(\frac{12}{d} \right) \left(\frac{5.125}{b} \right) \right)^{1/x} \leq 1.0$$

$$= \left(\left(\frac{21 \text{ ft}}{40 \text{ ft}} \right) \left(\frac{12 \text{ in}}{36 \text{ in}} \right) \left(\frac{5.125 \text{ in}}{6.75 \text{ in}} \right) \right)^{1/10}$$

$$= 0.82 \quad \text{[practically no change]}$$

Check total load deflection against limit of $L/240$. Superimpose case 5, AISC Table 3-22a, with dead and live loads of 11.1 kips, and case 26, AISC Table 3-23, with dead loads on the overhang.

$$\Delta_{\text{TL}} = \frac{ePL^3}{EI} - \frac{0.0208 P_D a L^2}{EI}$$

$$= \frac{\left(\begin{array}{c} (0.063)(11.1 \text{ kips})(40 \text{ ft})^3 \\ - (0.0208)(6.5 \text{ kips})(8 \text{ ft})(40 \text{ ft})^2 \end{array} \right) \left(12 \frac{\text{in}}{\text{ft}} \right)^3}{\left(1700 \frac{\text{kips}}{\text{in}^2} \right) (26{,}240 \text{ in}^4)}$$

$$= 1.6 \text{ in} \quad [L/240 = 2.0 \text{ in}]$$

For beam B,

$$S = \frac{M_{34}^+}{F_b'} = \frac{(177{,}000 \text{ ft-lbf}) \left(12 \frac{\text{in}}{\text{ft}} \right)}{2064 \frac{\text{lbf}}{\text{in}^2}} = 1029 \text{ in}^3$$

> Increase member size for beam B to $6\frac{3}{4}$ in $\times$ $31\frac{1}{2}$ in ($S_x = 1116 \text{ in}^3$, $I_x = 17{,}580 \text{ in}^4$).

19.3. Determine the camber for beam B. Use 1.5 times the dead load deflection (AISC Table 3-22a, case 4).

$$\Delta_{\text{DL}} = \frac{eP_nL^3}{EI} = \frac{(0.05)(4.3 \text{ kips})(32 \text{ ft})^3 \left(12 \frac{\text{in}}{\text{ft}} \right)^3}{\left(1700 \frac{\text{kips}}{\text{in}^2} \right) (17{,}580 \text{ in}^4)}$$

$$= 0.41 \text{ in}$$

$$\Delta_{\text{camber}} = 1.5 \Delta_{\text{DL}} = (1.5)(0.41 \text{ in})$$

$$= 0.62 \text{ in} \quad \text{[upward]}$$

19.4. The feasible camber diagram for beam A is simply a mirror image of the deflected shape under the dead load. Thus,

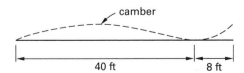

19.5. Design the connection at point 3. Compute the bearing length required for a total shear of $V = V_3 + P = 16.7 \text{ kips} + 11.1 \text{ kips} = 27.8 \text{ kips}$. Take $F_{c\perp}' = 500 \text{ psi}$ for a 24F-1.7E stress grade glulam.

$$N \geq \frac{R}{F_{c\perp}' b} = \frac{27{,}800 \text{ lbf}}{\left(500 \frac{\text{lbf}}{\text{in}^2} \right) (6.75 \text{ in})}$$

$$= 8.2 \text{ in} \quad \text{[say 8 in]}$$

Use $1/4$ in side plates of ASTM A36 steel, and use LRFD.

$$A_g = 2wt = (2)(0.25 \text{ in})(8 \text{ in}) = 4 \text{ in}^2$$

$$\phi P_n = \phi_t A_g F_y = (0.9)(4 \text{ in}^2) \left(36 \frac{\text{kips}}{\text{in}^2} \right)$$

$$= 130 \text{ kips} \quad \text{[more than adequate]}$$

The factored load on the bearing plate is

$$w_u = \frac{1.2 w_D + 1.6 w_L}{b} = \frac{1.2(\%D)R + 1.6(1 - \%D)R}{b}$$

$$= \frac{\begin{array}{c} (1.2)(0.39)(27.8 \text{ kips}) \\ + (1.6)(1 - 0.39)(27.8 \text{ kips}) \end{array}}{6.75 \text{ in}}$$

$$= 5.95 \text{ kips/in}$$

$$M_u = 0.125 w_u L^2 = (0.125) \left(5.95 \frac{\text{kips}}{\text{in}} \right) (6.75 \text{ in})^2$$

$$= 33.9 \text{ in-kips}$$

$$\phi M_n = \phi F_y \left(\frac{wt^2}{4} \right) = M_u = 33.9 \text{ in-kips}$$

$$t = \sqrt{\frac{(4)(33.9 \text{ in-kips})}{\phi F_y w}}$$

$$= \sqrt{\frac{(4)(33.9 \text{ in-kips})}{(0.9) \left(36 \frac{\text{kips}}{\text{in}^2} \right) (8 \text{ in})}}$$

$$= 0.72 \text{ in} \quad [\text{say } 3/4 \text{ in}]$$

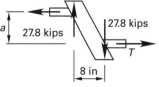

To resist the couple formed by the equal and opposite shear transfer, use bolts placed approximately 4 in from the top and bottom edges of the connected members; that is, from the depth $d = 31.5$ in of beam B, deduct two times 4 in to obtain a lever arm of $a = 31.5$ in $-(2)$ $(4$ in$) = 23.5$ in.

$$\sum M_{\text{top}} = Rw - Ta = 0$$

$$T = \frac{Rw}{a} = \frac{(27.8 \text{ kips})(8 \text{ in})}{23.5 \text{ in}} = 9.5 \text{ kips}$$

From NDS Table 11I, $^3/_4$ in bolts with steel side plates in double shear for a $6^3/_4$ in main member will furnish $Z'_{\parallel} = Z_{\parallel} = 3340$ lbf/bolt; therefore, use three $^3/_4$ in bolts on each side.

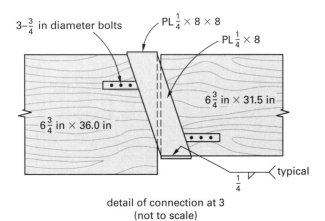

detail of connection at 3
(not to scale)

SOLUTION 20

20.1. Modify the existing ceiling framing to support the new equipment loads. Use ASD.

$$w = w_D s = \left(10 \ \frac{\text{lbf}}{\text{ft}^2}\right)(12 \text{ ft}) = 120 \text{ lbf/ft}$$

$$M_{\max} = 0.125wL^2 + 0.33PL$$
$$= (0.125)\left(120 \ \frac{\text{lbf}}{\text{ft}}\right)(12 \text{ ft})^2$$
$$\quad + (0.33)(1000 \text{ lbf})(12 \text{ ft})$$
$$= 6120 \text{ ft-lbf}$$

For a 4×8, no. 1 Douglas fir-larch, NDS Table 4A gives $F_b = 1000$ psi $(C_F = 1.3)$, $F_v = 180$ psi, $F_{c\perp} = 625$ psi, and $E = 1.7 \times 10^6$ psi. For the existing 4×8 beam, $b = 3.5$ in, $d = 7.25$ in, $I = 111$ in^4, and $S = 30.7$ in^3. Check the existing 4×8 under the proposed loading.

$$f_b = \frac{M_{\max}}{S} = \frac{(6120 \text{ ft-lbf})\left(12 \ \frac{\text{in}}{\text{ft}}\right)}{30.7 \text{ in}^3} = 2392 \text{ psi}$$

$$F'_b = C_D C_F F_b = (0.9)(1.3)\left(1000 \ \frac{\text{lbf}}{\text{in}^2}\right)$$
$$= 1170 \text{ psi} \quad [<f_b]$$

$$V = 0.5wL - wd + P$$
$$= (0.5)\left(120 \ \frac{\text{lbf}}{\text{ft}}\right)(12 \text{ ft})$$
$$\quad - \left(120 \ \frac{\text{lbf}}{\text{ft}}\right)(0.60 \text{ ft}) + 1000 \text{ lbf}$$
$$= 1648 \text{ lbf}$$

$$f_v = \frac{1.5 V}{bd} = \frac{(1.5)(1648 \text{ lbf})}{(3.5 \text{ in})(7.25 \text{ in})} = 97 \text{ psi}$$

$$F'_v = C_D F_v = (0.9)\left(180 \ \frac{\text{lbf}}{\text{in}^2}\right) = 162 \text{ psi} \quad [>f_v]$$

$$E' = E = 1,700,000 \text{ psi}$$

$$\Delta_{\text{TL}} = \frac{0.013wL^4}{E'I} + \frac{0.036PL^3}{E'I}$$
$$= \frac{(0.013)\left(120 \ \frac{\text{lbf}}{\text{ft}}\right)(12 \text{ ft})^4 \left(12 \ \frac{\text{in}^3}{\text{ft}}\right)}{\left(1,700,000 \ \frac{\text{lbf}}{\text{in}^2}\right)(111 \text{ in}^4)}$$
$$\quad + \frac{(0.036)(1000 \text{ lbf})\left((12 \text{ ft})\left(12 \ \frac{\text{in}}{\text{ft}}\right)\right)^3}{\left(1,700,000 \ \frac{\text{lbf}}{\text{in}^2}\right)(111 \text{ in}^4)}$$
$$= 0.30 \text{ in} + 0.57 \text{ in}$$
$$= 0.87 \text{ in}$$
$$\frac{L}{240} = \frac{144 \text{ in}}{240} = 0.6 \text{ in} \quad [<\Delta_{\text{TL}}]$$

Thus, the 4×8 is overstressed in bending and the deflection is excessive. Since the beam is accessible from above, try to reinforce the beam by adding a 4 in wide piece at top. Assume that the 4×8 is deflected by the uniformly distributed load (no increase applied for creep).

$$\Delta_{\text{unif}} = 0.30 \text{ in}$$

The beam stiffness must be sufficient to limit the deflection due to the 1000 lbf equipment loads.

$$\Delta_{\text{conc}} = L/240 - \Delta_{\text{unif}} = 0.60 \text{ in} - 0.30 \text{ in} = 0.30 \text{ in}$$

Therefore,

$$\Delta_{\text{conc}} = \frac{0.0417Pa(3L^2 - 4a^2)}{E'I_{\text{req}}} = 0.30 \text{ in}$$

$$I_{\text{req}} = \frac{0.0417Pa(3L^2 - 4a^2)}{(0.30 \text{ in})E'}$$
$$= \frac{\begin{array}{c}(0.0417)(1000 \text{ lbf})(4 \text{ ft}) \\ \times \left((3)(12 \text{ ft})^2 - (4)(4 \text{ ft})^2\right)\left(12 \ \frac{\text{in}}{\text{ft}}\right)^3\end{array}}{(0.30 \text{ in})\left(1,700,000 \ \frac{\text{lbf}}{\text{in}^2}\right)}$$
$$= 208 \text{ in}^4$$

Let h be the overall depth.

$$I = \frac{bh^3}{12}$$

$$\frac{bh^3}{12} \geq I_{\text{req}}$$

$$h \geq \sqrt[3]{\frac{12I_{\text{req}}}{b}} = \sqrt[3]{\frac{(12)(208 \text{ in}^4)}{3.5 \text{ in}}} = 9.0 \text{ in}$$

Try adding a 4×6 above the existing 4×8, which takes advantage of the available space and provides more than adequate stiffness.

$$h = h_8 + h_6 = 7.25 \text{ in} + 5.5 \text{ in} = 12.75 \text{ in}$$

$$I = \frac{bh^3}{12} = \frac{(3.5 \text{ in})(12.75 \text{ in})^3}{12} = 605 \text{ in}^4$$

$$S = \frac{I}{c} = \frac{605 \text{ in}^4}{6.375 \text{ in}} = 94.9 \text{ in}^3$$

Superimpose the stresses due to 120 lbf/ft acting on the existing 4×8 with the stresses due to the 1000 lbf equipment load on the built-up section.

$$f_b = \frac{M_1}{S_1} + \frac{M_2}{S_2} = \frac{(2160 \text{ ft-lbf})\left(12 \frac{\text{in}}{\text{ft}}\right)}{30.7 \text{ in}^3}$$

$$+ \frac{(4000 \text{ ft-lbf})\left(12 \frac{\text{in}}{\text{ft}}\right)}{94.9 \text{ in}^3}$$

$$= 1350 \text{ psi} \quad [> F'_b = 1170 \text{ psi}]$$

This is no good because stresses due to the ceiling load are too high. Shore up the 4×8 prior to installing the top piece and then release the built-up section. The deflection will obviously be acceptable. The built-up section will resist stresses of

$$f_b = \frac{M_1 + M_2}{S_2} = \frac{(2160 \text{ ft-lbf} + 4000 \text{ ft-lbf})\left(12 \frac{\text{in}}{\text{ft}}\right)}{94.9 \text{ in}^3}$$

$$= 780 \text{ psi} \quad [< F'_b = 1170 \text{ psi}]$$

Shear stresses are OK by previous calculations. Try connecting the two pieces with 16d toenails as shown.

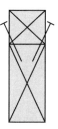

From NDS Table 11N for a 16d common nail ($D = 0.162$ in), toenailed ($C_{\text{tn}} = 0.83$),

$$Z' = C_D C_{\text{tn}} Z_{\parallel} = (0.9)(0.83)(141 \text{ lbf})$$

$$= 105 \text{ lbf} \quad [\text{per nail}]$$

Critical shear flow occurs at the interface between the 4×8 and 4×6 near the supports where shear is maximum.

$$V = R = 1720 \text{ lbf}$$

$$\overline{y} = 0.5(h - d_6) = (0.5)(12.75 \text{ in} - 5.5 \text{ in})$$

$$= 3.625 \text{ in}$$

$$Q = bd_6\overline{y} = (3.5 \text{ in})(5.5 \text{ in})(3.625 \text{ in}) = 70 \text{ in}^3$$

$$q = \frac{VQ}{I} = \frac{(1720 \text{ lbf})(70 \text{ in}^3)}{605 \text{ in}^4} = 199 \text{ lbf/in}$$

$$s = \frac{Z'_{\parallel}}{q} = \frac{105 \frac{\text{lbf}}{\text{nail}}}{199 \frac{\text{lbf}}{\text{in}}} = 0.5 \text{ in/nail}$$

Even with nails on opposite sides, which would increase the spacing along each side to 1 in, the spacing is too close to be practical. Therefore, try attaching the two pieces using lag screws installed from above and penetrating the existing 4×8. The resulting loss of cross section should not be critical because the bending stresses are only about 75% of that allowable in the built-up section.

From NDS Table 11J, try $^3/_4$ in diameter $\times 9$ in lag screws and assume a group factor, C_G, of 0.7.

$$Z'_{\parallel} = C_D C_G Z_{\parallel} = (0.9)(0.7)(960 \text{ lbf})$$

$$= 605 \text{ lbf} \quad [\text{per lag screw}]$$

Near the supports,

$$q = 199 \text{ lbf/in}$$

$$s = \frac{Z'_{\parallel}}{q} = \frac{605 \text{ lbf}}{199 \frac{\text{lbf}}{\text{in}}} = 3 \text{ in} \quad [= 4d]$$

Near the load point,

$$q = \frac{VQ}{I} = \frac{(1000 \text{ lbf})(70 \text{ in}^3)}{605 \text{ in}^4} = 116 \text{ lbf/in}$$

$$s = \frac{Z'_{\parallel}}{q} = \frac{605 \text{ lbf}}{116 \frac{\text{lbf}}{\text{in}}} = 5.25 \text{ in}$$

In the 48 in end regions, the spacing averages more than 4 in o.c., requiring 10 lag screws. Therefore, per NDS Table 10.3.6A, $A_s = (3.5 \text{ in})(5.5 \text{ in}) = 19 \text{ in}^2$; the group factor for 10 lag screws equals 0.7 and is conservative for the fastener diameter and spacings used. Start the first lag screw at 3 in from the end, then vary spacing linearly from 3 in o.c. to 5.25 in near the load point; no lag screws are required in the middle region (where V is zero). Minimum penetration into main member (NDS Sec. 11.1.3) is $4D = 3.0$ in; therefore, the overall length of 9 in is adequate.

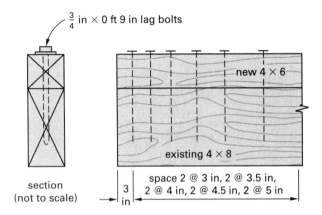

$\frac{3}{4}$ in $\times$ 0 ft 9 in lag bolts

new 4×6

existing 4×8

section
(not to scale)

space 2 @ 3 in, 2 @ 3.5 in,
2 @ 4 in, 2 @ 4.5 in, 2 @ 5 in

Recalculate the flexural stresses assuming that the portion of the lag bolt in the tension zone resists no tension (per the *AITC Manual*). Let c equal the distance from the bottom fiber to the neutral axis.

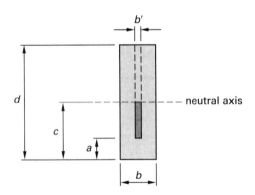

neutral axis

$$c = \frac{bd^2 - b'c^2 + b'a^2}{2(bd - b'c + b'a)}$$

$$c = \frac{\begin{array}{c}(3.5 \text{ in})(12.75 \text{ in})^2 - (0.75 \text{ in})c^2 \\ + (0.75 \text{ in})(3.75 \text{ in})^2\end{array}}{(2)\left(\begin{array}{c}(3.5 \text{ in})(12.75 \text{ in}) - (0.75 \text{ in})c \\ + (0.75 \text{ in})(3.75 \text{ in})\end{array}\right)}$$

$$0 \text{ in}^3 = (0.75 \text{ in})c^2 - (94.875 \text{ in}^2)c + 579.516 \text{ in}^3$$

$$c = 6.07 \text{ in}$$

Using the parallel axis theorem and treating the area of the hole below the neutral axis as negative contribution gives

$$I' = I + bd(0.5d - c)^2 - \frac{b'(c-a)^3}{12}$$
$$- b'(c-a)\big((0.5)(c-a)\big)^2$$
$$= 604.5 \text{ in}^4 + (3.5 \text{ in})(12.75 \text{ in})$$
$$\times \big((0.5)(12.75 \text{ in}) - 6.07 \text{ in}\big)^2$$
$$- \frac{(0.75 \text{ in})(6.07 \text{ in} - 3.75 \text{ in})^3}{12}$$
$$- (0.75 \text{ in})(6.07 \text{ in} - 3.75 \text{ in})$$
$$\times \big((0.5)(6.07 \text{ in} - 3.75 \text{ in})\big)^2$$
$$= 602.6 \text{ in}^4$$

The top fiber stress is more critical than the bottom.

$$c_{\text{top}} = 12.75 \text{ in} - 6.07 \text{ in} = 6.68 \text{ in}$$
$$S_{\text{top}} = \frac{I'}{c_{\text{top}}} = \frac{602.6 \text{ in}^4}{6.68 \text{ in}} = 90.2 \text{ in}^3$$

$$f_b = \frac{M_1 + M_2}{S_{\text{top}}}$$
$$= \frac{(2160 \text{ ft-lbf} + 4000 \text{ ft-lbf})\left(12 \frac{\text{in}}{\text{ft}}\right)}{90.2 \text{ in}^3}$$
$$= 820 \text{ psi} \quad [< F_b' = 1170 \text{ psi}]$$

Therefore, the built-up beam is OK.

The following illustration shows an alternative way to strengthen the beam and stay within the 5.5 in additional depth limit. This option may be feasible, but it is not investigated in this solution.

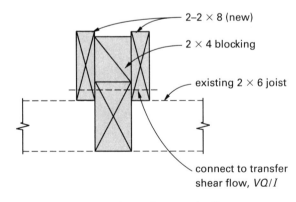

2–2 $\times$ 8 (new)

2 $\times$ 4 blocking

existing 2 $\times$ 6 joist

connect to transfer
shear flow, VQ/I

section—alternate detail

7 Masonry Design

PROBLEM 1

The north wall of a building contains a number of masonry piers. A portion of the wall elevation is shown. The seismic forces indicated were computed using ASCE/SEI7, and are given as strength-level loads.

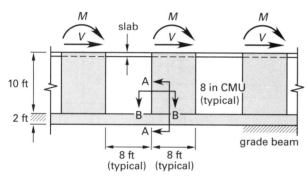

part north wall elevation

Design Criteria

- seismic design category D
- $S_{DS} = 1.0$
- redundancy factor $= 1.0$
- dead and live loads acting on wall to be neglected
- width of grade beam $= 1$ ft 4 in
- masonry units are 8 in concrete masonry units (CMU) solid grouted hollow concrete block, running bond
- $f'_m = 1.5$ ksi
- grade 60 reinforcement
- $f'_c = 3$ ksi standard weight concrete
- $V = \pm 28.6$ kips
- $M = \pm 57.1$ ft-kips

1.1. Design a typical wall and a typical grade beam. Use horizontal wall steel at 32 in on centers.

1.2. Provide detailed sketches of sections A-A and B-B.

PROBLEM 2

The retaining wall shown is to be constructed with concrete masonry units (CMU) in conformance with the IBC.

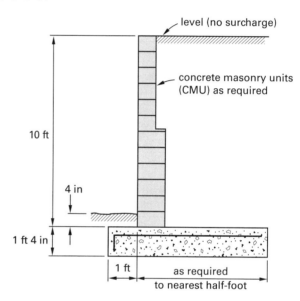

Design Criteria

- equivalent fluid unit weight, $k_a \gamma_s = 30$ lbf/ft^3
- resultant of soil pressure is to be in middle third of footing
- maximum soil bearing pressure $= 2500$ lbf/ft^2
- coefficient of friction (soil to concrete), $\mu_s = 0.4$
- $\gamma_s = 90$ lbf/ft^3
- 12 in (CMU) by 8 in high (unit weight $= 100$ lbf/ft^3)
- $f'_m = 1.5$ ksi
- grade 60 reinforcement
- $f'_c = 3.0$ ksi

2.1. Design the masonry cantilever retaining wall, and size the concrete footing. Do not design the footing reinforcement.

2.2. Provide a sketch detail indicating all dimensions, reinforcement, and any other pertinent data. Show reinforcement for the wall as designed. Indicate the placement of footing reinforcement, but do not design steel for footing.

PROBLEM 3

The uniform tributary roof load on a fully grouted exterior bearing brick masonry wall is shown. The brick wall is 13.5 in thick, and the wall weight is 135 lbf/ft².

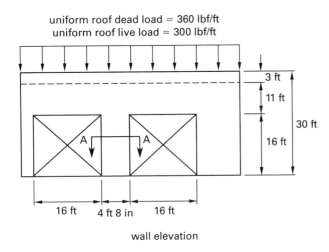

uniform roof dead load = 360 lbf/ft
uniform roof live load = 300 lbf/ft

wall elevation

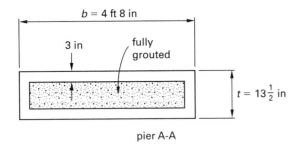

pier A-A

Design Criteria

- seismic design category D
- $S_{DS} = 1.0$
- redundancy factor = 1
- $f'_m = 2.5$ ksi
- $f_y = 60$ ksi
- importance factor = 1.0
- use strength design
- neglect eccentricity

3.1. Calculate reinforcement at pier A-A to satisfy combined vertical and seismic forces perpendicular to the wall.

3.2. Sketch a plan detail showing vertical and horizontal reinforcing steel.

PROBLEM 4

Shown is an exterior wall of regular weight hollow concrete block with all cells filled.

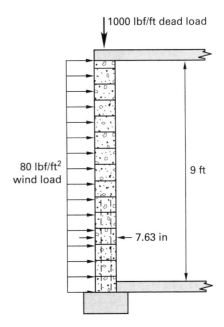

Design Criteria

- assume connection of slab to wall is adequate
- $f'_m = 1.4$ ksi
- $f_y = 60$ ksi
- weight of ungrouted masonry units $= 35$ lbf/ft²

4.1. Determine the adequate reinforcement to be placed in the center of the wall for the loads shown. Show all calculations.

SOLUTION 1

1.1. Design a typical grouted masonry shear wall for the given seismic forces. Neglect the gravity loads, and use grade 60 reinforcement and $f'_m = 1500$ psi, per the problem statement.

Use the working stress design method from ACI 530 Chap. 2. Convert the strength level seismic loads to working stress level. Use the basic load combinations of IBC Sec. 1605.3.1 (with no increase in allowable stresses, per IBC Sec. 1605.3.1.1). According to ACI 530 Sec. 1.18.3.2.6.1.2, shear must be increased by a factor of 1.5 when checking shear strength, but no increase is required for the overturning moment.

$$V = (0.7)(\pm 28.6 \text{ kips})$$
$$= \pm 20 \text{ kips} \quad [\text{working stress level}]$$
$$M = (0.7)(\pm 57.1 \text{ ft-kips})$$
$$= \pm 40 \text{ ft-kips} \quad [\text{working stress level}]$$

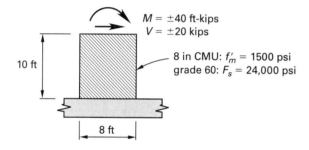

At the base of the wall,

$$M_o = M + Vh = \pm \big(40 \text{ ft-kips} + (20 \text{ kips})(10 \text{ ft})\big)$$
$$= \pm 240 \text{ ft-kips} \quad (240,000 \text{ ft-lbf})$$

Try $d \approx 0.9h = (0.9)(8 \text{ ft})(12 \text{ in/ft}) = 86.4$ in.

$$A_s \approx \frac{M_o}{F_s d} = \frac{(240,000 \text{ ft-lbf})\left(12 \frac{\text{in}}{\text{ft}}\right)}{\left(32,000 \frac{\text{lbf}}{\text{in}^2}\right)(86.4 \text{ in})}$$
$$= 1.04 \text{ in}^2 \quad [\text{say two no. 7 at each end}]$$

Two no. 7 bars would give an area of $A_s = 1.20$ in². Check stresses based on this trial value of A_s (see ACI 530 Chap. 2). Calculate a more accurate estimate of d. Position the bars 4 in and 6 in from the end of the wall. The centroid of the bars is $(4 \text{ in} + 6 \text{ in})/2 = 5$ in from the end of the wall.

$$d = (8 \text{ ft})\left(12 \frac{\text{in}}{\text{ft}}\right) - 5 \text{ in} = 91 \text{ in}$$

$$\rho = \frac{A_s}{bd} = \frac{1.20 \text{ in}^2}{(7.625 \text{ in})(91 \text{ in})} = 0.00173$$

$$n = \frac{E_s}{E_m} = \frac{29,000,000 \frac{\text{lbf}}{\text{in}^2}}{900 f'_m}$$
$$= \frac{29,000,000 \frac{\text{lbf}}{\text{in}^2}}{(900)\left(1500 \frac{\text{lbf}}{\text{in}^2}\right)}$$
$$= 21.5$$

$$\rho n = (0.00173)(21.5) = 0.037$$

$$k = \sqrt{(\rho n)^2 + 2\rho n} - \rho n$$
$$= \sqrt{(0.037)^2 + (2)(0.037)} - 0.037$$
$$= 0.238$$

$$j = 1 - 0.333k = 1 - (0.333)(0.238) = 0.92$$

$$f_s = \frac{M_o}{A_s j d} = \frac{(240,000 \text{ ft-lbf})\left(12 \frac{\text{in}}{\text{ft}}\right)}{(1.20 \text{ in}^2)(0.92)(91.0 \text{ in})}$$
$$= 28,700 \text{ psi} \quad [< F_s = 32,000 \text{ psi}]$$

$$f_c = \left(\frac{M_o}{bd^2}\right)\left(\frac{2}{jk}\right)$$
$$= \left(\frac{(240,000 \text{ ft-lbf})\left(12 \frac{\text{in}}{\text{ft}}\right)}{(7.625 \text{ in})(91.0 \text{ in})^2}\right)\left(\frac{2}{(0.92)(0.238)}\right)$$
$$= 417 \text{ psi}$$

$$F_c = 0.45 f'_m = (0.45)\left(1500 \frac{\text{lbf}}{\text{in}^2}\right)$$
$$= 675 \text{ psi} \quad [f_c < F_c]$$

Per IBC Sec. 2107.2.1, the required lap splice length is

$$l_d \geq \begin{cases} 0.002 d_b f_s = \left(0.002 \frac{\text{in}^2}{\text{lbf}}\right)(0.875 \text{ in})\left(28,700 \frac{\text{lbf}}{\text{in}^2}\right) \\ \qquad = 50 \text{ in} \quad [\text{controls}] \\ 40 d_b = (40)(0.875 \text{ in}) \\ \qquad = 35 \text{ in} \\ 12 \text{ in} \end{cases}$$

However, the stress in the no. 7 bars is $f_s = 28{,}700$ psi, which is greater than $0.8F_y = (0.8)(32{,}000 \text{ lbf/in}^2) = 25{,}600$ psi, so IBC Sec. 2107.3 also requires that this value for l_d be increased by at least 50%.

$$l_d \geq (1.5)(50 \text{ in}) = 75 \text{ in}$$

Use two no. 7 bars on each side.

Check the shear. According to ACI 530 Sec. 1.18.3.2.6.1.2, a 50% increase in shear is required for a special reinforced masonry wall, so use $1.5V$ in calculations instead of V.

$$\frac{M_o}{Vd} \leq \begin{cases} \dfrac{(240{,}000 \text{ ft-lbf})\left(12 \, \frac{\text{in}}{\text{ft}}\right)}{(20 \text{ kips})\left(1000 \, \frac{\text{lbf}}{\text{kip}}\right)(91 \text{ in})} = 1.58 \\[3mm] 1.0 \quad [\text{controls}] \end{cases}$$

$$A_{nv} = A_g = bL$$
$$= (7.625 \text{ in})(8 \text{ ft})\left(12 \, \frac{\text{in}}{\text{ft}}\right)$$
$$= 732 \text{ in}^2$$

$$F_{vm} = 0.25\left(\left(4 - 1.75\frac{M_o}{Vd}\right)\sqrt{f'_m}\right) + (0.25)\left(\frac{P}{A_{nv}}\right)$$
$$= (0.25)\left(\left(4 - (1.75)(1.0)\right)\sqrt{1500} \, \frac{\text{lbf}}{\text{in}^2}\right)$$
$$\quad + (0.25)\left(\frac{0 \text{ kips}}{732 \text{ in}^2}\right)$$
$$= 22 \text{ psi}$$

$$f_v = \frac{1.5V}{A_{nv}}$$
$$= \frac{(1.5)(20 \text{ kips})\left(1000 \, \frac{\text{lbf}}{\text{kip}}\right)}{732 \text{ in}^2}$$
$$= 41 \text{ psi}$$

$$F_{vm} + F_{vs} = f_v$$
$$F_{vs} = f_v - F_{vm}$$
$$= 41 \, \frac{\text{lbf}}{\text{in}^2} - 22 \, \frac{\text{lbf}}{\text{in}^2}$$
$$= 19 \text{ psi}$$

$$F_{vs} = (0.5)\left(\frac{A_v F_s d}{A_{nv} s}\right)$$

$$\frac{A_v}{s} = \frac{F_{vs}A_{nv}}{0.5F_s d}$$
$$= \frac{\left(19 \, \frac{\text{lbf}}{\text{in}^2}\right)(732 \text{ in}^2)}{(0.5)\left(32{,}000 \, \frac{\text{lbf}}{\text{in}^2}\right)(91 \text{ in})}$$
$$= 0.0096 \text{ in}$$

Try horizontal bars spaced at 32 in on centers.

$$A_v = s(0.0096 \text{ in}) = (32 \text{ in})(0.0096 \text{ in})$$
$$= 0.31 \text{ in}^2 \quad [\text{no. 5 rebars at 32 in on centers}]$$

According to ACI 530 Sec. 1.18.3.2.6, the area of vertical steel must be at least one-third of the required shear area and, for masonry laid in running bond, must provide at least 0.0007 times the gross area. The spacing should not exceed one-third the wall length (96 in/3 = 32 in). For bars spaced at 32 in on centers,

$$\frac{A_{s,\text{vert}}}{s} \geq \begin{cases} (0.0007)\dfrac{A_g}{s} = (0.0007)\left(\dfrac{732 \text{ in}^2}{32 \text{ in}}\right) = 0.016 \, \dfrac{\text{in}^2}{\text{in}} \\[3mm] \dfrac{0.096 \, \frac{\text{in}^2}{\text{in}}}{3} = 0.032 \, \dfrac{\text{in}^2}{\text{in}} \quad [\text{controls}] \end{cases}$$

$$A_{s,\text{vert}} = s\left(0.032 \, \frac{\text{in}^2}{\text{in}}\right) = (32 \text{ in})\left(0.032 \, \frac{\text{in}^2}{\text{in}}\right)$$
$$= 0.10 \text{ in}^2 \quad [\text{say, no. 3 at 32 in on centers}]$$

The combined vertical and horizontal steel must be at least 0.002 times the gross area.

$$A_v + A_{s,\text{vert}} \geq 0.002bs$$
$$0.31 \text{ in}^2 + 0.10 \text{ in}^2 \geq (0.002)(32 \text{ in})(7.625 \text{ in})$$
$$0.42 \text{ in}^2 \geq 0.49 \text{ in}^2 \quad [\text{false}]$$

Revise the vertical steel to no. 4 at 32 in on center.

$$A_v + A_{s,\text{vert}} \geq 0.002bs$$
$$0.31 \text{ in}^2 + 0.20 \text{ in}^2 \geq (0.002)(32 \text{ in})(7.625 \text{ in})$$
$$0.51 \text{ in}^2 \geq 0.49 \text{ in}^2 \quad [\text{OK}]$$

Use no. 5 bars at 32 in o.c. horizontally and no. 4 bars at 32 in o.c. vertically (anchor horizontal bars with standard hook).

1.2. Design the typical interior portion of the 16 in $\times$ 24 in reinforced concrete grade beam supporting the wall. Use the strength design method from ACI 318, with $f'_c = 3000$ psi and $f_y = 60{,}000$ psi.

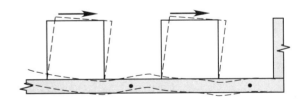

Points of inflection occur at the midspan of the grade beam.

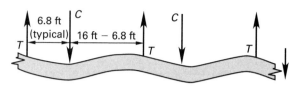

$$C = T = \frac{M}{jd}$$

$$= \frac{(240{,}000 \text{ ft-lbf})\left(12 \, \frac{\text{in}}{\text{ft}}\right)}{(0.91)(91 \text{ in})}$$

$$= 34{,}800 \text{ lbf} \quad [\text{working stress level}]$$

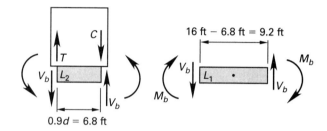

$$\sum M_{L1} = 2M_b - (9.2 \text{ ft})V_b = 0$$

$$V_b = \left(\frac{2M_b}{9.2 \text{ ft}}\right)$$

$$\sum M_{L2} = -240{,}000 \text{ ft-lbf} + V_b(6.8 \text{ ft}) + 2M_b = 0$$

$$= -240{,}000 \text{ ft-lbf} + \left(\frac{2M_b}{9.2 \text{ ft}}\right)(6.8 \text{ ft}) + 2M_b$$

$$M_b = \frac{240{,}000 \text{ ft-lbf}}{\left(\frac{2}{9.2 \text{ ft}}\right)(6.8 \text{ ft}) + 2}$$

$$= 69{,}000 \text{ ft-lbf}$$

$$V_b = \frac{(2)(69{,}000 \text{ ft-lbf})}{9.2 \text{ ft}} = 15{,}000 \text{ lbf}$$

Design flexural steel.

$$d = h - \text{cover} - 0.5d_b$$

$$= 24 \text{ in} - 3 \text{ in} - 1 \text{ in}$$

$$= 20 \text{ in}$$

$$M_u = \frac{M_b}{0.7} = \frac{69{,}000 \text{ ft-lbf}}{0.7}$$

$$= 98{,}600 \text{ ft-lbf}$$

$$M_u = \phi M_n = \phi \rho b d^2 f_y \left(1 - 0.59\rho\left(\frac{f_y}{f'_c}\right)\right)$$

$$(98{,}600 \text{ ft-lbf})\left(12 \, \frac{\text{in}}{\text{ft}}\right)$$

$$= 0.9\rho(16 \text{ in})(20.0 \text{ in})^2$$

$$\times \left(60{,}000 \, \frac{\text{lbf}}{\text{in}^2}\right)$$

$$\times \left(1 - 0.59\rho\left(\frac{60{,}000 \, \frac{\text{lbf}}{\text{in}^2}}{3000 \, \frac{\text{lbf}}{\text{in}^2}}\right)\right)$$

$$\rho = 0.0036$$

$$A_s = \rho b d = (0.0036)(16 \text{ in})(20.0 \text{ in})$$

$$= 1.15 \text{ in}^2$$

$$A_{s,\text{min}} = \frac{200 b_w d}{f_y}$$

$$= \frac{\left(200 \, \frac{\text{lbf}}{\text{in}^2}\right)(16 \text{ in})(20.0 \text{ in})}{60{,}000 \, \frac{\text{lbf}}{\text{in}^2}}$$

$$= 1.07 \text{ in}^2 \quad [\text{does not control}]$$

Therefore,

$$A_s^- = A_s^+ = 1.15 \text{ in}^2$$

Use two no. 7 continuous top and bottom bars.

Design for shear based on the probable flexural strength, treating the beam as singly reinforced for flexural calculations.

$$a_{\text{pr}} = \frac{1.25 f_y A_s}{0.85 f'_c b}$$

$$= \frac{(1.25)\left(60{,}000 \, \frac{\text{lbf}}{\text{in}^2}\right)(1.20 \text{ in}^2)}{(0.85)\left(3000 \, \frac{\text{lbf}}{\text{in}^2}\right)(16 \text{ in})}$$

$$= 2.21 \text{ in}$$

$$M_{\text{pr}} = 1.25 f_y A_s \left(d - \frac{a_{\text{pr}}}{2}\right)$$

$$= \frac{(1.25)\left(60{,}000 \, \frac{\text{lbf}}{\text{in}^2}\right)(1.20 \text{ in}^2)}{\times \left(20 \text{ in} - \frac{2.21 \text{ in}}{2}\right)}{12 \, \frac{\text{in}}{\text{ft}}}$$

$$= 142{,}000 \text{ ft-lbf}$$

Add the gravity load shear of the grade beam.

$$w_b = w_c b h$$

$$= \left(150 \, \frac{\text{lbf}}{\text{ft}^3}\right)\left(\frac{16 \text{ in}}{12 \, \frac{\text{in}}{\text{ft}}}\right)\left(\frac{24 \text{ in}}{12 \, \frac{\text{in}}{\text{ft}}}\right)$$

$$= 400 \text{ lbf/ft}$$

$$V_u = \frac{2M_{pr}}{l_t} + (1.2 + 0.2S_{DS})w_b(0.5L)$$

$$= \frac{(2)(142,000 \text{ ft-lbf})}{9.2 \text{ ft}}$$

$$+ \left(1.2 + (0.2)(1.0)\right)\left(400 \ \frac{\text{lbf}}{\text{ft}}\right)(0.5)(9.2 \text{ ft})$$

$$= 33,400 \text{ lbf}$$

More than 50% of the design shear is due to seismic effects; therefore, neglect V_c and design stirrups to resist the entire shear. Use no. 3 hoops.

$$A_v = (2)(0.11 \text{ in}^2) = 0.22 \text{ in}^2$$

$$V_s = \frac{V_u}{\phi} = \frac{33,400 \text{ lbf}}{0.75} = 44,500 \text{ lbf}$$

$$s \leq \begin{cases} \dfrac{d}{4} = \dfrac{20 \text{ in}}{4} = 5 \text{ in} \quad [\text{controls}] \\[2mm] 8d_b = (8)(0.875 \text{ in}) = 7 \text{ in} \\[2mm] 24d_h = (24)(0.5 \text{ in}) = 12 \text{ in} \\[2mm] \dfrac{A_v f_y d}{V_s} = \dfrac{(0.22 \text{ in}^2)\left(60,000 \ \dfrac{\text{lbf}}{\text{in}^2}\right)(20 \text{ in})}{44,500 \text{ lbf}} \\[2mm] \qquad = 5.9 \text{ in} \end{cases}$$

For simplicity, use 5 in spacing throughout. There is only about 1 ft over which the spacing would be allowed to increase to $d/2$.

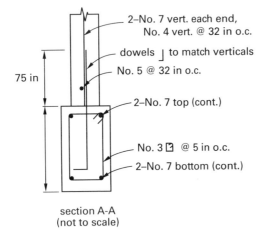

2–No. 7 vert. each end,
No. 4 vert. @ 32 in o.c.

dowels ⌐ to match verticals

No. 5 @ 32 in o.c.

2–No. 7 top (cont.)

No. 3 ⊡ @ 5 in o.c.

2–No. 7 bottom (cont.)

75 in

section A-A
(not to scale)

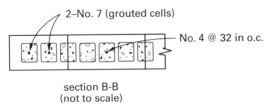

2–No. 7 (grouted cells)

No. 4 @ 32 in o.c.

section B-B
(not to scale)

SOLUTION 2

2.1. Design the masonry retaining wall and size the footing. Given: The unit weight of the soil is 90 lbf/ft³, the equivalent fluid weight is 30 lbf/ft³, the allowable soil bearing pressure is $F_p = 2500$ lbf/ft², $f'_m = 1500$ psi, and $f_y = 60,000$ psi.

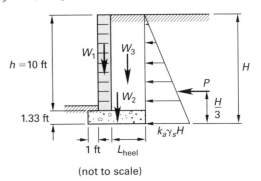

(not to scale)

Per the problem statement, the footing is to be reinforced concrete, but design of the footing reinforcement is not required. The stem is to be reinforced concrete masonry with $f'_m = 1500$ psi. The toe of the wall is to be 1 ft forward of the outside face of the stem. The compression resultant is to pass through the middle third of the footing (that is, within the kern so that there is compression beneath the footing throughout). Provide a factor of safety against sliding of 1.5 (friction coefficient is 0.4) and provide a factor of safety against overturning of 2.0. Try a 12 in wall uniform thickness, but vary the vertical reinforcement. The unit weight of the reinforced masonry is 100 lbf/ft³. Design on a unit length basis. The force of the compression resultant per foot of wall is

$$P = 0.5(k_a\gamma_s)H^2(1 \text{ ft})$$

$$= (0.5)\left(30 \ \frac{\text{lbf}}{\text{ft}^3}\right)(11.33 \text{ ft})^2(1 \text{ ft})$$

$$= 1926 \text{ lbf}$$

The overturning moment created by this force is

$$M_o = P\left(\frac{H}{3}\right)$$

$$= (1926 \text{ lbf})\left(\frac{11.33 \text{ ft}}{3}\right)$$

$$= 7267 \text{ ft-lbf}$$

Neglect the passive resistance to sliding and provide a factor of safety of 1.5. Find the minimum vertical force.

$$\mu_s W \geq (\text{FS})P$$

$$W \geq \frac{(\text{FS})P}{\mu_s}$$

$$\geq \frac{(1.5)(1926 \text{ lbf})}{0.4}$$

$$\geq 7223 \text{ lbf}$$

The stem weight per foot of wall is

$$W_1 = w_{\text{wall}} th(1 \text{ ft})$$

$$= \left(100 \ \frac{\text{lbf}}{\text{ft}^3}\right)(1 \text{ ft})(10 \text{ ft})(1 \text{ ft})$$

$$= 1000 \text{ lbf}$$

The weight of the backfill plus footing per foot of wall is

$$w = \gamma_s h L_{\text{toe}}(1 \text{ ft}) + w_c t L_{\text{toe}}(1 \text{ ft})$$

$$= \left(90 \ \frac{\text{lbf}}{\text{ft}^3}\right)(10 \text{ ft})(1 \text{ ft})(1 \text{ ft})$$

$$+ \left(150 \ \frac{\text{lbf}}{\text{ft}^3}\right)(1.33 \text{ ft})(1 \text{ ft})(1 \text{ ft})$$

$$= 1100 \text{ lbf}$$

To provide the required factor of safety against sliding,

$$W_1 + w\left(\frac{L_{\text{heel}}}{L_{\text{toe}}}\right) \geq W$$

$$L_{\text{heel}} \geq \frac{(W - W_1)L_{\text{toe}}}{w}$$

$$\geq \frac{(7223 \text{ lbf} - 1000 \text{ lbf})(1 \text{ ft})}{1100 \text{ lbf}}$$

$$\geq 5.7 \text{ ft}$$

The total required footing width is

$$B = L_{\text{toe}} + t + L_{\text{heel}}$$

$$= 1 \text{ ft} + 1 \text{ ft} + 5.7 \text{ ft}$$

$$= 7.7 \text{ ft}$$

Try 7 ft 6 in total width of footing. The total vertical force per foot of wall is then

$$W = W_1 + W_2 + W_3$$

$$= W_1 + \gamma_s L_{\text{heel}} h(1 \text{ ft}) + w_c t B(1 \text{ ft})$$

$$= 1000 \text{ lbf} + \left(90 \ \frac{\text{lbf}}{\text{ft}^3}\right)(5.7 \text{ ft})(10 \text{ ft})(1 \text{ ft})$$

$$+ \left(150 \ \frac{\text{lbf}}{\text{ft}^3}\right)(1.33 \text{ ft})(7.5 \text{ ft})(1 \text{ ft})$$

$$= 1000 \text{ lbf} + 5130 \text{ lbf} + 1496 \text{ lbf}$$

$$= 7626 \text{ lbf}$$

Calculate the factor of safety against overturning. The moment resisting overturning per foot of wall is

$$M_r = W_1 \overline{x}_1 + W_{\text{footing}} \overline{x}_2 + W_{\text{fill}} \overline{x}_3$$

$$= W_1 \overline{x}_1 + w_c t B(1 \text{ ft}) \overline{x}_2 + \gamma_s L_{\text{heel}} h(1 \text{ ft}) \overline{x}_3$$

$$= (1000 \text{ lbf})(1.5 \text{ ft}) + \left(150 \ \frac{\text{lbf}}{\text{ft}^3}\right)$$

$$\times (1.33 \text{ ft})(7.5 \text{ ft})(1 \text{ ft})\Big((0.5)(7.5 \text{ ft})\Big)$$

$$+ \left(90 \ \frac{\text{lbf}}{\text{ft}^3}\right)(5.7 \text{ ft})(10 \text{ ft})(1 \text{ ft})$$

$$\times \Big(7.5 \text{ ft} - (0.5)(5.5 \text{ ft})\Big)$$

$$= 31{,}480 \text{ ft-lbf}$$

$$\text{FS}_{\text{overturning}} = \frac{M_r}{M_o}$$

$$= \frac{31{,}480 \text{ ft-lbf}}{7267 \text{ ft-lbf}}$$

$$= 4.3 \quad [\text{OK}]$$

The resultant vertical force acting on the soil beneath the footing is located at

$$\overline{x} = \frac{M_r - M_o}{W}$$

$$= \frac{31{,}480 \text{ ft-lbf} - 7267 \text{ ft-lbf}}{7630 \text{ ft}}$$

$$= 3.2 \text{ ft}$$

One-third of the footing width is

$$\frac{B}{3} = \frac{7.5 \text{ ft}}{3} = 2.5 \text{ ft}$$

The resultant falls within the middle third of the footing, as required. Check the bearing pressure.

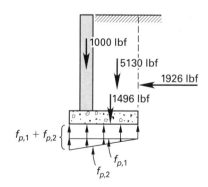

With the resultant within the middle third, the footing exerts compression throughout the width. Properties of the footing per foot of wall are

$$A = B(1 \text{ ft}) = (7.5 \text{ ft})(1 \text{ ft}) = 7.5 \text{ ft}^2$$

$$S = \frac{B^2(1 \text{ ft})}{6} = \frac{(7.5 \text{ ft})^2(1 \text{ ft})}{6} = 9.38 \text{ ft}^3$$

Take moments about the midwidth (7.5 ft/2 or 3.75 ft from the toe).

$$M = M_o + W_1 x_1 - W_2 x_2$$

$$= 7267 \text{ ft-lbf} + (1000 \text{ lbf})(2.25 \text{ ft})$$

$$\quad - (5130 \text{ lbf})(1 \text{ ft})$$

$$= 4387 \text{ ft-lbf}$$

$$f_{p,1} = \frac{W}{A} - \frac{M}{S} = \frac{7626 \text{ lbf}}{7.5 \text{ ft}^2} - \frac{4387 \text{ ft-lbf}}{9.38 \text{ ft}^3}$$

$$= 506 \text{ lbf/ft}^2 \quad [< F_p = 2500 \text{ lbf/ft}^2]$$

$$f_{p,2} = \frac{W}{A} + \frac{M}{S} = \frac{7626 \text{ lbf}}{7.5 \text{ ft}^2} + \frac{4387 \text{ ft-lbf}}{9.38 \text{ ft}^3}$$

$$= 1480 \text{ lbf/ft}^2 \quad [< F_p = 2500 \text{ lbf/ft}^2]$$

> Use a footing width of 7 ft 6 in.

2.2. Design the wall reinforcement at the base of stem.

$$P_{\text{stem}} = 0.5(k_a \gamma_s) H_{\text{stem}}^2 (1 \text{ ft})$$

$$= (0.5)\left(30 \frac{\text{lbf}}{\text{ft}^3}\right)(10.0 \text{ ft})^2 (1 \text{ ft})$$

$$= 1500 \text{ lbf}$$

$$M = P_{\text{stem}}\left(\frac{H_{\text{stem}}}{3}\right)$$

$$= (1500 \text{ lbf})\left(\frac{10 \text{ ft}}{3}\right)$$

$$= 5000 \text{ ft-lbf}$$

For a 12 in (nominal) block, allow 2 in from the back face to the centroid of steel. Use grade 60 rebar ($F_s = 32,000$ psi) with $d = 9.63$ in.

$$d = t_{\text{actual}} - 2 \text{ in}$$

$$= 11.63 \text{ in} - 2 \text{ in}$$

$$= 9.63 \text{ in}$$

$$A_s = \frac{M}{F_s j d} \approx \frac{(5000 \text{ ft-lbf})\left(12 \frac{\text{in}}{\text{ft}}\right)}{\left(32,000 \frac{\text{lbf}}{\text{in}^2}\right)(0.875)(9.63 \text{ in})}$$

$$\approx 0.22 \text{ in}^2$$

Try no. 5 bars (nominal area of 0.31 in^2) at 16 in on centers. Check stresses based on this trial value, $A_s = 0.23 \text{ in}^2$ per foot, using ACI 530 Chap. 2.

$$\rho = \frac{A_s}{bd} = \frac{0.23 \text{ in}^2}{(12 \text{ in})(9.63 \text{ in})} = 0.0020$$

$$n = \frac{E_s}{E_m} = \frac{E_s}{900 f_m'}$$

$$= \frac{29,000,000 \frac{\text{lbf}}{\text{in}^2}}{(900)\left(1500 \frac{\text{lbf}}{\text{in}^2}\right)}$$

$$= 21.5$$

$$\rho n = (0.0020)(21.5) = 0.043$$

$$k = \sqrt{(\rho n)^2 + 2\rho n} - \rho n$$

$$= \sqrt{(0.043)^2 + (2)(0.043)} - 0.043$$

$$= 0.253$$

$$j = 1 - 0.333k = 1 - (0.333)(0.253) = 0.92$$

$$f_s = \frac{M}{A_s j d} = \frac{(5000 \text{ ft-lbf})\left(12 \frac{\text{in}}{\text{ft}}\right)}{(0.23 \text{ in}^2)(0.92)(9.63 \text{ in})}$$

$$= 29,400 \text{ psi} \quad [< F_s = 32,000 \text{ psi}]$$

$$f_c = \left(\frac{M}{bd^2}\right)\left(\frac{2}{jk}\right)$$

$$= \left(\frac{(5000 \text{ ft-lbf})\left(12 \frac{\text{in}}{\text{ft}}\right)}{(12 \text{ in})(9.63 \text{ in})^2}\right)\left(\frac{2}{(0.92)(0.253)}\right)$$

$$= 463 \text{ psi}$$

$$F_c = 0.45 f_m' = (0.45)\left(1500 \frac{\text{lbf}}{\text{in}^2}\right)$$

$$= 675 \text{ psi} \quad [f_c' < F_c]$$

Check the shear using ACI 530 Sec. 2.3.6.

$$f_v = \frac{V}{b t_{\text{actual}}} = \frac{1500 \text{ lbf}}{(12 \text{ in})(11.63 \text{ in})}$$

$$= 11 \text{ psi}$$

$$\frac{M}{Vd} = \frac{(5000 \text{ ft-lbf})\left(12 \frac{\text{in}}{\text{ft}}\right)}{(1500 \text{ lbf})(9.63 \text{ in})} = 4.15 \quad [>1.0]$$

Since $4.15 > 1.0$, ACI 530 Eq. 2-27 applies.

$$F_v = F_{vm} = 2\sqrt{f_m'}$$

$$= 2\sqrt{1500 \frac{\text{lbf}}{\text{in}^2}}$$

$$= 77 \text{ psi} \quad [> f_v, \text{so OK}]$$

Per IBC Sec. 2107.2.1, the required lap splice length is

$$l_d \geq \begin{cases} 0.002 d_b f_s = \left(0.002 \ \dfrac{\text{in}^2}{\text{lbf}}\right)(0.625 \ \text{in})\left(29{,}400 \ \dfrac{\text{lbf}}{\text{in}^2}\right) \\ \qquad = 36.8 \ \text{in} \quad [\text{controls}] \\ 40 d_b = (40)(0.625 \ \text{in}) \\ \qquad = 25 \ \text{in} \\ 12 \ \text{in} \end{cases}$$

However, the stress in the no. 5 bars is $f_s = 29{,}400$ psi, which is greater than $0.8 F_s = (0.8)(32{,}000 \ \text{lbf/in}^2 = 25{,}600$ psi, so IBC Sec. 2107.3 also requires that this value for l_d be increased by at least 50%.

$$l_d \geq (1.5)(36.8 \ \text{in}) = 55 \ \text{in}$$

Provide horizontal bed joint reinforcement consistent with ACI 530 Sec. 1.18.4.3.1.

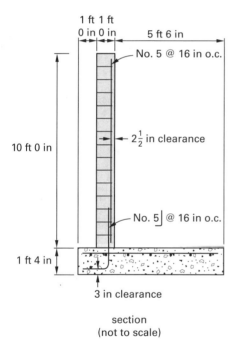

section
(not to scale)

SOLUTION 3

3.1. Calculate the reinforcement for the 4 ft 8 in wide pier under the dead gravity plus seismic loading. Per the problem statement, assume that the given dead and live loads are applied concentrically. Check load cases 5 and 6 of ASCE/SEI7 Sec. 12.4.2.3. For seismic design category D, ASCE/SEI7 Table 12.2-1 requires a special reinforced masonry bearing wall for which the out of plane forces are determined per ASCE/SEI7 Sec. 12.11.1.

$$w \geq \begin{cases} 0.4 S_{DS} I_e w_{\text{wall}} = (0.4)(1.0)(1.0)\left(135 \ \dfrac{\text{lbf}}{\text{ft}^2}\right) \\ \qquad = 54 \ \text{lbf/ft}^2 \quad [\text{controls}] \\ 0.1 w_{\text{wall}} = (0.1)\left(135 \ \dfrac{\text{lbf}}{\text{ft}^2}\right) = 14 \ \text{lbf/ft}^2 \end{cases}$$

For the cantilevered parapet, ASCE/SEI7 Table 13.5-1 gives $a_p = R_p = 2.5$, and the lateral force on the parapet is given by the equations in ASCE/SEI7 Sec. 13.3.1.

$$F_p \geq \begin{cases} \left(\dfrac{0.4 a_p S_{DS} I_p w_{\text{wall}}}{R_p}\right)\left(1 + \dfrac{2z}{h}\right) \\ = \left(\dfrac{(0.4)(2.5)(1.0)(1.0)}{\times \left(135 \ \dfrac{\text{lbf}}{\text{ft}^2}\right)}\Bigg/ 2.5\right) \\ \qquad \times \left(1 + \dfrac{(2)(27 \ \text{ft})}{27 \ \text{ft}}\right) \\ \qquad = 162 \ \text{lbf/ft}^2 \quad [\text{controls}] \\ 0.3 S_{DS} I_p w_{\text{wall}} = (0.3)(1.0)(1.0)\left(135 \ \dfrac{\text{lbf}}{\text{ft}^2}\right) \\ \qquad = 41 \ \text{lbf/ft}^2 \end{cases}$$

For design of the structural bearing wall, it is not necessary to include the peak loading from the parapet (a nonstructural component), but it is conservative to do so and that approach is used in the following analysis. For vibration in the fundamental mode, the force in the parapet acts in the opposite sense to the lateral force on the wall. This loading creates maximum bending moment in the wall pier, but it underestimates the reaction that must be transferred between the wall and the roof diaphragm, which would be calculated for a higher mode shape with wall and parapet forces in the same direction. Also, the special condition of IBC Sec. 1604.8.2 must be met. The tributary wall width is 4.67 ft in the lower 16 ft, and 4.67 ft + 16 ft = 20.67 ft in the upper region and parapet.

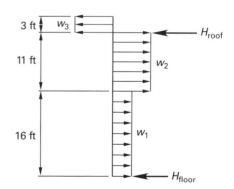

$$w_1 = wb_1 = \left(54 \ \frac{\text{lbf}}{\text{ft}^2}\right)(4.67 \ \text{ft}) = 252 \ \text{lbf/ft}$$

$$w_2 = wb_2 = \left(54 \ \frac{\text{lbf}}{\text{ft}^2}\right)(20.67 \ \text{ft}) = 1116 \ \text{lbf/ft}$$

$$w_3 = w_P b_3 = \left(162 \ \frac{\text{lbf}}{\text{ft}^2}\right)(20.67 \ \text{ft}) = 3350 \ \text{lbf/ft}$$

$$\sum M_{\text{roof}} = H_{\text{floor}} h - w_1 a(h - 0.5a)$$
$$- w_2 b(0.5b) - w_3 c(0.5c) = 0$$

$$H_{\text{floor}} = \frac{w_1 a(h - 0.5a) + w_2 b(0.5b) + w_3 c(0.5c)}{h}$$

$$= \frac{
\begin{aligned}
&\left(252 \ \frac{\text{lbf}}{\text{ft}}\right)(16 \ \text{ft})\left(27 \ \text{ft} - (0.5)(16 \ \text{ft})\right)\\
&+ \left(1116 \ \frac{\text{lbf}}{\text{ft}}\right)(11 \ \text{ft})(0.5)(11 \ \text{ft})\\
&+ \left(3350 \ \frac{\text{lbf}}{\text{ft}}\right)(3 \ \text{ft})(0.5)(3 \ \text{ft})
\end{aligned}
}{27 \ \text{ft}}$$

$$= 5896 \ \text{lbf}$$

$$\sum F_{\text{horiz}} = w_1 a + w_2 b - w_3 c - H_{\text{floor}} - H_{\text{roof}} = 0$$

$$H_{\text{roof}} = w_1 a + w_2 b - w_3 c - H_{\text{floor}}$$

$$= \left(252 \ \frac{\text{lbf}}{\text{ft}}\right)(16 \ \text{ft}) + \left(1116 \ \frac{\text{lbf}}{\text{ft}}\right)(11 \ \text{ft})$$

$$- \left(3350 \ \frac{\text{lbf}}{\text{ft}}\right)(3 \ \text{ft}) - 5896 \ \text{lbf}$$

$$= 362 \ \text{lbf}$$

Maximum bending moment occurs where shear is zero.

$$V_{16 \ \text{ft}} = H_{\text{floor}} - w_1(16 \ \text{ft})$$

$$= 5896 \ \text{lbf} - \left(254 \ \frac{\text{lbf}}{\text{ft}}\right)(16 \ \text{ft})$$

$$= 1832 \ \text{lbf}$$

$$V = V_{16 \ \text{ft}} - w_2(x - 16 \ \text{ft}) = 0$$

$$= 1832 \ \text{lbf} - \left(1116 \ \frac{\text{lbf}}{\text{ft}}\right)(x - 16 \ \text{ft})$$

$$x = 17.6 \ \text{ft}$$

Therefore, maximum bending moment occurs above the top of the pier, which is not critical. Check combined stresses at the top of the pier, 16 ft above the base.

$$M_E = 0.5(H_{\text{floor}} + V_{16 \ \text{ft}})a$$

$$= (0.5)(5896 \ \text{lbf} + 1832 \ \text{lbf})(16 \ \text{ft})$$

$$= 62{,}000 \ \text{ft-lbf}$$

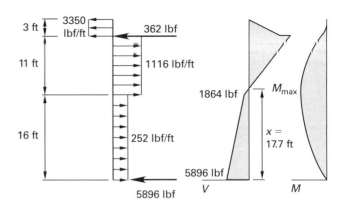

The service axial loads at the critical section are

$$P_D = (w_D + w_{\text{wall}} c)s$$

$$= \left(360 \ \frac{\text{lbf}}{\text{ft}} + \left(135 \ \frac{\text{lbf}}{\text{ft}^2}\right)(11 \ \text{ft})\right)(20.67 \ \text{ft})$$

$$= 38{,}140 \ \text{lbf}$$

$$P_L = w_L s = \left(300 \ \frac{\text{lbf}}{\text{ft}}\right)(20.67 \ \text{ft}) = 6200 \ \text{lbf}$$

With the redundancy factor, ρ, taken as 1.0, the load combinations can be simplified. For case 5,

$$(1.2 + 0.2S_{DS})D + \rho Q_E + L = \left(1.2 + (0.2)(1)\right)D$$
$$+ 1Q_E + L$$
$$= 1.4D + Q_E + L$$

$$P_u = 1.4P_D + P_L$$
$$= (1.4)(38{,}140 \ \text{lbf})$$
$$+ 6200 \ \text{lbf}$$
$$= 59{,}600 \ \text{lbf}$$

$$M_u = M_E = 62{,}000 \ \text{ft-lbf}$$

For case 7,

$$(0.9 - 0.2S_{DS})D + \rho Q_E = \left(0.9 - (0.2)(1.0)\right)D + 1Q_E$$
$$= 0.7D + Q_E$$

$$P_u = (0.7)(38{,}140 \ \text{lbf})$$
$$= 26{,}700 \ \text{lbf}$$

$$M_u = M_E = 62{,}000 \ \text{ft-lbf}$$

Place the reinforcement in two faces and consider only the far face to resist tension. Per ACI 530 Sec. 1.16.3, allow a minimum of 1 in grout cover between the inside shell and the reinforcement, and estimate the effective depth based on an assumed $d_b = 1$ in.

$$d = t - \text{shell} - \text{cover} - 0.5d_b$$
$$= 13.5 \ \text{in} - 3 \ \text{in} - 1.0 \ \text{in} - (0.5)(1 \ \text{in})$$
$$= 9.0 \ \text{in}$$

Try the minimum reinforcement as required by ACI 530 Sec. 1.14.1.3.

$$A_{st} = 0.0025 A_n = 0.0025 bt$$
$$= (0.0025)(56 \text{ in})(13.5 \text{ in})$$
$$= 1.89 \text{ in}^2$$
$$A_s \geq 0.5 A_{st} = (0.5)(1.89 \text{ in}^2) = 0.95 \text{ in}^2$$

Try four no. 5 bars in each face.

$$A_s = (4)(0.31 \text{ in}^2) = 1.24 \text{ in}^2$$

For a tension-controlled flexural failure, case 7 is critical.

$$a = \frac{A_s f_y + P_u}{0.8 f'_m b}$$

$$= \frac{(1.24 \text{ in}^2)\left(60,000 \dfrac{\text{lbf}}{\text{in}^2}\right) + 26,700 \text{ lbf}}{(0.8)\left(2.5 \dfrac{\text{kips}}{\text{in}^2}\right)\left(1000 \dfrac{\text{lbf}}{\text{kip}}\right)(56 \text{ in})}$$

$$= 0.90 \text{ in}$$

$$\phi M_n = \phi\left(A_s f_y (d - 0.5a) + P_u(0.5t - 0.5a)\right)$$

$$= \frac{(0.9)\begin{pmatrix}(1.24 \text{ in}^2)\left(60,000 \dfrac{\text{lbf}}{\text{in}^2}\right) \\ \times \left(9.0 \text{ in} - (0.5)(0.9 \text{ in})\right) \\ + (26,700 \text{ lbf})\left((0.5)(13.5 \text{ in}) \right. \\ \left. - (0.5)(0.9 \text{ in})\right)\end{pmatrix}}{12 \dfrac{\text{in}}{\text{ft}}}$$

$$= 60,325 \text{ ft-lbf} \quad [< M_u = 62,000 \text{ ft-lbf}]$$

Flexural strength is inadequate, so increase the area of steel. Try four no. 6 bars in each face.

$$A_s = (4)(0.44 \text{ in}^2) = 1.76 \text{ in}^2$$

Check case 7.

$$a = \frac{A_s f_y + P_u}{0.8 f'_m b}$$

$$= \frac{(1.76 \text{ in}^2)\left(60,000 \dfrac{\text{lbf}}{\text{in}^2}\right) + 26,700 \text{ lbf}}{(0.8)\left(2.5 \dfrac{\text{kips}}{\text{in}^2}\right)\left(1000 \dfrac{\text{lbf}}{\text{kip}}\right)(56 \text{ in})}$$

$$= 1.18 \text{ in}$$

$$\phi M_n = \phi\left(A_s f_y (d - 0.5a) + P_u(0.5t - 0.5a)\right)$$

$$= \frac{(0.9)\begin{pmatrix}(1.76 \text{ in}^2)\left(60,000 \dfrac{\text{lbf}}{\text{in}^2}\right) \\ \times \left(9.0 \text{ in} - (0.5)(1.18 \text{ in})\right) \\ + (26,700 \text{ lbf})\left((0.5)(13.5 \text{ in}) \right. \\ \left. - (0.5)(1.18 \text{ in})\right)\end{pmatrix}}{12 \dfrac{\text{in}}{\text{ft}}}$$

$$= 78,900 \text{ ft-lbf} \quad [> M_u = 62,000 \text{ ft-lbf}]$$

Verify that tension steel yields.

$$c = \frac{a}{\beta_1} = \frac{1.18 \text{ in}}{0.8} = 1.48 \text{ in}$$

$$\frac{\epsilon_s}{d - c} = \frac{\epsilon_{mu}}{c}$$

$$\epsilon_s = \frac{(d - c)\epsilon_{mu}}{c}$$

$$= \frac{(9.0 \text{ in} - 1.48 \text{ in})(0.0025)}{1.48 \text{ in}}$$

$$= 0.0127$$

$$\epsilon_y = \frac{f_y}{E_s} = \frac{60,000 \dfrac{\text{lbf}}{\text{in}^2}}{29,000,000 \dfrac{\text{lbf}}{\text{in}^2}}$$

$$= 0.00207 \quad [\epsilon_s > \epsilon_y]$$

Tension steel yields as assumed and the flexural strength is adequate. Calculate the nominal axial strength modified for slenderness in accordance with ACI 530 Sec. 3.3.4.1.

$$r = \sqrt{\frac{bt^3}{12}} = \sqrt{\frac{\dfrac{(56 \text{ in})(13.5 \text{ in})^3}{12}}{(56 \text{ in})(13.5 \text{ in})}} = 3.90 \text{ in}$$

Use the full unsupported height of wall to compute (conservatively) the axial compression capacity of the pier.

$$\frac{h}{r} = \frac{(27 \text{ ft})\left(12 \dfrac{\text{in}}{\text{ft}}\right)}{3.90 \text{ in}} = 83 \quad [< 99, \text{ so use ACI 530 Eq. 3-18}]$$

$$\phi P_n = \phi 0.8 \left(0.8 f'_m (A_n - A_{st}) + f_y A_{st} \right)$$

$$\times \left(1 - \left(\frac{h}{140r} \right)^2 \right)$$

$$= (0.9)(0.8) \begin{pmatrix} (0.8) \left(2.5 \ \frac{\text{kips}}{\text{in}^2} \right) \left(1000 \ \frac{\text{lbf}}{\text{kip}} \right) \\ \times \left((56 \ \text{in})(13.5 \ \text{in}) - 3.52 \ \text{in}^2 \right) \\ + (3.52 \ \text{in}^2) \left(60{,}000 \ \frac{\text{lbf}}{\text{in}^2} \right) \end{pmatrix}$$

$$\times \left(1 - \left(\frac{83}{140} \right)^2 \right)$$

$$= 801{,}300 \ \text{lbf} \quad [> P_u = 59{,}600 \ \text{lbf}]$$

Therefore, the design is OK. Provide minimum lateral reinforcement for a special column per ACI 530 Sec. 1.18.4.4.2.1; that is, no. 3 ties at 8 in on centers.

> Provide no. 3 ties at 8 in o.c. Use 4 no. 6 vertical bars in each face.

3.2. A plan detail showing the vertical and horizontal steel is shown.

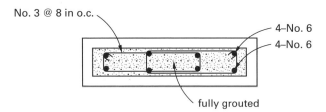

No. 3 @ 8 in o.c.
4–No. 6
4–No. 6
fully grouted

SOLUTION 4

4.1. Design the wall reinforcement for the given loading using grade 60 steel. f'_m is given as 1400 psi and end connections are adequate. Design according to ACI 530 Chap. 3, using strength design theory based on a unit width of wall spanning simply supported from the base to the floor above. Per ACI 530 Sec. 1.13.1, the span length of the wall is

$$\text{span} \leq \begin{cases} (1.15)(\text{clear height}) = (1.15)(9.0 \ \text{ft}) \\ \qquad = 10.4 \ \text{ft} \\ \text{center to center} \\ \text{of supports} \end{cases} \approx 9.5 \ \text{ft} \quad [\text{controls}]$$

For factored load combination 4 of ASCE/SEI7 Sec. 2.3.2,

$$U = 1.2D + W$$

Per foot of wall,

$$w_u = w(1 \ \text{ft}) = \left(80 \ \frac{\text{lbf}}{\text{ft}^2} \right)(1 \ \text{ft}) = 80 \ \text{lbf/ft}$$

$$P_{uf} = 1.2 P_D (1 \ \text{ft}) = (1.2) \left(1000 \ \frac{\text{lbf}}{\text{ft}} \right)(1 \ \text{ft}) = 1200 \ \text{lbf}$$

Per the problem statement, use a single layer of steel positioned at the middle of the wall.

$$d = 0.5t = (0.5)(7.63 \ \text{in}) = 3.8 \ \text{in}$$

For design of walls for out-of-plane loads, ACI 530 Sec. 3.3.5.3 applies. Per foot of wall,

$$\frac{h}{t} = \frac{(9.5 \ \text{ft}) \left(12 \ \frac{\text{in}}{\text{ft}} \right)}{7.63 \ \text{in}} = 15 \quad [< 30]$$

$$\frac{P_{uf}}{A_g} = \frac{P_{uf}}{bt} = \frac{1200 \ \text{lbf}}{(12 \ \text{in})(7.63 \ \text{in})}$$

$$= 13 \ \text{psi}$$

$$0.2 f'_m = (0.2) \left(1400 \ \frac{\text{lbf}}{\text{in}^2} \right)$$

$$= 280 \ \text{psi} \quad \begin{bmatrix} > P_{uf}/A_g, \\ \text{so ACI 530 Sec. 3.2.5.4 applies} \end{bmatrix}$$

Calculate the deflection at midheight based on the gross area and the factored wind loading. Per foot of wall,

$$E_m = 900 f'_m = (900) \left(1400 \ \frac{\text{lbf}}{\text{in}^2} \right)$$

$$= 1{,}260{,}000 \ \text{psi}$$

$$I = \frac{bt^3}{12}$$

$$= \frac{(12 \ \text{in})(7.63 \ \text{in})^3}{12}$$

$$= 444 \ \text{in}^4$$

$$\delta_u = \frac{0.013 w_u h^4}{E_m I}$$

$$= \frac{(0.013) \left(80 \ \frac{\text{lbf}}{\text{ft}} \right) (9.5 \ \text{ft})^4 \left(12 \ \frac{\text{in}}{\text{ft}} \right)^3}{\left(1{,}260{,}000 \ \frac{\text{lbf}}{\text{in}^2} \right) (444 \ \text{in}^4)}$$

$$= 0.026 \ \text{in}$$

Try no. 4 bars at 24 in o.c., so $A_s = 0.10 \ \text{in}^2/\text{ft}$. Per Table C3-1 in ASCE/SEI7, nominal 8 in ungrouted masonry units weighing 35 lbf/ft^2 have an average weight of approximately 50 lbf/ft^2 when cells are grouted at 24 in on centers. So, per foot of wall,

$$P_{uw} = 1.2 P_w (1 \ \text{ft}) = 1.2 w_c (0.5h)(1 \ \text{ft})$$

$$= (1.2) \left(50 \ \frac{\text{lbf}}{\text{ft}^2} \right)(0.5)(9.5 \ \text{ft})(1 \ \text{ft})$$

$$= 285 \ \text{lbf}$$

$$P_u = P_{uf} + P_{uw}$$

$$= 1200 \text{ lbf} + 285 \text{ lbf}$$

$$= 1485 \text{ lbf}$$

$$M_u = \frac{w_u L^2}{8} + P_{uf} e_u + P_u \delta_u$$

$$= \frac{\left(80 \ \frac{\text{lbf}}{\text{ft}}\right)(9.5 \text{ ft})^2}{8} + (1200 \text{ lbf})(0 \text{ ft})$$

$$+ (1485 \text{ lbf})\left(\frac{0.026 \text{ in}}{12 \ \frac{\text{in}}{\text{ft}}}\right)$$

$$= 906 \text{ ft-lbf}$$

$$a = \frac{A_s f_y + P_u}{0.8 f'_m b}$$

$$= \frac{(0.10 \text{ in}^2)\left(60 \ \frac{\text{kips}}{\text{in}^2}\right)\left(1000 \ \frac{\text{lbf}}{\text{kip}}\right) + 1485 \text{ lbf}}{(0.8)\left(1400 \ \frac{\text{lbf}}{\text{in}^2}\right)(12 \text{ in})}$$

$$= 0.56 \text{ in}$$

$$\phi M_n = \phi\big(A_s f_y(d - 0.5a) + P_u(0.5t - 0.5a)\big)$$

$$= \frac{(0.9)\begin{pmatrix}(0.10 \text{ in}^2)\left(60 \ \frac{\text{kips}}{\text{in}^2}\right)\left(1000 \ \frac{\text{lbf}}{\text{kip}}\right)\\[2mm] \times \left(3.8 \text{ in} - (0.5)(0.56 \text{ in})\right)\\[2mm] + (1485 \text{ lbf})\big((0.5)(7.63 \text{ in})\\[2mm] - (0.5)(0.56 \text{ in})\big)\end{pmatrix}}{12 \ \frac{\text{in}}{\text{ft}}}$$

$$= 1978 \text{ ft-lbf} \quad [> M_u = 906 \text{ ft-lbf}]$$

The flexural strength is more than twice the design moment. Try increasing the spacing of the no. 4 bars to 48 in on centers ($A_s = 0.05 \text{ in}^2$). There is negligible change in P_{uw}. Therefore, check capacity using the previously computed values of M_u and P_u. Per foot of wall,

$$a = \frac{A_s f_y + P_u}{0.8 f'_m b}$$

$$= \frac{(0.05 \text{ in}^2)\left(60 \ \frac{\text{kips}}{\text{in}^2}\right)\left(1000 \ \frac{\text{lbf}}{\text{kip}}\right) + 1485 \text{ lbf}}{(0.8)\left(1400 \ \frac{\text{lbf}}{\text{in}^2}\right)(12 \text{ in})}$$

$$= 0.33 \text{ in}$$

$$\phi M_n = \phi\big(A_s f_y(d - 0.5a) + P_u(0.5t - 0.5a)\big)$$

$$= \frac{(0.9)\begin{pmatrix}(0.05 \text{ in}^2)\left(60 \ \frac{\text{kips}}{\text{in}^2}\right)\left(1000 \ \frac{\text{lbf}}{\text{kip}}\right)\\[2mm] \times \left(3.8 \text{ in} - (0.5)(0.33 \text{ in})\right)\\[2mm] + (1485 \text{ lbf})\big((0.5)(7.63 \text{ in})\\[2mm] - (0.5)(0.33 \text{ in})\big)\end{pmatrix}}{12 \ \frac{\text{in}}{\text{ft}}}$$

$$= 1224 \text{ ft-lbf} \quad [> M_u = 906 \text{ ft-lbf}]$$

The wall is adequate with no. 4 bars at 48 in on centers.